W9-BZM-831

Mechanical Drawing

Mechanical Drawing

TENTH EDITION

Thomas E. French

Carl L. Svensen

Jay D. Helsel

Byron Urbanick

BUTLER PUBLIC LIBRARY
BUTLER, PENNSYLVANIA 16001
1710289

604.2
MEC

McGRAW-HILL BOOK COMPANY

New York   Atlanta   Dallas   St. Louis   San Francisco
Auckland   Bogotá   Guatemala   Hamburg   Lisbon
London   Madrid   Mexico   Montreal   New Delhi   Panama   Paris
San Juan   São Paulo   Singapore   Sydney   Tokyo   Toronto

Sponsoring Editor: D. Eugene Gilmore
Editing Supervisor: Larry Goldberg
Design and Art Supervisor: Patricia F. Lowy
Production Supervisor: Priscilla Taguer

Cover Photograph: Evans & Sutherland

*The computer-generated drawing on the cover
of this book was created through solid modeling
techniques using data from MBB GmbH.*

Library of Congress Cataloging in Publication Data

Main entry under title:

Mechanical drawing.

   Includes index.
   Summary: A textbook introducing the basic theory,
techniques, and uses of drafting for industrial arts
and vocational high school students.
   1. Mechanical drawing.   [1. Mechanical drawing]
I. French, Thomas Ewing, 1871–1944.
[T353.M48 1985]   604.2   84-19432
ISBN 0-07-022333-5

MECHANICAL DRAWING, Tenth Edition

Copyright © 1985, 1980, 1974, 1966, 1957 by McGraw-Hill, Inc. All rights reserved.
Copyright © 1948, 1940 by McGraw-Hill, Inc. All rights reserved. Copyright 1934,
1927, 1919 by McGraw-Hill, Inc. All rights reserved. Copyright renewed 1968, 1962 by
J. F. Houston and C. L. Svensen. Printed in the United States of America. Except as
permitted under the United States Copyright Act of 1976, no part of this publication may
be reproduced or distributed in any form or by any means, or stored in a data base or
retrieval system, without the prior written permission of the publisher.

3 4 5 6 7 8 9 0 VNHVNH 8 9 2 1 0 9 8

ISBN 0-07-022333-5

About the Authors

JAY D. HELSEL is a professor of industrial arts and technology at California University of Pennsylvania. He completed his undergraduate work in industrial arts at California State College and was awarded a master's degree from the Pennsylvania State University. He completed a doctoral degree in educational communications and technology at the University of Pittsburgh.

Dr. Helsel has worked in industry and has taught drafting, metalworking, woodworking, and a variety of laboratory and professional courses at the high school, college, and university levels. During the past 25 years, he has also owned and operated Technical Graphics, a technical art and illustrating company. His work appears in a great variety of technical publications.

Dr. Helsel is also coauthor of *Engineering Drawing and Design*, *Programmed Blueprint Reading*, *Reading Engineering Drawings through Conceptual Sketching*, and a variety of other technical publications.

BYRON URBANICK is chairman of industrial technology at Oak Park and River Forest High School in Oak Park, Illinois, where he teaches mechanical, architectural, and engineering drawing. He has also taught at the Illinois Institute of Technology, Northern Illinois University, and Chicago State University.

Mr. Urbanick earned his B.S. and M.S. degrees at Northern Illinois University, with additional studies at the University of Illinois, Illinois State University, and the Illinois Institute of Technology. His professional experiences include design and detailing in architectural and engineering offices. His present practice includes educational and design consultation.

Mr. Urbanick is a member of many professional organizations. He serves on the National Advisory Team for the Industrial Clubs of America (VICA) and is involved with the annual drafting competition for the United States Skill Olympics (USSO).

Contents

Preface

The first edition of *Mechanical Drawing* was written by Thomas E. French and Carl L. Svensen and was published by McGraw-Hill in 1919. They carefully prepared each unit to help students learn to visualize in three dimensions, to develop and strengthen their technical imagination, to think precisely, and to read and write the language of industry. While it was one of the first drafting textbooks written, it was so well done that many of the concepts remain as a very successful part of the book even into this tenth edition. However, in order to remain current with industrial and American National Standards Institute (ANSI) practices, each edition has been updated.

The latest ANSI drafting standards are once again covered in this edition. Particular attention should be given to Chap. 6, Dimensioning, where the latest standards from ANSI Y14.5M, *Dimensioning and Tolerancing for Engineering Drawing*, are covered. Notice especially the increased amount of "symbology" (drafting shorthand) that has been added as well as the increased emphasis on decimal-inch and metric dimensioning practices. However, some of the older practices also remain in this edition to accommodate those who will work in industries not yet entirely current with ANSI practices.

Beginning with the eighth edition, some new concepts were introduced. Many were covered in new chapters such as Systems for Graphic Communication, Basic Descriptive Geometry, and Drafting Media and Reproduction. Also, the addition of four-color illustrations helped show procedural steps and methods of projection in some of the more complex concepts.

In the tenth edition, a chapter on computer-aided design and drafting (CAD or CADD) has been added. This introduces the student to the basic elements of CAD systems as well as specific drafting techniques involved.

Once again, the student will notice the relatively low reading level throughout the text. Difficult words are carefully defined and illustrated when first used, and a separate vocabulary section has been added at the end of many chapters. Also, to add interest and to provide a variety of experiences, many of the chapters now have a new section called Learning Activities. The activities listed provide for enrichment and allow students to exercise their creative abilities.

A sincere effort has been made to accommodate the requests of the faithful users of this text. By popular demand, more of the problems from the seventh edition of *Mechanical Drawing* have been brought back, and new problems have been added. Teachers should find a sufficient number of problems to provide meaningful assignments over a number of years without duplication and at increasing levels of difficulty.

Conversion to the metric system continues to be a consideration in the tenth edition. Once again, enough flexibility has been provided to allow teachers to set the rate at which their programs move toward total metric conversion. By popular demand, U.S. Customary dimensions appear first, with metric equivalents following in parentheses. In most cases, however, a soft conversion is used, providing the nearest practical unit rather than a precise equivalent.

We again wish to thank those users of previous editions whose suggestions have helped shape this edition. Their interest and help are greatly appreciated. The authors are especially indebted to David Caldwell, Elizabeth Fowler, Carol Lees, Jim Mizener, Richard Smit, and Donald Voisinet for their help and encouragement. Credit is also given to the many companies who contributed technical assistance as well as illustrations. Comments and suggestions are welcomed and appreciated.

Jay D. Helsel
Byron Urbanick

1 Twentieth-Century Drafting

■ PLANS

For Industry—Your Career—Your Lifestyle

Choose a new product you would like to buy. Do you think a 10-speed bike, a dirt bike, a stereo sound system, a home computer game, or a pair of sports shoes could be produced without detailed plans? The people who work on the plans for a

Fig. 1-2. The industrial robot was planned to work around the clock. The plan emerged from the designer's drafting board. (General Motors Corp.)

Fig. 1-1. A drafting technician prepares a detail of a bike frame. (Schwinn Bicycle Co.)

new product can affect our style of living. Plans are needed for all new products (Fig. 1-1).

You will find various types of industrial plants in or near every major city in the United States (Fig. 1-2). The plants that make our new products will each have a design-engineering team to produce plans that

describe their products. The description, or plan, is generally called a *drawing*. The drawings define the shape of the product with bold lines and dimensions (Fig. 1-3).

There are technical devices or products for doing *ordinary* things like simple arithmetic or computer games (Fig. 1-4). Twentieth-century

1

Fig. 1-3. The drafting technician defines the shape of a complex project with numbers for assembly. (Bruning.)

Fig. 1-4. New ideas must be detailed for manufacturing. The design team works with up-to-date equipment. (Bruning.)

technology has produced useful things that are important parts of our daily lives. Each year, new discoveries and inventions help make our lives more interesting. The machines that are designed for *extraordinary* things like space travel will certainly influence the lifestyle of those who will live in the twenty-first century (Fig. 1-5).

All the new products you buy or see advertised on TV or in magazines, catalogs, and newspapers have been planned on the drawing board. An understanding of the role of the designer and drafting technician will help you to see how new ideas are created and detailed so that they work (Fig. 1-6). You will realize how hard the design team works to develop designs to meet ever-changing lifestyles.

There is an old saying that when something doesn't work, it's back to the drawing board (Fig. 1-7). That is a sure sign that the design has to be reworked, revised, or improved. Sometimes it means the design team has to develop additional new ideas for a changing marketplace. Another way of putting it would be that people want a faster bike, a better sound system, a better camera, a simpler computer language, or fancier and faster sporting shoes. The plans for these new products are sometimes called *prints* or *working drawings*. They are considered graphic communication, or *the language of industry*.

The design-engineering teams are responsible for all new achievements. Everyone, in every part of society, is a consumer of new products. No matter which direction your life takes, the designer and drafting technician will influence your future—at work, at play, in worship, and at home (Fig. 1-8).

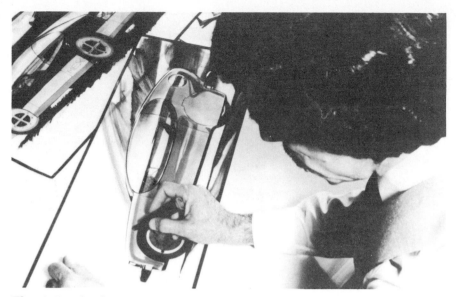

Fig. 1-5. The designer is creating new shapes for tomorrow's automobiles. The preliminary design is needed before details with instruments can be developed. (General Motors Corp.)

Fig. 1-7. Drafting technicians calculate dimensions when preparing detailed drawings. (Tobias Brown.)

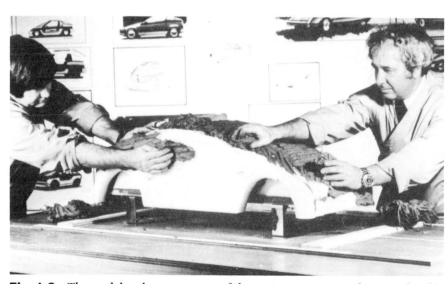

Fig. 1-6. The model makers are a part of the engineering team. They must be able to work with detailed drawings. (General Motors Corp.)

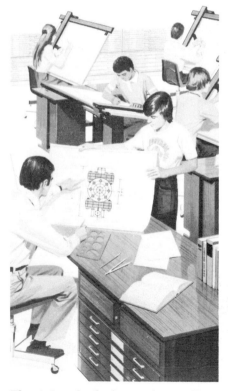

Fig. 1-8. The drafting technician illustrates a new design for your consideration and approval. (Bruning.)

◼ CAREER PLANS

You can make plans to serve on a design-engineering team or to have the team serve you. There are many design careers available at several technical levels of performance. Many businesses are in need of employees that can relate to the design team. All companies need design and drafting services for growth in technical development, construction, or production.

One of the most important jobs on any design team centers on the skills performed by the drafting technician. The career skills you can learn through hands-on experiences in junior and senior high school are taught in the beginning drafting courses. As noted earlier, drafting is referred to as the language of industry. You may develop a visual language that has an alphabet of lines (Fig. 1-9) and symbols for describing materials and technical information. The drafting technician is also referred to as a *drafter*.

CAREER OPPORTUNITIES

A career in the design world begins with courses in drafting, mechanical drawing, science, mathematics, and English. High school course work is the basis for starting a successful career (Fig. 1-10). The employer or design-engineering company offers young drafters good opportunities when they have an aptitude for prob-

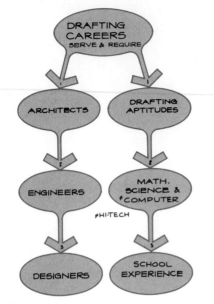

Fig. 1-10. The personal traits that are important in the development of a drafting technician.

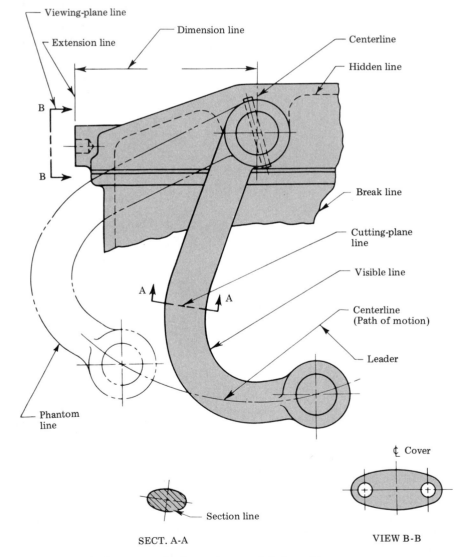

SECT. A-A   VIEW B-B

Fig. 1-9. The language of industry has an alphabet of lines. (ANSI.)

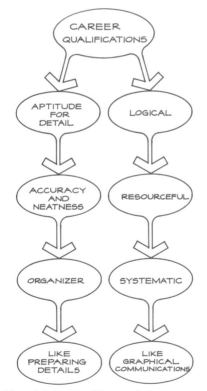

Fig. 1-11. Drafting technicians serve the professional and require technical skills.

lem solving along with drafting skills. This requires an understanding of applied science and mathematics (Fig. 1-11).

■ JOINING THE TEAM

The want-ads listing employment opportunities for beginners generally consist of the following three entry levels to the design-engineering team (Fig. 1-12):

1. *Junior drafter.* Must have good lettering and drafting skills and at least one year of high school drafting (Fig. 1-13).
2. *Drafting technician.* Must have at least one year of high school drafting and two years of work in a drafting design curriculum at a technical or community college. A mechanical aptitude with good drafting skills in an area of machine design or product design is also required.
3. *Engineer trainee.* Must have four years of college with a degree in a specialized area, for example, aerospace, architectural, me-

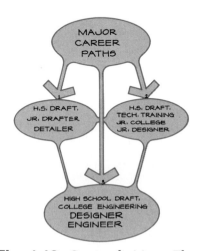

Fig. 1-12. Career decisions. Three ways to become professionally qualified.

DRAFTER

PANELTECH LTD has an immediate need for a qualified drafter. We are seeking a high school graduate with four semesters of drafting experience. You will prepare detail drawings from sketches, prepare charts and diagrams. Small precision machine parts and tooling mechanisms are custom designed. Eventually use CAD system for detailing. Good starting salary with fringe benefits.

Call for interview appointment.

Mr. William Weathers
354-5844

PANELTECH LTD.
701 North Catherine Avenue
LaGrange, IL. 60155

An Equal Opportunity Employer

Fig. 1-13. Want ads generally stress drafting skills learned in school; this one stresses the importance of sketching.

chanical, or biomedical engineering.

Design drafting technicians and engineers must specialize. There are so many technical advances that no one person or team can know all the areas of the new technology well.

■ THE DESIGN-ENGINEERING TEAM

The Major Jobs and the Creative Challenges

All the design-oriented jobs require creative drawing skills and the ability to interpret or read prints (Fig. 1-14).

Researchers People who are responsible for creating experiments that will uncover new theories. They propose processes for using new technologies through creative sketches and the use of science and mathematics.

Development managers Managers who engineer the research data until the data can be applied as new knowledge to produce new products.

Designers Creative people who use the latest materials and processes for creating new forms, structures, and design patterns.

Project coordinator Person who coordinates all the specialized areas of engineering and drawing for major production or construction projects.

Engineers People who play the major role on the design-engineering team. They apply math, science, and drawings to solve problems for production and construction in specialized areas. All of the many specialized areas of engineering will interact under the project coordinator's guidance as systems are needed.

Illustrators Specialized drafters who can make a three-dimensional pictorial from the details of an engineering drawing. The pictorial forms are picturelike and generally easier to understand.

Senior detailers People who are especially skilled in understanding the details of how things work and go together. They are capable of detailing complex parts and making the complicated details understandable.

Design drafting technicians Technicians who are capable of employing some design skills with drafting skills. This requires the interpretation of the designer's sketches and the engineer's details so that the plans can be prepared accurately for the senior detailer.

Drafters Junior detailers who use drafting skills to prepare single

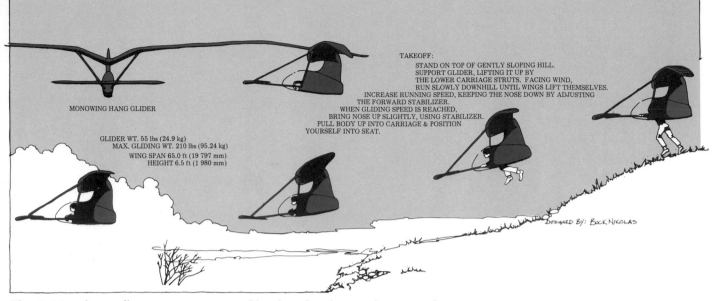

MONOWING HANG GLIDER

GLIDER WT. 55 lbs (24.9 kg)
MAX. GLIDING WT. 210 lbs (95.24 kg)
WING SPAN 65.0 ft (19 797 mm)
HEIGHT 6.5 ft (1 980 mm)

TAKEOFF:
STAND ON TOP OF GENTLY SLOPING HILL.
SUPPORT GLIDER, LIFTING IT UP BY
THE LOWER CARRIAGE STRUTS. FACING WIND,
RUN SLOWLY DOWNHILL UNTIL WINGS LIFT THEMSELVES.
INCREASE RUNNING SPEED, KEEPING THE NOSE DOWN BY ADJUSTING
THE FORWARD STABILIZER.
WHEN GLIDING SPEED IS REACHED,
BRING NOSE UP SLIGHTLY, USING STABILIZER.
PULL BODY UP INTO CARRIAGE & POSITION
YOURSELF INTO SEAT.

DESIGNED BY: BUCK NIKOLAS

Fig. 1-14. The mind's eye generates picturelike ideas, but the pencil captures the shape on paper.

technical drawings under the direction of a senior detailer.

Computer graphics programmers Programmers who become drafting specialists capable of preparing data to be detailed through a computer language. The computer can accept drafting instructions and give the data in turn to a drafting plotter which can prepare detailed drawings or illustrations.

Production clerks Clerks who serve the design-engineering team by developing prints and properly storing engineering drawings. The reproduction machines provide copies of the original drawings. Prints are used by the team for checking accuracy and for design interaction and discussion.

The design-engineering team serves most major corporations and is considered essential in preparing legal documents for contracting services. There are large offices with more than 100 engineers on a team, many small offices with 5 to 10 on the team, and medium-size offices with from 10 to 100 team members. Many specialized design offices have 4 to 12 on a team.

The universal graphic language is used by design-engineering teams all over the world. The stepped pyramid in Fig. 1-15 illustrates that the basis for all drawing is geometry.

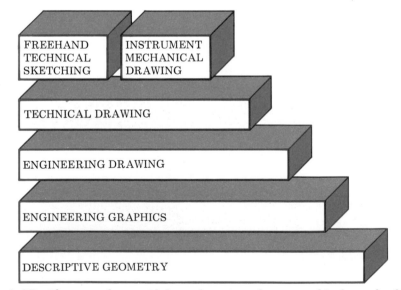

FREEHAND TECHNICAL SKETCHING

INSTRUMENT MECHANICAL DRAWING

TECHNICAL DRAWING

ENGINEERING DRAWING

ENGINEERING GRAPHICS

DESCRIPTIVE GEOMETRY

Fig. 1-15. The stepped pyramid shows the universal terms used in the study of the universal graphic language. The pyramid illustrates the hierarchy of graphic language. The base of all graphic language is descriptive geometry. You will start at the top.

This course will emphasize an introduction to freehand and mechanical drawing skills. The most common names associated with the graphic language are:

Drafting The common family name given to all the following types of graphic language.

Technical sketching Freehand technical drawing. You use a pencil to proportion the shape of ideas so that others can understand the shape of things.

Mechanical drawing A technical drawing made with drafting instruments.

Engineering drawing Drawing used by engineers and other members of the design-engineering team to describe the production of a part, shape, size, and material.

Technical drawing A broad term for any drawing which expresses technical ideas, including sketches, instrument drawings, charts, and illustrations.

Engineering graphics The graphic illustration or drawing which represents physical objects used in engineering and science.

Descriptive geometry The grammar of the graphic language; the basic principles set up to use geometric descriptions for solving two- and three-dimensional problems.

Computer graphics Today, the latest chapter in the truly universal language. It combines a computer with a plotter that can be programmed with information to prepare drawings automatically.

Basic drafting A course consisting of drafting experiences to acquaint the student with some of the processes, activities, and skills

Fig. 1-16. The drafting technician and the design engineer are at the hub of the engineering process. They make things happen. (Bruning.)

needed on the drafting table in the engineering office.

Before beginning your experience with the graphic language, let's examine the role of the drafting technician on the design-engineering team. The drafting technician, along with the engineer, forms the hub of the engineering process (Fig. 1-16). This means everything revolves around the engineer and his or her team. Engineering is the reason for the existence of drafting and the graphic language.

■ DRAFTING SERVES THE MANY BRANCHES OF ENGINEERING

Engineering has evolved into many specialized branches, including *aerospace, agricultural, architectural, chemical, civil, electrical, industrial, mechanical, mining* and *metallurgical, nuclear, petroleum,* *plastics,* and *safety*. Each branch uses, or employs, specialized drafting skills. The science and mathematics common to each branch helps convert natural resources into processes that provide new material for products and machines useful to humans.

The chart in Fig. 1-17 illustrates five common branches of engineering. Notice the different types of activities for the design-engineering team. The list of products they work with could be increased. Can you think of a few more activities to add to the list? Can you name more products? For example, could helicopters be added to aerospace engineering? And could houses be added to architectural engineering? How about civil, electrical, and mechanical engineering products? What is the most common activity that engineers engage in? You should have identified designing as the most common activity.

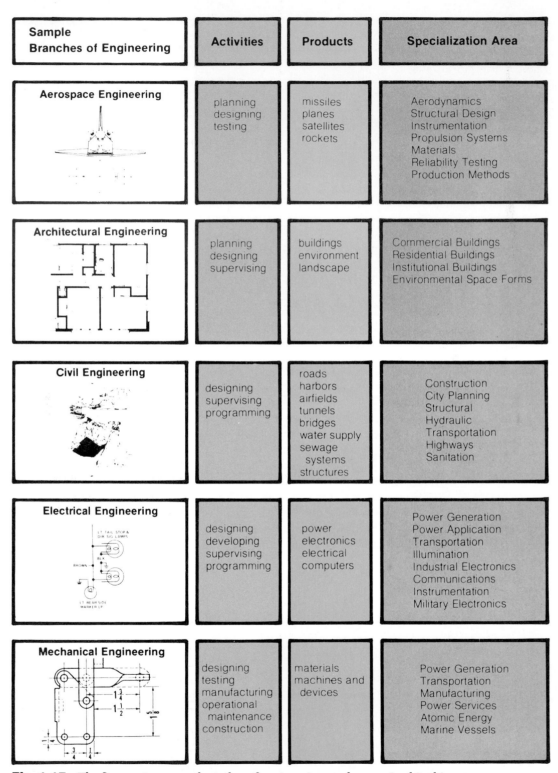

Sample Branches of Engineering	Activities	Products	Specialization Area
Aerospace Engineering	planning designing testing	missiles planes satellites rockets	Aerodynamics Structural Design Instrumentation Propulsion Systems Materials Reliability Testing Production Methods
Architectural Engineering	planning designing supervising	buildings environment landscape	Commercial Buildings Residential Buildings Institutional Buildings Environmental Space Forms
Civil Engineering	designing supervising programming	roads harbors airfields tunnels bridges water supply sewage systems structures	Construction City Planning Structural Hydraulic Transportation Highways Sanitation
Electrical Engineering	designing developing supervising programming	power electronics electrical computers	Power Generation Power Application Transportation Illumination Industrial Electronics Communications Instrumentation Military Electronics
Mechanical Engineering	designing testing manufacturing operational maintenance construction	materials machines and devices	Power Generation Transportation Manufacturing Power Services Atomic Energy Marine Vessels

Fig. 1-17. The five most common branches of engineering can be examined in this chart. Designing is the engineer's most common activity.

Even though there are many branches of engineering with areas of specialization, some activities are common to all engineers. Other activities may be limited to just a few engineers.

Let's examine some of these activities, also referred to as the "levels of engineering functions or services."

LEVELS OF ENGINEERING FUNCTIONS OR SERVICES

In each branch of engineering, the engineer performs services at one or more of the levels of engineering functions discussed below. The chart in Fig. 1-18 shows the levels in order of decreasing emphasis on mathematics and science.

Look at the working processes that the engineer engages in at each level. As less mathematics and science are required, the working processes become less theoretical and more practical. That is to say, research and development generally involves high-level thinking about things that haven't been achieved yet—those that don't work yet. Production, construction, and operation levels are developed around things we know will work—those that can be made to work successfully.

■ FOUR LEVELS OF GRAPHIC COMMUNICATION

Graphic communication is communication that is written or drawn. It may be talked about as having four separate levels: (1) *creative communication*, (2) *technical communication*, (3) *market communication*, and (4) *construction communication*. Each of these levels has its place in the field of drafting (Fig. 1-19).

Level One: Creative Communication

Graphic communication usually begins as an idea in the mind of a designer. The designer's first sketches are the beginnings of a product. This first work may be thought of as the *birth of an idea*. The creative communication of sketching is important to people who need to express ideas quickly. Sketching can capture ideas for further study.

Level Two: Technical Communication

When the designer gives a sketch to the other members of the design team, it is ready for level two. Engineers and technicians or architects and their assistants study and change the original design to make it more practical. The design is refined, or improved, through the ideas of several people. Two (or more) heads are better than one! The result is a set of drawings that fully explain the design.

Level Three: Market Communication

This level includes the evaluation of a design by a client or customer. This is especially used in architectural design. Before designs are made final, customers look at them to see if they like them. Clients must evaluate designs for style, form, and function. An example would be approving the designs of the floor plan and the outside of a house before actual construction work on the house begins.

Level Four: Construction Communication

This level includes all the details needed for manufacturing or construction. These drawings must be complete so that estimators can figure the exact costs of a project and factory superintendents can know exactly how a product is to be made. People who use the drafter's plans should not have to guess about details or ask the designer exactly what was meant.

Society needs students who can continue technical progress through drafting. If you come to thoroughly understand the first 13 chapters of this book, you will have the basic skills to accept the challenge of the advanced chapters.

■ COMMUNICATION

Each one of us spends much of our waking lives communicating. Mostly we talk or listen, but we also read and write, and sometimes we draw pictures. Drafting is a form of communication that is technical and very exact. Drafters the world over have agreed to use the same lines and symbols and the same methods for drawing. For this reason, drafting is called the *universal language of industry*. Drawings made in New York can be used by technicians in Chicago, London, Paris, or Tokyo.

Drawings are used for communication in industry because they are the clearest way to tell someone what to make and how to make it. Think how hard it would be to tell a friend in words about the shape and size of all the parts in an automobile! Figure 1-20 is an example of a "word picture" of a simple tool, a V-block. Read it and see if you can tell what a

Engineering Functions	Working Processes	Communication Forms
RESEARCH. The research engineer through experimentation and inductive reasoning employs the theories of science and mathematics in his or her attempts to find new processes.	Imagination Inductive Reasoning Ideation	**Genetic Doodling -** creating and evolving an idea. **Sketching -** creating and refining.
DEVELOPMENT. The development engineer utilizes the fruits of research by creatively applying new knowledge to produce ingenious products.	Creative Imagination Ingenious Clever Models	**Sketching -** refining **Delineation -** a graphic outline **Drawing -** construction
DESIGN. The design engineer creates products by determining construction methods, materials, and physical shapes to satisfy technical requirements, performance specifications, economic constraints, and marketing asthetics.	Creative Clever	**Sketching -** refining **Designing -** **Charts Reports Delineation**
PRODUCTION. The production engineer selects equipment and plans plant facilities to accommodate human and economic factors and allow for efficient material flow, testing, and inspection.	Plans Supervises Organizes	**Reports Flow Diagrams Architectural and Engineering Diagrams**
CONSTRUCTION. The construction engineer prepares the building environment to yield the safest, most economical, highest quality product possible.	Supervises Organizes	**Reports Charts Flow Diagrams Engineering Blueprints**
OPERATION. The operating engineer supervises the day-to-day control of machines and facilities providing power, transportation, and communication.	Supervises Operates Maintains	**Flow Diagrams Charts Architectural and Engineering Drawings**

Fig. 1-18. Examine the various services performed by the engineer in this chart. The drafting technician serves at all levels.

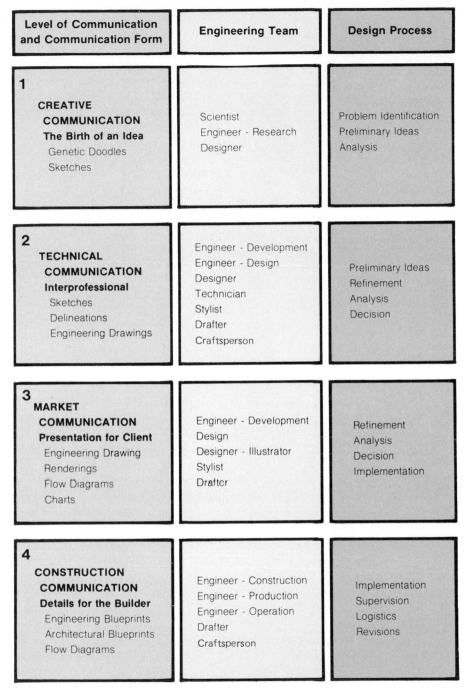

Level of Communication and Communication Form	Engineering Team	Design Process
1 **CREATIVE COMMUNICATION** **The Birth of an Idea** Genetic Doodles Sketches	Scientist Engineer - Research Designer	Problem Identification Preliminary Ideas Analysis
2 **TECHNICAL COMMUNICATION** **Interprofessional** Sketches Delineations Engineering Drawings	Engineer - Development Engineer - Design Designer Technician Stylist Drafter Craftsperson	Preliminary Ideas Refinement Analysis Decision
3 **MARKET COMMUNICATION** **Presentation for Client** Engineering Drawing Renderings Flow Diagrams Charts	Engineer - Development Design Designer - Illustrator Stylist Drafter	Refinement Analysis Decision Implementation
4 **CONSTRUCTION COMMUNICATION** **Details for the Builder** Engineering Blueprints Architectural Blueprints Flow Diagrams	Engineer - Construction Engineer - Production Engineer - Operation Drafter Craftsperson	Implementation Supervision Logistics Revisions

Fig. 1-19. Four levels of drawing or graphic communication.

THE V-BLOCK IS TO BE MADE OF CAST IRON AND MACHINED ON ALL SURFACES. THE OVERALL SIZES ARE TWO AND ONE-HALF INCHES HIGH, THREE INCHES WIDE, AND SIX INCHES LONG. A V-SHAPED CUT HAVING AN INCLUDED ANGLE OF 90° IS TO BE MADE THROUGH THE ENTIRE LENGTH OF THE BLOCK. THE CUT IS TO BE MADE WITH THE BLOCK RESTING ON THE THREE INCH BY SIX INCH SURFACE. THE V-CUT IS TO BEGIN ONE-QUARTER INCH FROM THE OUTSIDE EDGES. AT THE BOTTOM OF THE V-CUT THERE IS TO BE A RELIEF SLOT ONE-EIGHTH INCH WIDE BY ONE-EIGHTH INCH DEEP.

Fig. 1-20. A written description of a V-block.

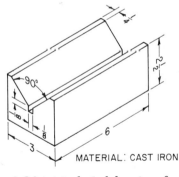

MATERIAL: CAST IRON

Fig. 1-21. A technical drawing of a V-block.

V-block looks like. Could one be made from this description? Figure 1-21 is a drawing of this same V-block. See how much easier it is to know the shape and size from the drawing than it is from the written description. It would be easy to make one from this drawing. Such a drawing is an example of the fourth level of graphic communication.

Notice how the communication forms, or drawing types, change from level to level as you move from the unknown in research to the more concrete in construction. At the highest level, in which mathematics and science are emphasized, the drawings have to be creative—they are unrefined and lack a final finished form. Ideas are in the formative stage.

SYSTEM OF GRAPHIC COMMUNICATION

There are various types of drawings used to communicate technical information in the engineer's office. Look back at the chart in Fig. 1-19 showing the four levels of drawings or graphic communication. Notice that the column to the right of each level shows who might be working at each level. The design-process column shows the stages in the design related to the four levels.

The successful drafting technician and design engineer must be able to communicate ideas accurately. Whether the ideas are for new structures, vehicles, or household products, the engineer must understand how to prepare and interpret plans drawn up by team members. The engineer will know the standards set up for his or her specialized area of design. The penmanship, so to speak, of the universal language is a very *necessary skill to learn*. The standards are established by the American National Standards Institute (ANSI) in New York. Large companies also maintain a set of standards for drafting and design. The standards help large design teams develop major achievements that would otherwise be difficult to organize.

ACHIEVEMENTS THROUGH DESIGN AND DRAFTING

Choose the world's most outstanding engineering achievements. What are the highlights of achievement in

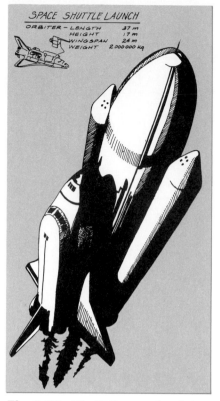

Fig. 1-22. The U.S. space shuttle is an extraordinary engineering achievement that required thousands of drawings before it could be built or placed in space.

America, Canada, Britain, France, and Japan?

Americans have designed a sleek, smooth form (an *aerodynamic* form) that can travel in space. Engineering teams have discovered high-technology fuels, materials, and controls. And what has been the result? A scientific space shuttle, with a crew and equipment, has been traveling into and out of space safely (Fig. 1-22). This kind of planning required a team of experts. One of the experts is the drafting technician on the design team.

Review the tall structures in the world (Fig. 1-23). Which one is the tallest? In Canada, the CN Tower in

Toronto is an outstanding achievement. Engineered in concrete and steel, it is a smooth, vertical form that serves as a communication tower. It has an activity center, a weather station, a restaurant, and an observation deck that pokes into the clouds (Fig. 1-24). The plan for a tall structure requires architects, structural engineers, mechanical engineers, and design drafting technicians for the final detail drawings.

Consider air travel between continents today. One way it is achieved is with the Concorde supersonic transport. This outstanding vehicle travels at speeds up to Mach 2. Aerospace designers shaped a form that can be controlled and moved smoothly through all types of weather in air travel patterns. The aerospace engineer employs a drafting technician as a member of the design group.

Smooth design forms are developed for land travel as well as space travel. All vehicles that travel smoothly must move with the least amount of resistance to air and friction of materials when in motion. French and Japanese designers have created supercontemporary trains that can travel in excess of 150 miles per hour (mph) between major cities (Fig. 1-25). The engineering team makes plans that have well-developed details so everything works safely. The drafting detailer is a key member of the design project during development.

The racing car travels in a very competitive world. Designers shape unusual forms to set new world records. Each year new achievements emerge with high-powered technology in America and on the international scene. Plans are on the drawing boards today for tomorrow's winner (Fig. 1-26).

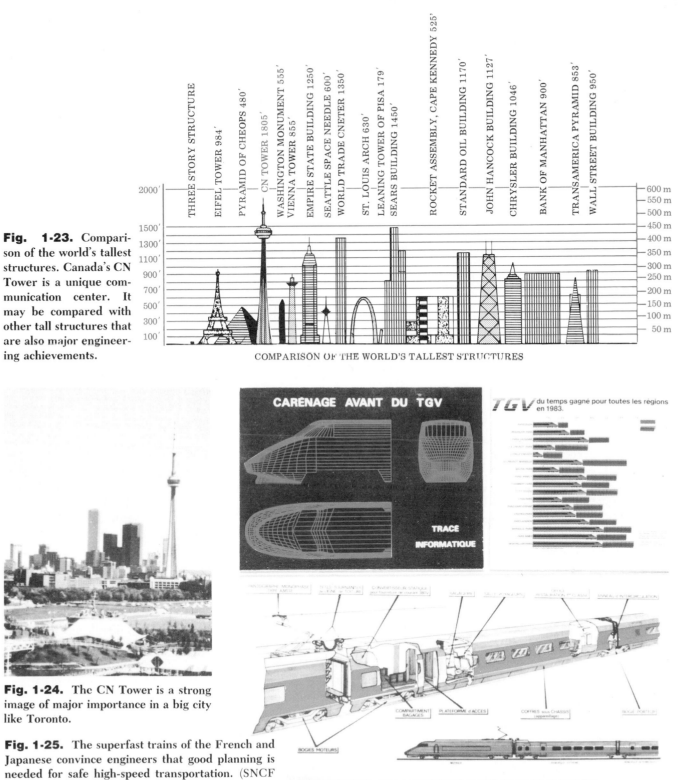

Fig. 1-23. Comparison of the world's tallest structures. Canada's CN Tower is a unique communication center. It may be compared with other tall structures that are also major engineering achievements.

THREE STORY STRUCTURE
EIFEL TOWER 984'
PYRAMID OF CHEOPS 480'
CN TOWER 1805'
WASHINGTON MONUMENT 555'
VIENNA TOWER 855'
EMPIRE STATE BUILDING 1250'
SEATTLE SPACE NEEDLE 600'
WORLD TRADE CNETER 1350'
ST. LOUIS ARCH 630'
LEANING TOWER OF PISA 179'
SEARS BUILDING 1450'
ROCKET ASSEMBLY, CAPE KENNEDY 525'
STANDARD OIL BUILDING 1170'
JOHN HANCOCK BUILDING 1127'
CHRYSLER BUILDING 1046'
BANK OF MANHATTAN 900'
TRANSAMERICA PYRAMID 853'
WALL STREET BUILDING 950'

COMPARISON OF THE WORLD'S TALLEST STRUCTURES

Fig. 1-24. The CN Tower is a strong image of major importance in a big city like Toronto.

Fig. 1-25. The superfast trains of the French and Japanese convince engineers that good planning is needed for safe high-speed transportation. (SNCF French National Railroads.)

CARENAGE AVANT DU TGV

TRACE INFORMATIQUE

TGV du temps gagné pour toutes les régions en 1983.

PERSPECTIVE DE LA REMORQUE EXTRÉME DE 1re CLASSE (R1)

Fig. 1-26. Designers of racing cars on the international scene and engineers from around the world are planning tomorrow's winner on the drawing board today. (D. R. Shuck, Designer.)

all who are on the engineering team.

This textbook provides course work that assumes you are not experienced or skilled in drafting. It will provide opportunities to learn some of the basic drafting standards and techniques through hands-on experience (Fig. 1-28).

Start to build the best possible portfolio of drawings for a successful venture into drafting. You have a chance to add professional dimensions to your career plans for tomorrow. Without a command of the

If you could examine the plans of all types of vehicles more closely, you would learn about the importance of drafting details. The plans for vehicles, buildings, electronic communications, sound systems, or things needed to live are essential to the manufacturing process. Most products are made (manufactured) in factories, from plans sometimes called *prints*.

Products can add new dimensions to your life. Whether it is for your car, home, office, school, church, recreation center, or factory, technical progress is taking shape in the form of new products every year. What are some of the products you consider "new" this year? Do they require new technical advances? Research design? Inventions? Do we generally accept new technical advances without understanding their inner workings? Drafting technicians are the people who detail the inner working parts of new products.

Behind every major product or outstanding achievement are men and women of exceptional ability. They serve on design teams that are planning for your tomorrow (Fig. 1-27).

For a design project to go right and finish right, it has to start right. Drafting is an important part of starting any project. That's the reason drafting "know-how" is essential to

Fig. 1-27. A high-technology design team uses a microcomputer with a plotter, tablet, and cursor. (Bruning.)

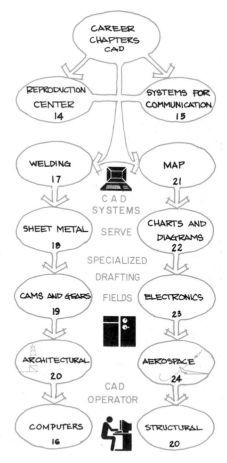

Fig. 1-28. A number of chapters in this book specialize in career-oriented subjects such as aerospace.

graphic language, engineers would be unable to communicate their ideas to members of their teams. All people should develop some technical knowledge of this universal language. It could be the most important course of study for an engineering or technical career.

For information about your future career plans, write to a professional society in your field of interest.

LEARNING ACTIVITIES

1. Write a short note or letter to a professional organization. Tell the public relations director you are studying drafting. Ask whether you can receive information or a brochure about special careers. Study the information to see what courses are recommended in high school and college. Also ask for posters or pamphlets that can be used for "engineering week" in February.

2. Each class member should try to bring at least one set of prints to class. (Some prints are called *blueline prints*.) Ask your parents, relatives, or neighbors to share information they may have for technical communications. Compare the drafting performed by the architect and the machine designer. Can you recognize the difference in techniques? How many kinds of prints can your class collect for an exhibit? Can anyone bring in a printed plan prepared by a computer with a plotter? Are unique techniques used by the computer? Can anyone bring in a microfilm card used for storing technical information?

3. Check the Yellow Pages of your local telephone directory for the number of architects, interior designers, and engineers in or near your community. Visit an architect's office and discuss the architect's responsibility as a designer. How big is the office? How many people are employed? Are there people in the office who specialize in other fields, such as a mechanical engineer? The average small office will have between 4 and 12 workers. Does the architect have a broad range of design work or a specialized area (for example, schools or churches)? Can you survey the type of work the other designers perform in your neighborhood?

4. Collect a set of working drawings and specifications for a building that was constructed in the last year in the community. Assemble a set of drawings that would represent the four levels of graphic communication. Select them from the catalogs or magazines in your home.

5. List what you consider to be the top five products developed for teenagers' recreational use. What are their important features? Compile a list of the top five achievements in American technology developed in the last five years. List the designer and the major features of each.

VOCABULARY

1. Plans
2. Drafting technician
3. Engineer
4. Lifestyle
5. Achievement
6. Want-ads
7. Employment
8. Development
9. Illustrator
10. Programmer
11. Reproduction
12. Specialization
13. Technology
14. Techniques
15. Prints
16. Designer
17. Architect
18. Universal
19. Drafter
20. Opportunity
21. Research
22. Coordinator
23. Detailer
24. Computer
25. Function
26. Aerodynamics
27. Processes
28. Skills

REVIEW

Answer True or False on a separate sheet of paper.

1. Detailed plans are needed before any new product is manufactured in an industrial plant.

2. There are six levels of graphic communication for the design-engineering team.

3. The drafting technician is not an important member of the engineering research team.

4. There are only five branches of engineering that use the drafting technician's services.

5. Drafting is a common family name for mechanical drawing and technical drawing.

6. The computer graphics programmer must have drafting experience and drafting skills.

7. There are only four major engineering achievements in the world.

8. Research is the top creative level of service performed by the engineer.

9. Mathematics and science are important skills for the drafting technician.

10. An illustrator is a specialized drafter who prepares picturelike drawings.

2 Sketching and Lettering

■ A LANGUAGE OF VISUAL SYMBOLS

When you see sketches bring ideas alive, you know that the saying *progress begins on paper* is true.

When words alone cannot describe new or futuristic forms, sketching is needed to show the thoughts that cannot be said (Fig. 2-1). The language of sketching has four basic *visual symbols* (things that can be

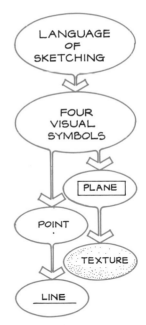

Fig. 2-2. Simple visual symbols are the basic elements of sketching.

seen). These are a point, a line, a plane, and a texture, or surface quality (Fig. 2-2). Any idea, no matter how simple or complicated or how plain or spectacular, can be sketched by using these four visual symbols. Good sketching often means using as

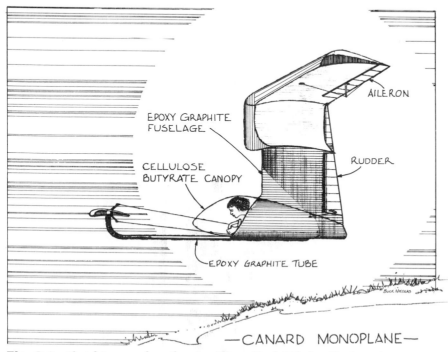

AILERON

EPOXY GRAPHITE FUSELAGE

CELLULOSE BUTYRATE CANOPY

RUDDER

EPOXY GRAPHITE TUBE

—CANARD MONOPLANE—

Fig. 2-1. Sketching can describe new ideas. (Buck Nikolas, Designer.)

17

few details as possible to show an idea fully.

REASONS FOR SKETCHING

There can be many reasons for doing technical sketching. The following are the most important:

1. To persuade people who make decisions about a project that an idea is good.
2. To develop a refined sketch of a proposed solution to a problem so that a client can respond to it (Fig. 2-3).
3. To show a complicated detail of a *multiview* (more-than-one-view) drawing enlarged or in a simple *pictorial* (picturelike) sketch.
4. To give design ideas to drafters so that they can do the detail drawings.
5. To develop a series of ideas for refining a new product or machine part.
6. To develop and analyze the best methods and materials for making a product.
7. To record permanently a design improvement on a project that already exists. The change may result from the need to repair a part that breaks over and over again. It may result from the discovery of an easier and cheaper way to make a part.
8. To show that there are many ways to look at or to solve a problem.
9. To spend less time in drawing. It is quicker to make a sketch, which takes only a pencil and a sheet of paper, than to do a mechanical drawing.

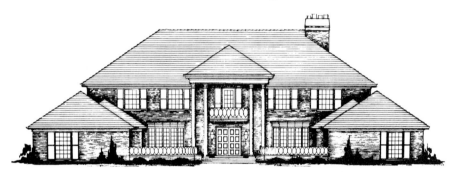

Fig. 2-3. A refined sketch for the client. (A. W. Wendell and Sons, Architects.)

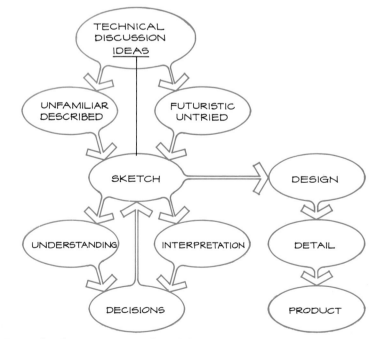

Fig. 2-4. Sketches can assist technical discussions.

SKETCHING AND TECHNICAL DISCUSSIONS

Freehand sketching is the simplest form of drawing. It is one of the quickest ways to express ideas. A sketch can help simplify a technical discussion. Designers, drafters, technicians, engineers, and architects will often explain complicated or unclear thoughts with a freehand sketch. Ideas imagined in the mind can be caught in sketches and thus held in simple lines for further study. The pencil is an instrument that aids clear thinking and creative communication. Figure 2-4 shows when sketching can become an important part of technical discussions and design decisions.

DESIGN SKETCHES NEED LETTERING

Freehand drawings generally need some freehand lettering to explain features of a new idea or how a new product works. A drawing well planned may be worth a thousand words, as an old saying goes, but a few choice words well organized can explain some details. Can you imagine what kind of notes could be used

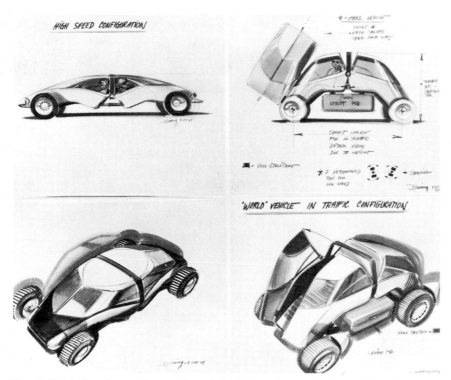

Fig. 2-5. A sketch can shape a futuristic design. (General Motors Corp.)

to assist in explaining the sketch of the foldable vehicle (Fig. 2-5)?

The notes lettered in Fig. 2-6 describe some features on a rough sketch that are functional and important to operation. On most of the sketches you have prepared or studied in the past, can you recall a need for notes to explain an idea? Simple freehand lettering will complement an idea that is captured in a sketch, especially if it is neat and carefully placed on the drawing.

LETTERING ON MECHANICAL DRAWINGS

To be sure that the notes and directions are clear, the drafter generally uses lettering instead of writing. Even on drawings made with instruments, notes are hand-lettered (Fig. 2-7). Good lettering is always needed by drafters throughout their careers. Lettering is one of the first things students of drafting study. This is because you do not learn how to do good lettering all at once. You learn it by practicing it little by little for a long time. Your lettering should get better with every sketch or drawing you do.

THE USE OF LETTERING

Simple freehand lettering, quickly made and perfectly *legible* (readable), is important to business, industry, and engineering. Single-stroke Gothic lettering (see Fig. 2-8) is used on technical drawings to tell the kinds of materials, sizes, distances, and amounts; to identify units; and to give other necessary information.

COMPOSITION

In lettering, *composition* means arranging words and lines with letters of the right style and size. Letters in words are not placed at equal distances from each other. They are placed so that the spaces between the letters *look equal*. The distance

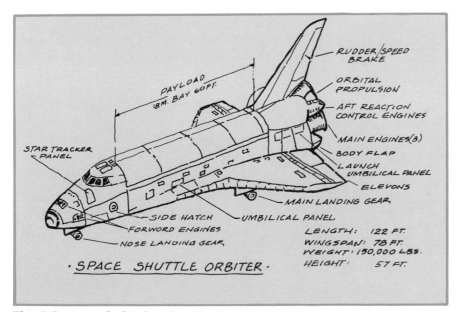

Fig. 2-6. A rough sketch with notes about important features.

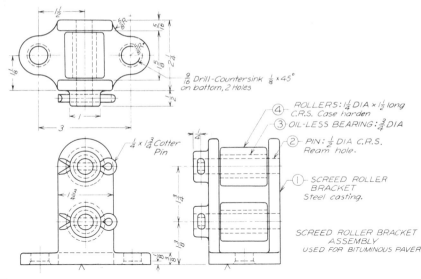

Fig. 2-7. Three-view drawing with good freehand lettering.

ABCDEFGHIJKLMNOPQRSTUVWXYZ

abcdefghijklmnopqrstuvwxyz

1234567890

ABCDEFGHIJKLMNOPQRSTUVWXYZ

abcdefghijklmnopqrstuvwxyz

1234567890

Fig. 2-8. Single-stroke Gothic letters, vertical and inclined.

LE TTERIN G CO MPOSITI O N
INVOLVESTHE SPACING OF LETTERS,
WORDS, ANDLINES AND THE CHOICE
OF APPROPRIATE STYLES AND SIZES.

INCORRECT LETTER, WORD, AND LINE SPACING

LETTERING COMPOSITION
INVOLVES THE SPACING OF LETTERS,
WORDS, AND LINES AND THE CHOICE
OF APPROPRIATE STYLES AND SIZES.

CORRECT LETTER, WORD, AND LINE SPACING

Fig. 2-9. Single-stroke lettering. Study the word spacing.

between words, called *word spacing*, should be about equal to the height of the letters. Figure 2-9 shows examples of proper and improper letter and word spacing.

GUIDELINES

In order to make letters the same height and in line, you must *rule* (draw with a straightedge) guidelines for the top and bottom of each line of letters. Draw guidelines lightly with a sharp pencil. The *clear distance* (open space) between lines of letters is from ½ to 1½ times the height of the letters. Use a lettering triangle or an Ames lettering instrument to space guidelines accurately (Fig. 2-10). Guidelines are spaced ⅛ in. (3 mm) apart for regular letters and *numerals* (numbers) on a drawing. Guidelines for fractions are usually twice the height of whole numbers, as shown in Fig. 2-11.

FRACTIONS

Fractions are always made with a division line that is *horizontal* (side-to-side). The whole fraction is usually twice the height of regular numbers. The numbers in the fraction are about three fourths the height of regular numbers. *Fraction numbers must never touch the division line* (Fig. 2-11).

SINGLE-STROKE INCLINED CAPITAL LETTERS AND NUMERALS

Inclined (slanted) letters and numerals should slope at an angle of about 67½° with the horizontal. Figure 2-12 shows some ways of laying out inclined guidelines. These may help you develop skill in hand lettering.

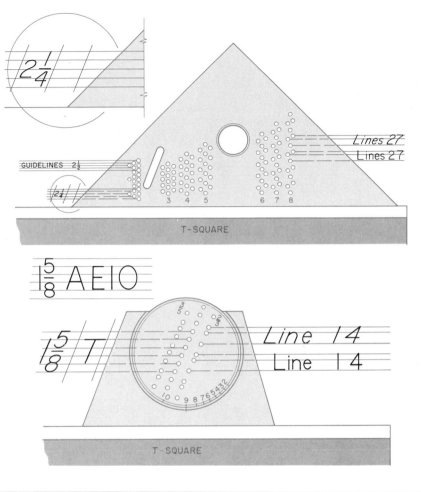

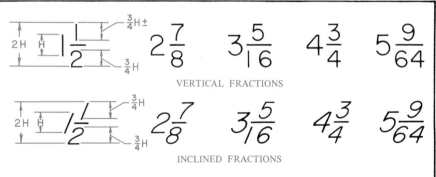

Fig. 2-10. Lettering guidelines should be evenly spaced. Use a lettering triangle or an Ames lettering guide.

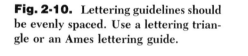

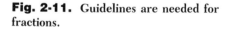

VERTICAL FRACTIONS

INCLINED FRACTIONS

Fig. 2-11. Guidelines are needed for fractions.

Fig. 2-12. There are various methods for laying out inclined guidelines for lettering.

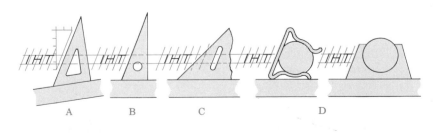

A   B   C   D

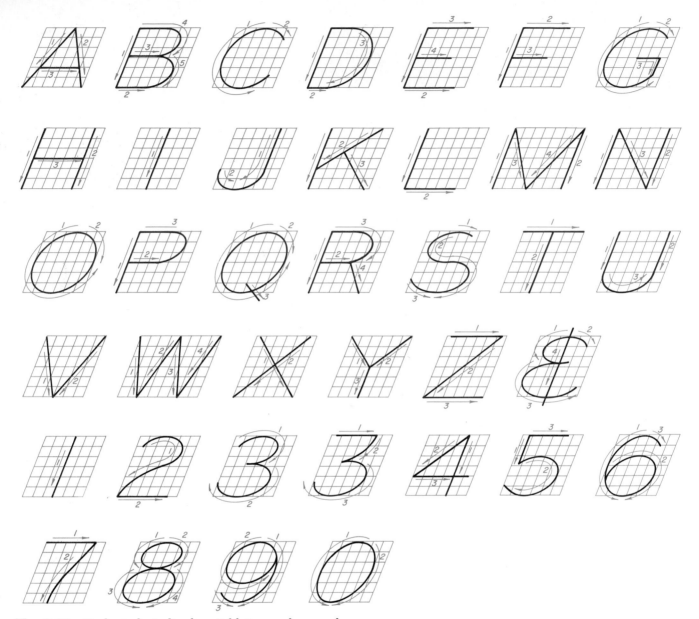

Fig. 2-13. Single-stroke inclined capital letters and numerals.

Guidelines should be very thin and light so that they will not detract from the letters.

Figure 2-13 shows single-stroke inclined letters and numerals on an inclined grid. Notice that the only difference between *vertical* (straight) and inclined letters is the slant.

■ SINGLE-STROKE VERTICAL CAPITAL LETTERS AND NUMERALS

The shapes and proportions of letters and numerals are shown in Fig. 2-14. Study these characters carefully until

you understand completely how each is made. Each character is shown in a square 6 units high. The squares are divided into unit squares. By following these, you can easily learn the right shapes, proportions, and strokes. You can vary your lettering to make it more individual. Figure 2-15 shows some of the possible variations. They are common styles for designers and architects.

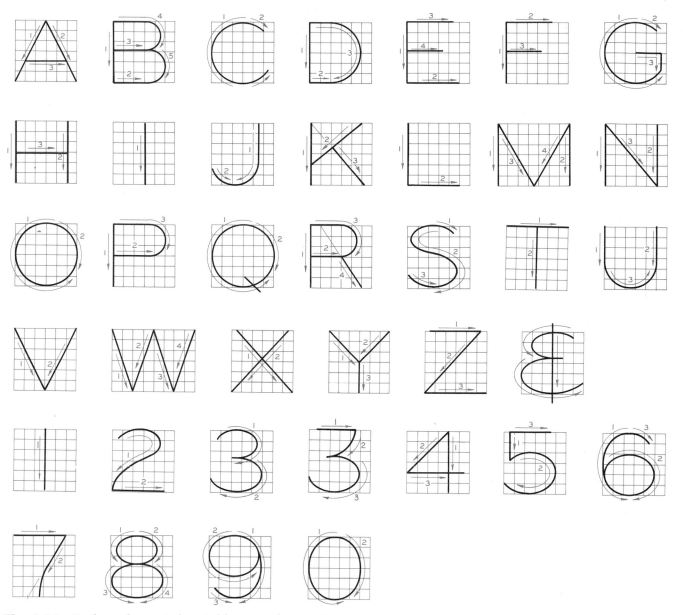

Fig. 2-14. Single-stroke vertical capital letters and numerals.

ABCDEFGHIJKLMNOPQRSTUVWXYZ 1234567890

ABCDEFGHIJKLMNOPQRSTUVWXYZ 1234567890

ABCDEFGHIJKLMNOPQRSTUVWXYZ
1234567890

Fig. 2-15. Variations in lettering styles for the designer.

■ VARIOUS STYLES OF LETTERING

There are various styles of letters, a few of which are shown in Fig. 2-16. Each style is right for a particular use. Be careful in choosing one. For example, it is not practical to use fancy roman-style lettering for notes and dimensions on technical drawings. Such lettering would be more

VERTICAL
lower case
INCLINED
COMPRESSED
EXTENDED
BOLD FACE
SLANT *Italics*
𝔒𝔩𝔡 𝔈𝔫𝔤𝔩𝔦𝔰𝔥
MODERN
OLD ROMAN
MICROFONT
ABCDEFGHIJKLMNOPQR
STUVWXYZ1234567890

Fig. 2-16. Various styles of lettering. Microfont is a style used for microfilm reproduction.

costly in time and money than the single-stroke styles in Figs. 2-13 and 2-14. Yet it would serve the purpose of the drawing no better.

The lettering style most commonly used on working drawings is *single-stroke commercial Gothic*. This style is best because it is easy to read and easy to hand letter. It is made up of *uppercase* (capital) letters, *lowercase* (small) letters, and numerals. Nearly all companies now use only uppercase lettering. As a result, this book stresses uppercase lettering. Letters and numerals may be either vertical or inclined. However, the same style should be followed throughout a set of drawings.

■ DISPLAY LETTERING

Display lettering is often used for title sheets, posters, and other places where large, easy-to-read lettering is needed. Although any style may be used for display, the drafter most often uses Gothic, the most legible of all styles. Figure 2-17 shows a block-style and a regular-style Gothic.

The block style is the easiest to make. It has no curved strokes and only takes a T-square and triangle. A

light-line grid may be drawn on the sheet and later erased. If you are using a tracing *medium* (drawing material), you can put a grid sheet under the drawing sheet. Block in the letters in pencil first. Then, if desired, trace them in ink. Letters can be either outlined or filled in, as shown in Fig. 2-17. The thickness of the strokes can vary from one fifth to one tenth the height of the letter. In Fig. 2-17, one seventh is used. In Fig. 2-18, one fifth through one tenth are shown.

Regular Gothic capital letters are made in much the same way. The only difference is that you have to make curved strokes. Do this with circle templates or a compass, or draw the curves freehand.

You can change the size and proportion of letters by using the grid as shown in Fig. 2-19. Notice that letters can be made larger or smaller overall by using this method. In addition, they can be stretched or squeezed.

FREEHAND DRAWING

■ TYPES OF SKETCHES— ROUGH (UNREFINED) AND REFINED

Any image drawn on paper *freehand* (without a straightedge or other tools) may be called a *sketch*. The sketching may be *rough*, or *unrefined*. That is, the sketch may be drawn quickly with jagged lines. The rough sketch can quickly express thoughts. Figure 2-20 shows rough sketches that were used to develop *preliminary* (early) designs of a two-position automobile mirror. The sketches show several design

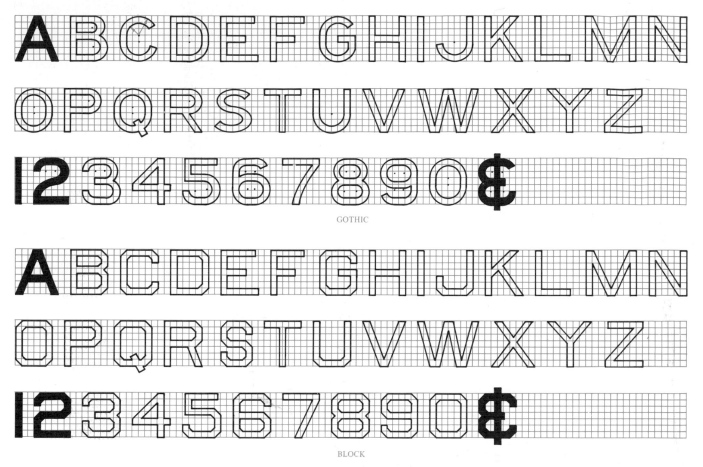

GOTHIC

BLOCK

Fig. 2-17. Block-style and regular-style Gothic letters.

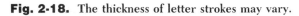

ONE - FIFTH   ONE - SIXTH   ONE - SEVENTH   ONE - EIGHTH   ONE - NINTH   ONE - TENTH

Fig. 2-18. The thickness of letter strokes may vary.

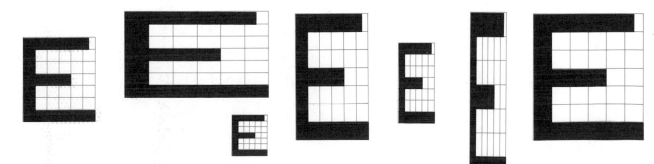

Fig. 2-19. A grid method for changing the size and proportion of letters.

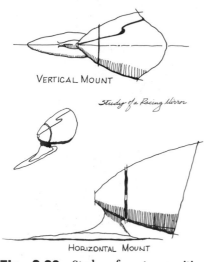

VERTICAL MOUNT

Study of a Racing Mirror

HORIZONTAL MOUNT

Fig. 2-20. Study of a two-position mirror for a racing car.

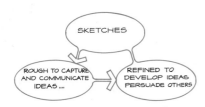

Fig. 2-21. Sketching—rough to refined.

choices. The other way of sketching is *refined*. That is, the sketch is neat and looks finished. A refined sketch is carefully drawn. It shows good proportion and excellent line values. It may be more persuasive than an unrefined sketch. Many refined sketches are based on a rough sketch that has captured the general idea. Figure 2-21 shows the usual way sketching is used to communicate. A very good way to *refine* (improve) a sketch is to use an overlay.

SKETCHING RULES

Rule One
Rough sketching

Never use instruments or straightedges when you are preparing rough sketches. Instru-

ments will restrict the creative expressions developed with good pencil techniques. Avoid a mechanical, hard line look and work on good proportions with a few choice notes.

Rule Two
Refined sketching

Straightedges may be used for controlling long lines when refined sketches are being prepared. However, never allow the line to look or appear to be mechanically drawn. Sketched lines should have some irregular character.

THE OVERLAY

Sketches are often drawn on paper that can be seen through. This paper is called *translucent paper* or *tracing paper*. The best parts of a sketch may be quickly traced by putting a new piece of this paper, called the *overlay*, on top (Fig. 2-22). Thus, refining ideas means sketching over and over again on tracing paper until the design is right.

TWO USES FOR OVERLAYS

The overlay is used in two important ways that may seem very much alike. The first use is reshaping an idea. This might include refining the pro-

portions of the parts of an object or changing its shape entirely. Second, an overlay can be used to refine, or improve, the drawing itself without really changing the design. These two things can be, and usually are, done at the same time (Fig. 2-23).

NATURE OF A SKETCH

A sketch is an important form of *graphic* (drawn) communication. There are many levels of sketches for different uses. (Fig. 2-24). If you know the language of mechanical drawing well, you will know how to sketch your ideas quickly, clearly, and accurately.

Temporary

Many technical sketches have short lives. They are done merely to solve an immediate problem. Then they are thrown away. Other technical sketches are kept longer. It may take weeks or even months to study some sketches and make mechanical drawings from them. However, these sketches too may be thrown away some day.

Permanent

Sometimes the engineering department or the management of a company will include a sketch in a notice to other employees. Such a sketch is an important record and should be kept. Therefore, some sketches are

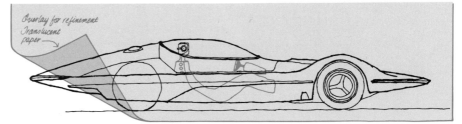

Overlay for refinement
Translucent paper

Fig. 2-22. The overlay can speed up the design process.

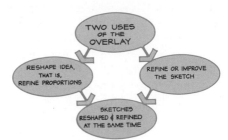

Fig. 2-23. Two uses for the overlay process.

filed as part of a company's permanent records.

Presentation

The sketch that is refined for presentation is generally *pictorial* (picturelike). It is used to convince a client or management to accept and approve the ideas presented. The pictorial sketch has a three-dimensional view that can be understood easily by nontechnical people. Such sketches are generally drawn so that they look glamorous, artistic, or eye-appealing.

■ VIEWS NEEDED FOR A SKETCH

In Chap. 5, multiview projection will be discussed in full. However, you need to know some basic things about views and how they are placed in order to do sketches.

There are two types of drawings that you can sketch easily. One is called *pictorial* (picturelike) *drawing*. In this type of drawing, the width, height, and depth of an object are shown in one view (Fig. 2-25). The other is called *multiview projection* or *orthographic projection*. In this type of drawing, an object is usually shown in more than one view. You do this by drawing sides of the object and relating them to each other, as shown in Fig. 2-26.

■ ONE-VIEW SKETCH

If an object can be fully shown in only one view, this means that its shape can be described well in two *dimensions* (sizes). These are height and width. Things shown in one-view drawings generally have a depth or thickness that is *uniform* (the same throughout). This third dimension can be given in a note rather than being drawn. A typical one-view drawing is shown in Fig. 2-27 at A. The thickness of the stamping at B is shown by a note on the

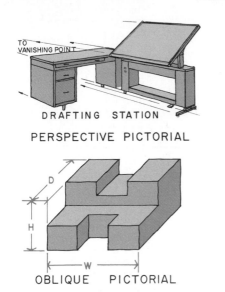

Fig. 2-25. Comparing oblique and perspective pictorials.

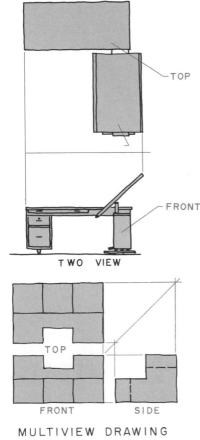

Fig. 2-26. Typical multiview drawings (two view and three view).

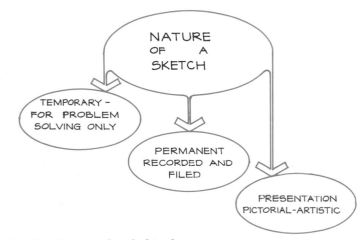

Fig. 2-24. Sketches are classified in three ways.

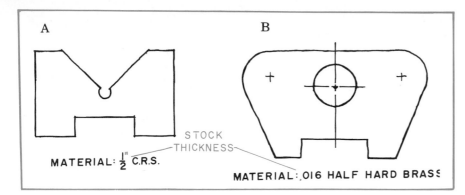

Fig. 2-27. Typical one-view drawings.

sketch. Many objects that are shaped like cylinders can also be shown in single views if the diameter of the cylindrical part is noted, as in Fig. 2-28.

TWO-VIEW SKETCHES

There are many objects that can be described in only two views, such as the one shown in Fig. 2-29. If two views that describe the object well are carefully selected, this can help simplify the drawing. Figure 2-30 shows 5 three-view drawings in which the objects could have been described well with only two views. Which view is not needed at A, B, C, D, and E? At A, is it the top view?

MULTIVIEW SKETCHES

A pictorial drawing shows how the object looks in three-dimensional form. Three directions are suggested for viewing the residence in Fig. 2-31. From in front, the width and height would show. This is the front view. From the side, the depth and height would show. This is the right-side view. From above, the depth and width would show. This is the top view. However, a three-view

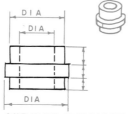

Fig. 2-28. A cylindrical object may require only one view.

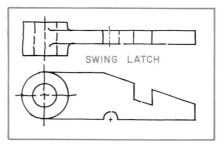

Fig. 2-29. A two-view sketch.

pictorial does not fully describe the residence. There are lines and details that do not show completely (Fig. 2-32).

Multiview projection, also called *orthographic projection*, is the sys-

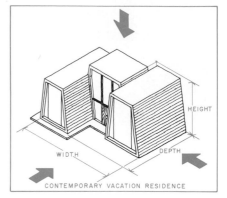

Fig. 2-31. A-frame residence in pictorial showing three-dimensional form.

tem that arranges views in relation to each other (Fig. 2-33).

THE GLASS BOX

The residence in Fig. 2-32 can be thought of as being inside a *transparent* (clear) glass box. By looking at each side of the building through the glass box, you can see the five possible views of the object. When the glass box is opened up into one plane (Fig. 2-33), the views are placed as they would be drawn on paper.

MATERIALS FOR SKETCHING

The good things about sketching are that only a few materials are required and that it can be done anywhere. You are ready to sketch with a pencil, an eraser, and a pad of paper. If you need more equipment than that, you are probably not as good a drafter as you could be.

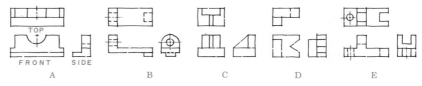

Fig. 2-30. Select the two required views at A, B, C, D, and E.

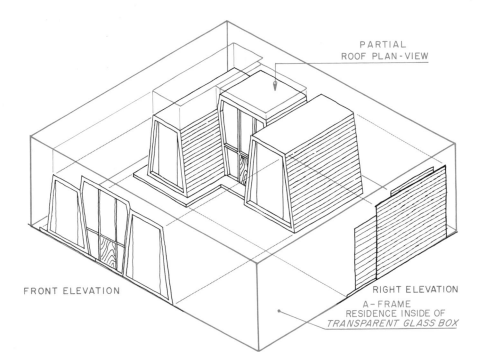

Fig. 2-32. Two elevations can be projected from the pictorial drawing to the transparent glass box.

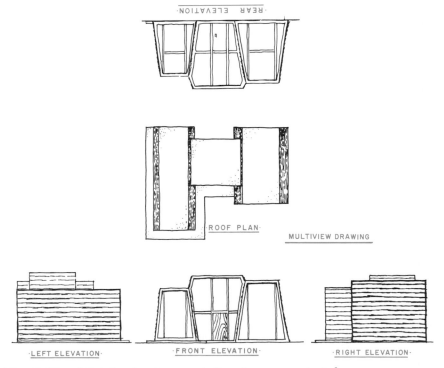

Fig. 2-33. If the glass box is opened, five views are projected.

Paper

You can use plain paper for sketching. If you need to refine the sketch, use tracing paper. Control proportions while sketching by using *cross-section paper*, also called *graph paper* or *squared paper*. The graph paper that is used most often has heavily ruled 1-inch (in.) squares. The 1-in. squares are then subdivided into lightly ruled $\frac{1}{10}$-, $\frac{1}{8}$-, $\frac{1}{4}$-, or $\frac{1}{2}$-in. squares. This paper is called 10 to the inch, 8 to the inch, and so on. Graph paper is also ruled in millimeters (mm). There are many specially ruled graph papers for particular kinds of drawing such as isometric and perspective. These kinds of drawings are explained in Chap. 12.

You can sketch on any convenient size of paper. However, standard 8½ × 11 in. (216 × 279 mm) letter paper is the best for making small sketches quickly. You can hold the paper on stiff cardboard or on a clipboard while working on it. If you use cross-section paper, put it under tracing paper to help guide line spacing.

Pencils and Erasers

Most drafters like to use soft lead pencils (grades F, H, or HB), properly sharpened. They also use an eraser that is good for soft leads, such as a plastic eraser or a kneaded-rubber eraser.

Use a drafter's pencil sharpener to remove the wood from the *plain* end of a pencil. This is done so that the grade mark (F, H, or HB) on the other end will show. Sharpen the lead to a point on a sandpaper block or on a file. If you are using a lead holder, use a lead pointer. Do not forget to adjust the grade mark in the window of the lead holder if it has one. Be careful to remove the needle point that the sharpener leaves by touching it gently on a piece of scrap paper. Then the pencil will not groove or tear the drawing paper.

Four types of points are used for sketching: sharp, near-sharp, near-dull, and dull (Fig. 2-34 at A). The points should make lines of the following kinds:

- *Sharp point*—a thin black line for center, dimension, and extension lines
- *Near-sharp point*—visible or object lines
- *Near-dull point*—cutting-plane and border lines
- *Dull point*—construction lines

Lines drawn freehand have a natural look (see Fig. 2-34 at B). They show freedom of movement because of their slight changes in direction. Hold the pencil far enough from the point that you can move your fingers easily and yet can put enough pressure on the point to make dense, black lines when you need to. Draw light construction lines with very little pressure on the point. They should be light enough that they need not be erased.

STRAIGHT LINES

You can sketch lines in the following ways: (1) Draw them continuously. (2) Draw short dashes where the line should start and end. Then place the pencil point on the starting dash. Keeping your eye on the end dash, draw toward it. (3) Draw a series of strokes that touch each other or are separated by very small spaces. (4) Draw a series of overlapping strokes (Fig. 2-35 at A).

Before you try to draw objects, practice sketching straight lines to improve your line technique. Draw horizontal lines from left to right.

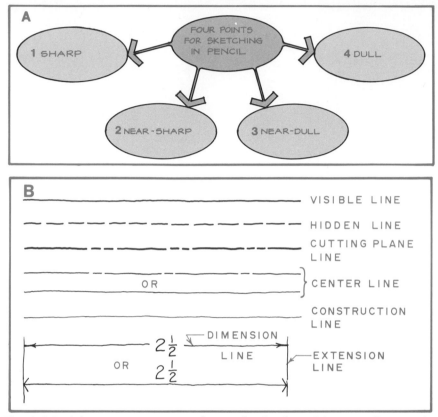

Fig. 2-34. (A) Four convenient pencil points for sketching. (B) Types of lines used in sketching.

Draw vertical lines from the top down (Fig. 2-35 at B).

SLANTED LINES AND SPECIFIC ANGLES

Sketch slanted, or inclined, lines from left to right. It might be easiest to turn the paper and draw an inclined line the same way as a horizontal one. When trying to sketch a specific angle, first draw a vertical line and a horizontal line to form a *right angle* (90°). Divide the right angle in half to form two 45° angles (Fig. 2-36). Or divide it in thirds to form three 30° angles. By starting with these simple angles, you can

estimate (guess) other angles more exactly. Note the direction of the inclined lines drawn to form a desired angle in Fig. 2-37.

PROPORTIONS FOR SKETCHING

Sketches are not usually made to *scale* (exact measure). Nonetheless, it is important to keep sketches in *proportion* (similar to exact measure). In preparing the layout, look at the largest overall dimension, usually width, and estimate the size. Next, determine the proportion of the height to the width. Then, as the front view with the width and height takes shape, compare the smaller de-

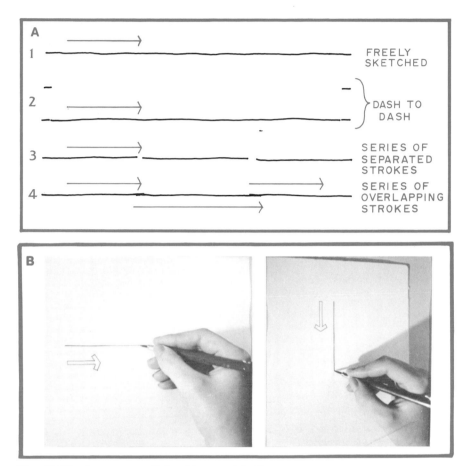

A

1 ——————————→   FREELY SKETCHED

2 ——————————→   } DASH TO DASH

3 ——————————→   SERIES OF SEPARATED STROKES

4 ——————————→   SERIES OF OVERLAPPING STROKES

Fig. 2-35. Some ways of sketching straight lines.

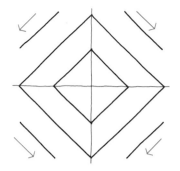

Fig. 2-36. Draw horizontal and vertical lines before sketching slanted lines.

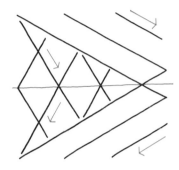

Fig. 2-37. Sketching slanted lines and angles.

tails with the larger ones and fill them in too (Fig. 2-38).

It is important that the design drafter, in sketching an object, have a good sense of how distances relate to each other. This will allow the drafter to show the width, height, and depth of an object in the right proportions.

For example, suppose that the design drafter plans a contemporary stereo cabinet to be 4 units wide. The height of the cabinet is 2 units. The depth of the cabinet is 1 unit. This is a proportion of 2 to 1. If you were designing this cabinet in customary units, you might choose a 15-in. width, height, and depth unit.

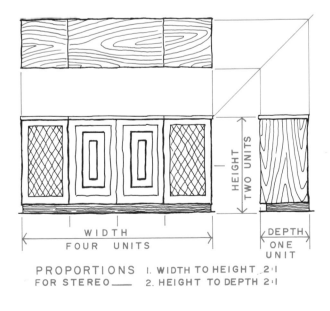

HEIGHT TWO UNITS

DEPTH ONE UNIT

WIDTH FOUR UNITS

PROPORTIONS FOR STEREO —— 1. WIDTH TO HEIGHT 2:1   2. HEIGHT TO DEPTH 2:1

Fig. 2-38. Sketching a contemporary stereo with proportional units.

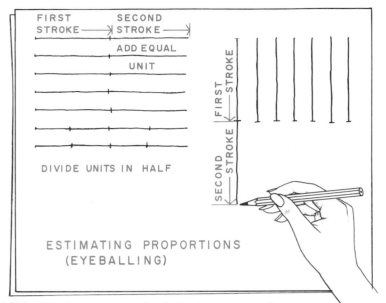

Fig. 2-39. Practice estimating (eyeballing) proportional units.

The width overall would be 60 in. The height would be 30 in., and the depth 15 in. The proportions are developed in 2-to-1 units in Fig. 2-38.

TECHNIQUE IN DEVELOPING PROPORTION

In order to sketch well, the design drafter must be able to divide a line in half by *eyeballing* (estimating by eye). The halves can be divided again to give fourths. Through practice, you can train your eye to work in at least two directions. Start by drawing a line of 1 unit. Increase it by one equal unit so that it is twice as long as at first. Practice adding an equal unit and dividing a unit equally in half. Practice developing units on parallel horizontal lines. Then develop them vertically. By learning to compare distances, you can get better and better at estimating (Fig. 2-39).

Design drafters can use scrap paper or a rigid card as a straightedge when they do not have *scales* (rulers) at hand. Fold the paper in half to find the length of units on the marked edge, as in Fig. 2-40. In this way, you can draw lines of the same length several times in different directions. This will help maintain the right proportions.

Figure 2-41 shows how to use *diagonal* (corner-to-corner) lines to

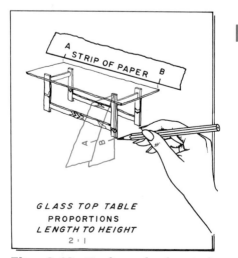

Fig. 2-40. Drafters develop techniques for blocking out sketches.

multiply proportional units. At A, the square *ABCD* is sketched using diagonals. At B, a horizontal centerline is sketched through the *intersection* (crossing point) of the diagonals. The centerline is then extended. At C, the large diagonal lines *AL* and *BN* are sketched. The rectangle with a length 3 times the height is then completed. The side view (at D) is sketched using diagonals, as at A.

EXAMPLE: SKETCH OF A CHAIR

The chair is shown in picture form in Fig. 2-42. Note that three views are enough to sketch all the important features. After looking at the chair carefully, estimate the width, height, and depth. Then block in the major dimensions as shown. Large areas are first drawn in to the right proportion. Next, the smaller features are added in the right places. Never try to finish one view at a time. Block them all in, then work out the same details in each view. By sketching this way, you can develop views of an object that have the right proportions to each other.

STEPS IN MAKING A SKETCH

You have to practice carefully to gain the skills needed to make good sketches. The steps in making a sketch are shown in Fig. 2-42.

1. Observe the chair at A.
2. Select the views needed to show all shapes.
3. Estimate the proportions carefully. Mark off major distances for width, height, and depth in all three views at B.
4. Block in the enclosing rectangles as at C.

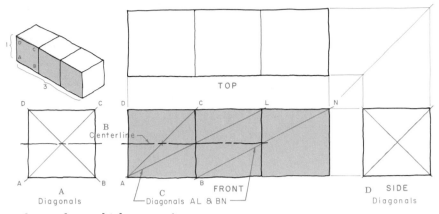

Fig. 2-41. The diagonal is used to multiply proportions.

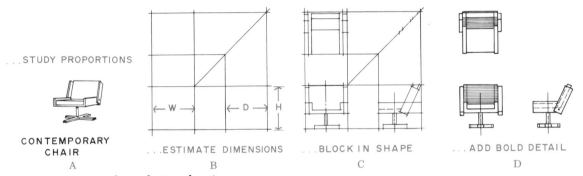

Fig. 2-42. The development of a multiview drawing.

5. Locate the details in each of the views. Block them in as at D.
6. Finish the sketch by darkening the object lines.
7. The blocking-in lines that you start with should be light enough that you do not have to erase them. You can easily draw the dashes for hidden lines over the light blocking-in lines.
8. Add the dimensions and notes that you need.

■ CIRCLES AND ARCS

There are two ways to sketch circles. For the first way, draw very light horizontal and vertical lines. Next, estimate the length of the *radius* (the distance from the center of the circle to its edge; plural, *radii*) and mark it off. Then draw a square in which you can sketch the circle (Fig. 2-43). For the second way, first draw very light centerlines. Then draw *bisecting* (halving) lines through the center at convenient angles (Fig. 2-44). Next, mark off estimated radii of the same length on all lines (Fig. 2-45). The bottom of the curve is generally easier to form, so draw it first. Then turn the paper so that the rest of the circle is on the bottom. Finish drawing it (Fig. 2-46). You can sketch *arcs* (parts of a circle); *tangent arcs* (parts of two circles that touch); and *concentric circles* (circles of different diameters that have the same center) in the same way you sketch circles (Fig. 2-47). Use light, straight, construction lines to block in the area of the figure.

For large circles and arcs and for *ellipses* (ovals), use a scrap of paper

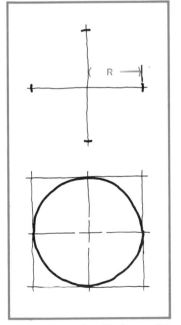

Fig. 2-43. Mark off the radii and draw a square in which to sketch a circle.

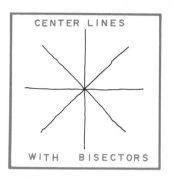

Fig. 2-44. For a circle, draw centerlines with bisecting lines.

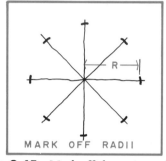

Fig. 2-45. Mark off the estimated radii on all lines before sketching the circle.

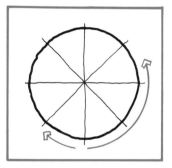

Fig. 2-46. The bottom, or concave, side of a curve is the easiest to form.

with the radius marked off along one edge. Put one end of the marked-off radius on the center of the circle. Draw the arc by placing a pencil at the other end of the radius and turning the scrap paper (Fig. 2-48). Note in Fig. 2-48 that two radii are needed for an ellipse. To draw the ellipse, sketch both centerlines. Keep both radius points on the centerlines as

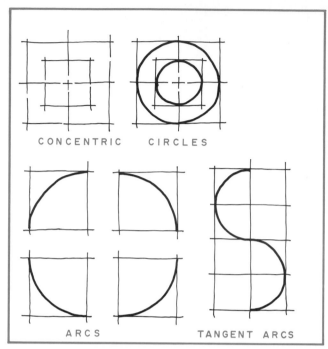

Fig. 2-47. Arcs and concentric circles are controlled by sketching squares.

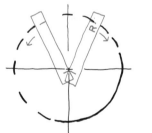 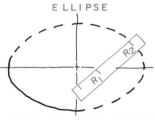

Fig. 2-48. Large circles, large arcs, and ellipses are easily sketched with the aid of a strip of paper.

you draw with the pencil at the other end.

You can also use your hand as a compass. To do this, use your little finger as a *pivot* (turning point) at the center of the circle. Use your thumb and forefinger to hold the pencil rigidly at the radius you want. Turn the paper carefully under your hand, thereby drawing the circle (Fig. 2-49). You can also draw circles with two crossed pencils. Hold them rigidly with the two points as far apart as the length of the desired radius. Put one pencil point at the center.

Fig. 2-49. The hand as a compass, using the little finger as a pivot point.

Fig. 2-50. Two pencils can serve as a compass.

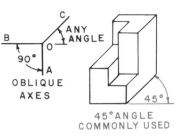

Fig. 2-51. Oblique drawings always have one right-angle corner.

UNFOLDING OBLIQUE PICTORIAL

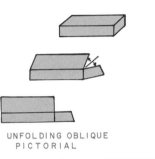

OBLIQUE RENDERING

Fig. 2-52. Oblique drawings can be developed with rendering techniques.

Hold it there firmly and turn the paper, drawing the circle with the other point (Fig. 2-50).

PICTORIAL DRAWING

There are several kinds of pictorial drawings. All types will be discussed in Chap. 12. For sketching, we will consider only two kinds of pictorial drawings, oblique and isometric. Making oblique and isometric sketches will help you learn how to "see" objects in your mind. You must be able to do this in order to draw *multiview projections* (three-view drawings). Pictorial drawings help people who are not trained to read multiview drawings understand basic shapes.

OBLIQUE SKETCHING

One of the easiest pictorial drawings to make is the *oblique* (inclined) sketch. In an oblique sketch, the depth of an object is drawn at any angle. Every object has three *dimensions* (sizes or directions): *width*, *height*, and *depth*. Each of these dimensions is called an *axis* (plural, *axes*). In oblique drawings, two of the axes are at *right angles* (90°) to each other. The third axis is drawn at any angle to the other two (Fig. 2-51). You may make any side of an object the front view. The front shows the width and height of the object. You usually make the side with the most detail the front view. In Fig. 2-52, a digital clock radio is used as an example. It is shown in an oblique pictorial view. The dial side has been made the front view. This is because it has the most detail and shows the width and height of the radio. The front view is sketched just

as the clock radio would appear when you look at it.

OBLIQUE SKETCHING ON GRAPH PAPER

Graph paper is useful for oblique sketching. This is so because the front view of the oblique sketch is like the front view of a multiview sketch, as shown in Fig. 2-53 at A and B. If you develop the oblique pictorial drawing on graph paper from a multiview drawing on graph paper, simply transfer the dimensions from one to the other by counting the graph-paper squares:

1. Block in lightly the front face of the object by counting squares.
2. Sketch lightly the *receding* (going-away) axis by drawing a line diagonally through the squares. You find the depth by using half as many squares, as shown on the side view of Fig. 2-53 at A.
3. Sketch in any arcs and circles.
4. If you have drawn all the layout lightly, darken the final object lines.

OBLIQUE LAYOUT

To make an oblique sketch, as shown in Fig. 2-54, always follow a good layout procedure:

1. Block in lightly the front face of the object with the estimated units for the width and height (Fig. 2-55).
2. Sketch in lightly the receding lines at any angle, for example, about 45° with the horizontal. You can choose an angle that will show as much as desired of the top and side. Thus, if you draw the third axis at a small angle like 30°, the side will show more

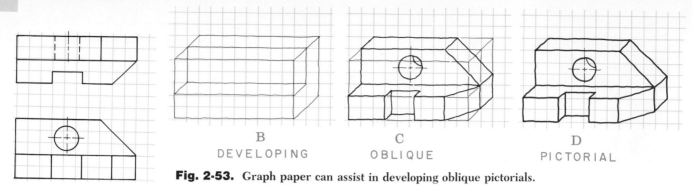

A GIVEN VIEWS

B
DEVELOPING

C
OBLIQUE

D
PICTORIAL

Fig. 2-53. Graph paper can assist in developing oblique pictorials.

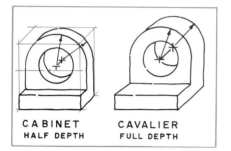

CABINET
HALF DEPTH

CAVALIER
FULL DEPTH

Fig. 2-54. Oblique drawings can vary in depth.

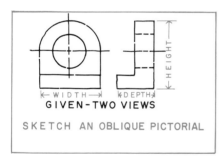

GIVEN-TWO VIEWS

SKETCH AN OBLIQUE PICTORIAL

Fig. 2-55. Check the width, height, and depth on the views given.

clearly (Fig. 2-56 at A). On the other hand, if you choose 60°, the top will show more clearly (Fig. 2-56 at B).

3. The receding axis may have the same proportioned units as the front axis. Or you can reduce its size by up to one half. When the depth dimension of a drawing is exactly one half the true dimension, it is called a *cabinet sketch*. Using a full-depth dimension produces a *cavalier sketch* (Fig. 2-54).

4. If you have drawn all the layout lightly, darken the final object lines.

OBLIQUE CIRCLES

In oblique sketching, circles in the front view can be drawn in their true shape. However, oblique circles drawn in the top or side views appear *distorted* (not in their true shape). Indeed, you must draw an ellipse to show such circles. Ellipses are not good in oblique pictorial sketching. It is better to show the circular shapes of important parts in the front view (Fig. 2-54).

ISOMETRIC AXES

An *isometric* (equal-measure) sketch is based on three lines called *axes*. These are used to show the three basic dimensions: width, height, and depth. A *cube* is an object with six equal square sides (Fig. 2-57). The isometric cube will have three equal sides. Thus, there are three equal angles at the Y axis at B (Fig. 2-57). The height *OA* is laid off on the vertical leg of the Y axis. The width *OB* is laid off to the left on a line 30° above the horizontal. The depth *OC* is laid off to the right on a line 30° above the horizontal. The 30° lines receding to the left and right can be located by estimating one third of a right angle, as shown in Fig. 2-57.

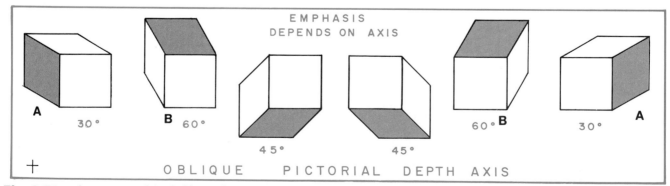

EMPHASIS
DEPENDS ON AXIS

A   30°

B   60°

45°

45°

60° B

30°   A

OBLIQUE   PICTORIAL   DEPTH   AXIS

Fig. 2-56. The many angles of oblique drawing.

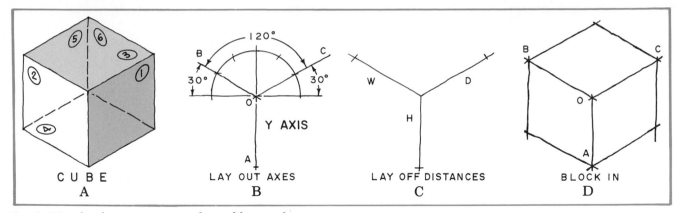

Fig. 2-57. Sketching isometric angles and layout of Y axis.

Lines parallel to the axes are called *isometric lines*. The estimated distances are laid off on them only as shown for the cube at C (Fig. 2-57).

The sketched lines for isometric axes tend to become steeper than 30° if you do not carefully prepare the layout. A better pictorial sketch will result when the angle is at 30° or a little less. Using isometric graph paper with 30° ruling lets you make sketches quickly and easily (Fig. 2-58). Figure 2-59 shows the steps in making an isometric sketch.

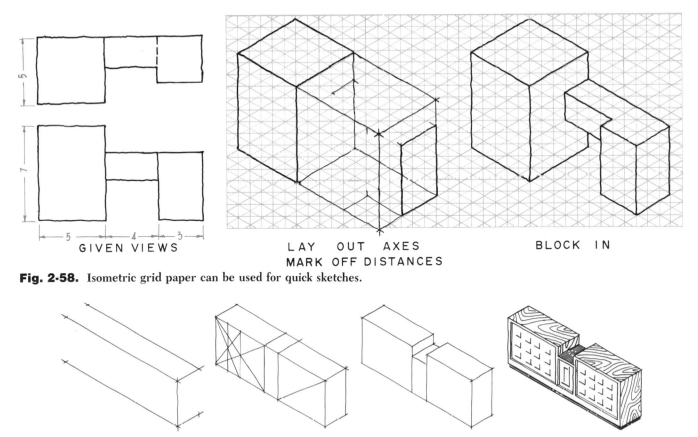

Fig. 2-58. Isometric grid paper can be used for quick sketches.

Fig. 2-59. The steps in making an isometric sketch.

■ NONISOMETRIC LINES ON ISOMETRIC SKETCHES

As we have seen, lines parallel to isometric axes are called isometric lines. Therefore, lines that are not parallel to the isometric axes are labeled *nonisometric lines* as in Figs. 2-60 and 2-61.

Any object may be sketched in a box, as suggested in Fig. 2-62. Note, however, that the objects in Figs. 2-63 and 2-64 have some lines that are not parallel to the isometric axes.

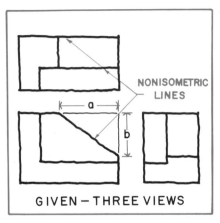

Fig. 2-60. Identifying the nonisometric lines on a three-view drawing.

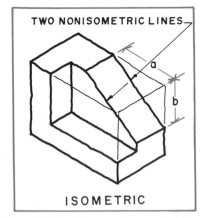

Fig. 2-61. The nonisometric lines form an inclined plane on the isometric drawing.

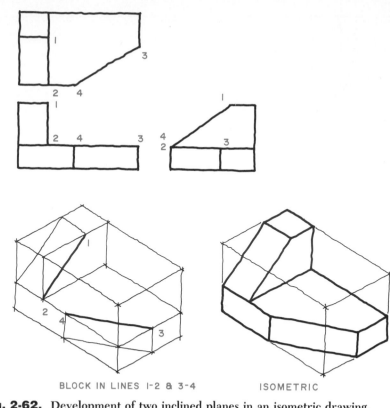

BLOCK IN LINES 1-2 & 3-4   ISOMETRIC

Fig. 2-62. Development of two inclined planes in an isometric drawing.

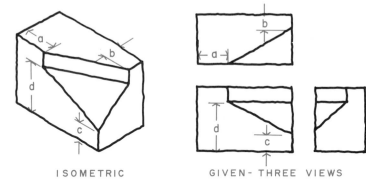

ISOMETRIC   GIVEN-THREE VIEWS

Fig. 2-63. Developing an oblique plane in an isometric drawing.

Lines not parallel to the axes can be drawn by extending their ends to touch the blocked-in box. Locate points at the ends of slanted lines by estimating measurements along or parallel to isometric lines. Having located the ends of the nonisometric lines, you can sketch the lines from point to point. Lines that are parallel to each other will also show parallel on the sketch (Fig. 2-64). Note how the ends have been located on lines 1-2 and 1-3 in Fig. 2-64. Distances *a* and *b* are estimated and transferred from the figure at A to B. Any inclined line, plane, or specific angle must be found by locating two points of intersection on isometric lines.

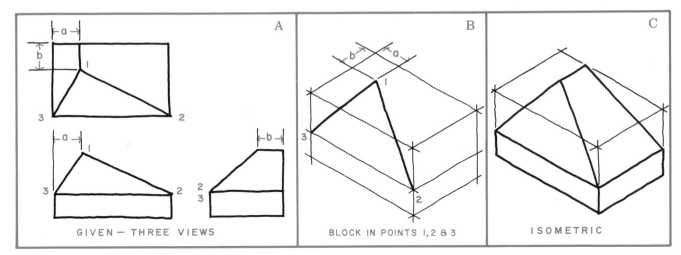

Fig. 2-64. Development of three inclined surfaces in an isometric drawing.

CIRCLES AND ARCS ON ISOMETRIC SKETCHES (ELLIPSES)

To sketch a circle in an isometric view (Fig. 2-65), sketch an isometric square first. The small-end arcs are sketched *tangent to* (touching) the square. Then sketch the larger arcs tangent at points *T* to finish the ellipse. Note that the *major diameter* (long axis) is longer than the true diameter of the circle. The *minor diameter* (short axis) is shorter. This difference is caused by the isometric angle. Figure 2-65 shows an ellipse for a top view only.

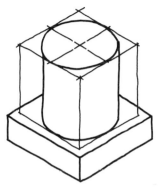

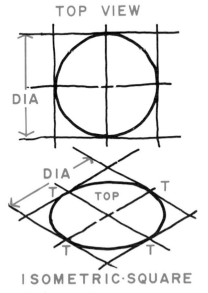

Fig. 2-65. The isometric square with arcs tangent to form an ellipse (isometric circle).

Fig. 2-66. Isometric circles (ellipses) sketched on the front, top, and sides of a cube.

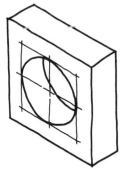

CYLINDRICAL SHAPES

Fig. 2-67. Isometric circles assist in describing cylindrical forms.

Circles on the three faces of an isometric cube are sketched in Fig. 2-66. Some ways to block in *cylindrical* (cylinderlike) shapes are shown in Fig. 2-67. Some ways to block in *con-*

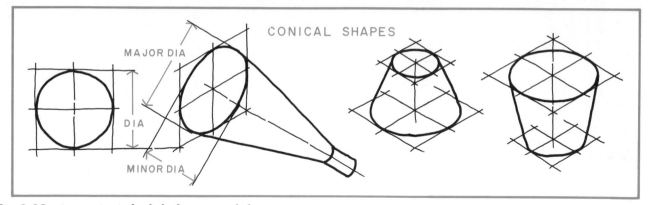

Fig. 2-68. Isometric circles help form conical shapes.

ical (conelike) shapes are shown in Fig. 2-68.

Arcs developed in an isometric view are shown in Fig. 2-69. A semicircular opening and rounded corners appear in the front view. The object is blocked in at A. Only partial circles are needed here. The outline of the object is darkened in at B. The rounded corners take up only a quarter of the full isometric circle that was plotted.

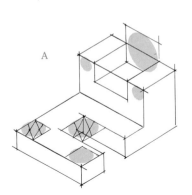

BLOCK IN ARCS

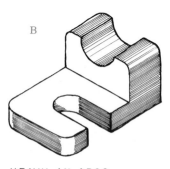

HEAVY IN ARCS

Fig. 2-69. Arcs are blocked in on a sketch similar to isometric circles.

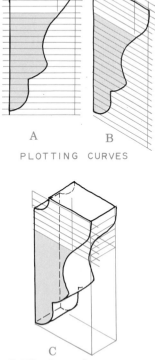

PLOTTING CURVES

Fig. 2-70. Irregular curves are plotted with a coordinated grid.

■ SKETCHING CURVES AND ARCS

You can draw curves in pictorial sketches by plotting *coordinates* (points). To locate the curve, you generally transfer the coordinates from a multiview sketch. In Fig. 2-70 at A, coordinates are plotted that stand for the width from the left edge of the front view and the height from the top edge of the front view. The series of lines at B drawn parallel to the vertical and horizontal axes locates points of intersection as shown. Similar coordinates are plotted on the pictorial at C. The intersections serve as points for sketching the pictorial curve.

■ OTHER ISOMETRIC AXES

Figure 2-71 shows a regular Y-axis situation. You can draw other isometric axes in any position as long as there is 120° between the axes, as shown in Fig. 2-72. The reverse axis is shown for objects that can be seen better from the bottom. The position of the axes in Fig. 2-73 is good for long objects. In preparing a pictorial sketch that must clearly describe the shape of an object, you have to decide how to use the isometric axes.

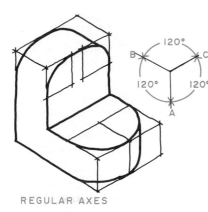

REGULAR AXES

Fig. 2-71. The regular isometric axes show as a *Y* axis.

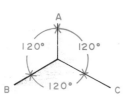

REVERSED AXES

Fig. 2-72. The isometric axes are reversed to emphasize the bottom view (inverted *Y* axis).

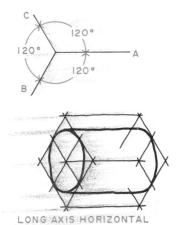

LONG AXIS HORIZONTAL

Fig. 2-73. The horizontal isometric axis.

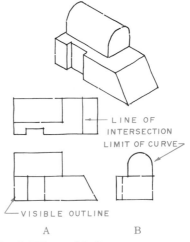

Fig. 2-74. The six views repeat geometric details.

SIX VIEWS VS. THREE VIEWS

The purpose of a sketch is to give a description of the shape of an object.

Six views are shown for the movie camera in Fig. 2-74. Studying the pictorial sketch and the six views will help you see which views are needed. The top, front, and side views describe the shape best, as they have the fewest hidden lines. Sketch only the views actually needed to show an object fully. Usually, fewer than six views need to be drawn. One, two, or three views are often enough.

LINE INTERPRETATION

Sketched lines that are *visible* (seen) have three important roles:

1. The visible line forms the outline of the drawing.
2. The visible line shows how two

Fig. 2-75. Visible line interpretation.

surfaces *intersect* (meet), as shown in Fig. 2-75 at A.
3. The visible line shows the limit of a curved surface, as in Fig. 2-75 at B.

Sketched lines that are *invisible*, or *hidden*, are used to show parts of an object that you normally cannot see. The right ways to draw hidden lines are shown in Fig. 2-76. Invisi-

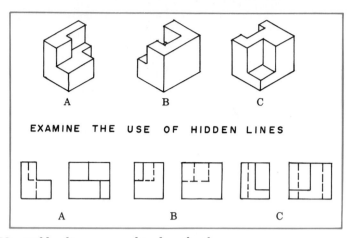

EXAMINE THE USE OF HIDDEN LINES

Fig. 2-76. Hidden lines are used to describe the unseen.

78 at *A*) describe the overall geometric elements that give an object form. Location dimensions (Fig. 2-78 at *B*) relate these geometric elements to each other.

Definitions

A *dimension line* is used to show the direction of a dimension. Dimension lines have an *arrowhead* at each end to show where the dimension begins and ends. They have a break for the dimension numbers. Dimension lines generally stop at centerlines or extension lines. *Extension lines* are thin lines used to extend the shape of the object. An extension line starts ⅟₁₆ to ⅛ in. (1.5 to 3 mm) away from the object. It extends ⅛ in. (3 mm) beyond the arrowhead on the dimension line. Centerlines can serve as extension lines. *Leaders* are thin lines drawn from a note or dimension to the place where it applies. A leader starts with a ⅛- to ¼-in. (3- to 6-mm) horizontal dash. It then angles off to the part featured, usually at 30°, 45°, or 60°. It ends with an arrowhead.

ble lines are shown by sketching dashes about ⅛ in. (3 mm) long with ⅟₁₆-in. (1.5-mm) spaces between. Hidden lines can help show the shape of an object. When a visible line *coincides* (lines up) with a hidden line, only the visible line shows.

You must use centerlines when sketching objects that have symmetrical parts. An object is *symmetrical* when its shape is the same on both sides of a line drawn through its center. The centerline (symbol ₵) can help develop proportions for cylinders and other curved surfaces. The centerline is shown in Fig. 2-77. The long dash of the centerline can be from ¾ to 1½ in. (19 to 38 mm) or more in length, depending on the size of the object. The short dashes are usually ⅛ in. (3 mm) long, with spaces on either side of about ⅟₁₆ in. (1.5 mm). When doing a sketch, make the centerline thinner than hidden and visible lines. When a centerline coincides with visible or invisible lines, omit it (Fig. 2-77).

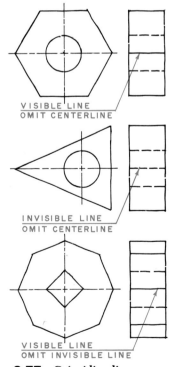

VISIBLE LINE
OMIT CENTERLINE

INVISIBLE LINE
OMIT CENTERLINE

VISIBLE LINE
OMIT INVISIBLE LINE

Fig. 2-77. Coinciding lines.

■ INTRODUCTION TO DIMENSIONING

A sketch must fully describe the object. Generally the sketch is made before the measurements of an object have been decided. After the needed dimensions are determined, they can be recorded on the sketch. Two types of dimensions are *specified* (given). Size dimensions (Fig. 2-

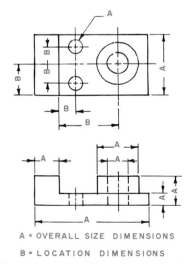

A = OVERALL SIZE DIMENSIONS

B = LOCATION DIMENSIONS

Fig. 2-78. Size dimensions at *A*, location dimensions at *B*.

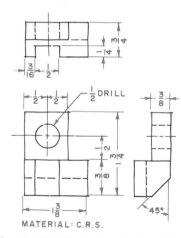

MATERIAL: C.R.S.

Fig. 2-79. Typical multiview dimensioning.

TYPICAL MULTIVIEW DIMENSIONING

Dimensions can be used to describe linear distances, angles, and notes. A typical dimensioned sketch is shown in Fig. 2-79. Chapter 6 discusses where to put dimensions in multiview drawings.

TYPICAL PICTORIAL DIMENSIONING

Pictorial dimension lines are always parallel to the major axes. The lettering that goes with them is always vertical. The American National Standards Institute (ANSI) approves the aligned and the unidirectional systems that are shown in Fig. 2-80. Chapter 6 discusses where to put dimensions in pictorial drawings.

CONCLUSION

Drafting begins with an idea. The idea develops with a sketch. Many jobs depend on the ability of the drafter to put ideas on paper with sketches and notes. The successful drafter is able to sketch technical details accurately and to render pictorial sketches artistically. Engineers, architects, and designers need drafters and artists with these skills. Drafting is an exciting challenge for those with the skills and techniques that industry needs.

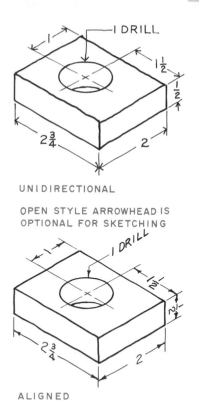

UNIDIRECTIONAL

OPEN STYLE ARROWHEAD IS OPTIONAL FOR SKETCHING

ALIGNED

Fig. 2-80. Dimensioning on a pictorial sketch.

LEARNING ACTIVITIES

1. Prepare a collection of six sketches and six mechanical drawings for classroom display that will show the various types of lettering used by the professional drafting technician. Compare and evaluate the quality of the work with your teacher. Rate the items as excellent, good, and fair, if possible. Use magazines and prints where possible.

2. Students are to bring prints for architectural studies to class and to review the architectural lettering styles on several residential or commercial prints from local architects. Are there freehand and mechanically drawn letters and numbers? Which represents the professional designer the best?

3. Check the styles of lettering used in Chap. 11 on working drawings with the styles used in Chap. 20 on architectural drawings. Which style has the creative form? Is it easy to tell the mechanical lettering from the freehand lettering?

4. Prepare a sketch of the Canard Monoplane in Fig. 2-1. Create the drawing twice the size shown, and letter your own notes and title.

5. Prepare a collection of six sketches from your favorite subject, such as racing cars or bicycles, that do not have notes. Prepare at least four lines of lettering on the sketches to describe special, important features. Try to make the lettering look like it was done by the designer.

VOCABULARY

1. Tracing
2. Translucent
3. Concentric
4. Multiview
5. Pictorial
6. Legible
7. Composition
8. Guidelines
9. Styles
10. Isometric
11. Tangent
12. Oblique
13. Cube
14. Trammel
15. Proportions
16. Ellipse
17. Refine
18. Overlay
19. Estimate
20. Distorted
21. Symmetrical
22. Recede
23. Coincide
24. Permanent
25. Radii
26. Analyze
27. Visual
28. Imagined
29. Futuristic
30. Interpretation
31. Gothic lettering
32. Single stroke
33. Perspective
34. Orthographic
35. Temporary
36. Artistic
37. Transparent
38. Glass box
39. Proportional
40. Technique
41. Diagonal
42. Customary

REVIEW

1. What kind of paper is used in the overlay method?

2. Why do the drafter, designer, and engineer who must take part in highly technical discussions consider sketching a valuable skill?

3. Of all the reasons for sketching, which three seem most important?

4. Name the three terms that describe the nature of a sketch.

5. Why are the four visual symbols in the language of sketching essential in graphic communication?

6. What qualities would be good for the paper used for refining a sketch?

7. What is the term applied to estimating proportions? Why is estimating important to the drafter and designer?

8. List the uses for the four kinds of pencil points used in freehand sketching.

9. Straight lines can be developed easily by drawing vertical lines from the _____ down and horizontal lines from _____ to _____.

10. Name two advantages of pictorial sketching.

11. When preparing sketches of objects, what is the drafter's first important decision?

12. What is another name for multiview drawing?

13. Name two basic types of pictorial sketching.

14. What are the three principal planes of projection? What view appears on each plane?

15. When a visible line and a hidden line coincide, which line should be drawn? Which should be drawn if the lines are a hidden line and a centerline?

Problems

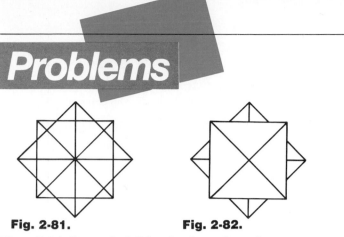

Fig. 2-81.

Fig. 2-82.

Fig. 2-83.

Fig. 2-84.

Figs. 2-81 through 2-84. Sketch in 2″ overlapping squares as creative visual studies. Create two of your own designs.

Fig. 2-85.

Fig. 2-86.

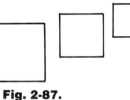

Fig. 2-87.

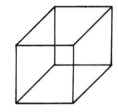

Fig. 2-88.

Figs. 2-85 through 2-88. Sketch the squares overlapping, diminishing, and as a transparent cube. Sizes are about 1½″, 1⅛″, and ¾″.

Fig. 2-89. Sketch a cube with the rectangular shape as shown. Observe the optical illusion.

Fig. 2-90. Sketch a rectangular solid with 2-to-1 proportions. Use ½″, 1″, and 2″.

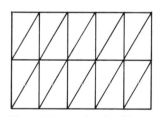

Fig. 2-91. Sketch the apparent two-dimensional form using six diagonals. How many times will Fig. 2-90 be found in Fig. 2-91?

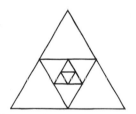

Fig. 2-92. Sketch a 3″ equilateral triangle with diminishing triangles at midpoints. Note the proportions. How many triangles can you make diminish inside?

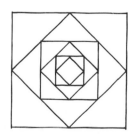

Fig. 2-93. Sketch a 3″ square with diminishing squares at midpoints. Note the proportions.

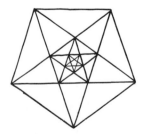

Fig. 2-94. Sketch a pentagon using 2″ sides with diminishing five-pointed stars. Note the proportions.

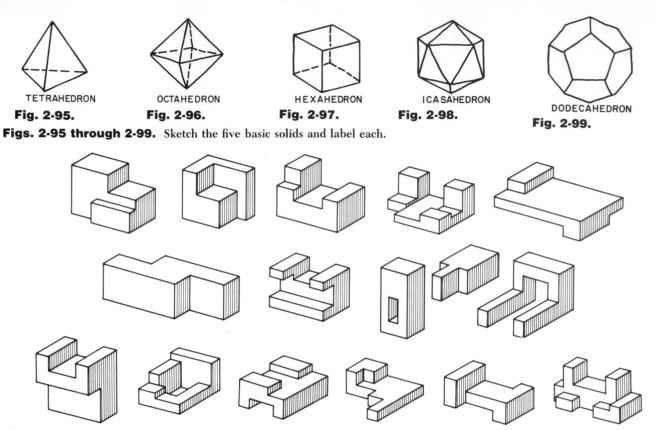

TETRAHEDRON   OCTAHEDRON   HEXAHEDRON   ICASAHEDRON

Fig. 2-95.   **Fig. 2-96.**   **Fig. 2-97.**   **Fig. 2-98.**

DODECAHEDRON
Fig. 2-99.

Figs. 2-95 through 2-99. Sketch the five basic solids and label each.

Fig. 2-100. Sketch the three views needed for a multiview drawing. Select approximate proportions and use methods as shown in Figs. 2-39 through 2-41 (straight lines).

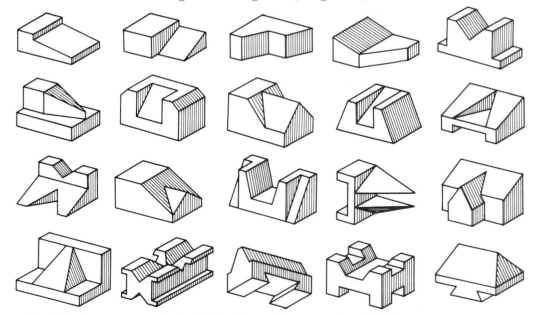

Fig. 2-101. Sketch the three views needed to define the machine part assigned. Select approximate sizes. Use methods as shown in Figs. 2-39 through 2-41 (inclined lines).

Fig. 2·102. Sketch the three views required to define the problems assigned with curved lines.

Fig. 2·103. Sketch the three views required to define the problems assigned.

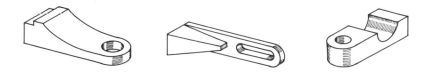

Figs. 2-104 through 2-107. Lettering is best learned in short unhurried periods. The problems are planned for a working space of 7 × 5 in. One problem may be worked on an 8½ × 11 in. sheet. Use an H or a 2H pencil with a conical point. The lettering should be black in order to reproduce, as most drawings are made on a translucent medium. Use a sharp 4H pencil to rule very light guidelines for the heights of the letters. Rotate the pencil in the fingers after each few strokes to keep the point in shape; sharpen the pencil as often as necessary to keep a proper point.

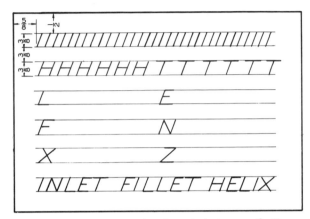

Fig. 2-104. Letter ⅜″ inclined capitals in pencil. Draw a few very light inclined guidelines.

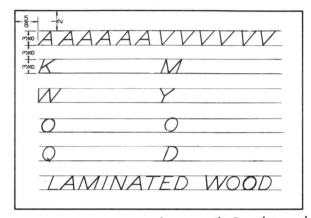

Fig. 2-105. Letter ⅜″ capitals in pencil. Complete each line.

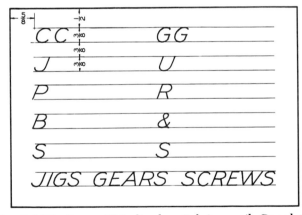

Fig. 2-106. Letter ⅜″ inclined capitals in pencil. Complete each line.

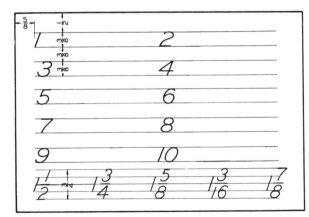

Fig. 2-107. Letter ⅜″ inclined numerals in pencil. Make the total height of the fractions 2 times the height of the numerals.

Figs. 2-108 through 2-111. Prepare a lettering sheet as you did for the four inclined lettering assignments and practice the same letters with vertical capitals for the four assignments.

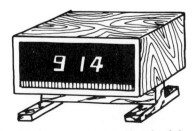

Fig. 2-112. Prepare a multiview sketch of the digital clock. Prepare an oblique sketch and dimension pictorial.

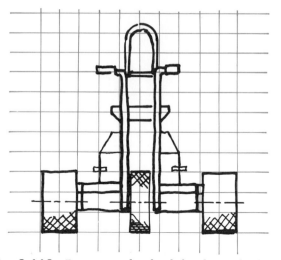

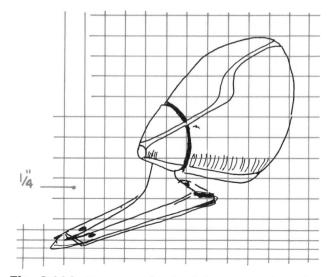

Fig. 2-113. Prepare a sketch of the three-wheel vehicle, showing three views. Given front, develop right side and top view. Create centerlines and structure.

Fig. 2-114. Prepare a sketch of the racing mirror from grid. Develop two or three views as per instructor.

Fig. 2-115. Prepare a sketch of the monowing hang glider. Refer to Fig. 2-1, and place notes on your sketch.

Fig. 2-116. Prepare a sketch of the space shuttle from Fig. 25-32 in Chap. 25. Two views are required, with notes.

Fig. 2-117. Prepare a sketch of the racing car in Fig. 1-26.

Fig. 2-118. Prepare your name and your school name on a cover sheet for your drawings. Use block lettering as shown in Fig. 2-17.

3 The Use and Care of Drafting Equipment

THE DRAFTING ROOM

The plans and directions that are needed for doing engineering work of all kinds are prepared in the drafting room (Fig. 3-1). There, the designs are worked out and checked. Technical drawing is really a language. We have seen why drawings are used in industrial, engineering, and scientific work. Sometimes, designers or scientists make freehand sketches to help them study new ideas or show ideas to other people.

Usually, though, technical drawings must be more accurate than any sketch can be. Such accurate drawings are made with drafting *instruments* (tools). In learning to read and write the drafting language, you must learn which instruments to use on a particular problem. You need to know how to use drafting equipment skillfully, accurately, and quickly.

BASIC EQUIPMENT

Figure 3-2 shows a variety of basic drafting equipment. Many of these items are included in a student drafting kit. This equipment and other common items are listed below.

- Drafting board
- T-square, or parallel-ruling straightedge, or drafting machine
- Drawing sheets (paper, cloth, or film)
- Drafting tape
- Drafting pencils
- Pencil sharpener
- Eraser
- Erasing shield
- Triangles, 45° and 30°-60° (not required with drafting machine)
- Architect's or engineer's scale
- Irregular curve
- Drawing instrument set
- Lettering instruments
- Black drawing ink
- Technical fountain pens
- Brush or dust cloth
- Protractor
- Cleaning powder

Your teacher can tell you exactly what equipment will be needed for your course.

Fig. 3-1. An industrial drafting room.

DRAWING BOARDS

The drawing sheet is attached to a drawing board. Drawing boards used in school or at home usually measure 9 × 12 in. (230 × 300 mm), 16 × 21 in. (400 × 530 mm), or 18 × 24 in. (460 × 600 mm). Boards used to make engineering or architectural drawings are typically larger and may be any size needed. Boards are generally made of soft pine or basswood. They are made so that they will stay flat and so that the guiding edge (or edges) will be straight. Hardwood or metal (steel or aluminum) strips are used on some boards to provide truer guiding edges.

T-SQUARES

T-squares (Fig. 3-4) are made of various materials. However, most have plastic-edged wood blades, or clear plastic blades, and heads of wood or plastic. Where extreme accuracy is needed, stainless steel or hard aluminum blades with metal heads are used. The blade (straightedge) must be very straight. It must be attached securely to the top surface of the T-square head.

You can easily find how accurate your T-square is, as shown in Fig. 3-5. First, draw a sharp line along the drawing edge of the T-square on a sheet of paper. Second, turn the drawing sheet around and line up the drawing edge of the T-square with the other side of the line. If the drawing edge and the pencil line do not match, the T-square edge is not accurate. When this condition is found, the T-square is generally replaced.

Some T-squares are made with an adjustable head that allows the blade to be lined up with the drawing. If the head of such a T-square has a

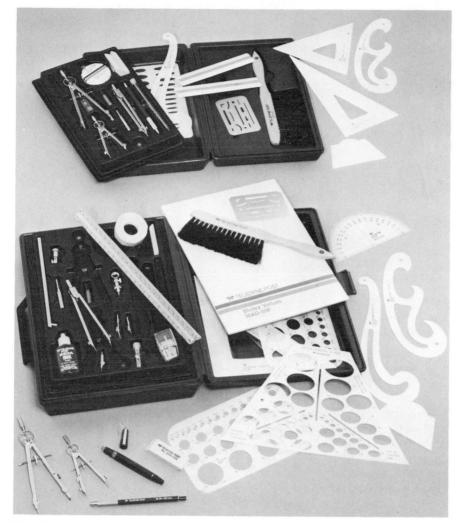

Fig. 3-2. Drafting equipment.

DRAWING TABLES AND DESKS

These items of equipment come in many different sizes and types (Fig. 3-3). One kind of table has a fixed top and a separate drawing board that you can move around. Another has a fixed top that is made to be used as a drawing board itself. A third has an adjustable top that holds a separate drawing board at the slope you want. Still another has an adjustable top that is made to be used as a drawing board.

Some tables are made to be used while you are standing up or sitting on a high stool. Other drawing tables are the same height as regular desks. There are also combined tables and desks that you can use either standing up or sitting down. The type that combines a drafting table, desk, and regular office chair is the most comfortable and efficient. This kind is replacing the high drawing table in drafting rooms. The cost of drawing tables and desks will vary according to their size, style, and number of components.

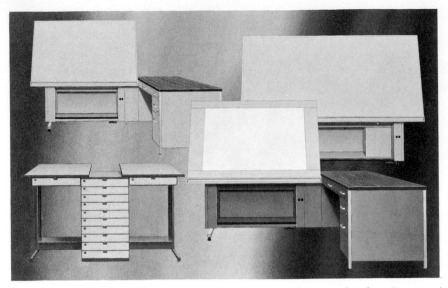

Fig. 3-3. Drafting tables are available in a variety of sizes and styles. (Bruning.)

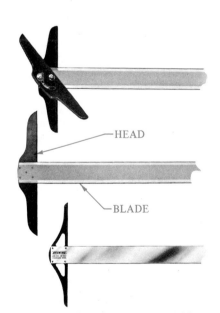

HEAD

BLADE

Fig. 3-4. T-squares are available in various styles and materials. (Bruning.)

protractor, the blade can be set at any angle. The size of a T-square is found by measuring along the blade from the contact surface of the head to the end of the blade.

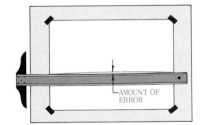

AMOUNT OF ERROR

Fig. 3-5. Check to see that the T-square is accurate.

AMERICAN NATIONAL STANDARD TRIMMED SIZES OF DRAWING SHEETS

Three series of sizes are listed in Fig. 3-6. One is based on the standard-size $8\frac{1}{2} \times 11$ in. letter paper. A second is based on a 9×12 in. sheet. The third is based on the International Standards Organization (ISO) standard. The ISO standard is developed downward in size from a base sheet with an area of about 1 square meter (m^2). Sheet sizes are based on a length-to-width ratio of one to the

square root of two $(1:\sqrt{2})$. Each smaller size has an area equal to half the preceding size. Multiples of these sizes are used for larger sheets also.

DRAWING SHEETS

Drawings are made on many different materials. Papers may be white or tinted cream or pale green. They are made in many thicknesses and qualities. Most drawings are made directly in pencil or ink on tracing paper, vellum, tracing cloth, glass cloth, or film. When such materials are used, copies can be made by printing or other reproduction methods (Chap. 14).

Vellum is tracing paper that has been treated to make it more *transparent* (clear). Tracing cloth is a finely woven cloth. It is treated to provide a good working surface and good transparency. Polyester (plastic) films are widely used in industrial drafting rooms. They are very transparent, strong, and lasting. Drawing films are made with a *matte* (dull and rough) surface. They are suitable for both pencil and ink work.

FASTENING THE DRAWING SHEET TO THE BOARD

The sheet may be held in place on the board in several ways (Fig. 3-7). Drafting tape can be put across the corners of the sheet and, if needed, at other places. Small, precut circular pieces of tape, called *dot tape*, are also commonly used. These ways of attaching the sheet let you move the T-square and triangles freely over the whole sheet. In addition, they do not damage the corners or the edges of the sheet. They also can be used

U.S. CUSTOMARY SERIES			ISO STANDARD	
Size	First series	Second series	Size	Third series
A	8½ × 11 in.	9 × 12 in.	A0	841 × 1189 mm
B	11 × 17 in.	12 × 18 in.	A1	594 × 841 mm
C	17 × 22 in.	18 × 24 in.	A2	420 × 594 mm
D	22 × 34 in.	24 × 36 in.	A3	297 × 420 mm
E	34 × 44 in.	36 × 48 in.	A4	210 × 297 mm

Fig. 3-6. Standard drawing sheet sizes.

on composition boards or other boards with hard surfaces. As a result, most drafters prefer to use one of these methods. Drafters in industry often use fine-wire staples for this purpose. However, care must be taken not to damage the sheet.

To fasten the paper or other drawing sheet (Fig. 3-8), place it on the drawing board with the left edge 1 in. (25 mm) or so away from the left edge of the board. (Left-handed students should work from the right edge.) Put the lower edge of the sheet at least 4 in. (100 mm) up from the bottom of the board so that you can work on it comfortably. Then line up the sheet with the T-square blade, as shown at A. Hold the sheet in position. Move the T-square down, as at B, keeping the head of the T-square against the edge of the board. Then fasten each corner of the sheet with drafting tape.

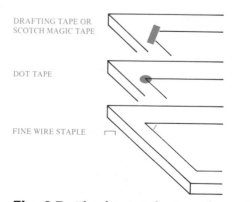

Fig. 3-7. The drawing sheet can be fastened to the board in several ways.

■ DRAWING PENCILS

Both regular wooden pencils and mechanical pencils are used for technical drawing. Four kinds of lead are now used in drawing pencils. One pencil lead is made from *graphite*, a form of the element carbon. It also

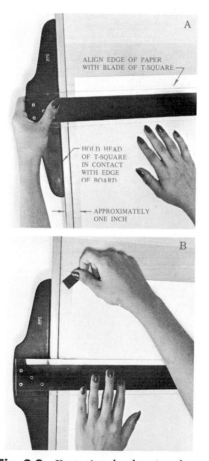

Fig. 3-8. Fastening the drawing sheet to the board.

contains clay and *resins* (sticky substances). Graphite pencils have been used for over 200 years. They are still the most important kind. Drafting pencils of this type are usually made in 17 degrees of hardness, or grades. These grades are listed below.

6B (softest and blackest)
5B (extremely soft)
4B (extra soft)
3B (very soft)
2B (soft, plus)
B (soft)
HB (medium soft)
F (intermediate, between soft and hard)
H (medium hard)
2H (hard)
3H (hard, plus)
4H (very hard)
5H (extra hard)
6H (extra hard, plus)
7H (extremely hard)
8H (extremely hard, plus)
9H (hardest)

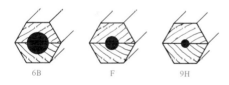

Graphite pencils are mostly used on paper or vellum. They may also be used on cloth.

Which grade of pencil you use depends on what kind of surface you are drawing on. It also depends on how *opaque* (dark) and how thick you want the line you are drawing to be. To lay out views on fairly hard-surfaced drawing paper, use grades 4H and 6H. When you use tracing paper or cloth and draw finished views that are to be *reproduced* (copied by a machine), use an H or 2H pencil. Grades HB, F, H, and 2H are sometimes used for sketching and lettering and for drawing arrowheads, symbols, border lines, and so on. The exact grade you use depends on the drawing and the surface. Very hard and very soft leads are seldom used in ordinary drafting.

Since film has come into use for drawings, new kinds of pencil lead have been developed. Three types are described below, based on information furnished by the Joseph Dixon Crucible Company. The first is a *plastic pencil*. This type is a black crayon. Its lead is *extruded* (pushed out) in a "plasticizing" process. Drawings made with this lead reproduce well on microfilm. The second type of lead is a "combination." It is part plastic and part graphite and is made by heating. This kind stays sharp, draws a good opaque line, does not smear easily, erases well, and microfilms well. It can be used on paper or cloth as well as on film. The third type is also a combination of plastic and graphite. This type is extruded, like the first kind, but not heated, like the second. The third type does not remain as sharp. However, it draws a fairly opaque line, erases well, does not smear easily, and microfilms well. It is used mostly on film. These three types of lead are made in only five or six grades. Their grades are not the same as those used for regular graphite leads. The companies that make these pencils use different systems of letters and numbers to tell what kind of lead is in each pencil and how hard it is. The drafter must experiment with various grades to determine which best suits specific needs.

■ SHARPENING THE PENCIL

A wooden pencil is sharpened by cutting away the wood at a long slope, as at A in Fig. 3-9. Always

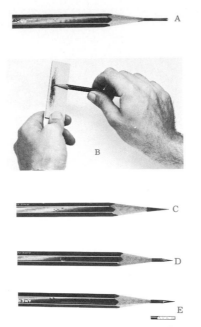

Fig. 3-9. Sharpening the pencil properly is important.

sharpen the end opposite the grade mark. Be careful not to cut the lead. Leave it exposed for about ⅜ to ½ in. (10 to 13 mm). Then shape the lead to a long *conical* (cone-shaped) point. Do this by rubbing the lead back and forth on a sandpaper pad (or on a fine file), as at B, while turning it slowly to form the point, as at C or D. Some drafters prefer the flat point (or chisel point) shown at E.

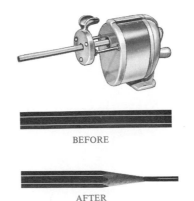

BEFORE

AFTER

Fig. 3-10. A drafter's pencil sharpener cuts the wood, not the lead.

Keep the sandpaper pad at hand so that you can sharpen the point often. Also, many drafters like to shape and smooth the point further by "burnishing" it on a piece of rough paper such as drafting paper.

Mechanical sharpeners (Fig. 3-10) are made with special drafter's cutters that remove the wood as shown. Special pointers are made for shaping the lead, as in Fig. 3-11. Such devices may be hand-operated or electrically powered.

Mechanical pencils, also called *lead holders*, are widely used by drafters. They hold plain sticks of lead in a chuck that allows the exposed lead to be extended to any length desired. The lead is shaped as with wooden pencils. However, some refill pencils have a built-in sharpener that shapes the lead. Still other mechanical drafting pencils use leads made to specific thicknesses for desired line widths and require no sharpening. These, however, have not generally been found to be very practical for the beginning drafting student.

Pencil lines must be clean and sharp. They must be dark enough for the views to be seen when the standard lines in Fig. 3-12 are used. If you use too much pressure, you will

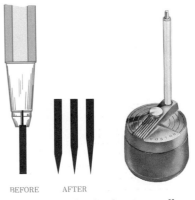

BEFORE   AFTER

Fig. 3-11. This lead pointer allows a choice of point shapes.

groove the drawing surface. You can avoid this if you use the correct grade of lead. Develop the habit of turning the pencil between your thumb and forefinger as you draw a line. This will help make the line uniform and keep the point from wearing down unevenly. *Never sharpen a pencil over the drawing board.* After you sharpen a pencil, wipe the lead with a cloth to remove the dust. Being careful in these ways will help keep the drawing clean and bright. This is important when the original pencil drawing is used to make copies.

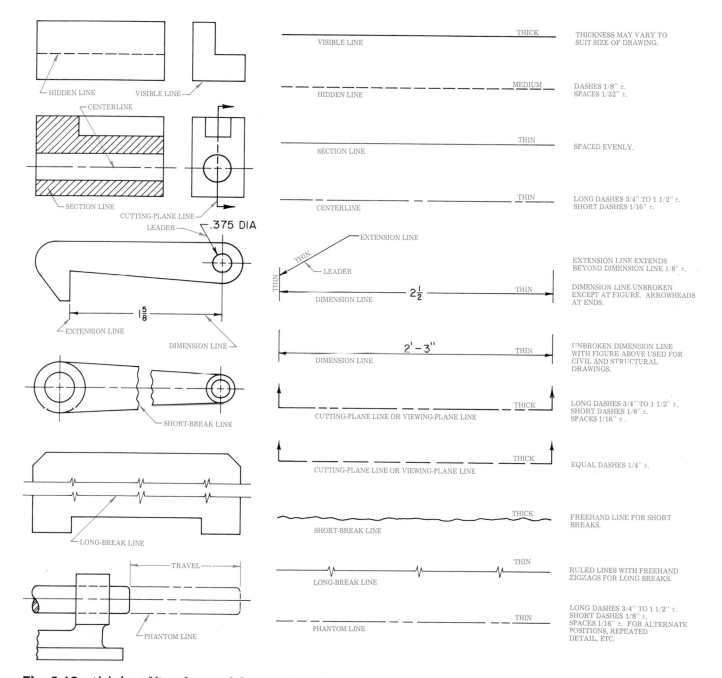

Fig. 3-12. Alphabet of lines for pencil drawing. (See Chap. 13 for ink lines.)

ALPHABET OF LINES

The different lines, or line symbols, used on drawings are a kind of graphical alphabet. The line symbols that the American National Standards Institute (ANSI) recommends for pencil drawings are shown in Fig. 3-12. Two widths of lines—thick and thin—are generally used. (Formerly, a medium line was used as a hidden line.) Drawings are much easier to read when there is good contrast between the different kinds of lines. All lines must be uniformly sharp and black.

ERASERS AND ERASING SHIELDS

Soft erasers, such as the vinyl type, the Pink Pearl, or the Artgum, are used for cleaning soiled spots or light pencil marks from drawings. Rubkleen, Ruby, or Emerald erasers are generally good for removing pencil or ink. Ink erasers contain grit. If they are used at all, they have to be used very carefully to keep from damaging the drawing surface. Electric erasing machines (Fig. 3-13) are in common use in drafting rooms.

Erasing shields, made of metal or plastic, have openings of different sizes and shapes. They are useful for protecting lines that are not to be erased (Fig. 3-14). Lines on paper or cloth are erased *along* the direction of the work. Lines on film are erased *across* the direction of the work. On film, use a vinyl eraser made especially for this material.

TO DRAW HORIZONTAL LINES

The upper edge of the T-square blade is used as a guide for drawing

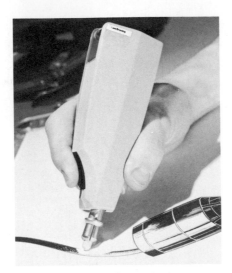

Fig. 3-13. An electric erasing machine saves time. (Bruning.)

Fig. 3-14. Using an erasing shield.

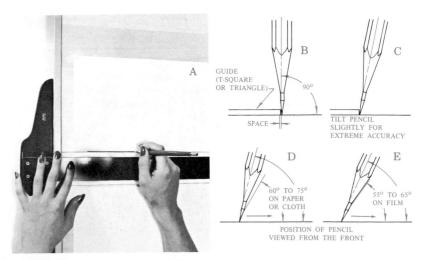

Fig. 3-15. Drawing a horizontal line.

horizontal (side-to-side) lines (Fig. 3-15). With your left hand, place the head of the T-square in contact with the left edge of the board. Keeping the head in contact, move the T-square to the place you want to draw the line. Slide your left hand along the blade to hold it firmly against the drawing sheet. Hold the pencil about 1 in. (25 mm) from its point. Slant it in the direction in which you are drawing the line. (This direction should be left to right for right-handers and right to left for left-handers.) While drawing the line, rotate the pencil slowly and slide your little finger along the blade of the T-square. This will help you to better control the pencil. On film, keep the pencil at the same angle (55° to 65°) all along the line. You must also use less pressure than on paper or other material. Always keep the point of the lead a little distance away from the corner between the guiding edge and the drawing surface, as shown in Fig. 3-15. This will let you see where you are drawing the line. It will also help you keep from making a poor or smudged line. *Be careful to keep the line parallel to the guiding edge.*

TO DRAW VERTICAL LINES

Use a triangle and a T-square to draw *vertical* (up-and-down) lines (Fig. 3-16). Place the head of the T-square in contact with the left edge of the board. Keeping the T-square in contact, move it to a position below the start of the vertical line. Place a triangle against the T-square blade. Move the triangle to the place you want to draw the line. Keeping the vertical edge of the triangle toward the left, draw upward. Slant the pencil in the direction in which you are drawing the line. Be sure to keep this angle the same when drawing on film. Keep the point of the lead far enough out from the guiding edge so that you can see where you are drawing the line. Be careful to keep the line parallel to the guiding edge.

ANGLES

An *angle* is formed when two straight lines meet at a point. The point where the lines meet is called the *vertex* of the angle. The lines are called the *sides* of the angle. Angles are measured in a unit called a *degree*. Degrees are marked by the

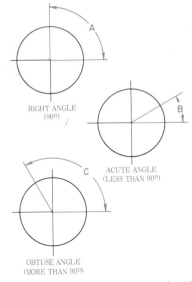

Fig. 3-17. Angles are measured in degrees, minutes, and seconds.

Decimal degree	Degrees, minutes, and seconds
10° ± 0.5°	10° ± 0°30′
0.75°	0°45′
0.004°	0°0′15″
90° ± 1.0°	90° ± 1°
25.6° ± 0.2°	25°36′ ± 0°12′
25.51°	25°30′40″

Fig. 3-18. Decimal-degree equivalents of degrees, minutes, and seconds.

symbol °. If the *circumference* (rim) of a circle is divided into 360 parts, each part is an angle of 1°. Then, one fourth of a circle would be 360/4=90°, as at A in Fig. 3-17. A part of a circle is called an *arc*. Two lines drawn to the center of a circle from the ends of a 90° arc on that circle form a *right angle*. An *acute angle* (B) is less than 90°. An *obtuse angle* (C) is greater than 90°. For closer measurements, a degree is divided into 60 equal parts called *minutes* (′). Minutes in turn, are divided into 60 equal parts called *seconds* (″). Notice that the size of an angle does not depend on the length of its sides. Therefore, the number of degrees in a given angle will be the same, whatever the size of the circle and arc that define it.

While angles are commonly given in degrees, minutes, and seconds, the use of *decimal degrees* is now growing in popularity. For example, 50.5° means the same as 50°30′. Figure 3-18 shows some additional examples.

An instrument used in measuring, or laying out, angles is called a *protractor*. A semicircular protractor is shown in Fig. 3-19, where an angle of 43° is measured. Drafting

Fig. 3-16. Drawing a vertical line.

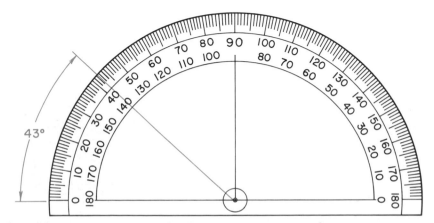

Fig. 3-19. A protractor is used to lay out, or measure, angles.

machines and adjustable-head T-squares have an adjustable protractor for laying out angles of any degree.

Angles of 45°, 30°, and 60° can be drawn directly, using the triangles. Angles varying by 15° can also be drawn with the triangles, as described in the following section.

■ TO DRAW LINES INCLINED 30°-60° AND 45°

The 30°-60° triangle (Fig. 3-20) has angles of 30°, 60°, and 90°. Lines inclined at 30° and 60° from horizontal or vertical lines are drawn with the triangle held against the T-square blade, as shown in Fig. 3-20, or against a horizontal straightedge. The 30°-60° triangle can be used to lay off equal angles, 6 at 60° or 12 at 30°, about a center and for many constructions.

The 45° triangle (Fig. 3-21) has two angles of 45° and one of 90°. Lines *inclined* (slanted) 45° from horizontal or vertical lines are drawn with the triangle held against the T-square blade, as shown at A, B, and C in Fig. 3-21, or against a horizontal straightedge. The 45° triangle can be used to lay off eight equal angles of 45° about a center and for many constructions.

■ TO DRAW LINES AT ANGLES VARYING BY 15°

The 45° and 30°-60° triangles, alone or together and combined with the T-square, can be used to draw angles increasing by 15° from the horizontal or vertical line. Some ways of placing the triangle to draw angles of 15° and 75° are shown at A, B, C, and D in Fig. 3-22. Ways of getting the different angles are shown in Fig. 3-23. In space O, the lines have been drawn for all the positions possible. All the angles are 15°. By placing the triangles as in spaces B and M or in spaces F and I, you get two angles of 30° and two angles of 15°. It is a good idea to try placing the triangles in all the different ways until you are used to them. (Remember, any angle may be obtained by using a protractor.)

■ TO DRAW PARALLEL LINES

You can draw *parallel* (side-by-side) horizontal lines with the T-square. You can draw parallel vertical lines

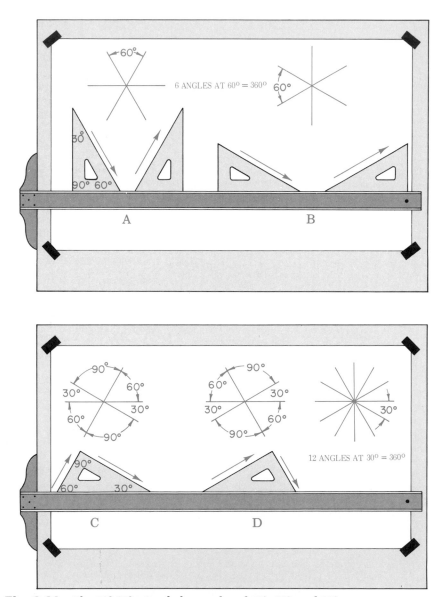

Fig. 3-20. The 30°-60° triangle has angles of 30°, 60°, and 90°.

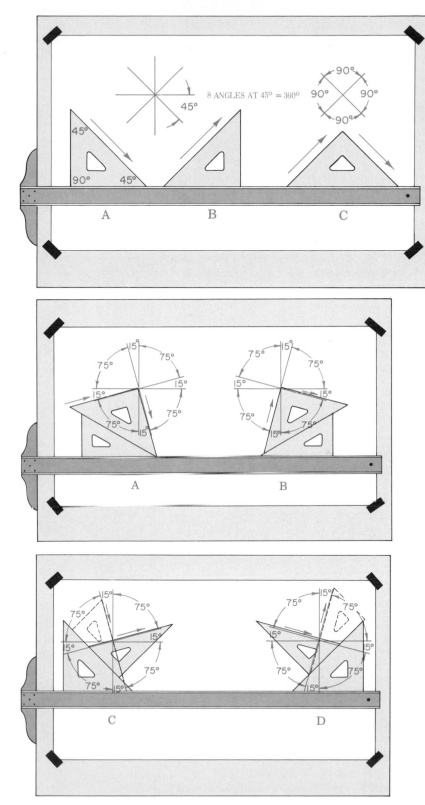

Fig. 3-21. The 45° triangle has angles of 45° and 90°.

Fig. 3-22. Drawing lines at 15° and 75° using the two triangles.

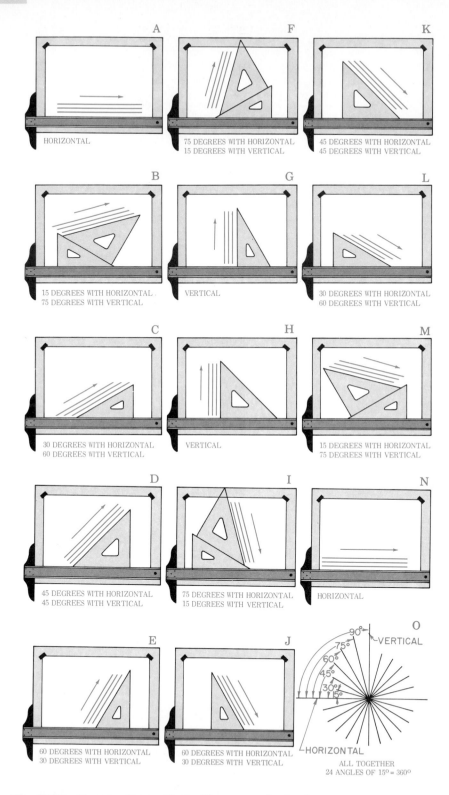

Fig. 3-23. Drawing lines with the T-square and triangles.

by using a triangle along with the T-square. Parallel lines can be drawn at regular angles with the triangles, as shown in Fig. 3-23. You can draw parallel lines in any place by using a triangle along with the T-square or another triangle, as shown in Fig. 3-24. Parallel lines at any angle can also be drawn directly with a drafting machine.

■ TO DRAW A LINE PARALLEL TO A GIVEN LINE

Place a triangle against the T-square blade and move them together until one edge of the triangle matches the

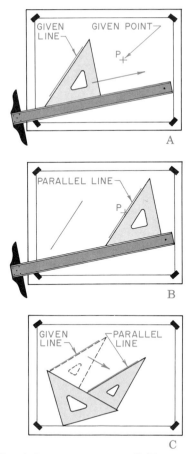

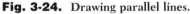

Fig. 3-24. Drawing parallel lines.

given line (Fig. 3-24, space A). Holding the T-square firmly, slide the triangle along its blade until it reaches the desired location. Then draw the parallel line (space B). In Fig. 3-24, the *hypotenuse* (the edge opposite the right angle) of the triangle is used to draw the parallel line. However, other edges of the 30°-60° triangle can be used instead. In space C, a triangle is used in place of the T-square. Parallel lines can also be drawn directly with a drafting machine.

■ TO DRAW PERPENDICULAR LINES

Lines at right angles (90°) with each other are *perpendicular*. A vertical line is perpendicular to a horizontal line. Perpendicular lines can be drawn using triangles and the T-square. They can also be drawn directly with a drafting machine.

■ TO DRAW A LINE PERPENDICULAR TO ANY LINE

Two methods can be used.

First Method (Fig. 3-25). Place a triangle and T-square together so that one edge of the triangle matches the given line, as in space A. Hold the T-square firmly. Slide the triangle along the blade to the desired position, as in space B. Draw the perpendicular line. The 45° triangle can be used instead of the 30°-60° triangle, as in space C. A triangle can be used instead of the T-square, as in space D.

Second Method (Fig. 3-26). Place a triangle and T-square together so that the hypotenuse of the triangle matches the given line, as in space A. Turn the triangle about its right-angled corner, as in space B. Slide the triangle along the T-square blade until it is in the desired location, as in space C, and draw the per- pendicular line. The 45° triangle can be used instead of the 30°-60° triangle. A triangle can be used instead of the T-square, as in space D.

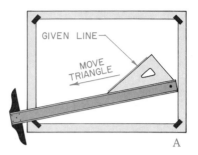
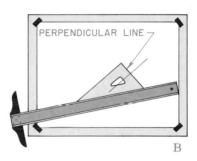
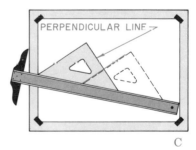
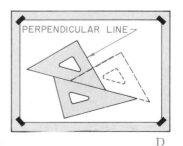

Fig. 3-25. Drawing a perpendicular line, first method.

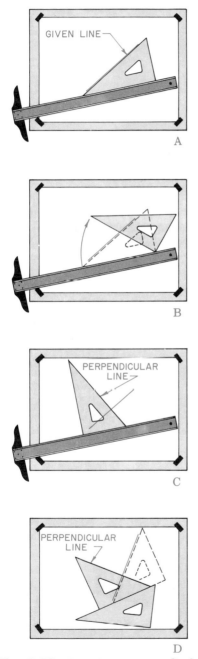

Fig. 3-26. Drawing a perpendicular line, second method.

Fig. 3-27. An arm-, or elbow-, type drafting machine. (Teledyne Post.)

DRAFTING MACHINES

There are two kinds of drafting machines in wide use. The *arm,* or *elbow,* type, shown in Fig. 3-27, uses an anchor and two arms to hold a movable protractor head with two scales. The scales are ordinarily at right angles to each other. The arms allow the scales to be moved to any place on the drawing that is parallel to the starting position. The *track*

type, shown in Fig. 3-28, uses a horizontal guide rail at the top of the board and a moving arm rail at right angles to the top rail. An adjustable protractor head and two scales, ordinarily at right angles, move up and down on the arm. The scales may be moved to any place on the drawing that is parallel to the starting position. This type of drafting machine is easy to use on large boards or on boards placed vertically or at a steep angle.

Most industrial drafting departments now use drafting machines. Many school drafting departments teach drawing with drafting machines. Drafting machines combine the functions of the T-square, triangles, scales, and protractor. You can draw lines as long as you want, where you want, and at any angle you want just by moving the scale ruling edge to the desired location. This lets you draw faster and with less work. You should learn how to use the drafting machine in all the possible ways and how to take care of it. If you do so, you will soon see how valuable a tool the drafting machine is.

PARALLEL-RULING STRAIGHTEDGES

Many drafters prefer to use parallel-ruling straightedges when working on large boards placed vertically or nearly so. A guide cord is clamped to the ends of the straightedge. The cord runs through a series of pulleys on the back of the board. This allows the straightedge to slide up and down on the board in parallel positions (Fig. 3-29).

SCALES

Scales are made in different shapes (Fig. 3-30). They are made of different materials, such as boxwood, plastic, plastic on boxwood, and metal. They are also made with different divisions so that they can be used to make different kinds of drawings. The customary-inch system uses separate scales for architectural, mechanical engineering, and civil engineering drawings. Thus, customary-inch scales that are commonly used include the architect's scale (Fig. 3-

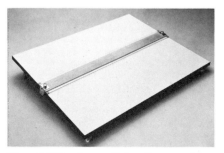

Fig. 3-29. A parallel-ruling straightedge is another convenient instrument used to save time. (Staedtler-Mars.)

Fig. 3-30. Scales are made in various shapes.

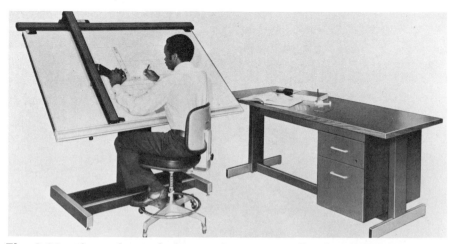

Fig. 3-28. The track-type drafting machine is especially adapted for wide drawings, in addition to its use for regular-size drawings. (Teledyne Post.)

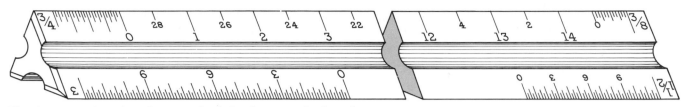

Fig. 3-31. Architect's scale, open divided. The triangular form has many proportional scales.

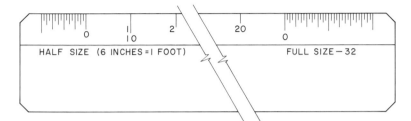

Fig. 3-32. Mechanical engineer's scale, open divided.

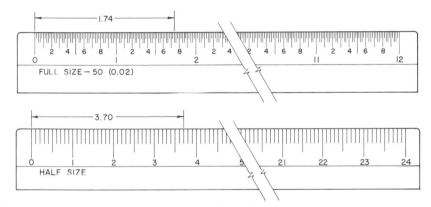

Fig. 3-33. Civil engineer's scale, divided into decimals.

31), the mechanical engineer's scale (Fig. 3-32), and the civil engineer's scale (Fig. 3-33). By contrast, the proportional metric scale can be used on all types of engineering drawings. Both customary and metric scales are made in any of the shapes shown in Fig. 3-30.

Scales are used for laying off distances and for making measurements. Measurements can be full size or in some exact proportion to full size. Scales can be *open-divided* (Figs. 3-31 and 3-32), with only the end units subdivided. Or they can be *full-divided* (Figs. 3-33 and 3-34),

with subdivisions over their entire length.

THE ARCHITECT'S SCALE

The architect's scale (Fig. 3-31) is divided into proportional feet and inches. The triangular form shown is used in many schools and in some drafting offices because it has many scales on a single stick. However, many drafters prefer flat scales, especially when they do not have to change scales often.

The usual proportional scales are listed in the following table.

Proportion	Gradations	Ratio
Full size	12″ = 1′-0″	1:1
¼ size	3″ = 1′-0″	1:4
⅛ size	1½″ = 1′-0″	1:8
1/12 size	1″ = 1′-0″	1:12
1/16 size	¾″ = 1′-0″	1:16
1/24 size	½″ = 1′-0″	1:24
1/32 size	⅜″ = 1′-0″	1:32
1/48 size	¼″ = 1′-0″	1:48
1/64 size	3/16″ = 1′-0″	1:64
1/96 size	⅛″ = 1′-0″	1:96
1/128 size	3/32″ = 1′-0″	1:128

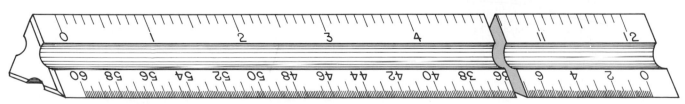

Fig. 3-34. Decimal-inch scales are used in drawing machine parts.

The symbol ′ is used for feet and ″ for inches. Thus, three feet four and one-half inches is written 3′-4½″. When all dimensions are in inches, the symbol is usually left out.

Proportional scales are used in drawing buildings and in making mechanical, electrical, and other engineering drawings. They are also often used in drafting in general. The proportional scale to which the views are drawn should be given on the drawing. This is done in the title block if only one scale is used. If different parts of a drawing are in different scales, the scales are given near the views in this way:

- Scale: 6″ = 1′-0″
- Scale: 3″ = 1′-0″
- Scale: 1½″ = 1′-0″

THE MECHANICAL ENGINEER'S SCALE

The mechanical engineer's scale (Fig. 3-32) has inches and fractions of an inch divided to stand for inches. The usual divisions are the following:

- *Full size*—1 in. divided into 32nds
- *Half size*—½ in. divided into 16ths
- *Quarter size*—¼ in. divided into 8ths
- *Eighth size*—⅛ in. divided into 4ths

These scales are used for drawing parts of machines or where larger reductions in scale are not needed. The proportional scale to which the views are drawn should be given on the drawing. This is done in the title block. If different parts of a drawing are in different scales, the scales are given near the views (as, for exam-

ple, full size or 1″ = 1″; half size or ½″ = 1″).

The decimal-inch system uses a scale divided into *decimals* (parts divisible by 10) of an inch (Fig. 3-33). For full size, 1 in. is divided into 50 parts (each part of 1/50 = .02″). In this way, you can easily measure to hundredths of an inch by sight. Some usual divisions are given below:

- *Full size*—1 in. divided into 50ths
- *Half size*—½ in. divided into 10ths
- *Three-eighths size*—⅜ in. divided into 10ths
- *One-quarter size*—¼ in. divided into 10ths

The decimal-inch system has been used in the automotive industry for many years. It is now in widespread use in other engineering fields, as well, and is rapidly replacing the fractional-inch system.

THE CIVIL ENGINEER'S SCALE

The civil engineer's scale (Fig. 3-34) has inches divided into decimals. The usual divisions follow:

- 10 parts to the inch
- 20 parts to the inch
- 30 parts to the inch
- 40 parts to the inch
- 50 parts to the inch
- 60 parts to the inch

With this scale, 1 in. may stand for feet, rods, miles, and so forth. It may also stand for quantities, time, or other units. The divisions may be single units or multiples of 10, 100, and so on. Thus, the 20-parts-to-an-inch scale may stand for 20, 200, or 2000 units. This scale is used for civil engineering work. This includes maps and drawings of roads and

other public projects. It is also used where decimal-inch divisions are needed. These uses include plotting data and drawing graphic charts.

The scale used should be given on the drawing or work as follows:

- Scale: 1″ = 500 pounds
- Scale: 1″ = 100 feet
- Scale: 1″ = 500 miles
- Scale: 1″ = 200 pounds

For some uses, a graphic scale is put on a map, drawing, or chart, as shown below.

METRIC SCALES

Metric scales (Fig. 3-35) are divided into millimeters. The usual proportional scales in the metric system are listed as a ratio in the following table.

Enlarged	Same size	Reduced
2000:1		
1000:1	1:1	
500:1		1:2
200:1		1:5
100:1		1:10
50:1		1:20
20:1		1:50
10:1		1:100
5:1		1:200
2:1		1:500
		1:1000
		1:2000

Metric architectural proportions are used in drawing buildings and in

Fig. 3-35. Metric scales are divided into millimeters.

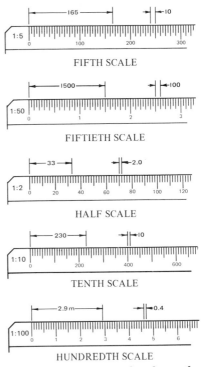

Fig. 3-36. Metric scales for reduction.

making many mechanical, electrical, and other engineering drawings. They are also often used in drafting in general. The proportional scale to which the views are drawn should be given on the drawing. This is done in the title block if only one scale is used. If different parts are in different scales, the scales are given near the views in this way:

- Scale: 1:2
- Scale: 1:5
- Scale: 1:10
- Scale: 1:50
- Scale: 1:100

To reduce the size of an object being drawn, a drafter uses one of

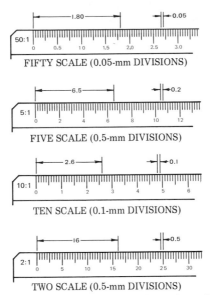

Fig. 3-37. Metric scales for enlarged objects.

the scales in Fig. 3-36. The proportional scales used to enlarge drawings of machine parts are shown in Fig. 3-37.

FULL-SIZE DRAWINGS

When an object is not too large for the paper on which you are drawing it, draw it full size (1:1). *To make a measurement*, put the scale on the paper in the direction in which you are measuring. Make a short, light dash next to the zero on the scale and another dash next to the division at the distance you want. Do not make a dot or punch a hole in the drawing sheet. Figure 3-38 shows a distance of 1⁷⁄₁₆″ laid off on a scale.

DRAWING TO SCALE

If an object is large or has little detail, you can draw it in a *reduced* (smaller) *proportion*. In the customary system, the first reduction is to

the scale of 6″ = 1′-0″, commonly called *half size*. You can use a full-size scale to draw a half size by letting each half inch stand for 1 in. and each 12-in. scale stand for a 24-in. scale. This is shown in Fig. 3-39, where 3⅝″ is laid off by three half inches and five eighths of the next half inch. Always think full size. A scale divided and marked for half size (Fig. 3-32) is easier to use.

If smaller views are needed, the next reduction that can be used is the scale of 3″ = 1′-0″, called *quarter size*. Find this scale on the architect's scale. The actual length of 3 in. represents 1 foot (ft) divided into 12 parts. Each part stands for 1 in. and is further divided into eighths. Learn to think of the 12 parts as standing for real inches. Fig. 3-40 shows how to lay off the distance of 1′-3½″. Notice where the zero mark is. It is placed so that inches can be measured in one direction from it and feet in the other direction.

You can use a regular mechanical engineer's scale with a scale of ¼″ = 1″. For other reductions, the proportional scales listed in the sections on

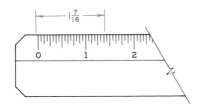

Fig. 3-38. Making a measurement with the full-size scale.

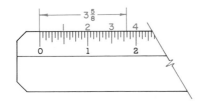

Fig. 3-39. Measuring to half size on a full-size scale.

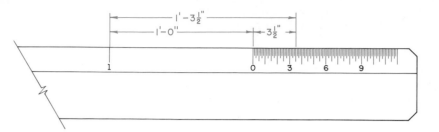

Fig. 3-40. Reading the scale of 3″ = 1′-0″, called quarter size.

the architect's scale and the mechanical engineer's scale are used. For small parts, you can use enlarged scales, such as 24″ = 1′-0″ for double-size views. Very small parts can be drawn 4 or 8 times full size or, for some purposes, 10, 20, or more times full size. The views of large parts and projects have to be drawn to a very small scale.

When drawing to scale using the metric system, you can get reduced or enlarged proportions with the scales shown in Figs. 3-36 and 3-37.

■ CASE INSTRUMENTS

A full set of instruments usually includes compasses with pen part, pencil part, lengthening bar, dividers, bow pen, bow pencil, bow dividers, and one or two ruling pens.

Large-bow sets (Fig. 3-41) are favored by most drafters. They are known as *master,* or *giant,* bows and are made in several patterns. With large bows, 6 in. (152 mm) or longer, circles can be drawn up to 13 in. (330 mm) in diameter or, with lengthening bars, up to 40 in. (1016 mm) in diameter. Large-bow sets let you use one instrument in place of the regular compasses, dividers, and small-bow instruments. Large-bow instruments let you hold the *radius* (one half the diameter; plural, *radii*) securely at any desired distance, up to their largest possible radius.

Drafting instrument sets are also available with technical fountain pens and attachments, if preferred (Fig. 3-42).

■ THE DIVIDERS

Lines are divided and distances *transferred* (moved from one place to another) with dividers. Dividers are held in the right hand. They are adjusted as in space A, Fig. 3-43.

To divide a line into three equal parts, adjust the points of the dividers until they seem to be about one third the length of the line. Put one point on one end of the line and the other point on the line (space B). Turn the dividers about the point that rests on the line, as in space C. Then turn it in the alternate direction, as in space D. If the last point falls short of the end of the line, increase the distance between the points of the dividers by an amount about one third the distance *mn.* Then start at the beginning of the line again. You may have to do this a few times. If the last point overruns the end of the line, decrease the distance between the points by one third the extra distance. For four, five, or more spaces, follow the same rules, but correct by one fourth, one fifth, and so forth, of the overrun or underrun. You divide an arc or the circumference of a circle in the same way.

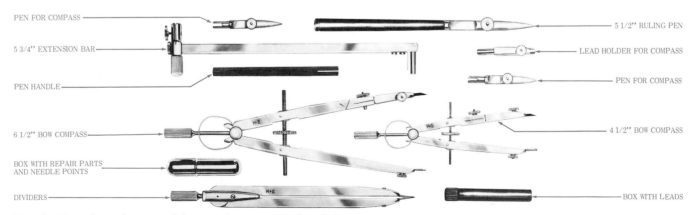

PEN FOR COMPASS

5 3/4″ EXTENSION BAR

PEN HANDLE

6 1/2″ BOW COMPASS

BOX WITH REPAIR PARTS AND NEEDLE POINTS

DIVIDERS

5 1/2″ RULING PEN

LEAD HOLDER FOR COMPASS

PEN FOR COMPASS

4 1/2″ BOW COMPASS

BOX WITH LEADS

Fig. 3-41. A large-bow set of drawing instruments. (Keuffel & Esser Co.)

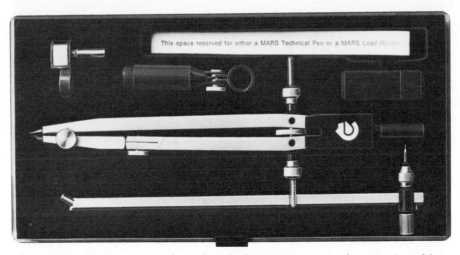

Fig. 3-42. Instruments with technical fountain pen attachments. (Staedtler-Mars.)

■ *THE COMPASS*

Views on drawings are made up of straight lines and curved lines. Most of the curved lines are circles or parts of circles (arcs). Draw these with the compass (Fig. 3-44). Leave the legs of the compass straight for radii under 2 in. (50 mm). For larger radii, make the legs perpendicular to the paper (Fig. 3-45). Put in a lengthening bar (Fig. 3-46) when you need a very large radius, usually one over 8 in. (200 mm).

To get the compass ready for use, sharpen the lead as in Fig. 3-47, allowing it to extend about ⅜ in. (10

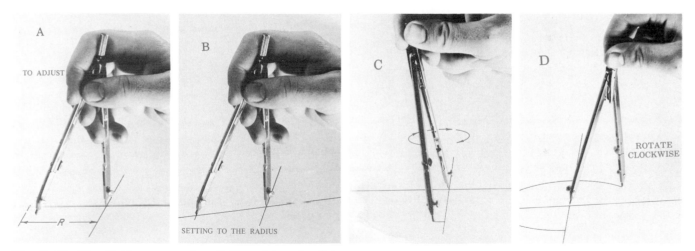

Fig. 3-43. The dividers are used to divide and transfer distances.

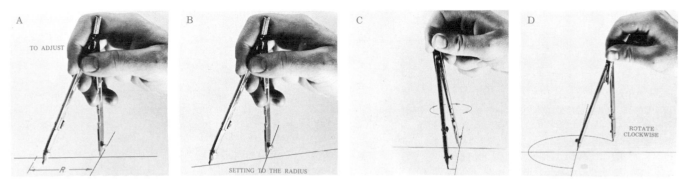

Fig. 3-44. The compass is used to draw circles and arcs.

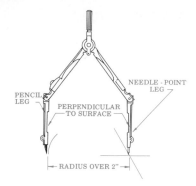

Fig. 3-45. Adjusting the compass for large circles.

Fig. 3-46. The lengthening bar is used in compasses for large radii.

mm). Using a long *bevel* (slant) on the outside of the lead will keep the edge sharp when you increase the radius. Then adjust the shouldered end of the needle point until it extends slightly beyond the lead point, as shown in Fig. 3-47. You cannot use as much pressure on the lead in the compass as you can on a pencil. Therefore, use lead one or two degrees softer in the compass to get the same line *weight* (thickness and darkness).

To draw a circle or an arc with the compass, locate the center of the arc or circle by two *intersecting* (crossing) lines. Lay off the radius by a short, light dash (Fig. 3-44 in space A). Open the compass by pinching it between your thumb and second finger (space A). Set it to the desired

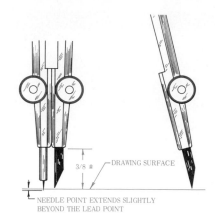

Fig. 3-47. Adjusting the point and shaping the lead of the compass.

radius by putting the needle point at the center and moving the pencil leg with your first and second fingers (space B). When the radius is set, raise your fingers to the handle (space C). Turn the compass by twirling the handle between your thumb and finger. Start the arc near the lower side and turn clockwise (space D). In doing this, slant the compass a little in the direction of the line. *Do not force* the needle point deep into the paper. Use only enough pressure to hold the point in place. Long radii may be drawn by

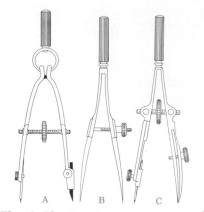

Fig. 3-48. Bow instruments are used for drawing small circles and arcs.

using a lengthening bar in the compass to make the pencil leg longer. For extra-long radii, a beam compass is used (Fig. 3-52).

■ BOW INSTRUMENTS

A set of bow instruments (Fig. 3-48) is made up of the bow pencil (A), the bow dividers (B), and the bow pen (C). Any of them can be adjusted with either a center wheel, as at A and C, or a side wheel, as at B. They may be of the hook-spring type, as at

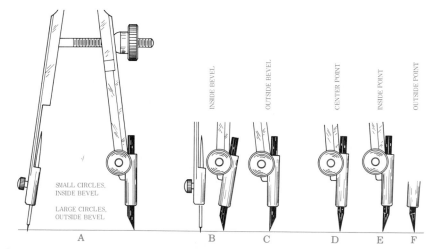

Fig. 3-49. Adjusting the lead for the bow pencil.

A, or the fork-spring type, as at B and C. They are usually about 4 in. (102 mm) long.

Use the bow dividers for taking off (transferring) small distances or for marking off a series of small distances. Also use them for dividing a line into small spaces. The bow pencil is used for drawing small circles. Whether you use instruments with center wheels or with side wheels is up to you. Sharpen and adjust the lead for the bow pencil as shown at A in Fig. 3-49. The inside bevel holds an edge for small circles and arcs, as shown at B. For larger radii, the outside bevel at C is better. Some drafters prefer a conical center point or an off-center point, as at D, E, and F. The bow instruments are easy to use and accurate for small distances or radii (less than 1¼ in. or 32 mm). They hold small distances better than the large instruments. You can make large adjustments quickly with the side-wheel bows by pressing the fork and spinning the adjusting nut. Some center-wheel bows are built

for making large, rapid adjustments. This is done by holding one leg in each hand and either pushing to close or pulling to open. Small adjustments are made with the adjusting nut on both the side-wheel and the center-wheel bows. Use the bow pencil (Fig. 3-50) with one hand. Set the radius as in space A. Start the circle near the lower part of the vertical centerline (space B). Turn clockwise, as in Space C.

DROP-SPRING BOW COMPASS

The drop-spring bow compass (Fig. 3-51) is good for drawing very small circles. It is very useful for drawing many small circles of the same size, such as when drawing rivets. You attach the marking point (pencil or pen) to a tube that slides on a pin.

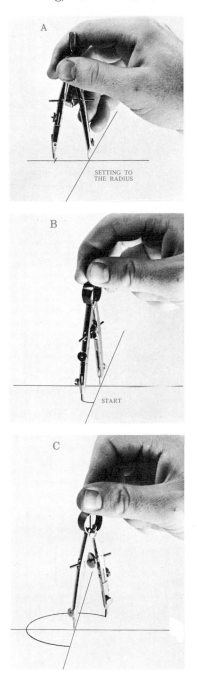

Fig. 3-50. Adjusting the radius for the bow-pencil compass.

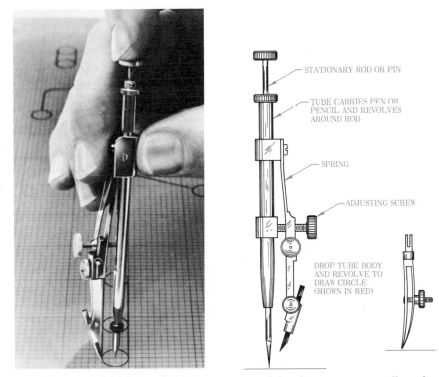

Fig. 3-51. The drop-spring bow compass is used for drawing very small circles, especially where there are many to be drawn. (Keuffel & Esser Co.)

When the bow is used, the pin stays still while the pencil point turns around it. Set the radius with the spring screw. Hold the marking point up while putting the pin on the center. Then drop it and turn it. The circles drawn will all•be the same size.

BEAM COMPASS

Beam compasses (Fig. 3-52) are used to draw arcs or circles with large radii. The beam compass is made up of a bar (beam) on which movable holders for a pencil part and a needle part can be put and fixed as far apart as desired. A pen part may be used in place of the pencil point. If a needle point is put in place of the pencil point, a beam compass can be used as a divider or to set off long distances. The usual bar is about 13 in. (330 mm) long. However, by using a coupling to add extra length, circles of almost any size can be drawn.

IRREGULAR CURVES

Irregular, or French, curves (Fig. 3-53) are used to draw noncircular curves (involutes, spirals, ellipses, and so forth). Irregular curves are also used for drawing curves on graphic charts. In addition, they can be used to plot motions and forces and to make some engineering and scientific graphs. Irregular curves are made of sheet plastic. They come in many different forms, some of which are shown. Sets are made for ellipses, parabolas, hyperbolas, and many special purposes.

To use an irregular curve (Fig. 3-54), find the points through which a curved line is to pass. Then set the path of the curve by drawing a light line, freehand, through the points.

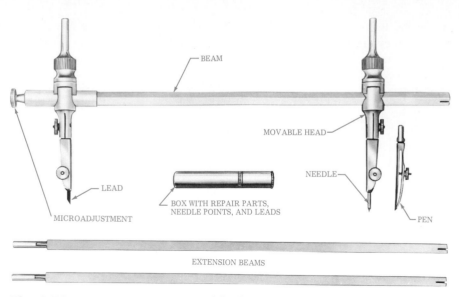

Fig. 3-52. Beam compasses are used for large radii.

Adjust it as needed to make the curve smooth. Next, match the irregular curve against a part of the curved line and draw part of the line. Move the irregular curve to match the next part, and so on. Each new position should fit enough of the part just drawn to make the line smooth. Note whether the radius of the curved line is increasing or decreasing and place the irregular curve in the same way. Do not try to draw too

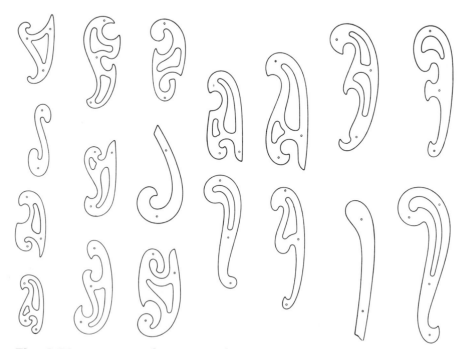

Fig. 3-53. Some irregular, or French, curves. They are made in a variety of forms. (Staedtler-Mars,)

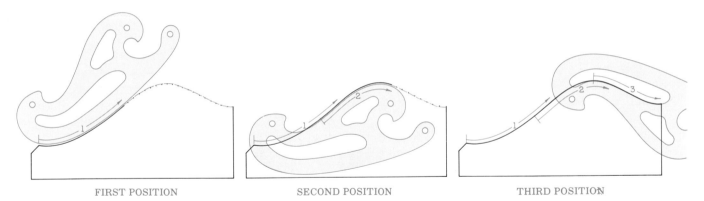

FIRST POSITION   SECOND POSITION   THIRD POSITION

Fig. 3-54. Steps in drawing a smooth curve.

much of the curve with one position. If the curved line is *symmetrical* (even) around an *axis* (center point), mark the position of the axis on the irregular curve with a pencil for one side. Then turn the irregular curve around to match and draw the other side. There are adjustable or flexible curves that you can use for certain special kinds of work (Fig. 3-55).

◼ *TEMPLATES*

Templates (Fig. 3-56) are an important part of the equipment of engineers and professional drafters They save a great deal of time in drawing shapes of details. These include boltheads, nuts, and electrical, architectural, and plumbing symbols.

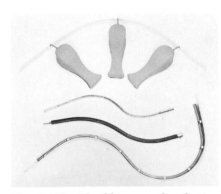

Fig. 3-55. Flexible curves for plotting smooth curves. (Bruning.)

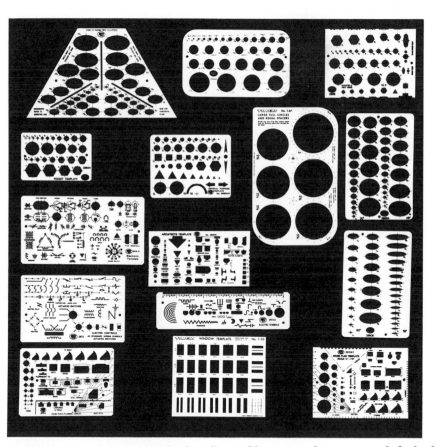

Fig. 3-56. Templates are made for all possible uses and save a good deal of time. (Teledyne Post.)

LEARNING ACTIVITIES

1. Visit an industrial drafting room and pay special attention to the types of drafting tables, equipment, and instruments being used.

2. Collect technical drawings done on various types of drafting sheets and compare them for such things as transparency and durability. Make prints of several and compare the quality of reproduction.

3. Draw sample lines on different types of tracing sheets using as many different grades of leads as available and make prints. Compare the line qualities.

4. Test different types of erasers on various types of drafting sheets to determine which works best on each.

VOCABULARY

1. Protractor
2. T-square
3. Vellum
4. Opaque
5. Transparent
6. Alphabet of lines
7. Horizontal
8. Vertical
9. Inclined
10. Vertex
11. Circumference
12. Acute angle
13. Obtuse angle
14. Parallel
15. Perpendicular
16. Hypotenuse
17. Bevel
18. French curve
19. Symmetrical
20. Axis

REVIEW

1. Explain how you can check the accuracy of a T-square.

2. Name three kinds of drawing sheets.

3. Drafting pencils are made in 17 degrees of hardness from _____ (softest and blackest) to _____ (hardest).

4. The shape of a drafting pencil point may be _____ or _____.

5. How many widths, or thicknesses, of lines are generally used in drafting?

6. You can draw angles in _____° intervals by using the 45° and 30°-60° triangles with the T-square.

7. What is another name for *smoothing* the pencil point?

8. What name is given to the different lines, or line symbols, used on drawings?

9. An instrument used in measuring, or laying out, angles is called a(n) _____.

10. Name two types of drafting machines.

Problems

■ GENERAL INSTRUCTIONS

The trim sizes of sheets recommended by the American National Standards Institute (ANSI) and the International Standards Organization (ISO) are in almost universal use in industry and are, therefore, desirable for use in drawing courses. Standard drawing sheet sizes are given in Fig. 3-6. Most of the drawing problems throughout this book are planned for working on A or B (ANSI) or A4 or A3 (ISO) sheets. It is possible, of course, to use other sizes and arrangements where necessary or where desired by the instructor. To assist in such cases, Fig. 3-57 is given with letters to indicate the dimensions. The desired numerical values can be filled in after the equal signs (=) on the figure. When this special layout is used, your instructor will supply the sizes.

Problems designed for B-size (11 × 17 in.) or A3-size (297 × 420 mm) sheets are designed for the layouts shown in Figs. 3-58 and 3-59. Slight adjustments and conversions are necessary when the metric-size sheets are used. Other layouts for B- and A3-size sheets are suggested in Fig. 3-60.

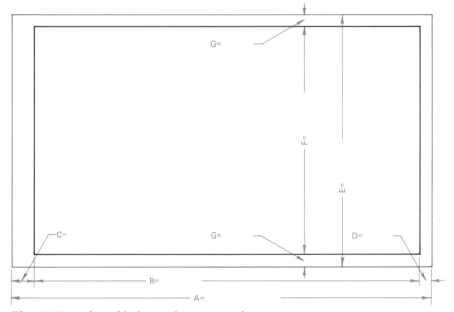

Fig. 3-57. Adjustable layout for any size sheet.

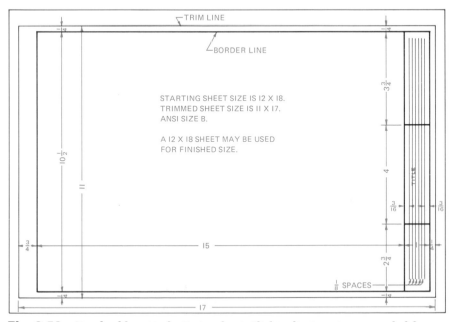

Fig. 3-58. Standard layout of a B-size sheet. Slight adjustments are needed for an A3-size (metric) sheet.

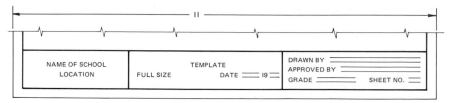

Fig. 3-59. Suggested title block for a B-size sheet. Slight adjustments are needed for an A3-size (metric) sheet.

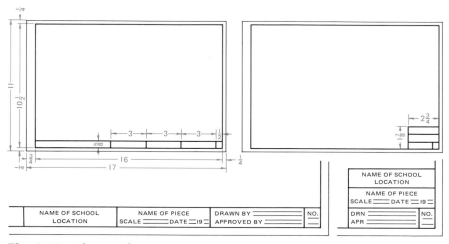

Fig. 3-60. Alternate layouts.

Sheet Layout: Two-Minute Method

If the method illustrated in progressive steps in Fig. 3-61 is followed, the layout of a sheet should not take more than two minutes.

Begin with a 12 × 18 in. sheet and tape it to the board as described earlier in this chapter. With the scale, measure 17″ near the bottom of the sheet (or on the bottom of the sheet) making short vertical marks, not dots. Measure and mark ¾″ in from the left-hand mark and ¼″ in from the right-hand mark (space 1). From this last mark measure 1″ toward the left and mark. Lay the scale vertically near the left of the paper (or on the left edge) and make short horizontal marks 11″ apart. Make short marks ¼″ up from the bottom mark and ¼″ down from the top mark.

With the T-square draw horizontal lines through the four marks last made (space 2). Next, draw vertical lines through the five vertical marks (space 3). Then brighten up the border lines and the sheet will appear as in space 4. It is now ready to be used for drawing. Lettering in the title block should be done after the assigned drawing problem is com-

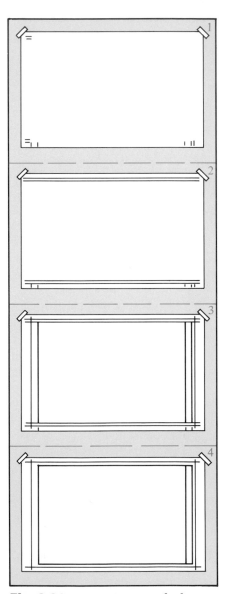

Fig. 3-61. Two-minute method.

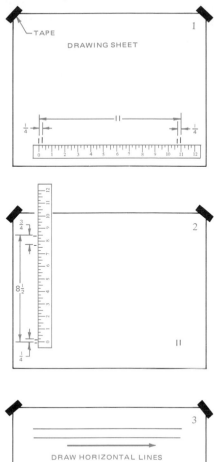

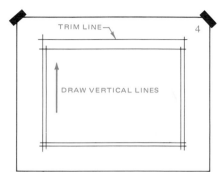

Fig. 3-62. Layout for an A-size sheet. Slight adjustments are needed for an A4-size (metric) sheet.

pleted. This will help to avoid smudging of the lettering.

The order of work for an A- or A4-size sheet (8½ × 11 in. or 210 × 297 mm, respectively) is given in Fig. 3-62. The method is the same for other sizes except for the dimensions. If sheets measuring 8½ × 11 in. or 210 × 297 mm trimmed size are used, the method is the same except that measurements are made from the edges of the sheet. Also, since the sizes given are in inches, slight adjustments will be needed for metric-size sheets.

Begin with a 9 × 12 in. sheet and tape it to the board. Lay the scale horizontally near the bottom of the sheet and make short vertical marks (not dots) 11″ apart (space 1). Hold the scale in position and make short vertical marks ¼″ from the left-hand and right-hand marks.

Lay the scale vertically near the left-hand edge of the sheet and make short horizontal marks (not dots) 8½″ apart (space 2). Hold the scale in position and make short marks (not dots) ¾″ down from the top mark and ¼″ up from the bottom mark.

Place the T-square in position and draw light horizontal lines through the four horizontal marks (space 3).

Place the T-square and triangle in position and draw light vertical lines through the four vertical marks (space 4). Go over the border lines to make them clear and black. Use an F or HB pencil. The sheet is now ready to be used for making a drawing.

The sheet may be used with the long measurement horizontal, as at A in Fig. 3-63, or vertical, as at B. A title strip may be used, as suggested. The height of the title strip may be ⅝″, or more, to suit the information required by the instructor. Four problems may be worked on a B- or an A3-size sheet, as at C.

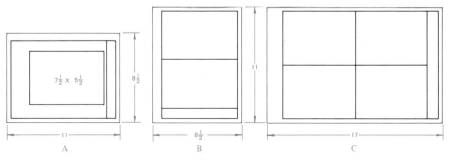

Fig. 3-63. Three arrangements of sheets.

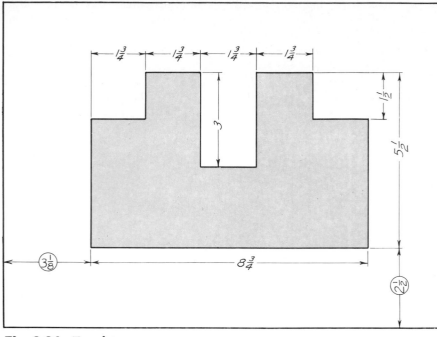

Fig. 3-64. Template.

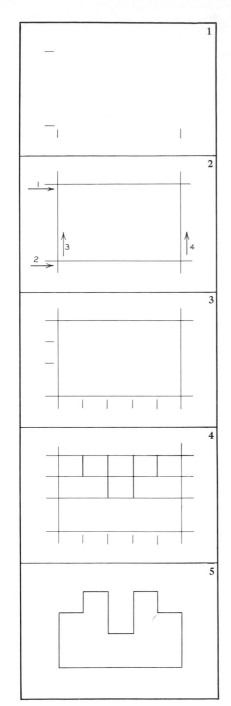

Fig. 3-64. Make a drawing of the template shown in Fig. 3-64. In this and several of the one-view drawings, the order of working is shown in progressive steps. These steps should be followed carefully, because they represent the drafter's procedure in making drawings. You should not consider the explanations as applying only to the particular problem. Instead, try to understand the system and apply it to all drawings.

1. On an 11 × 17 in. drawing sheet, lay out a working space 15″ wide and 10½″ high.

2. Measure 3⅛″ from left border line, and from this mark measure 8¾″ toward the right.

3. Lay the scale on the paper vertically near (or on) the left edge, make a mark 2½″ up, and from this measure 5½″ more. The sheet will appear as in space 1.

4. Draw horizontal lines 1 and 2 with the T-square, and vertical lines 3 and 4 with T-square and triangle (space 2).

5. Lay the scale along the bottom line of the figure with the measuring edge on the upper side and make marks 1¾″ apart. Then with the scale on line 3, with its measuring edge to the left, measure from the bottom line two vertical distances, 2½″ and 1½″ (space 3).

6. Through the two marks draw light horizontal lines.

7. Draw the vertical lines with T-square and triangle by setting the pencil on the marks on the bottom line and starting and stopping the lines on the proper horizontal lines (space 4).

8. Erase the lines not wanted and darken the lines of the figure to get the finished drawing (space 5). Do not add dimensions unless instructed to do so.

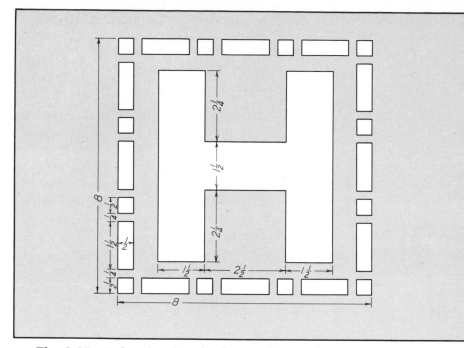

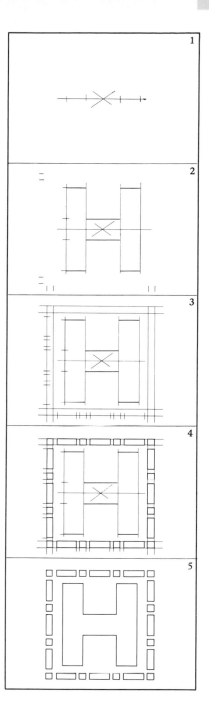

Fig. 3-65. Order of working for drawing the stencil.

Fig. 3-65. Make a drawing of the stencil shown in Fig. 3-65. This drawing gives practice in accurate measuring with the scale and making neat corners with short lines. The construction shown in the order of working should be drawn very lightly with a well-sharpened 3H or 4H pencil.

1. On an 11 × 17 in. (B-size) drawing sheet, lay out a working space 15″ wide and 10½″ high. Find the center of this space by laying the T-square blade face down across the opposite corners and drawing short lines where the diagonals intersect (space 1).

2. Through the center draw a horizontal centerline and on it measure and mark off points for the four vertical lines. The drawing will appear as in space 1.

3. Draw the vertical lines lightly with T-square and triangle. On the first vertical line, at the extreme left, measure and mark off points for all horizontal lines. The drawing will now appear as it does in space 2.

4. Draw the horizontal lines as finished lines. Measure points for the stencil border lines on the left side and bottom. The drawing will now appear as in space 3.

5. Draw the border lines. On the lower and left-hand border lines, measure the points for the ties.

6. Complete the border by drawing the cross lines as finished lines and darkening the other lines as in space 4.

7. Darken the vertical lines and finish as in space 5.

8. Fill in the title block.

Figs. 3-66 through 3-74. Make an instrument drawing of the figure assigned by your instructor. Include all centerlines. Do not add dimensions unless instructed to do so. All fractional-inch sizes may be converted to decimal-inch or metric (millimeter) sizes, if desired. Scale and sheet size are suggested. Your instructor may wish to change these.

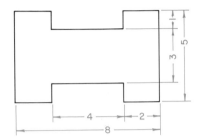

Fig. 3-66. Sheet-metal pattern. A- or A4-size sheet. Scale: full size.

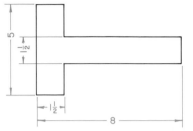

Fig. 3-67. Template for letter T. A- or A4-size sheet. Scale: full size.

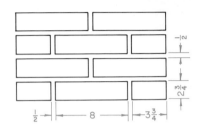

Fig. 3-68. Brick pattern. B- or A3-size sheet. Scale: three-quarter size.

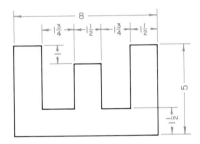

Fig. 3-69. Template for letter E. A- or A4-size sheet. Scale: full size.

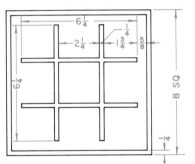

Fig. 3-70. Tic-tac-toe board. A- or A4-size sheet. Scale: three-quarter size.

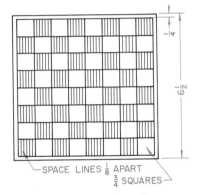

Fig. 3-71. Checker board. A- or A4-size sheet. Scale: full size.

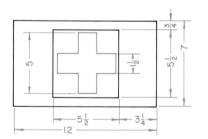

Fig. 3-72. First-aid station sign. B- or A3-size sheet. Scale: full size.

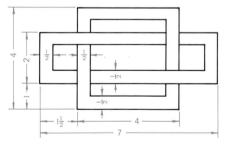

Fig. 3-73. Inlay. A- or A4-size sheet. Scale: full size.

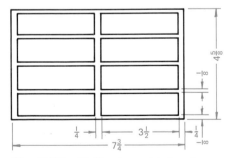

Fig. 3-74. Grille. A- or A4-size sheet. Scale: full size.

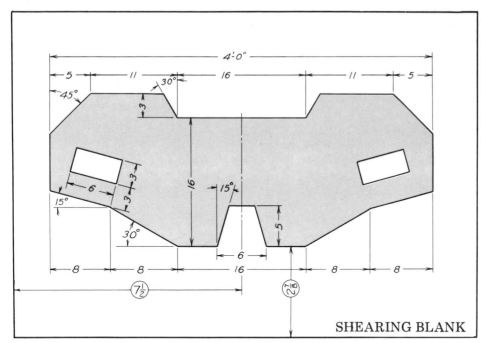

SHEARING BLANK

Fig. 3-75. Order of working for drawing the shearing blank.

Fig. 3-75. Make a drawing of the shearing blank shown in Fig. 3-75. When a view has inclined lines, it should first be blocked in with square corners. Angles of 15°, 30°, 45°, 60°, and 75° are drawn with the triangles after locating one end of the line.

1. Locate vertical centerline and measure 2′ on each side (space 1). Note that this drawing must be made to the scale of 3″ = 1′.

2. Locate vertical distances for top and bottom lines.

3. Draw main blocking-in lines as in space 2.

4. Make measurements for starting points of inclined lines (space 3).

5. Draw inclined lines with T-square and triangles (space 4).

6. Finish as in space 5.

7. Fill in the title block.

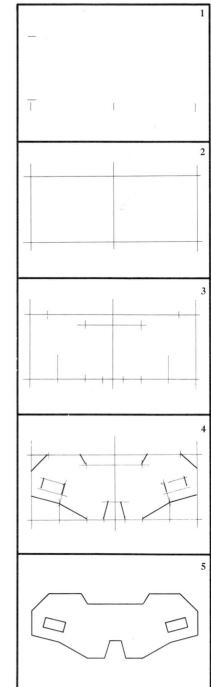

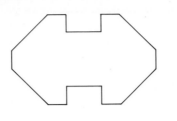

Fig. 3-76. Template. A- or A4-size sheet. Scale: full size.

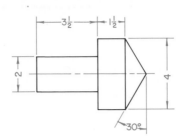

Fig. 3-77. Pivot. A- or A4-size sheet. Scale: full size.

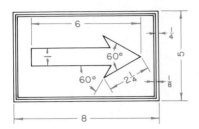

Fig. 3-78. Direction sign. A- or A4-size sheet. Scale: full size.

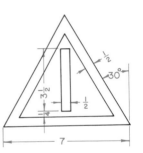

Fig. 3-79. International danger road sign. A- or A4-size sheet. Scale: full size.

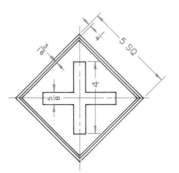

Fig. 3-80. Highway warning sign. A- or A4-size sheet. Scale: three-quarter size.

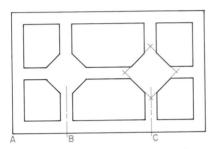

Fig. 3-81. Grill plate. Scale: full size or as assigned. Make all ribs ½″ wide. The distance $AB = 2¼″$, $BC = 3½″$, $AD = 2½″$. The diamond shapes are 1½″ square. A- or A4-size sheet. Scale: 1:1.

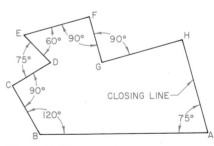

Fig. 3-82. Metric measurement. Scale: as assigned. Draw horizontal line AB 180 mm long. Work clockwise around the layout. *Remember!* Angular dimensions are the same in both the customary and metric systems. $BC = 60$ mm, $CD = 48$ mm, $DE = 42$ mm, $EF = 74$ mm, $FG = 50$ mm, $GH = 90$ mm. Measure the closing line and measure and label the angle at H. A- or A4-size sheet. Scale: 1:1.

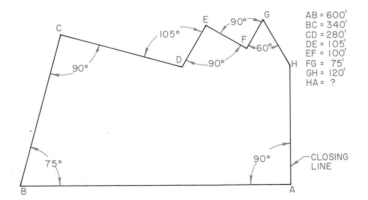

AB = 600'
BC = 340'
CD = 280'
DE = 105'
EF = 100'
FG = 75'
GH = 120'
HA = ?

Fig. 3-83. Civil engineer's scale. Draw a figure similar to the one shown above. Use a scale of 1″ = 40′-0″. Measure the length of the closing line to the nearest tenth of a foot and note it on your drawing. B- or A3-size sheet.

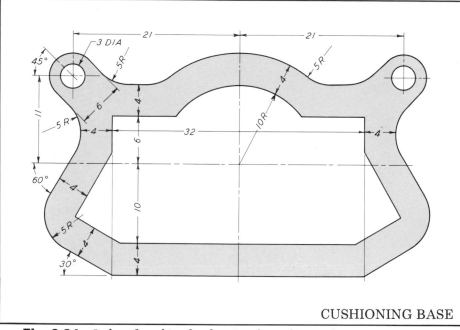

CUSHIONING BASE

Fig. 3-84. Order of working for drawing the cushioning base.

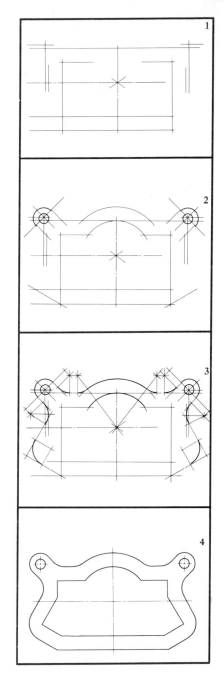

Fig. 3-84. Make a drawing of the cushioning base (Fig. 3-84). The drawing is for practice with triangles, compasses, and scale. Centers of arcs and tangent points must be carefully located.

Order of working for drawing the cushioning base in Fig. 3-84:

1. Through the center of the working space, draw horizontal and vertical centerlines. Measure horizontal and vertical distances. This drawing must be made to the scale of 3″ = 1′. Then draw horizontal and vertical lines (space 1).

2. Draw inclined lines with 45° and 30°-60° triangles. Then draw large arcs and two semicircles with tangents at 45° (space 2).

3. Locate centers and tangent points for the 5″ radius tangent arcs. To do this, measure 5″ perpendicularly from each tangent line and draw lines parallel to the tangent lines. The inter-

section of these lines will be the required centers. To find the points of tangency, draw lines from the centers perpendicular to the tangent lines. To find the centers for the 5″ arcs tangent to the middle arc, proceed as follows: Increase the radius of the larger arc by 5″ and draw two short arcs cutting lines parallel to and 5″ above the top horizontal tangent line. These points will be the centers. Lines joining these centers with the center of the large arc will locate the points of tangency of the arcs (space 3).

Draw all the 5″ tangent arcs above the horizontal centerline. Locate points of tangency and draw the two 60° tangent lines. Locate centers and draw the 5″ tangent arcs below the centerline.

4. Complete the view by drawing the lines for the opening. Darken the lines and finish the drawing as shown in Fig. 3-84 (space 4).

Figs. 3-85 through 3-96. Make an instrument drawing of the figure assigned. Include all centerlines. Do not dimension unless instructed to do so. Drawing sheet size and scale optional.

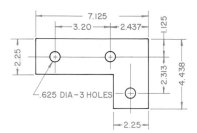

Fig. 3-85. Angle bracket.

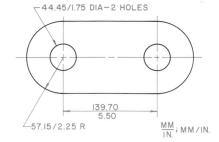

Fig. 3-86. Link plate.

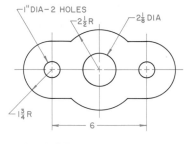

Fig. 3-87. Armature support.

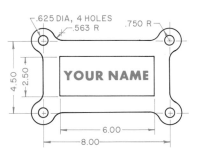

Fig. 3-88. Identification plate.

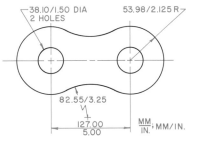

Fig. 3-89. Bicycle chain link.

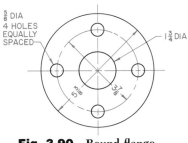

Fig. 3-90. Round flange.

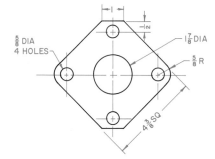

Fig. 3-91. Bronze shim.

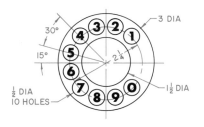

Fig. 3-92. Telephone dial.

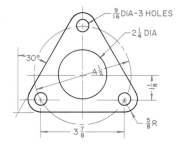

Fig. 3-93. Base plate.

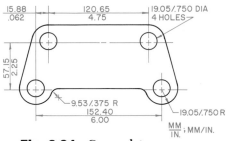

Fig. 3-94. Cover plate.

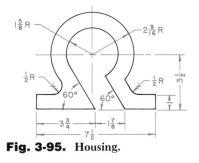

Fig. 3-95. Housing.

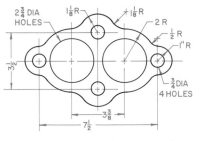

Fig. 3-96. Carburetor gasket.

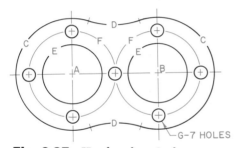

Fig. 3-97. Head gasket. Scale: as assigned. Use metric scale. The distance between center points *A* and *B* is 45.50 mm. Radius of arc *C* is 30 mm. Radius of arc *D* is 43 mm. Diameter of hole *E* is 32 mm. Diameter of circle *F* is 45.50 mm. Holes labeled *G* are 7-mm DIA.

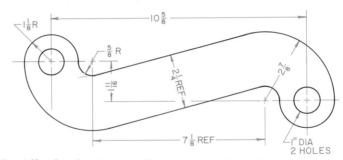

Fig. 3-98. Offset bracket. Locate all center points before beginning to draw circles and arcs.

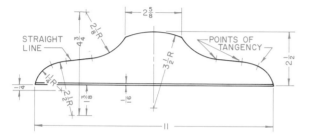

Fig. 3-99. T-square head. Draw to a scale of ¾″ = 1″. Be sure to locate points of tangency.

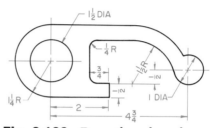

Fig. 3-100. Draw the release lever. Scale: full size. Use compass and bow pencil. Make neat tangent joints.

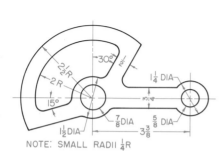

Fig. 3-101. Draw the adjustable sector. Scale: full size. Make neat tangent joints. Use compass and bow pencil.

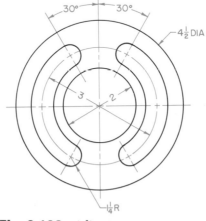

Fig. 3-102. Adjusting ring.

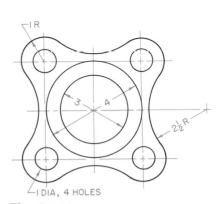

Fig. 3-103. Anchor plate.

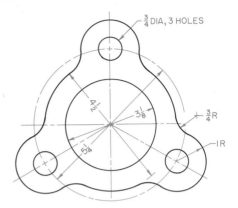

Fig. 3-104. Gasket.

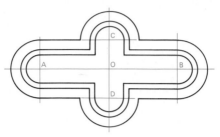

Fig. 3-105. Make a drawing of the frame. Scale: 6″ = 1′-0″. Draw intersecting lines at right angles through the center of the working space. Lay off $AO = OB = 5\frac{1}{4}″$ and $OC = OD = 2\frac{1}{4}″$. With centers at A, B, C, and D, draw semicircles with radii of $1\frac{1}{8}″$, $1\frac{1}{2}″$, and $2\frac{1}{4}″$. Draw horizontal and vertical tangent lines. Brighten lines necessary to show the view.

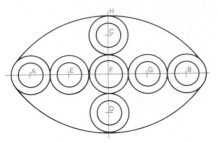

Fig. 3-106. Make a drawing of the multiple dial plate. Scale: 3″ = 1′-0″. Draw centerlines at right angles. Lay off $FC = FD = FG = FE = EA = GB = 6″$. With centers at A, B, C, D, E, F, and G, draw circles with a diameter of 6″. With center at F, draw a circle with a diameter of $4\frac{1}{2}″$. With centers at A, B, C, D, E, and G, draw circles with a diameter of 4″. With centers at H and I, draw tangent arcs with a radius of 18″. Brighten lines necessary to show the view.

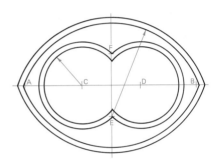

Fig. 3-107. Make a drawing of the double dial plate. Scale: full size. Draw line $AB = 7″$ and divide it into three equal parts. With centers at C and D, draw arcs with radii of $1\frac{1}{2}″$ and $1\frac{3}{4}″$. With centers at E and F, draw arcs with radii of $3\frac{3}{4}″$ and 4″ to complete the view.

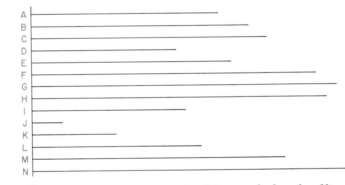

Fig. 3-108. Measuring practice. Measure the lengths of lines A through N at full size, ¾″ = 1″, ½″ = 1″, 1″ = 40′-0″, 1″ = 1′-0″, etc., as assigned by your instructor.

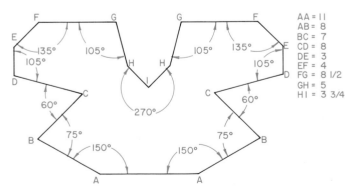

AA = 11
AB = 8
BC = 7
CD = 8
DE = 3
EF = 4
FG = 8 1/2
GH = 5
HI = 3 3/4

Fig. 3-109. Irregular polygon. Construct as shown. Use a scale of ½″ = 1″. Begin by drawing line AA near the bottom of the sheet and centered horizontally. The length of each line is given at the right of the figure above. All angles may be drawn with the T-square and a combination of triangles.

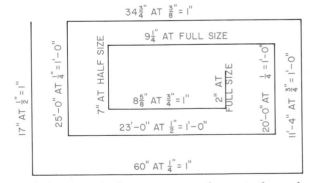

Fig. 3-110. Scale drawing. Draw a figure similar to the one shown above. Draw lines to the lengths and at the scales indicated.

4
Geometry for Technical Drawing

■ IMPORTANCE OF GEOMETRY

Geometry has always been important to people. It was used in ancient times for measuring land and making right-angle (90°) corners for buildings and other kinds of construction. The Egyptians used people called *rope stretchers* for this purpose. They used rope with marks or knots having 12 equal spaces. It was divided into 3-, 4-, and 5-space sections, as shown in Fig. 4-1. A square (right-angle) corner was made by stretching the rope and driving pegs into the ground at the 3-, 4-, and 5-space marks. This was one way an ancient people used geometry.

The use of the 3-4-5 triangle for making a right angle was proved by the mathematician Pythagoras in the sixth century B.C. This proof is called the pythagorean theorem. The theorem is shown graphically and mathematically in Fig. 4-2.

This method also works well for triangles that have the same proportions such as 6, 8, and 10 units.

$$6^2 + 8^2 = 10^2$$
$$36 + 64 = 100$$
$$100 = 100$$

Fig. 4-1. Egyptian rope stretchers.

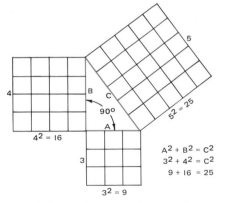

Fig. 4-2. Pythagorean theorem shown graphically and mathematically.

The units may be millimeters, meters, inches, fractions of an inch, or any other units of measure.

Geometry is the study of the size and shape of things. The relationship of straight and curved lines in drawing shapes is also a part of geometry. Some geometric figures used in drafting include circles, squares, triangles, hexagons, and octagons. Many other shapes and lines are shown in Figure 4-3. Study the "dictionary of drafting geometry" (Fig. 4-3) before beginning the geometric

constructions (drawings) on the following pages.

Geometric constructions are made of individual lines and points drawn in proper relationship to one another. Accuracy is extremely important.

Geometric constructions are very important to drafters, surveyors, engineers, architects, scientists, mathematicians, and designers. Therefore, nearly everyone in all technical fields needs to know the constructions explained in this chapter.

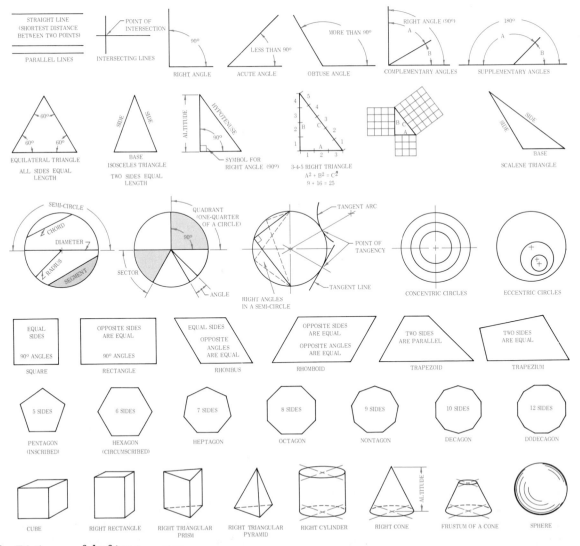

Fig. 4-3. Dictionary of drafting geometry.

PROBLEM Bisect line *AB* or arc *AB*. NOTE: *Bisect* means to divide into two equal parts.

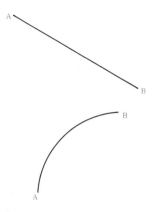

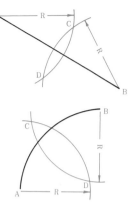

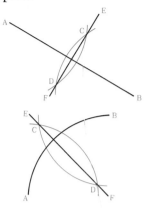

Given line *AB* or arc *AB*.

With *A* and *B* as centers and any radius *R* greater than one half of *AB*, draw arcs to intersect (cut across) at *C* and *D*.

Draw line *EF* through intersections *C* and *D*.

Fig. 4-4. To bisect a straight line or an arc.

PROBLEM Divide line *AB* into eight equal parts.

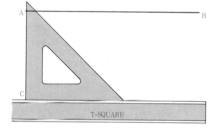

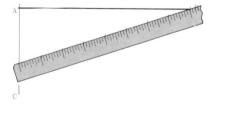

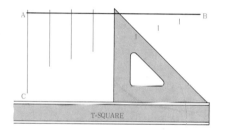

Draw a line of any length at *A* perpendicular (at right angles, 90°) to line *AB*.

Place scale with zero at point *B* and adjust it along line *AC* until any eight equal divisions are included between points *B* and *C*. (In this case, 8 in.) Mark the divisions.

Draw lines parallel (side-by-side) to *AC* through the division marks to intersect line *AB*.

Fig. 4-5. To divide a straight line into any number of equal parts (first method).

PROBLEM Divide line *AB* into five equal parts.

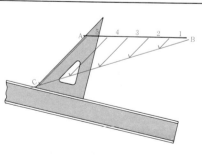

Draw line *BC* from point *B* at any convenient angle and of any length.

Use dividers or a scale to step off five equal spaces on line *BC* beginning at point *B*.

Draw a line connecting points *A* and *C*. Draw lines through each point on *BC* parallel to line *AC* to intersect line *AB*.

Fig. 4-6. To divide a straight line into any number of equal parts (second method).

PROBLEM Draw a perpendicular line at point *O* on line *AB*.

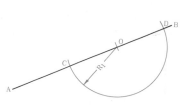

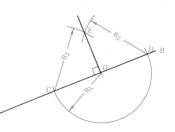

Given line *AB* and point *O*.

With *O* as the center and any convenient radius R_1, construct an arc cutting line *AB*, locating points *C* and *D*.

With *C* and *D* as centers and any radius R_2 greater than *OC*, draw arcs intersecting at *E*. Draw a line connecting points *E* and *O* to form the perpendicular.

Fig. 4-7. To draw a perpendicular line to a given line at a given point on the line (first method).

PROBLEM Draw a perpendicular line at *O* near the end of the line *AB*.

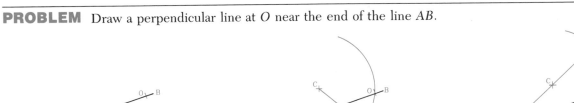

Given line *AB* and point *O*.

From any point *C* above line *AB*, construct an arc using *CO* as the radius and passing through line *AB* to locate point *D*.

Draw a line through points *D* and *C*, extending it through the arc to locate point *E*. Connect points *E* and *O* to form the perpendicular line.

Fig. 4-8. To draw a perpendicular line to a given line at a given point on the line (second method).

PROBLEM Draw a perpendicular line to line *AB* through point *O*.

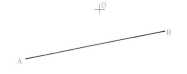

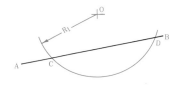

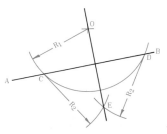

Given line *AB* and point *O*.

With *O* as the center, draw an arc with radius R_1 long enough to intersect line *AB* to locate points *C* and *D*.

With *C* and *D* as centers and radius R_2 greater than one half of *CD*, draw intersecting arcs to locate point *E*. A line drawn through points *O* and *E* is the perpendicular line.

Fig. 4-9. To draw a perpendicular line to a given line through a point outside the line (first method).

PROBLEM Draw a perpendicular line to line *AB* through point *O*.

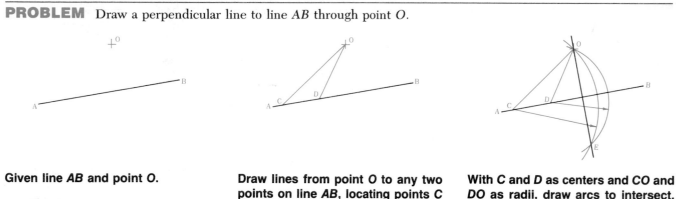

Given line *AB* and point *O*.

Draw lines from point *O* to any two points on line *AB*, locating points *C* and *D*.

With *C* and *D* as centers and *CO* and *DO* as radii, draw arcs to intersect, locating point *E*. Connect points *O* and *E* to form the perpendicular line.

Fig. 4-10. To draw a perpendicular line to a given line through a point outside the line (second method).

PROBLEM Draw a perpendicular line to line *AB* through point *O*.

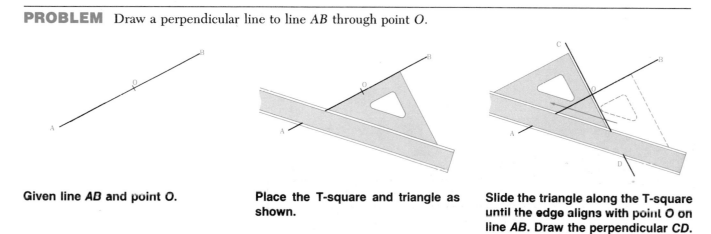

Given line *AB* and point *O*.

Place the T-square and triangle as shown.

Slide the triangle along the T-square until the edge aligns with point *O* on line *AB*. Draw the perpendicular *CD*.

Fig. 4-11. To draw a perpendicular line through a point outside the line (third method).

PROBLEM Draw a line parallel to line *AB* through point *P*. NOTE: *Parallel* lines are always the same distance apart.

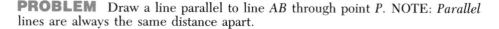

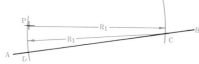
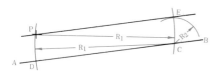

Given line *AB* and point *P*.

With point *P* as the center and any convenient radius R_1, draw an arc cutting line *AB* to locate point *C*. With point *C* as the center and the same radius R_1, draw an arc through point *P* and line *AB* to locate point *D*.

With *C* as the center and radius R_2, equal to chord *PD*, draw an arc to locate point *E*. Draw a line through points *P* and *E* that is parallel to *AB*.

Fig. 4-12. To construct (draw) a line parallel to a given line (first method).

PROBLEM   Construct (draw) a line parallel to *AB* at a distance you need from *AB*.

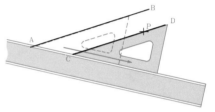

Given line *AB*.

Draw two arcs with centers anywhere along line *AB*. The arcs should have a radius *R* equal to the distance you need between the parallel lines.

Draw a parallel line *CD* tangent to (just touching) the arcs.

Fig. 4-13. To construct a line parallel to a given line (second method).

PROBLEM   Construct a line parallel to *AB* through point *P*.

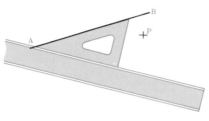

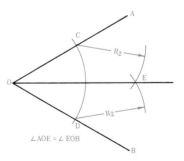

Given line *AB* and point *P*.

Place the T-square and triangle as shown.

Slide the triangle until the edge aligns with point *P*. Draw the parallel *CD*.

Fig. 4-14. To construct a line parallel to a given line (third method).

PROBLEM   Construct a line to bisect angle *AOB*.

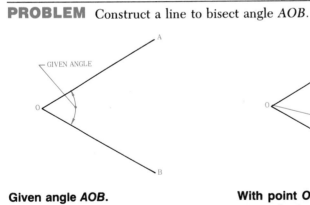

Given angle *AOB*.

With point *O* as the center and any convenient radius R_1, draw an arc to intersect with *AO* and *OB* at *C* and *D*.

With *C* and *D* as centers and any radius R_2 greater than one half of arc *CD*, draw arcs to intersect, locating point *E*. Draw a line through points *O* and *E* to bisect angle *AOB*.

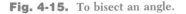

Fig. 4-15. To bisect an angle.

PROBLEM Construct angle *AOB* in a new location.

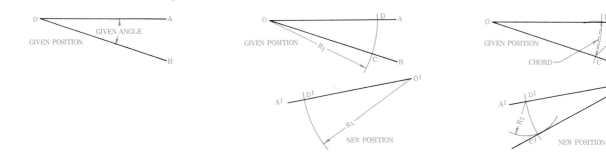

Given angle *AOB*.

Draw one side *O¹A¹* in the new position. With *O* and *O¹* as centers and any convenient radius R_1, construct arcs to cut *BO* and *AO* at *C* and *D* and *A¹O¹* at *D¹*.

With *D¹* as the center and radius R_2 equal to chord *DC*, draw an arc to locate point *C¹* at the intersection of the two arcs. Draw a line through points *O¹* and *C¹* to complete the angle.

Fig. 4-16. To copy a given angle.

PROBLEM Construct an isosceles triangle. NOTE: An *isosceles* triangle has two equal sides.

Given base line *AB*.

Fig. 4-17. To draw an isosceles triangle.

With points *A* and *B* as centers and a radius *R* equal to the length of the sides you want, draw intersecting arcs to locate the vertex (top point) of the triangle.

Draw lines through point *A* and the vertex and through point *B* and the vertex to complete the triangle.

PROBLEM Construct an equilateral triangle. NOTE: An *equilateral* triangle has all sides equal and all angles equal.

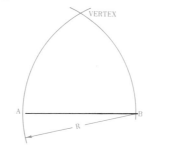

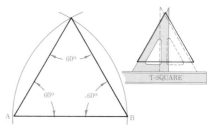

Given base line *AB*.

With points *A* and *B* as centers and a radius *R* equal to the length of line *AB*, draw intersecting arcs to locate the vertex.

Draw lines through point *A* and the vertex and through point *B* and the vertex to complete the triangle. NOTE: An equilateral triangle may also be constructed by drawing 60° lines through the ends of the base line with the 30°-60° triangle, as shown at right.

Fig. 4-18. To draw an equilateral triangle.

PROBLEM Construct a right triangle with two sides given.
NOTE: A *right* triangle has one right angle (90°).

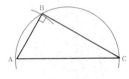

Given sides *AB* and *BC*.

Draw side *AB* in the desired position. Draw a perpendicular line to *AB* at *B* equal to *BC*. NOTE: Use the method of Fig. 4-7 or the method of Fig. 4-11 to construct the perpendicular line.

Draw a line connecting points *A* and *C* to complete the right triangle. NOTE: Line *AC* is called the *hypotenuse*. It is always the side opposite the 90° angle.

Fig. 4-19. To draw a right triangle with two sides given.

PROBLEM Construct a right triangle with the hypotenuse and one side given.

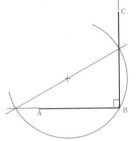

Given hypotenuse *AC* and side *AB*.

Draw the hypotenuse in the desired location. Draw a semicircle on *AC* using ½*AC* as the radius.

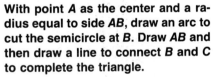

With point *A* as the center and a radius equal to side *AB*, draw an arc to cut the semicircle at *B*. Draw *AB* and then draw a line to connect *B* and *C* to complete the triangle.

Fig. 4-20. To draw a right triangle with the hypotenuse and one side given.

PROBLEM Construct a right triangle on a base line 3 units long.

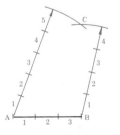

Given base line *AB* 3 units long.

With *A* and *B* as centers and radii 4 and 5 units long, draw intersecting arcs to locate point *C*.

Draw lines *AC* and *BC* to complete the triangle.

Fig. 4-21. To draw a right triangle by the 3-4-5 method.

PROBLEM Construct a triangle with three sides given.

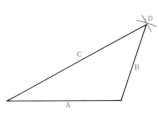

Given triangle sides *A*, *B*, and *C*.

Fig. 4-22. To draw a triangle with three sides given.

Draw base line *A* in the desired location. Construct arcs from the ends of line *A* with radii equal to lines *B* and *C* to locate point *D*.

Connect the ends of line *A* with point *D* to complete the triangle.

PROBLEM Draw a square within a circle with corners tangent to (that touch but do not cross) the circle.

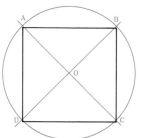

Given a circle with center point *O*. Construct 45° diagonals through the center point *O* to locate points *A*, *B*, *C*, and *D*. Connect *A* and *B*, *B* and *C*, *C* and *D*, and *D* and *A* to complete the square.

Fig. 4-23. To draw a square within a circle.

PROBLEM Draw a square outside the circle with the circle tangent to the midpoints of the sides of the square.

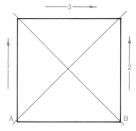

Given a circle with center point *O*. Construct 45° diagonals through the center point *O*. Draw sides tangent to the circle, intersecting at the 45° diagonals to complete the square.

Fig. 4-24. To draw a square about a circle.

PROBLEM Construct a square.

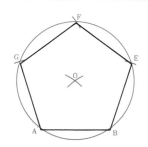

Given the length of the side *AB*. Construct 45° diagonals from ends of line *AB*. Complete the square by drawing the sides in the order shown by the numbered arrows.

Fig. 4-25. To draw a square.

PROBLEM Construct a regular pentagon. NOTE: *Regular* means the object has equal sides and equal angles.

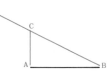

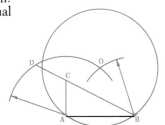

Given line *AB*, construct a perpendicular line *AC* equal to one half of *AB*. Draw line *BC* and extend it to make line *CD* equal to *AC*.

Fig. 4-26. To construct a regular pentagon given one side.

With radius *AD* and points *A* and *B* as centers, draw intersecting arcs to locate point *O*. With the same radius and *O* as the center, draw a circle.

Step off *AB* as a chord to locate points *E*, *F*, and *G*. Connect the points to complete the pentagon.

PROBLEM Draw a regular pentagon in a given circle.

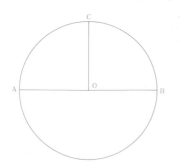

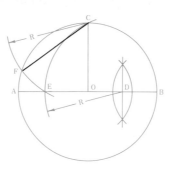

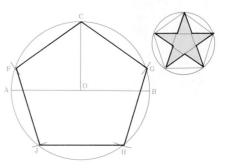

Given a circle with diameter *AB* and radius *OC*, draw a pentagon within the circle with points touching the circle.

Bisect radius *OB* to locate point *D*. With *D* as center and radius *DC*, draw an arc to locate point *E*. With *C* as center and radius *CE*, draw an arc to locate point *F*. Chord *CF* is one side of the pentagon.

Step off chord *CF* around the circle to locate points *G, H,* and *J*. Draw the chords to complete the pentagon. NOTE: Another method for constructing a pentagon in a circle is to use dividers and locate the points by trial and error.

Fig. 4-27. To draw a regular pentagon in a given circle.

PROBLEM Construct a hexagon.

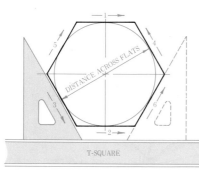

Given the distance across the flats of a regular hexagon. Draw centerlines and a circle with the diameter equal to the distance across the flats. With the T-square and 30°-60° triangle, draw the tangents in the order shown.

Fig. 4-28. To draw a regular hexagon given the distance across the flats.

PROBLEM Construct a hexagon.

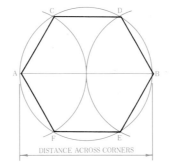

Given the distance *AB* across the corners. Draw a circle with *AB* as the diameter. With *A* and *B* as centers and the same radius, draw arcs to intersect the circle at points *C, D, E,* and *F*. Connect the points to complete the hexagon.

Fig. 4-29. To draw a regular hexagon given the distance across the corners (first method).

PROBLEM Construct a hexagon.

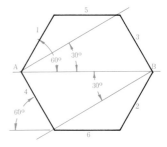

From points *A* and *B* draw lines of indefinite (any) length at 30° to line *AB*. With the T-square and 30°-60° triangle, draw the sides of the hexagon in the order shown.

Fig. 4-30. To draw a regular hexagon given the distance across the corners (second method).

PROBLEM Construct an octagon outside a circle that touches the midpoints of each of its sides.

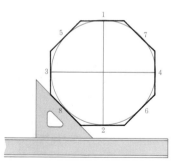

Given the distance across the flats, draw centerlines and a circle with the diameter equal to the distance across the flats. With the T-square and 45° triangle, draw lines tangent to the circle in the order shown to complete the octagon.

Fig. 4-31. To draw a regular octagon about a circle.

PROBLEM Construct an octagon within a circle so that the corners touch the circle.

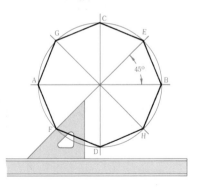

Given the distance across the corners, draw centerlines AB and CD and a circle with the diameter equal to the distance across the corners. With the T-square and 45° triangle, draw diagonals EF and GH. Connect the points to complete the octagon.

Fig. 4-32. To draw a regular octagon within a circle.

PROBLEM Construct an octagon within a square.

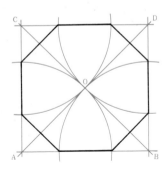

Given the distance across the flats, construct a square having the sides equal to AB. Draw diagonals AD and BC with their intersection at O. With A, B, C, and D as centers and radius R = AO, draw arcs to cut the sides of the square. Connect the points to complete the octagon.

Fig. 4-33. To draw a regular octagon in a square.

PROBLEM Construct a circle through points A, B, and C.

Given points A, B, and C. Draw lines AB and BC.

Draw perpendicular bisectors of AB and BC to intersect at point O.

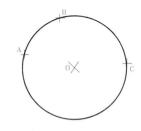

Draw the required circle with point O as the center, and radius R = OA = OB = OC.

Fig. 4-34. To construct a circle through any three points not in a straight line.

PROBLEM Draw a tangent line through point *P* on the circle.

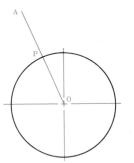

Given circle with center point *O* and tangent point *P*. Draw line *OA* from the center of the circle to extend beyond the circle through point *P*.

Draw a perpendicular line to *OA* at *P*. The perpendicular line is the tangent line.

Fig. 4-35. To draw a tangent to a circle at a given point *P* on the circle (first method).

PROBLEM Draw a tangent line through point *P* on the circle.

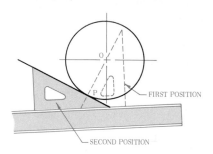
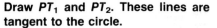

Given a circle with center point *O* and tangent point *P*. Place a T-square and triangle so that the hypotenuse of the triangle passes through points *P* and *O* (first position). Hold the T-square, turn the triangle to the second position at point *P*, and draw the tangent line.

Fig. 4-36. To draw a tangent to a circle at a given point *P* on the circle (second method).

PROBLEM Draw a line from point *P* tangent to the circle.

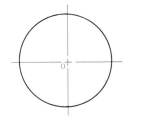

Given a circle with center point *O* and point *P* outside the circle.

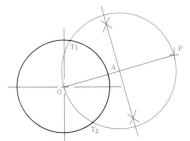

Draw line *OP* and bisect it to locate point *A*. Draw a circle with center *A* and radius *R* = *AP* = *AO* to locate tangent points T_1 and T_2.

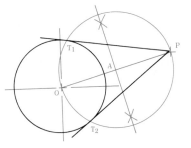

Draw PT_1 and PT_2. These lines are tangent to the circle.

Fig. 4-37. To draw a tangent to a circle from a point outside the circle.

PROBLEM Construct an arc tangent to two straight lines.

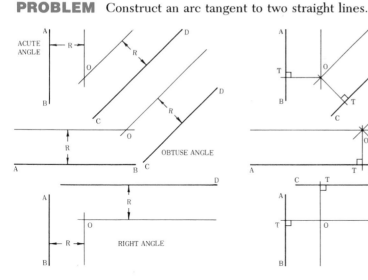
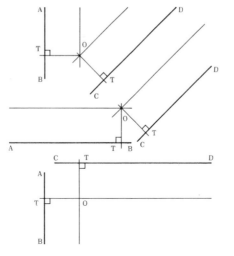
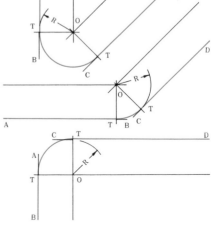

Given lines *AB* and *CD*, draw lines parallel to *AB* and *CD* at a distance *R* from them on the inside of the angle. The intersection *O* will be the center of the arc you need.

Draw perpendicular lines from *O* to *AB* and *CD* to locate the points of tangency *T*.

With *O* as the center and radius *R*, draw the needed arc.

Fig. 4-38. To construct an arc tangent to two straight lines at an acute angle, an obtuse angle, and a right angle.

PROBLEM Draw a reverse curve, or an ogee, between two straight lines.

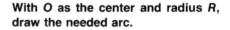

Given lines *AB* and *CD*, draw line *BC*. Select a point *E* on line *BC* through which the curve is to pass.

Draw perpendicular bisectors of *BE* and *EC*. Draw perpendicular lines to *AB* at *B* and to *CD* at *C*. They must cross the bisectors of *BE* and *EC* at *O_1* and *O_2*, respectively.

Draw one arc with center *O_1* and radius *O_1E* and the other with center *O_2* and radius *O_2E* to complete the required curve.

Fig. 4-39. To draw a reverse, or ogee, curve.

PROBLEM Construct an arc tangent to two given arcs.

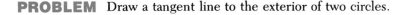

Given two arcs having radii R_1 and R_2 (radii may be equal or unequal) and radius R of the tangent arc.

Draw an arc with center O_1 and radius $= R + R_1$. Draw an arc with center O_2 and radius $= R + R_2$. The intersection at O is the center of the tangent arc.

Draw lines O_1O and O_2O to locate tangent points T_1 and T_2. With point O as the center and radius R, draw the tangent arc needed.

Fig. 4-40. To draw an arc of a given radius tangent to two given arcs.

PROBLEM Draw a tangent line to the exterior of two circles.

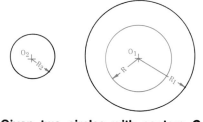

Given two circles with centers O_1 and O_2 and radii R_1 and R_2. $R = R_1 - R_2$. Using radius R and point O_1 as the center, draw a circle.

From center point O_2 draw a tangent O_2T to the circle of radius R. Draw radius O_1T. Extend it to locate tangent point T_1. Draw O_2T_2 parallel to O_1T_1.

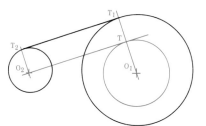

Draw the needed tangent T_1T_2 parallel to TO_2.

Fig. 4-41. To draw an exterior common tangent to two circles of unequal radii.

PROBLEM Draw a tangent line to the interior of two circles.

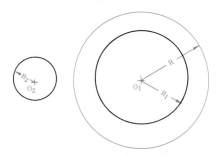

Given two circles with centers O_1 and O_2, and radii R_1 and R_2. $R = R_1 + R_2$. Using radius R and point O_1 as the center, draw a circle.

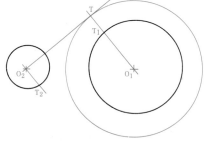

From center point O_2, draw a tangent O_2T to the circle of radius R. Draw radius O_1T to locate tangent point T_1. Draw O_2T_2 parallel to O_1T.

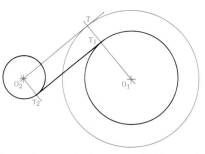

Draw the needed tangent T_1T_2 parallel to TO_2.

Fig. 4-42. To draw an interior common tangent to two circles of unequal radii.

PROBLEM Construct an arc tangent to line *AB* and arc *CD*.

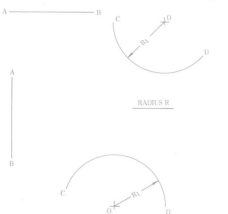

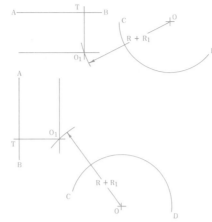

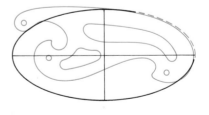

Given line *AB*, arc *CD*, and radius *R*.

Fig. 4-43. To draw an arc of given radius tangent to an arc and a straight line.

Draw a line parallel to *AB*, at distance *R*, toward arc *CD*. Use radius $R_1 + R$ to locate point O_1. A perpendicular line from O_1 to *AB* locates tangent point *T*.

Draw a line from *O* to O_1 to locate tangent point T_1 on *CD*. With point O_1 as the center and radius *R*, draw the tangent arc.

PROBLEM Construct an ellipse (oval) by the pin-and-string method.

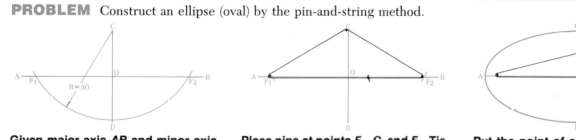

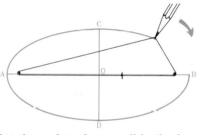

Given major axis *AB* and minor axis *CD* intersecting at *O*. With *C* as center and radius $R = AO$, an arc locates F_1 and F_2.

Place pins at points F_1, *C*, and F_2. Tie a string around the three pins and remove pin *C*.

Put the point of a pencil in the loop and draw the ellipse. Keep the string taut (tight) when moving the pencil.

Fig. 4-44. To draw an ellipse by the pin-and-string method.

PROBLEM Construct an ellipse by the trammel method.

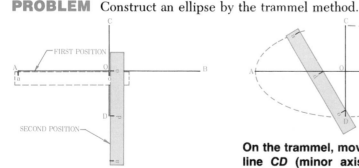

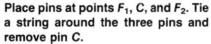

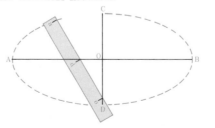

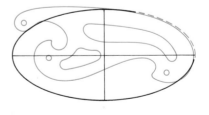

Given major axis *AB* and minor axis *CD* intersecting at point *O*. Cut a strip of paper or plastic (*trammel*). Mark off distances *AO* and *OD* on trammel.

On the trammel, move point *o* along line *CD* (minor axis) and point *d* along line *AB* (major axis) and mark points at *a*.

Use a French curve or flexible curve to connect the points to draw the ellipse.

Fig. 4-45. To draw an ellipse by the trammel method.

PROBLEM Draw an ellipse using axes *AB* and *CD*.

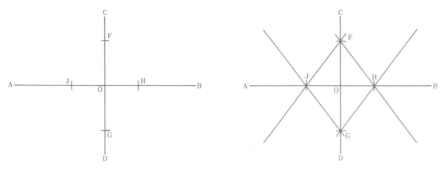 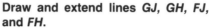 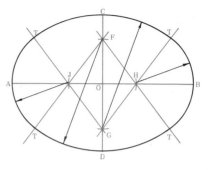

Given major axis *AB* and minor axis *CD* intersecting at point *O*. Lay off *OF* and *OG*, each equal to *AB* minus *CD*. Lay off *OJ* and *OH*, each equal to three fourths of *OF*.

Draw and extend lines *GJ*, *GH*, *FJ*, and *FH*.

Draw arcs with centers *F* and *G* and radii *FD and GC* to the points of tangency. Draw arcs with centers *J* and *H* and radii *JA* and *HB* to complete the ellipse. The points of tangency are marked *T*.

Fig. 4-46. To draw an approximate ellipse when the minor axis is at least two thirds the size of the major axis.

PROBLEM Draw an ellipse using axes *AB* and *CD*.

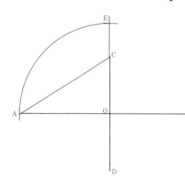 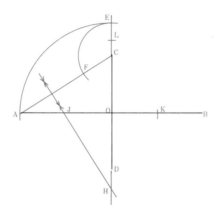

Given major axis *AB* and minor axis *CD* intersecting at point *O*. Draw line *AC*. Draw an arc with point *O* as the center and radius *OA* and extend line *CD* to locate point *E*.

Draw an arc with point *C* as the center and radius *CE* to locate point *F*. Draw the perpendicular bisector of *AF* to locate points *J* and *H*. Locate points *L* and *K*. *OL = OH* and *OK = OJ*.

Draw arcs with *J* and *K* as centers and radii *JA* and *KB*. Draw arcs with *H* and *L* as centers and radii *HC* and *LD* to complete the ellipse.

Fig. 4-47. To draw an approximate ellipse when the minor axis is less than two thirds of the major axis.

PROBLEM Reduce or enlarge the drawing of the sailboat shown at *A*.

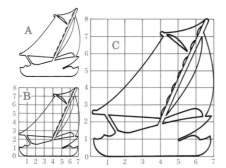

Lay out a grid over the drawing. Use squares of an approximate size. Draw a larger or smaller grid on a separate sheet of paper. The size of the grid depends upon the amount of enlargement or reduction needed. Use dots to mark key points on the enlarged or reduced grid corresponding to points on the original drawing. Connect the points to complete the new drawing.

Fig. 4-48. To reduce or enlarge a drawing.

PROBLEM Change the proportion of the drawing shown at *A*.

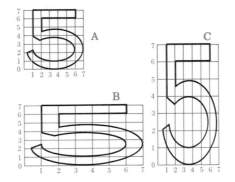

Draw a grid over the original drawing. Draw a grid on a separate sheet of paper in the needed proportion, as at *B* or *C*. Use dots to mark key points from the original drawing. Connect the points to complete the new drawing.

Fig. 4-49. To change the proportion of a drawing.

PROBLEM Enlarge or reduce the original rectangle.

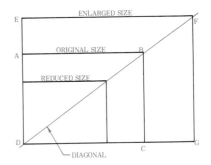

Draw a diagonal through corners *D* and *B*. Measure the width or height you need along line *DC* or *DA* (example: *DG*). Draw a perpendicular line from that point (*G*) to the diagonal. Draw a line perpendicular to *DE* intersecting at point *F*. Reductions are made in the same way.

Fig. 4-50. To enlarge or reduce a square or rectangular area.

VOCABULARY

1. Adjacent	11. Complementary angle	21. Concentric circles	31. Bisect
2. Hypotenuse	12. Supplementary angle	22. Eccentric circles	32. Perpendicular
3. Geometry	13. Equilateral triangle	23. Rhombus	33. Intersect
4. Hexagon	14. Isosceles triangle	24. Rhomboid	34. Arc
5. Octagon	15. Scalene triangle	25. Trapezoid	35. Tangent
6. Pentagon	16. Radius, radii	26. Heptagon	36. Vertex
7. Parallel	17. Diameter	27. Nontagon	37. Right triangle
8. Right angle	18. Chord	28. Decagon	38. Ogee
9. Acute angle	19. Quadrant	29. Dodecagon	39. Ellipse
10. Obtuse angle	20. Sector	30. Cube	40. Drawing grid

REVIEW

1. What kind of triangle has one right angle (90°)?

2. A hexagon has how many sides?

3. An octagon has how many sides?

4. What name is given to a plane figure having equal sides and equal opposite angles?

5. What is another name for one quarter of a circle?

6. What is an angle greater than 90° called?

7. What is an angle less than 90° called?

8. What name is given to a triangle with all three sides of equal length?

9. What is the name of the point at which a line touches an arc or a circle?

10. A pentagon has how many sides?

Problems

Each of the problems is to be drawn three times the size shown below. Use dividers to pick up the dimensions from Figs. 4-51 to 4-71 as assigned, and step off each measurement three times.

Problems 4-51 to 4-70 are designed for working four problems on an A- or A4-size sheet, laid out as shown in the accompanying figure.

Problems 4-71 to 4-83 are single-view drawings. These are designed to give additional practice in geometric constructions. Each should be drawn using the assigned sheet size and scale. Dimensions may be omitted unless required by your instructor.

Nearly all the problems for Chap. 3 may be used as problems for Chap. 4 and vice versa.

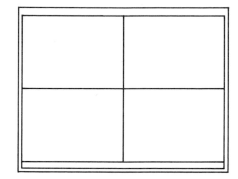

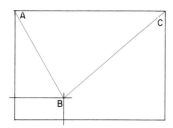

Fig. 4-51. Bisect line *AB*.

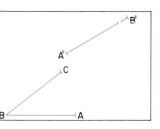

Fig. 4-52. Construct a perpendicular at point *P*.

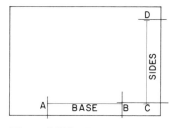

Fig. 4-53. Divide line *AB* into five equal parts.

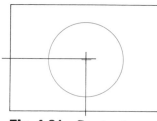

Fig. 4-54. Construct line *CD* parallel to *AB* and equal in length to *AB* through *P*.

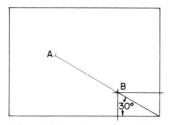

Fig. 4-55. Bisect angle *ABC*.

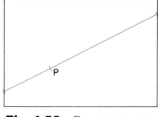

Fig. 4-56. Copy angle *ABC* in a new location, beginning with *A'B'*.

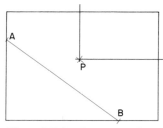

Fig. 4-57. Construct an isosceles triangle on base *AB* with sides equal to *CD*.

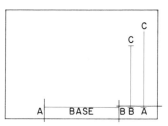

Fig. 4-58. Construct a triangle on base *AB* with sides equal to *BC* and *AC*.

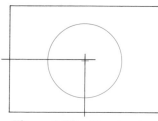

Fig. 4-59. Construct a square in the 3″ DIA circle with corners touching the circle.

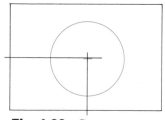

Fig. 4-60. Construct a regular pentagon in the 3″ DIA circle.

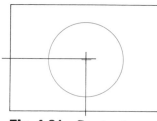

Fig. 4-61. Construct a regular hexagon around the 3″ DIA circle.

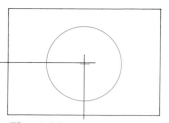

Fig. 4-62. Construct a regular octagon around the 3″ DIA circle.

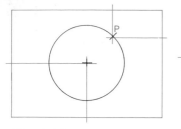

Fig. 4-63. Construct a tangent line through point *P* on the 3″ DIA circle.

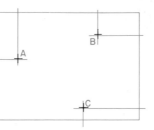

Fig. 4-64. Construct a circle through points *A*, *B*, and *C*.

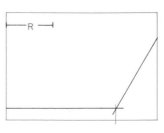

Fig. 4-65. Construct an arc having a radius *R* tangent to the two lines.

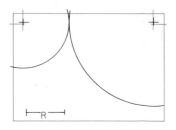

Fig. 4-66. Construct an arc having a radius *R* tangent to two given arcs.

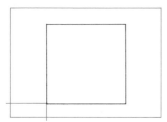

Fig. 4-67. Construct a regular octagon within the 3″ square.

Fig. 4-68. Construct an ellipse on the 4″ major and 2½″ minor axes.

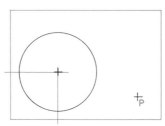

Fig. 4-69. Construct a line from point *P* tangent to the 2½″ DIA circle.

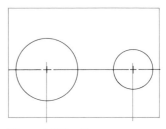

Fig. 4-70. Construct two tangent lines to the exterior of two circles.

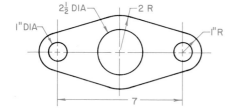

Fig. 4-71. Draw the view of the gasket full size or as assigned. Mark all points of tangency. Do not dimension.

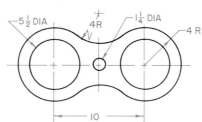

Fig. 4-72. Pipe support. Scale: as assigned. Locate and mark all centers and all points of tangency. Do not dimension.

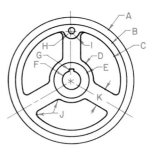

Fig. 4-73. Handwheel. Scale: as assigned. *A* = 7″ DIA, *B* = 6⅛″ DIA, *C* = 2¾″R, *D* = 1¼″R, *E* = 2″ DIA, *F* = 1″ DIA, *G* (keyway) = ³⁄₁₆″ wide × ³⁄₃₂″ deep, *H* = ⅜″ DIA, *I* = ⅜″R, *J* = ³⁄₁₆″R, *K* = 1″.

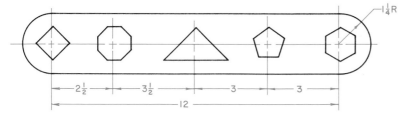

Fig. 4-74. Combination wrench. Scale: as assigned. Square: 1″; octagon: 1⅜″ across flats; isosceles triangle: 2¾″ base, 2″ sides; pentagon: inscribed within 1⅜″-diameter circle; hexagon: 1¼″ across flats. Use the method of your choice for constructing geometric shapes. Do not erase construction lines.

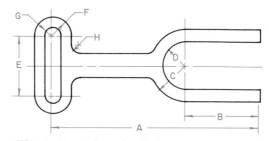

Fig. 4-75. Adjustable fork. Scale: as assigned. *A* = 220 mm, *B* = 80 mm, *C* = 40 mm, *D* = 26 mm, *E* = 64 mm, *F* = 20 mm, *G* = 8 mm, *H* = 10 mm.

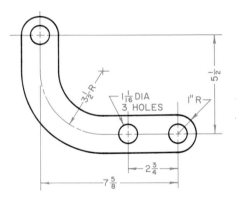

Fig. 4-76. Rod support. Scale: as assigned by instructor.

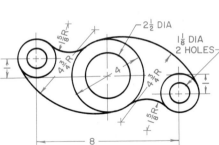

Fig. 4-77. Rocker arm. Scale: full size or as assigned.

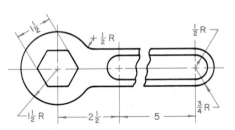

Fig. 4-78. Hex wrench. Scale: as assigned. Mark all tangent points. Do not erase construction lines.

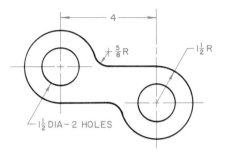

Fig. 4-79. Offset link. Scale: full size or as assigned. Locate and mark all points of tangency.

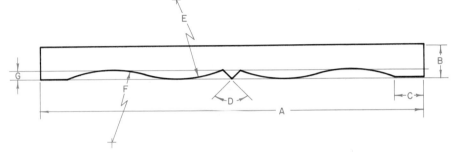

Fig. 4-80. Valance board. Scale: 1″ = 1′-0″ or as assigned. A = 8′-0″, B = 0′-8″, C = 0′-7″, D = 90°, E = 2′-6″, F = 2′-6″, G = 0′-2″. Locate and mark points of tangency. Do not erase construction lines.

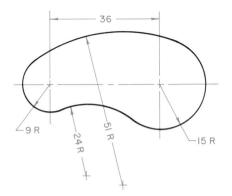

Fig. 4-81. Kidney-shaped table top. Scale: full size or as assigned.

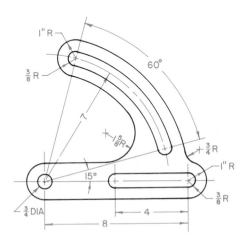

Fig. 4-82. Adjustable table support. Scale: as assigned.

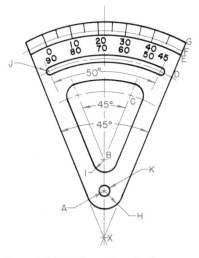

Fig. 4-83. Tilt scale. Scale: as assigned. *AB* = 1¾″, *AX* = 2⅝″, *AC* = 5½″, *AD* = 7¼″, *AE* = 8½″, *AF* = 8¾″, *AG* = 9¼″, *H* = 1″R, *I* = ⅝″R, *J* = 3/16″R, *K* = ½″ DIA.

5 Multiview Drawing

INTRODUCTION TO MULTIVIEW DRAWING

Technical drawing is a way of communicating ideas. People communicate by verbal and written language and by *graphic* (pictorial) means. One of the graphic means is technical drawing. It is a language used and understood in all countries. When accurate *visual* (sight) understanding is necessary, technical drawing is the most exact method that can be used.

Technical drawing involves two things: (1) visualization and (2) implementation. *Visualization* is the ability to see clearly what a machine, device, or other object looks like in the mind's eye. *Implementation* is the drawing of the object that has been visualized. In other words, the designer, engineer, or drafter first visualizes the object and then explains it pictorially by a technical drawing. Thus, the idea is recorded in a form that can be used as a means of communication.

A technical drawing, properly made, will give a more accurate and clearer description of an object than a photograph or written explanation. Technical drawings made according to standard *principles* (rules) result in views that give an exact visual description of an object (Fig. 5-1).

MULTIVIEW DRAWING

A photograph of a V-block is shown in Fig. 5-2. It shows the object as it appears to the eye. Notice that three sides of the V-block are shown in a single view. In Figure 5-3 the same photograph is shown with the three sides labeled according to their *relative* (related) positions. If all sides could be shown in a single photo-

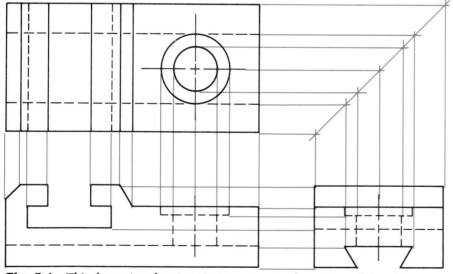

Fig. 5-1. This three-view drawing gives an accurate description of the object.

Fig. 5-2. Photograph of a V-block.

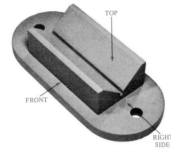

Fig. 5-3. Photograph of a V-block with front, top, and side views labeled.

graph, it would also include a left-side view, a rear view, and a bottom view. Nearly all objects have six sides.

An object cannot be photographed before it has been built. Therefore it is necessary to use another kind of graphic representation. One possibility is to make a *pictorial drawing*, as shown in Fig. 5-4. It shows, just as a photograph, the way the object looks generally. However, it does not show the exact forms and relationships of the parts of the object. It shows the V-block as it appears, not as it really is. For example, the holes in the base appear as ellipses, not as true circles.

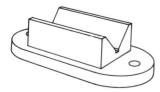

Fig. 5-4. Pictorial drawing.

Fig. 5-5. Front view of a V-block.

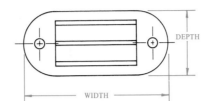

Fig. 5-6. Top view of a V-block.

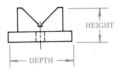

Fig. 5-7. Right-side view of a V-block.

The problem, then, is to represent an object on a sheet of paper in a way that will describe its exact shape and proportions. This can be done by drawing views of the object as it is seen from different positions. These views are then arranged in a particular order.

In order to describe accurately the shape of each view, imagine a position directly in front of the object, then above it, and finally at the right side of it. This is where the ability to

visualize is important. Figure 5-5 shows the exact shape of the V-block when viewed from the front. The dashed lines are used to show the outline of details behind the front surface (hidden details). Notice that this view shows the width and the height of the object.

Figure 5-6 is a top view of the V-block. It shows the width and the depth. Since the view is taken directly from above, the exact shape of the top is shown. Notice that the holes are true circles and that the rounded ends of the base are true radii. In the photograph and in the pictorial drawing, these appeared as elliptical shapes.

Figure 5-7 is a right-side view of the V-block. It shows the depth and the height. Notice that the shape of the V appears to be balanced (symmetrical) in the drawing. It appears distorted or misshapen in the photograph and in the pictorial drawing.

The Relationship of Views

Views must be placed in proper relationship to one another. Only in this way can technical drawings be read and understood properly. Figure 5-8 shows the V-block and how its three normal views have been *revolved*

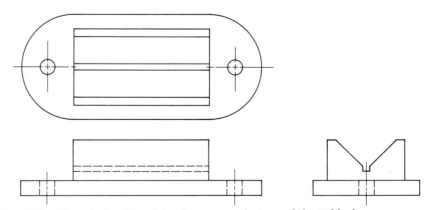

Fig. 5-8. The relationship of the three normal views of the V-block.

(turned) into their proper places. Notice that the top view is directly above the front view. The right-side view is directly to the right of the front view. Each of the normal views is where it logically belongs. When the normal views are placed in proper relationship to one another, the result is a *multiview drawing*. Multiview drawing is the exact representation of two or more views of an object on a *plane* (flat surface). These three views will usually give a complete description of the object. Figure 5-8 is an example of a multiview drawing.

Other Views

Most objects have six sides or six views. In most cases two or three views will completely describe the shape and size of all parts of an object. However, in some cases it may be necessary to show views other than the front, top, and right sides. In Fig. 5-9 dice are shown in the upper left-hand corner. Since the detail is different on each of the six sides, six views are needed for a complete graphic description. The six views are shown in their proper locations in the lower part of Fig. 5-9. Only in unusual cases are six views necessary.

■ ORTHOGRAPHIC PROJECTION

Earlier we said that multiview drawing is the exact representation of two or more views of an object on a plane (flat surface). These views are developed through the *principles* (rules) of orthographic projection. *Ortho-* means "straight or at right angles." *Graphic* means "written or drawn." *Projection* comes from two Latin words: *"pro,"* meaning *"forward,"* and *"jacere,"* meaning *"to throw."* Thus, orthographic projection literally means "thrown forward, drawn at right angles." *Orthographic projection* is the method of representing the exact form of an object in two or more views on planes, usually at right angles to each other, by lines drawn *perpendicular* (at right angles) from the object to the planes.

Angles of Projection

Orthographic projection involves the use of three planes. They are the *vertical* (up-and-down) plane, the *horizontal* (side-to-side) plane, and the *profile* (side-view) plane. These are shown in Fig. 5-10. In technical drawing, a *plane* is an imaginary flat surface that has no thickness. A view of an object is then projected and drawn upon this plane. Notice that the vertical and horizontal planes divide space into four *quadrants* (quarters of a circle). In orthographic projection, quadrants are usually called *angles*. Thus, we get the names *first-angle projection* and *third-angle projection*. First-angle projection is used in European countries. Third-

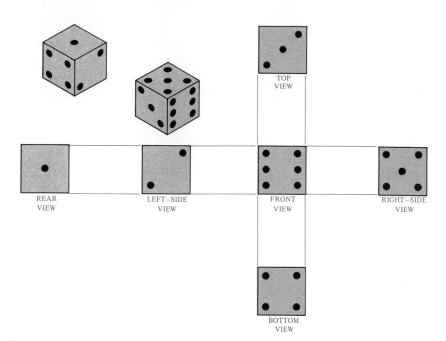

Fig. 5-9. Pictorial drawing and six views of dice.

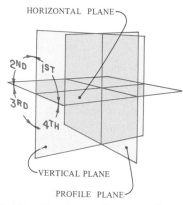

Fig. 5-10. The three planes used in orthographic projection.

angle projection is used in the United States and Canada. Second- and fourth-angle projections are not used in technical drawing.

First-Angle Projection

Figure 5-11 shows an object within the three planes of the first quadrant for developing a three-view drawing in first-angle projection. The front view is projected to the vertical plane. The top view is projected to the horizontal plane. The *left-side*

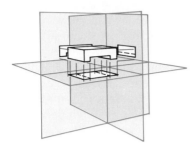

Fig. 5-11. The position of the three planes used in first-angle projection.

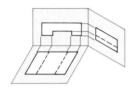

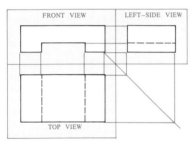

Fig. 5-12. Three views in first-angle projection.

view is projected to the profile plane. The horizontal and profile planes are rotated into a single plane. *In this case the front view is above the top*

view. The left-side view is to the right of the front view (Fig. 5-12).

Third-Angle Projection

Third-angle projection uses the same basic *principles* (rules) as first-angle. The main difference is in the *relative* (related) positions of the three planes. Figure 5-13 shows the same object placed within the third quadrant for developing a three-view drawing in third-angle projection. In this case the front view is projected to the vertical plane. The top view is projected to the horizontal plane. The *right-side view* is projected to the profile plane. The horizontal and profile planes are rotated into a single plane. *Thus, the top view is above the front view*. The right-side view is to the right side of the front view (Fig. 5-14).

■ THE GLASS BOX

The three views of an object have been developed by using imaginary *transparent* (see-through) planes. The views are projected onto these planes. We mentioned earlier that most objects have six sides. Therefore, six views may result. To explain the theory of projecting all six views,

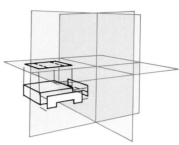

Fig. 5-13. The position of the three planes used in third-angle projection.

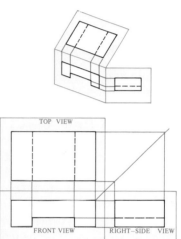

Fig. 5-14. Three views in third-angle projection.

an imaginary glass box will be used.

Figure 5-15 shows the glass box partially opened with the six views labeled. When the box is fully

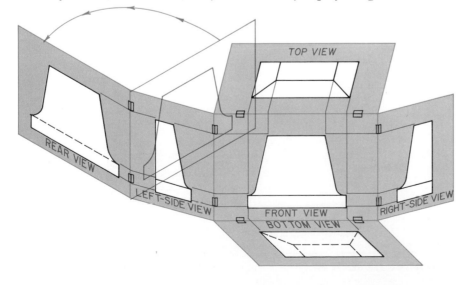

Fig. 5-15. Opening the glass box.

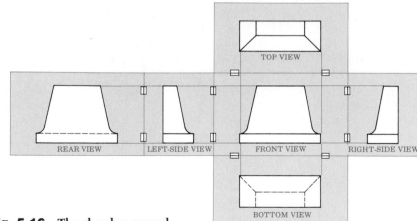

Fig. 5-16. The glass box opened.

opened up into one plane (Fig. 5-16), the views are in their relative positions as they would be if they had been drawn on paper. These views are arranged according to proper order for the six views. Notice that the rear view is located to the left of the left-side view.

Also notice that some views give the same information that is in other views. The views may also be mirror images of one another. Thus, it is not necessary to show all six views for a complete description of the object. The three normal views, as ordinarily drawn, are shown in Fig. 5-17.

■ HIDDEN LINES

It is necessary to describe every part of an object. Therefore, everything

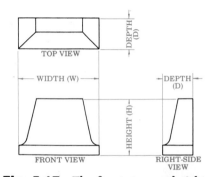

Fig. 5-17. The front, top, and right-side views.

must be represented in each view, whether or not it can be seen. *Interior* (inside) features and *exterior* (outside) features are both projected in the same way. Parts that cannot be

Fig. 5-18. Hidden lines.

seen in the views are drawn with hidden lines that are made up of short dashes (Fig. 5-18). Notice that the first dash of a hidden line touches the line where it starts (Fig. 5-18 at A). If a hidden line is a continuation of a visible line, space is left between the visible line and the first dash of the hidden line, as at B. If the hidden lines show corners, the dashes touch at the corners, as at C.

Dashes for hidden arcs (Fig. 5-19 at A) start and end at the tangent points. When a hidden arc is tangent to a visible line, a space is left, as at B. When a hidden line and a visible line project at the same place, show the visible line (Fig. 5-19 at C). When a centerline and a hidden line project at the same place (Fig. 5-20

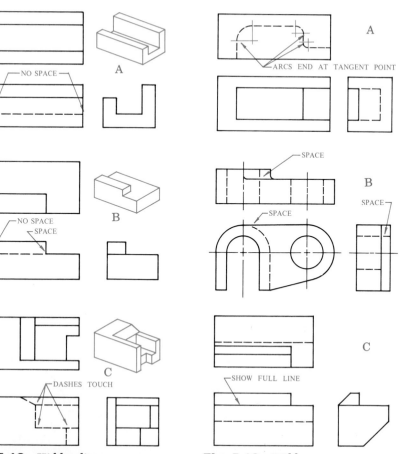

Fig. 5-19. Hidden arcs.

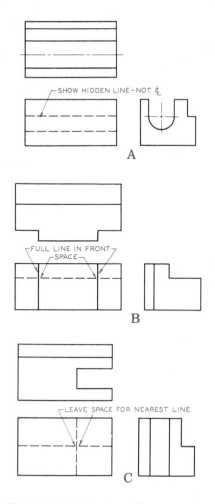

Fig. 5-20. Technique of representing hidden and visible lines.

at A), draw the hidden line. When a hidden line crosses a visible line (Fig. 5-20 at B), do not cross the visible line with a dash. When hidden lines cross (Fig. 5-20 at C), the nearest hidden line has the "right of way." Draw the nearest hidden line through a space in the farther hidden line.

■ CENTERLINES

Centerlines are used to locate views and dimensions. (See the alphabet of lines, Fig. 3-12.) Primary centerlines, marked P in Fig. 5-21, locate the center on *symmetrical* (balanced) views where one part is a mirror image of another. Primary centerlines are used as major locating lines to help in making the views. They are also used as base lines for dimensioning. Secondary centerlines, marked S in Fig. 5-21, are used for drawing details of a part. Primary centerlines are the first lines to be drawn. The views are developed from them. Note that centerlines represent the axes (singular, axis; turning point) of cylinders in the side view. The centers of circles or arcs are located first so that measurements can be made

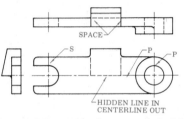

Fig. 5-22. Centerlines and hidden lines.

from them to locate the lines on the various views. Show a hidden line instead of a centerline (Fig. 5-22).

■ CURVED SURFACES

Some curved surfaces, such as cylinders and cones, do not show as curves in all views. This is illustrated in Fig. 5-23. A cylinder with its *axis* (centerline) perpendicular to a plane will show as a circle on that plane. It will show as a rectangle on the other two planes. Three views of a cylinder in different positions are shown at B, C, and D. The holes may be thought of as *negative cylinders*. (In mathematics, *negative* means an amount less than zero. A hole is a "nothing" cylinder. However, it has size. Thus, in a sense, it is negative.) A cone appears as a circle in one view. It appears as a triangle in the others, as shown at E. One view of a frustum of a cone appears as two circles, as at F. In the top view, the conical surface is represented by the space between the two circles.

Cylinders, cones, and frustums of cones have single curved surfaces. They are represented by circles in one view and straight lines in the other. The handles in Fig. 5-24 at A have double curved surfaces that are represented by curves in both views. The ball handle has spherical ends. Thus, both views of the ends are circles because a sphere appears as a circle when viewed from any direction. The slotted link in Fig. 5-24 at B and C is an example of tangent

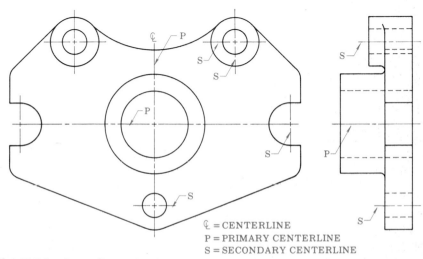

℄ = CENTERLINE
P = PRIMARY CENTERLINE
S = SECONDARY CENTERLINE

Fig. 5-21. Centerlines.

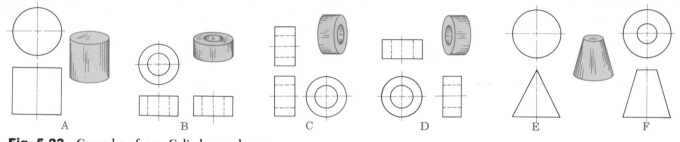

Fig. 5-23. Curved surfaces. Cylinders and cones.

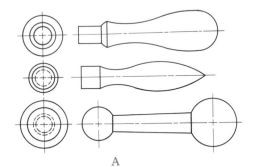

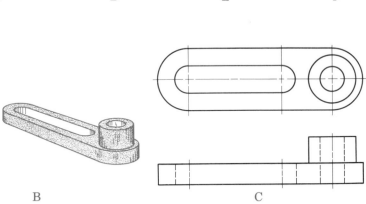

Fig. 5-24. Curved surfaces.

plane and curved surfaces. The rounded ends are tangent to the sides of the link and the ends of the slot are tangent to the sides. Therefore, the surfaces are smooth. There is no line of separation.

▪ *WHAT VIEWS TO DRAW*

Six views are not always needed to describe most objects. Usually three views are sufficient. The general *characteristics* (different features) of an object will suggest the views that are required to describe its shape. Three properly selected views will describe most shapes. Sometimes, however, there are features that can be more clearly described by using more views or parts of extra views.

Most pieces can be recognized because they have a characteristic view. This is the first view to consider. Usually, it is the first view to draw. Next, consider the normal position of the part when it is in use. It

is often desirable to draw the part in its normal position. However, it is not always necessary. For example, tall parts, such as vertical shafts, can be drawn in a horizontal position more easily. Views with the fewest hidden lines are easiest to read. They also take much less time to lay out and draw.

The practical purpose for drawing views is to describe the shape of something. Therefore, it is a waste of time to make more views than are necessary to describe an object. In fact, some objects require only one view in order to be described adequately. Some things that can often be described in one view are the following: turned parts, such as the handles shown at A in Fig. 5-24; sheet material; plywood; plate material; and parts of uniform thickness, such as the latch and the stamping in Fig. 5-25 at A and B. For the handles in Fig. 5-24 at A, give the diameter in a note. For the latch or the stamping in Fig. 5-25 at A and B, give the

thickness in a note. Parts, such as the bushing shown at D and the sleeve shown at E, are often shown in one view. Dimensions for diameters are marked ∅ or DIA. The two views of the bushing at C are not necessary, as shown at D.

Many things can be described in two views. These are shown in Fig. 5-26. When two views are used, they must be selected carefully so that they describe the shape of the object accurately. For the parts at A and B in Fig. 5-26, there would be no question about what views to make. For the part at A, the top view and either front or side view would be enough. A third view would add nothing to the description of the placer cone at C. There should be no question about the selection of views for the rod guide at D or for the wedge cam at E. Figure 5-27 shows some objects that can be described in two views. The top and front views at A and B are the same. Since the side views are necessary, the front and side

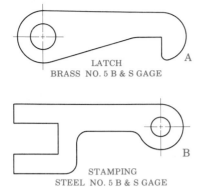

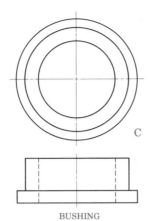

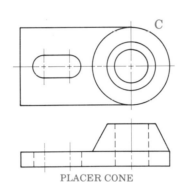

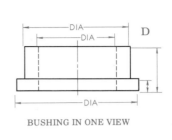

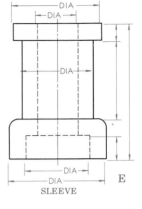

LATCH
BRASS NO. 5 B & S GAGE

STAMPING
STEEL NO. 5 B & S GAGE

BUSHING

BUSHING IN ONE VIEW

SLEEVE

Fig. 5-25. One-view drawings.

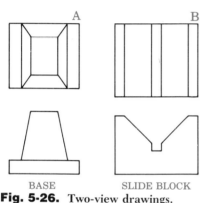

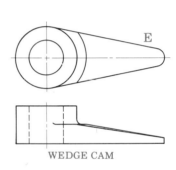

BASE   SLIDE BLOCK

PLACER CONE

ROD GUIDE

WEDGE CAM

Fig. 5-26. Two-view drawings.

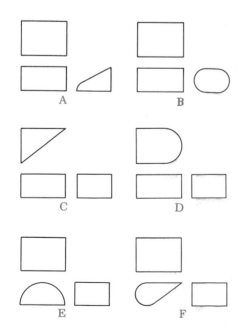

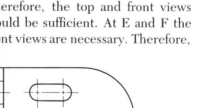

Fig. 5-27. Selection of two views.

views would be adequate. At C and D the top views are necessary. Therefore, the top and front views would be sufficient. At E and F the front views are necessary. Therefore,

the front and top views or the front and side views would be sufficient.

Some things, such as the angle in Fig. 5-28, require three views. Three views are needed because it is

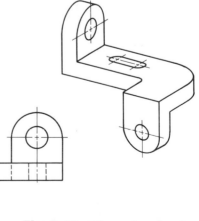

Fig. 5-28. Three-view drawing.

necessary to describe shapes in each of the views.

Six views are shown for the sliding base in Fig. 5-29. As you look at the pictorial view you will see that the top, front, and right-side views will give the best shape description. These views also have the fewest hidden lines. The six views are shown here simply to illustrate and explain the selection of views. You would draw only the necessary views in real practice.

Careful thought about your "mind's-eye picture" of an object will help you decide which views best describe its shape.

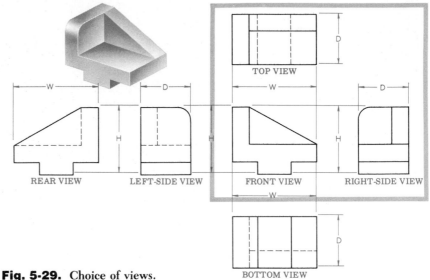

Fig. 5-29. Choice of views.

SECOND POSITION OF THE SIDE VIEW

The proportions of an object or the size of the sheet sometimes makes you want to show the side view in the second position. This position is directly across from the top view, as shown in Fig. 5-30. This view can be made by revolving the side plane around its intersection with the top plane.

PLACING VIEWS

The size of the drawing sheet should allow enough space for the number of views that you need to give a clear description of the part. Working space is suggested by the sheet layout in Fig. 5-31. Working space may also be specified by your instructor. The method for working out the positions of the views is the same for any space.

In Fig. 5-32 at A, a pictorial drawing of a slide stop is shown with its overall width, height, and depth dimensions. Some simple arithmetic is needed to place the three normal views properly. It may also be helpful to make a rough layout on scrap paper, as shown at B. This layout need not be made to scale.

A working space of $10\frac{1}{2} \times 7$ in. (267×178 mm) is used to explain how to place the views of the slide

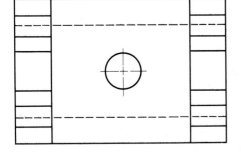
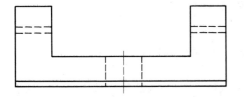
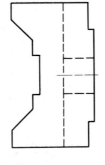
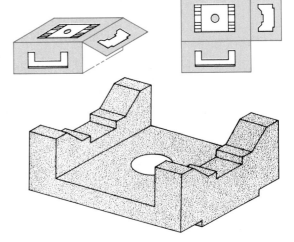

Fig. 5-30. Second position of the side view.

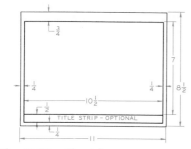

Fig. 5-31. Sheet layout.

stop in Fig. 5-32 at A. The overall dimensions are: width W = 5¼ in. (133 mm); depth D = 1¾ in. (45 mm); height H = 3 in. (76 mm).

The width, depth, and height dimensions are given in turquoise on the sketch (Fig. 5-32 at B). The dimensions in blue indicate the spacing at the top, bottom, side, and between views. Use the following procedure to determine spacing (Fig. 5-32).

1. Add the width 5¼ in. (133 mm) and the depth 1¾ in. (45 mm): 5¼ + 1¾ = 7 in. (133 + 45 = 178 mm). Subtract 7 in. (178 mm) from the width of the drawing space 10½ in. (268 mm): 10½ − 7 = 3½ in. (268 − 178 = 90 mm). The remaining 3½ in. (90 mm) is the amount left for the space at the left, right, and between views. If a space of about 1 in. (25 mm) is used between the front

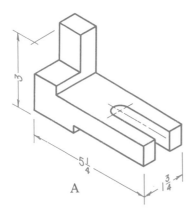

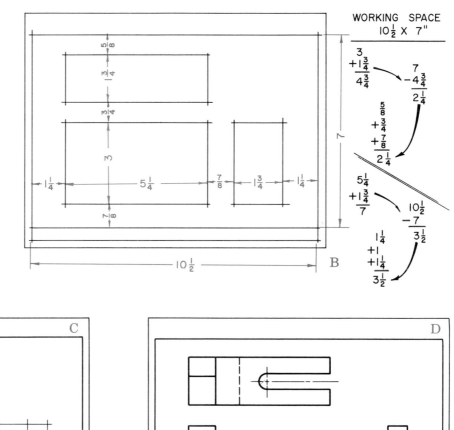

Fig. 5-32. Calculations for the placement of three views in the customary-inch system.

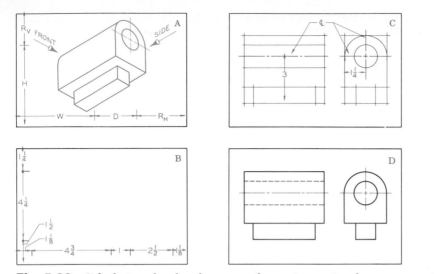

Fig. 5-33. Calculations for the placement of two views using the customary-inch system.

and side views, it will allow 1¼ in. (32 mm) on the left and right sides. These spaces may be larger or smaller, depending upon the shapes of the views, the space available, and the space needed for dimensions and notes when added.

2. Next, add the height, 3 in. (76 mm) and the depth, 1¾ in. (45 mm): 3 + 1¾ = 4¾ in. 76 + 45 = 121 mm). Subtract 4¾ in. (121 mm) from the height of the drawing space, 7 in. (178 mm): 7 − 4¾ = 2¼ in. (178 + 121 = 57 mm). The remaining 2¼ in. (57 mm) is the amount left for the space at the bottom, top, and between views. If a space of ¾ in. (19 mm) is used between the front and top views, it will allow 1½ in. (38 mm) for spaces above and below the views. These could be ¾ in.

(19 mm) each, but a better visual balance will result if ⅞ in. (22 mm) is used below and ⅝ in. (16 mm) above.

After calculations are made, proceed with the layout on the final drawing sheet, as shown at C in Fig. 5-32. Notice that the views are blocked in with light construction lines until all details have been added. Figure 5-32 at D shows all necessary visible, hidden, and centerlines darkened.

Figures 5-33 and 5-34 show the same procedure being used for a two-view drawing. Regardless of the number of views, the basic procedure does not change. The views can be arranged as in Fig. 5-33 at D or as in Fig. 5-34 at D.

■ LOCATING MEASUREMENTS

After lines have been drawn to locate the views, make measurements for details. Then draw the views (Fig. 5-35). Measurements made on one view can be transferred to another to save the time of making them again.

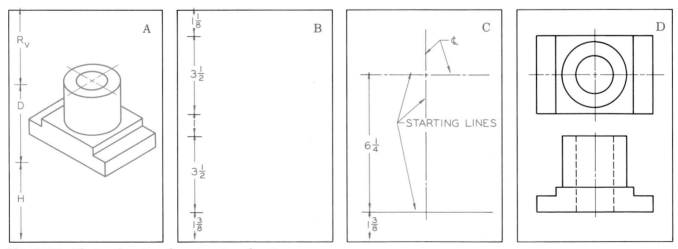

Fig. 5-34. Placing the views for a two-view drawing.

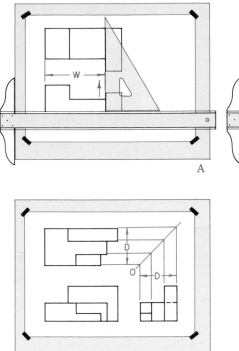

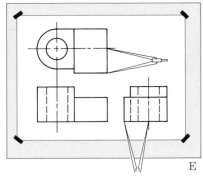

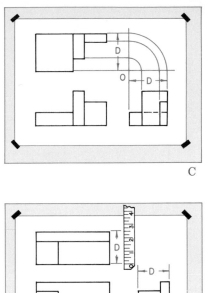

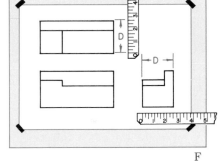

Fig. 5-35. Locating measurements.

This procedure will also ensure accuracy and correctness. Distances in the three directions, width *W*, height *H*, and depth *D*, can be easily transferred, as seen in Fig. 5-35.

1. Width *W* (horizontal) measurements made on the front view can be located on the top view by drawing up from the front view. In the same way, measurements can be projected down from the top view to the front view.
2. Height *H* (vertical) measurements on the front view can be located on the side view by drawing a light line across to the side view. Measurements can also be projected to the front view from the side view.
3. Depth *D* measurements show as vertical distances in the top view and as horizontal distances in the side view. Such measurements

can be taken from the top view to the side view by the following methods: by drawing arcs from a center O (at C), by using a 45° triangle through O (at D), by using the dividers (at E), or by using the scale as shown at F.

◼ TO MAKE A DRAWING

Follow a step-by-step method of working to be sure your drawing is accurate and easy to understand (Fig. 5-36). All views should be carried along together. Do not attempt to finish one view before starting the others. Use a hard lead pencil (4H or 6H) and light, thin lines for preliminary lines. Use a soft lead pencil (F, HB, or H) for final lines. The grade of pencil you use depends partly upon the surface of the paper, cloth,

or film you use. The following order of working is suggested:

1. Consider the characteristic view (at A; the front view).
2. Determine the number of views (at A; three views needed).
3. Locate the views (at B).
4. Block in the views with light, thin lines (at C).
5. Lay off the principal measurements (at D).
6. Draw the principal lines (at E).
7. Lay off the measurements for details (at F; centers for arcs, circles, and triangular ribs).
8. Draw the circles and arcs (at G).
9. Draw any additional lines needed to complete the views.
10. Brighten the lines where necessary to make then sharp and black and of the proper thickness (at H).

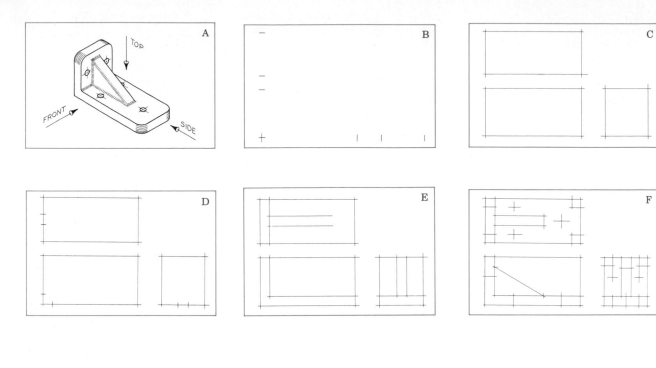

Fig. 5-36. Making a drawing.

LEARNING ACTIVITIES

1. Using thin sheets of Plexiglas, construct a glass box to be used as a visual aid in helping to better understand orthographic projection.

2. Collect various shapes and sizes of objects. Practice selecting views and determining the number of views needed. Sketch or draw views of objects assigned.

3. Obtain multiview drawings from industry. Analyze them to determine whether proper views have been selected and whether all views drawn are necessary. Also, determine whether any views are missing.

VOCABULARY

1. Visualization
2. Implementation
3. Multiview drawing
4. Orthographic projection
5. Quadrant

6. First-angle projection
7. Third-angle projection
8. Vertical`plane
9. Horizontal plane
10. Profile plane

11. Negative cylinder
12. Spherical
13. Front view
14. Top view
15. Right-side view

REVIEW

1. Describe multiview drawing.

2. Most objects have six sides or six views. Name them.

3. There are four angles of projection. Which two are used in technical drawing? Which one is used in the United States?

4. What type of line is used to represent *interior* (inside) details not seen on the outside of an object?

5. On which plane is the front view projected?

6. On which plane is the top view projected?

7. How many views does a cylindrical object usually require?

8. If a centerline and a hidden line fall in the same place, which is drawn?

9. On which plane is the right-side view projected?

10. Which type of drawing gives the most accurate shape description of an object?

Problems

The following problems provide practice in representing objects by views so that the student may gain a thorough understanding of the theory of shape description. Chapters 3 and 5 should be carefully studied before attempting the following problems.

Views and pictures are given for some of the problems to assist in visualizing the object. In others, two views are given from which to work out the third view. In still others, pictures are given from which to determine which views are necessary, to plan and arrange the views in the working space, and then to work out the views.

Sheet layouts and title blocks are covered at the end of Chap. 3. Chapter 5 describes a systematic method for placing the views on a drawing sheet. The student should review this material carefully before beginning to solve the drawing problems.

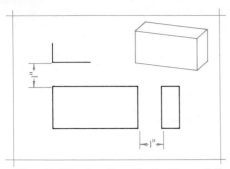

Fig. 5-37. Sanding block. Draw full size the two views shown and complete the third (top) view. Do not draw the pictorial view. The block is ¾″ × 1¾″ × 3½″.

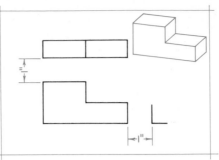

Fig. 5-38. Step block. Scale: full size. Draw the front and top views, but not the pictorial view. Complete the right-side view in its proper location. The step block is ¾″ × 1¾″ × 3½″. The notch is ⅞″ × 1¾″.

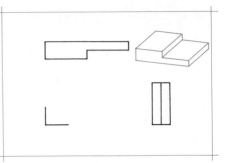

Fig. 5-39. Half lap. Scale: full size. Draw the top and right-side views, but not the pictorial. Complete the front view in its proper shape and location. The half lap is ¾″ × 1¾″ × 3½″. The notch is ⅜″ × 1¾″.

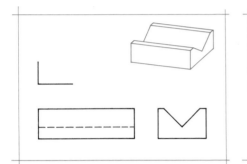

Fig. 5-40. V-block. Scale: full size. Draw the front and right-side views as shown. Complete the top view in its proper location. Do not draw the pictorial view. The overall sizes are 1¼″ × 2″ × 4″. The V-cut has a 90° included angle and is ¾″ deep.

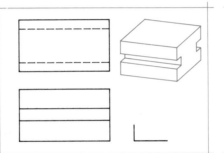

Fig. 5-41. Slide. Scale: full size. Draw the front and top views as shown. Complete the right-side view in its proper location. Do not draw the pictorial view. The overall sizes are 2⅛″ square × 3¾″. The slots are ⅜″ deep and ½″ wide.

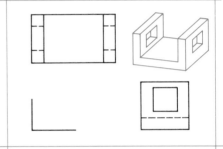

Fig. 5-42. Rod support. Scale: full size. Draw the top and right-side views, but not the pictorial. Complete the front view. The overall sizes are 2″ square × 3½″. Bottom and ends are ½″ thick. The holes are 1″ square and are centered on the upper portions.

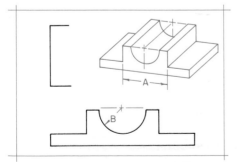

Fig. 5-43. Cradle. Scale: full size. Draw the front view, but not the pictorial. Complete the top view in the proper shape and location. Height = 2″, width = 6″, depth = 2½″. Base = ½″ thick. A = 3″, B = 1″R.

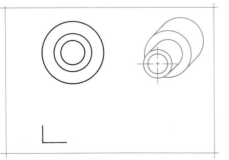

Fig. 5-44. Spacer. Scale: as assigned. Draw the top view, but not the pictorial. Complete the front view. Base = 2½″ DIA × 1″. Top = 1½″ DIA × ¾″. Hole = 1″ DIA. A vertical sheet will permit a larger scale.

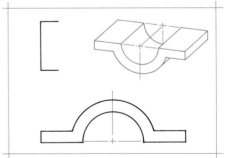

Fig. 5-45. Strap. Scale: full size. Draw the front view as shown. Complete the top view in the proper shape and location. Do not draw the pictorial view. Overall width is 6″.

Figs. 5-46 through 5-57. Two- and three-view problems. Each problem has one view missing. Draw the view or views given and complete the remaining view in the proper shape and location. Scale: full size or as assigned. Do not dimension unless instructed to do so.

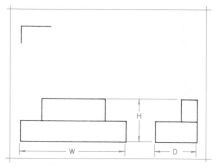

Fig. 5-46. Stop. W = 5″, H = 2″, D = 2″, base = 1″ × 2″ × 5″, top = ¾″ × 1″ × 3″.

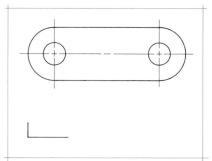

Fig. 5-47. Link. W = 7½″, D = 2½″, H = 1¼″, holes = 1¹⁄₁₆″ DIA.

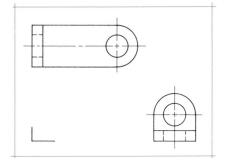

Fig. 5-48. Angle bracket. W = 5″, D = 2″, H = 2¼″, matl thk = ½″, holes = 1″ DIA.

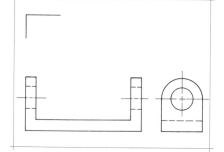

Fig. 5-49. Saddle. W = 5½″, D = 2″, H = 2½″, matl thk = ½″, hole = 1″ DIA.

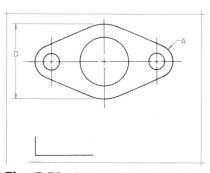

Fig. 5-50. Spacer. W = 6½″, D = 3¼″, thk = 1″, holes = 2⅜″ DIA, ¾″ DIA, A = ¾″R.

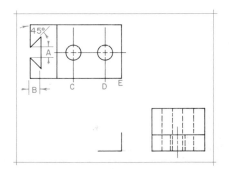

Fig. 5-51. Dovetail slide. W = 4¼″, D = 2½″, H = 2″, base thk = ¾″, upright thk = 1¼″, holes = ⅝″ DIA, A = ½″, B = ½″, CD = 1½″, DE = ¾″.

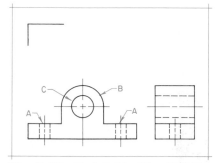

Fig. 5-52. Rod Guide. W = 5⅛″, D = 1⅞″, H = 2½″, C = 1″ DIA, A = ½″ DIA —3⅝″ apart, base thk = ¾″, B = 1″R.

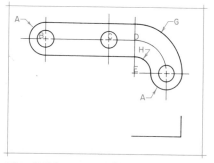

Fig. 5-53. Hinge plate. A = ¾″ R, BC = 3″, CD = 1¼″, DE = 1½″, EF = 1½″, G = 2¼″R, H = ¾″R, holes = ¾″ DIA, thk = 1″.

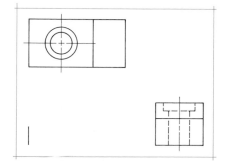

Fig. 5-54. Offset lug. W = 4½″, D = 2¼″, H = 2″, notch = ¾″ × 1½″, hole = 1″ DIA, Cbore = 1½″ DIA × ⅜″ deep.

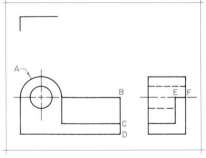

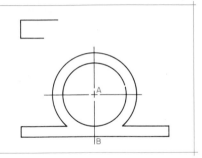

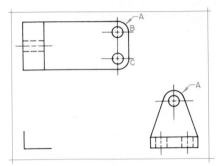

Fig. 5-55. Pin holder. *W* = 4¾″, *D* = 1¾″, *H* = 2¾″, hole = 1″ DIA, *A* = 1″R, *BC* = 1¼″, *BD* = 1¾″, *EF* = ½″.

Fig. 5-56. Ring. Base = ½″ × ⅞″ × 7″, ring = 4″ OD, 3″ ID, *AB* = 2″.

Fig. 5-57. Bracket. *L* = 5″, *W* = 2¼″, *H* = 2¾″, base thk = ½″, upright = 1″, *A* = ½″R, *BC* = 1¼″, holes = ½″ DIA.

Figs. 5-58 through 5-66. Make two- or three-view drawings of objects with instruments. Scale: as assigned. Do not dimension unless instructed to do so. Include all centerlines.

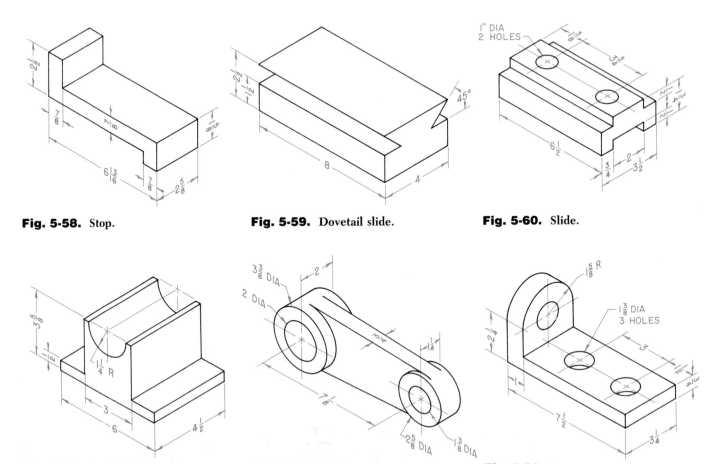

Fig. 5-58. Stop.

Fig. 5-59. Dovetail slide.

Fig. 5-60. Slide.

Fig. 5-61. Cradle block.

Fig. 5-62. Pivot arm.

Fig. 5-63. Base.

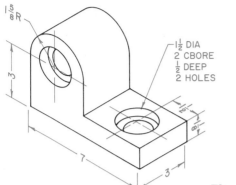

Fig. 5-64. Shaft support.

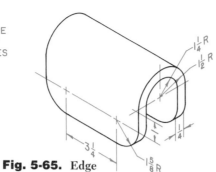

Fig. 5-65. Edge protector.

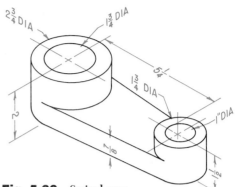

Fig. 5-66. Swivel arm.

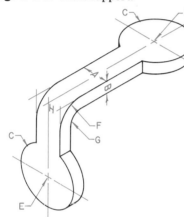

Fig. 5-67. Offset ring. *A* = 1½″, *B* = ⅝″, *C* = 1⅝″R, *D* = 1″ DIA, *E* = 1⅛″ DIA, *F* = 1½″R, *G* = ⅞″R, *EH* = 4″, *HD* = 6¾″.

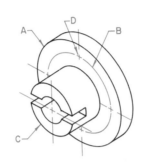

Fig. 5-68. Socket. Scale: as assigned. All dimensions are in millimeters. *A* = 50.50-mm DIA × 7 mm thick, *B* = 38 mm, *C* = 25.2-mm DIA × 17 mm long with 13-mm-DIA hole through, slots = 4.5 mm wide × 8 mm deep, *D* = 6-mm DIA, 4 holes equally spaced.

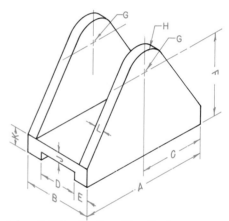

Fig. 5-69. Shaft guide. Scale: as assigned. *A* = 7¼″, *B* = 3⅞″, *C* = 3⅝″, *D* = 2¼″, *E* = ¹³⁄₁₆″, *F* = 4½″, *G* = 1″ DIA, 2 holes, *H* = 1¼″R, *J* = ½″, *K* = 1″, *L* = ½″.

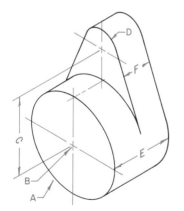

Fig. 5-70. Cam. *A* = 2⅝″ DIA, *B* = 1¼″ DIA, 1¾″ Cborc, ¼″ deep, both ends, *C* = 2⅛″, *D* = ⁹⁄₁₆″R, *E* = 1¾″, *F* = ⅞″.

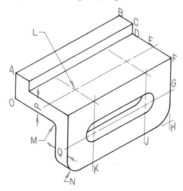

Fig. 5-71. Adjustable stop. Scale: as assigned. *AB* = 10″, *BC* = 1⅜″, *CD* = ¾″, *DE* = 1⅞″, *EF* = 2″, *FG* = 2½″, *CH* = 2½″, *FH* = 5″, *HJ* = 2½″, *JK* = 5″, *L* = 1″ DIA, 2 holes, *M* = ½″R, *N* = 1¼″R, *AO* = 2″, *P* = 1¼″, *Q* = 1¼″, slot = 1½″ wide.

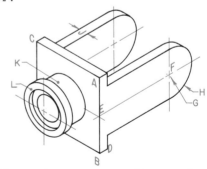

Fig. 5-72. Camera swivel base. Scale: as assigned. *AB* = 40 mm, *AC* = 38 mm, *BD* = 5.5 mm, *BE* = 20 mm, *EF* = 45 mm, *H* = 12-mm R, *G* = 10-mm DIA, *J* = 6 mm. Boss: *K* = 24-mm DIA × 9 mm long, *L* = 28-mm DIA × 4 mm long, hole = 12-mm DIA × 14 mm deep, Cbore = 18-mm DIA × 3 mm deep.

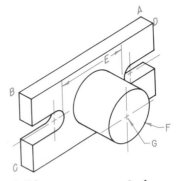

Fig. 5-73. Pipe support. Scale: as assigned. $AB = 8''$, $BC = 4''$, $AD = \frac{3}{4}''$, $E = 3\frac{7}{8}''$, $F = 2\frac{3}{4}''$ DIA × $2\frac{1}{4}''$ long, $G = 1\frac{1}{8}''$ DIA hole through, slots = $1''$ wide.

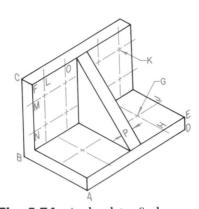

Fig. 5-74. Angle plate. Scale: as assigned. $AB = 6''$, $BC = 6\frac{1}{2}''$, $AD = 9\frac{3}{4}''$, $DE = 1''$, $CF = 1''$, $G = \frac{3}{4}''$ DIA, 2 holes, $EH = 2''$, $EJ = 2\frac{1}{2}''$, $FL = 1\frac{1}{8}''$, $LO = 2''$, $FM = 1\frac{1}{2}''$, $MN = 2\frac{1}{2}''$, $P = 1\frac{1}{8}''$, $K = \frac{1}{2}''$ DIA, 8 holes.

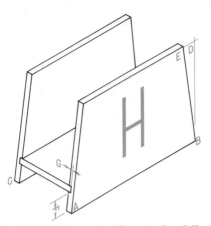

Fig. 5-75. Letter holder. Scale: full size or as assigned. Draw all necessary views. All material is $\frac{3}{16}''$ thick (plastic or wood). AB is $4''$, AC is $2''$, BD is $3''$, DE is $\frac{3}{8}''$, F is $\frac{3}{8}''$, G is $\frac{3}{32}''$. Add a design to the front view. See Chap. 2 for layout of block-style lettering. Dimension only if instructed to do so.

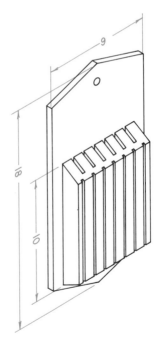

Fig. 5-76. Knife rack. Scale: half size or as assigned. Draw all necessary views. Back is $\frac{1}{2}'' \times 9'' \times 18''$. Front is $1\frac{1}{2}'' \times 7'' \times 10''$ with 30°-angle bevels on each end. Slots for knife blades are $\frac{1}{8}''$ wide × $1''$ deep. Grooves on front are $\frac{1}{8}''$ wide × $\frac{1}{8}''$ deep. Estimate all sizes not given. Redesign as desired.

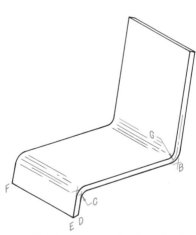

Fig. 5-77. Desktop book rack. Scale: as assigned. Draw all necessary views. Material is laminated wood, plastic, or aluminum. AB is $8''$, BC is $9\frac{1}{2}''$, CD is $2''$, DE is $\frac{3}{8}''$, EF is $6''$, G is $1\frac{1}{4}''$R. All bends are 90°.

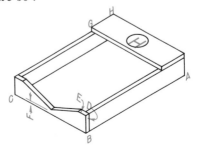

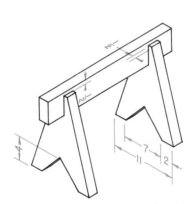

Fig. 5-78. Mini sawhorse. Scale: $\frac{1}{4}'' = 1''$ or as assigned. Top rail is $2'' \times 4'' \times 24''$. Legs are cut from 2 × 12, $14\frac{1}{2}''$ long. Dimension only if instructed to do so.

Fig. 5-79. Note-paper box. Scale: full size or as assigned. All stock is $\frac{1}{4}''$ thick. AB is $6\frac{5}{8}''$, BC is $4\frac{5}{8}''$, BD is $1''$, DE is $\frac{1}{2}''$, F is $\frac{1}{4}''$, GH is $1\frac{1}{2}''$. Draw all necessary views. Do not dimension unless instructed to do so. Initial inlay is optional. Redesign as desired.

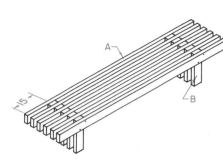

Fig. 5-80. Garden bench. Scale: 1″ = 1′-0″ or as assigned. *A* is 2 × 4 × 6′-0″, *B* is 2 × 4 × 1′-4″. Draw front, top, and right-side views of the bench. Use six or more top rails.

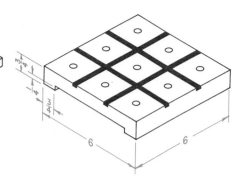

Fig. 5-81. Tic-tac-toe board. Scale: full size or as assigned. Material: hardwood with black Plexiglas inlays. Inlays are ⅛″ × ¼″ and are located 2″ OC. Holes are ⁵⁄₁₆″ DIA × ¼″ deep and are centered within squares. Game board is designed to use marbles.

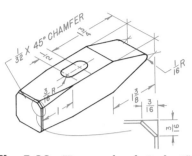

Fig. 5-82. Hammer head. Scale: 2″ = 1″ or as assigned. Draw all necessary views. Overall sizes are ⅞″ square × 3½″ long. Do not dimension unless instructed to do so.

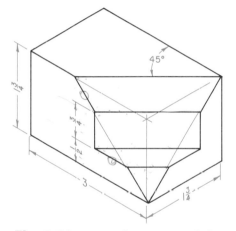

Fig. 5-83. Draw three views of the corner lock.

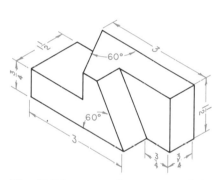

Fig. 5-84. Draw three views of the wedge spacer.

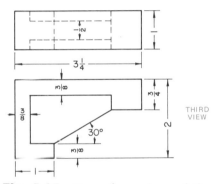

Fig. 5-85. Draw three views of the bracket.

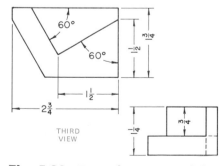

Fig. 5-86. Draw three views of the locator.

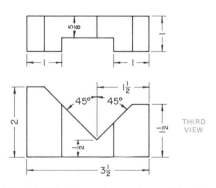

Fig. 5-87. Draw three views of the V-slide.

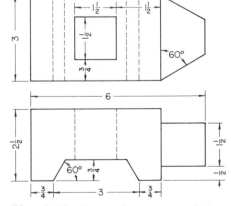

Fig. 5-88. Draw three views of the locating support.

Fig. 5-89. Draw three views of the support guide.

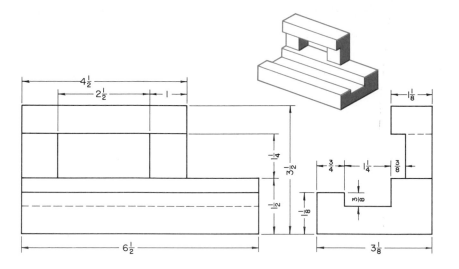

Fig. 5-90. Draw three views of the pivot.

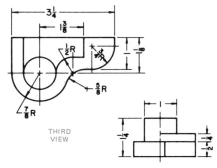
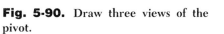

THIRD VIEW

Fig. 5-91. Draw three views of the link.

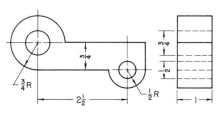

Fig. 5-92. Draw three views of the locating plate.

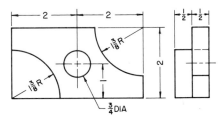

Fig. 5-93. Draw three views of the holder.

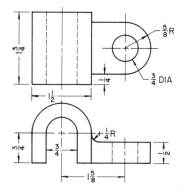

Fig. 5-94. Draw three views of the slide.

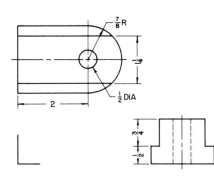
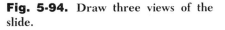

Fig. 5-95. Draw three views of the base.

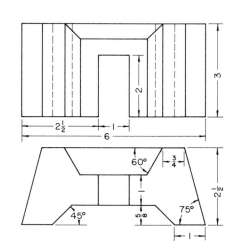

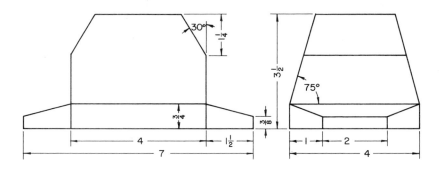

Fig. 5-96. Draw three views of the base.

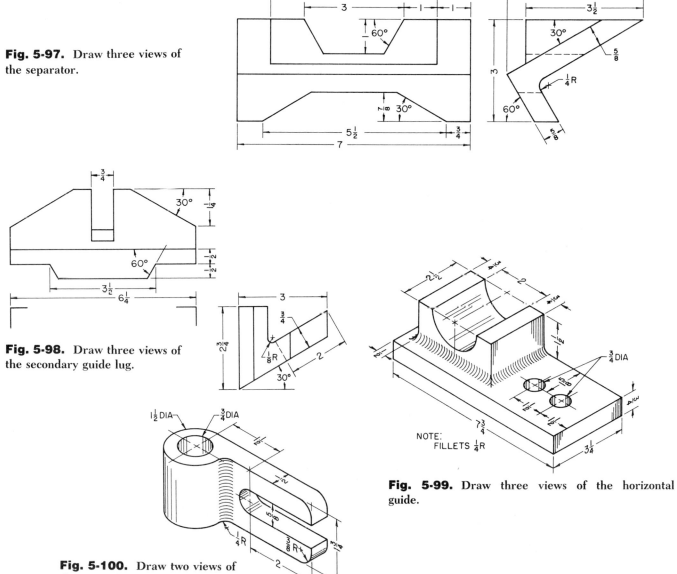

Fig. 5-97. Draw three views of the separator.

Fig. 5-98. Draw three views of the secondary guide lug.

Fig. 5-99. Draw three views of the horizontal guide.

NOTE:
FILLETS ¼R

Fig. 5-100. Draw two views of the vertical stop.

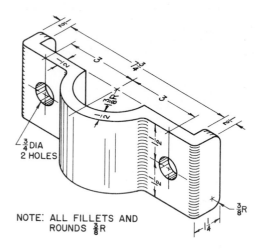

Fig. 5-101. Draw three views of the side cap.

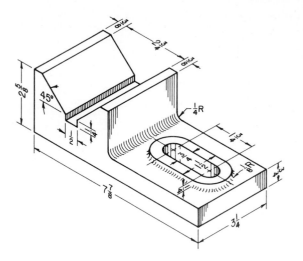

Fig. 5-102. Draw three views of the V-block base.

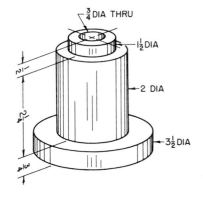

Fig. 5-103. Draw two complete views of the cast-iron collar.

6 Dimensioning

■ SIZE DESCRIPTION

To describe an object completely, two things are needed. One is its shape. The other is its size. Most of this book deals with the ways of describing shape. In this chapter, however, you will learn how to show the size of the objects that you draw. It is very important to clearly understand the rules and principles of size description. After all, a machinist cannot make parts correctly unless all the sizes on the drawing are accurate.

Another name for size description is *dimensioning. Dimensions* (sizes) are measured in either U.S. customary or metric units. In the customary system, measurements are given in feet and inches, or in inches and fractions, or in decimal divisions of inches. Decimal divisions and metric units are now most commonly used throughout industry and are used exclusively in ANSI Y14.5M (drafting standard on dimensioning).

Metric units, such as the millimeter, may be used for engineering drawings. Civil engineering drawings may be dimensioned in meters.

Architects may use both meters and millimeters, depending on the size of the item they are drawing.

Notes and symbols that show the kind of finish, materials, and other information needed to make a part are also part of dimensioning. A complete *working drawing* includes shape description, measurements, notes, and symbols (Fig. 6-1).

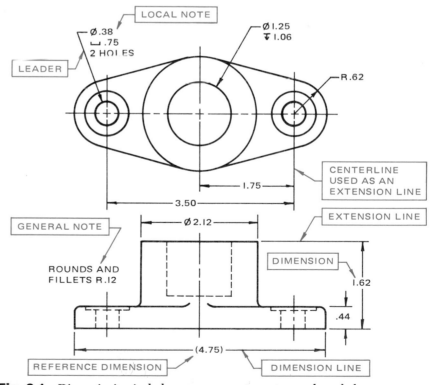

Fig. 6-1. Dimensioning includes measurements, notes, and symbols.

Size description is an important part of a working drawing. Sometimes, all a drafter has to do is give nominal and ordinary sizes in fractional inches, decimal inches, or millimeters. On other drawings, a note is added stating that the dimensions can be plus or minus a specific amount, such as ± ¹⁄₆₄ in., ± .02 in., or ± 0.4 mm. For large castings, ± ¹⁄₁₆ in., ± .06 in., or ± 1.6 mm or more would be close enough. Such a note may be placed on the drawing with the views. It can also be placed in the title block, usually in a space provided for it.

When dimensions must be exact, they are given in hundredths, thousandths, or ten-thousandths of an inch. If the metric system is being used, the measurements may be in tenths, hundredths, or even thousandths of a millimeter.

DIMENSIONING

The views on drawings describe the shape of an object. In theory, size could be found by measuring the drawing and applying a scale. In reality, though, this would not be practical, even when the views are drawn to actual or full size. The measuring would simply take too much time. More important, it is impossible to measure a drawing accurately enough for many interchangeable parts that must fit closely together. To ensure accuracy and efficiency, size information is added to the drawing. This is done through a system of lines, symbols, and numerical values.

LINES AND SYMBOLS

Lines and symbols are used on drawings to show where the dimensions

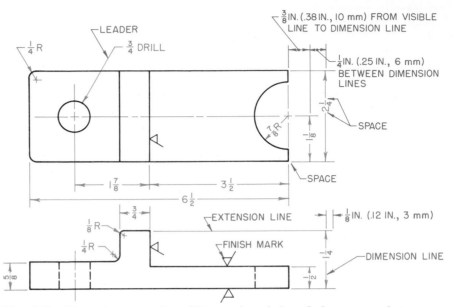

Fig. 6-2. Dimensioning consists of lines and symbols and placement techniques.

apply (see Fig. 6-2). Professional and trade associations, engineering societies, and certain industries have agreed upon the symbols to be used. Therefore, the lines and symbols are recognized by the people who use the drawings. The latest standards information on drawings and symbols can be found in publications from the American National Standards Institute (ANSI), the Society of Automotive Engineers (SAE), the Military Standards, and the International Standards Organization (ISO).

To make a correct drawing, drafters must know these symbols as well as the principles of dimensioning. Drafters must also know the shop processes that are used to build or make the products they draw. Sometimes drafters include symbols in the drawing to show which processes are needed.

DIMENSION LINES

A *dimension line* is a thin line that shows where a measurement begins

and where it ends (Fig. 6-3). It is also used to show the size of an angle. The dimension line should have a break in it for the dimension numbers. To keep the numbers from getting crowded, dimension lines should be at least ⅜ (.38) in. from the lines of the drawing. They should also be at least ¼ (.25) in. from each other. On metric drawings, dimension lines should be at least 10 mm from the lines of the drawing and 6 mm from each other.

ARROWHEADS

Arrowheads are used at the ends of dimension lines. They show where a dimension begins and ends (Fig. 6-3). They are also used at the end of a leader (see Figs. 6-2 and 6-8) to show where a note or dimension applies to a drawing.

Arrowheads can be open or solid. Their shapes are shown enlarged in Fig. 6-4 at A and reduced to actual size at B. Draw arrowheads carefully. In any one drawing, they should all be the same size and

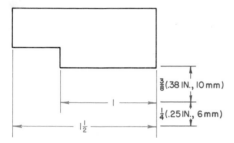

Fig. 6-3. Dimension lines must be spaced to provide clearness.

shape. However, in a small space you may have to vary the size somewhat.

Some industries use other means to point out the end point of a dimension line or leader. Figure 6-4 at C shows examples of these. These methods do the same job as arrowheads. Nevertheless, the arrow-

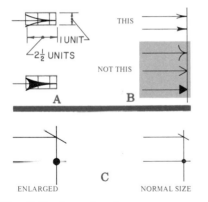

Fig. 6-4. Arrowheads.

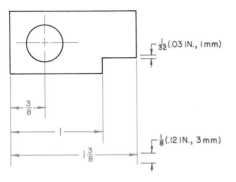

Fig. 6-5. Extension (witness) lines. A centerline may be used as an extension line.

heads shown in Fig. 6-4 at A and B are preferred.

EXTENSION LINES

Extension lines are thin lines that extend the lines or edges of the views. They are used to show center points and to provide space for dimension lines (Fig. 6-5). Extension lines are not part of the views, so they should not touch the outline. They should start after a visible space of about ¹⁄₃₂ (.03) to ¹⁄₁₆ (.06) in. (1 to 1.5 mm) and extend about ⅛ (.12) in. (3 mm) beyond the last dimension line.

NUMERALS AND NOTES

Numerals (numbers) and notes have to be easy to read, and therefore, they must be made carefully. Do not make them unnecessarily large, however. Capital letters are preferred on most drawings. They can be either *vertical* (straight up and down) or *inclined* (slanted). However, vertical letters are becoming more commonly used. In general, make numerals about ⅛ (.12) in. (3 mm) high. When fractions are used, make the fractions about ¼ (.25) in. (6 mm) and the fraction numerals about ³⁄₃₂ (.09) in. (2.5 mm) high. Always make the fraction bar (division line) in line with the dimension, never at an angle. Also, allow some space between the fraction numerals and the fraction bar.

Sometimes drawings are made to be microfilmed or to be reduced photographically and used at a smaller size. When this is the case, the numerals must be made larger and with heavier strokes so that they will be clear when reduced.

Light guidelines for figures and fractions may be drawn quickly and

easily with a lettering triangle or an Ames lettering instrument (Fig. 6-6).

THE FINISH MARK

The finish mark, or surface-texture symbol, shows that a surface is to be machined (finished), as shown in Fig. 6-7. The old symbol form ∨ is still used occasionally, but it is being replaced. An even older symbol form, ⨍, may be found on some very old drawings. The symbol ∇ is now in general use. The point of the symbol ∇ should touch the edge view of the surface to be finished. Modified forms of this symbol can be used to show that allowance for machining is needed, a certain surface condition is needed, and other conditions. These are describe in ANSI B46.1, *Surface Texture*.

LEADERS

Leaders are thin lines drawn from a note or a dimension to the place where it applies (Fig. 6-8). Leaders are drawn at an angle to the horizontal. An angle of 60° is preferred, but 45°, 30°, or other angles may be used. A leader starts with a dash, or short horizontal line. This line should be about ⅛ (.12) in. (3 mm) long, but it may be longer if needed. It generally ends with an arrowhead. However, a dot is used if the leader is pointing to a surface rather than to an edge (Fig. 6-8).

When a number of leaders must be drawn close together, it is best to draw them parallel. A leader to a circle or an arc should point to its center. Do not cross leaders. Do not draw long leaders. Do not draw leaders horizontally, vertically, at a small angle, or parallel to dimension, extension, or section lines.

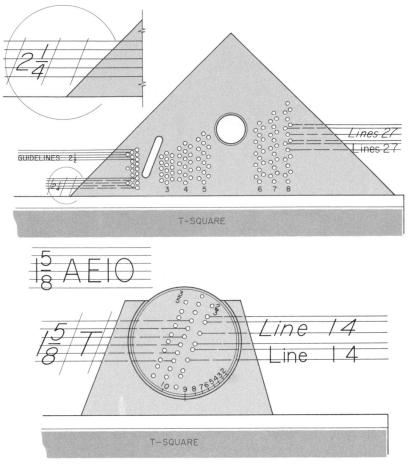

Fig. 6-6. Guidelines for letters, whole numbers, and fractions.

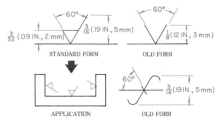

Fig. 6-7. The finish mark tells which surfaces are to be machined.

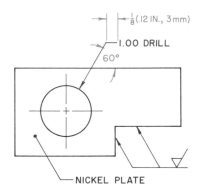

Fig. 6-8. Leaders point to the place where a note or dimension applies.

SCALE OF A DRAWING

Scales used in making drawings are described in Chap. 3. The scale used should be given in or near the title. If a drawing has views of more than one part and different scales are used, the scale should be given close to the views. Usual scales are stated as full, or full size, 1:1; half size, 1:2; and so forth. If enlarged views are used, the scale would be shown as 2 times full size, 2:1; 5 times full size, 5:1; and so forth.

The scales used on metric drawings are usually based on divisions of 10. Scales such as 100:1, 1:50, and 1:100 are examples.

UNITS AND PARTS OF UNITS

When the customary system is used, measurements are given in feet and decimals of a foot, feet and inches, inches and fractions of an inch, or inches and decimals of an inch. When customary dimensions are in inches, the inch symbol (″) is omitted. When feet and inches are used, the standard practice is to show the symbol for feet but *not* for inches: 7′-5, 7′-0, and so forth.

Feet and inches are used with common fractions, such as ½, ¼, ⅛, and so on, when particular accuracy is not necessary. They are used, for example, when measurements do

not have to be closer than ± ¹⁄₆₄ in.

When the metric system is used, dimensions on drawings are given in millimeters, meters, and, for special applications, micrometers.

Sometimes, parts must fit together with extreme accuracy. In that case, the machinist must work within specified limits. If the measurements are customary, the decimal inch is used. This is called *decimal dimensioning*. Such dimensions are used between finished surfaces, center distances, and pieces that must be held in a definite relationship to each other.

With customary measures, decimals to two places are used where limits of ± .01 in. are close enough (Fig. 6-9 at A, B, and C). Decimals to three or more places are used where limits smaller than ± .01 in. are required, as in Fig. 6-10 at A and B. For two-place decimals, fiftieths, such as .02, .04, .24 (even numbers), are preferred over decimals such as

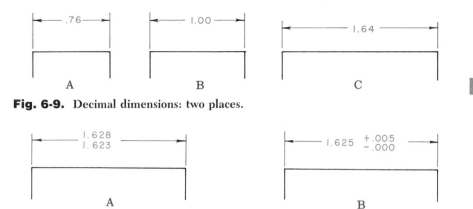

Fig. 6-9. Decimal dimensions: two places.

Fig. 6-10. Decimal dimensions: three places.

.03 and .05 (odd numbers). Fiftieths can be divided by 2 and still end up as two-place decimals. This is useful when, for example, a diameter is divided by 2 to get a radius. The decimal point should be clear and placed on the bottom guideline in a space about the width of a 0.

Decimal dimensioning is used in most industries. It is the preferred method in drafting with customary measure. In recent years, a dual dimensioning system has been used in industries involved in international trade. This system uses both the decimal inch and the millimeter (Fig. 6-11). However, in many industries it is becoming more common to use the metric system alone.

With metric dimensions, decimals to one place are used where limits of ± .01 mm are close enough. Deci-

mals to two places or more are used where limits smaller than ± 0.1 mm are required.

■ PLACING DIMENSIONS FOR READING

There are two methods in use: the *aligned system* and the *unidirectional system*.

In the *aligned system* of dimensioning (Fig. 6-12) the dimensions are placed in line with the dimension lines. Horizontal dimensions always read from the bottom of the sheet. Vertical dimensions read from the right-hand side of the sheet. Inclined dimensions read in line with the inclined dimension line. If possible, all dimensions should be kept outside the shaded area in Fig. 6-13. The

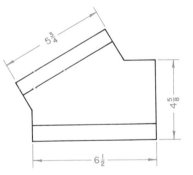

Fig. 6-12. Dimensioning: the aligned system.

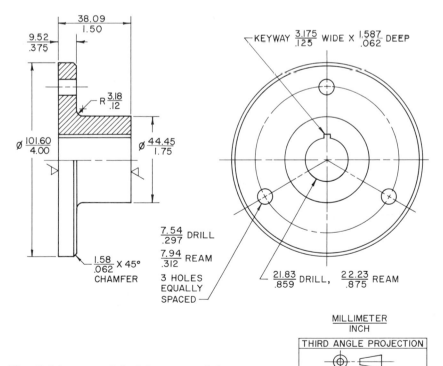

Fig. 6-11. Typical dual-dimensioned drawing.

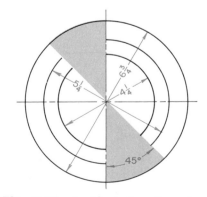

Fig. 6-13. Avoid placing dimensions in the shaded area.

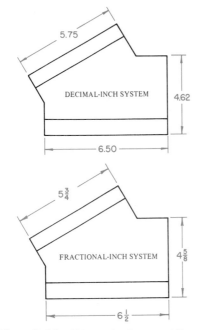

Fig. 6-14. Dimensioning: unidirectional system.

aligned system was once the only system in use.

In the *unidirectional system* of dimensioning (Fig. 6-14), all the dimensions read from the bottom of the sheet, no matter where they appear. In both systems, notes and dimensions with leaders should read from the bottom of the drawing. The unidirectional system has now replaced the aligned system in many industries. ANSI Y14.5M uses this system exclusively.

■ THEORY OF DIMENSIONING

There are two basic kinds of dimensions: *size dimensions* and *location dimensions*. Size dimensions define each piece. Giving size dimensions is really a matter of giving the dimensions of a number of simple shapes. Every object is broken down into its geometric forms, such as prisms, cylinders, pyramids, cones, and so forth, or into parts of such shapes. This is shown in Fig. 6-15, where the bearing is separated into simple parts. A hole or hollow part has the same outlines as one of the geometric shapes. Think of such open spaces in an object as *negative* (not solid) shapes.

The idea of open spaces is especially valuable to certain industries. Drafters in the aircraft industry need to know the weights of parts. These weights are worked out from the volumes of the parts as solids. From these solids, the volumes of holes and hollow or open spaces (negative or minus shapes) are subtracted. To get the total weight, the result is then multiplied by the weight per cubic inch or per cubic millimeter of the material.

When the object being dimensioned has a number of pieces, the positions of each piece must also be given. These are given by location dimensions. Each piece is first considered separately and then in relation to the other pieces. When the size and location dimensions of each piece are given, the size description is complete. Dimensioning a whole machine, a piece of furniture, or a

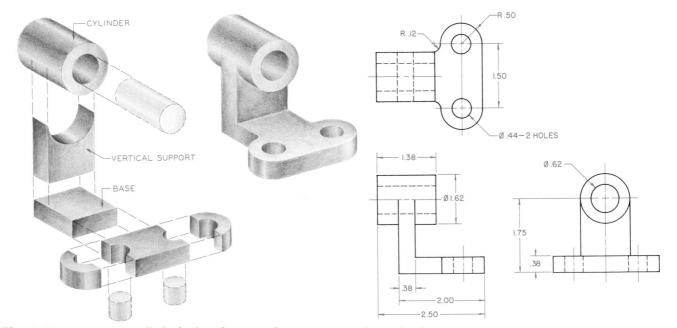

Fig. 6-15. Parts can usually be broken down into basic geometric shapes for dimensioning.

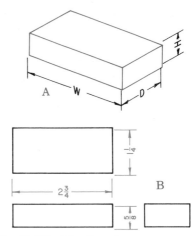

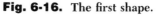

Fig. 6-16. The first shape.

building is just a matter of following the same orderly pattern that is used for a single part.

■ SIZE DIMENSIONS

The first shape is the *prism*. For a rectangular prism (Fig. 6-16), the width *W*, the height *H*, and the depth *D* are needed. This basic

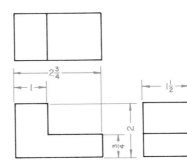

Fig. 6-17. The first rule applied.

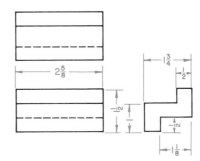

Fig. 6-18. The first rule applied.

Fig. 6-19. An irregular flat shape.

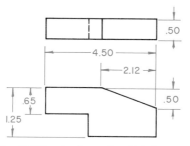

Fig. 6-20. An irregular flat shape.

shape may appear in a great many ways. A few of these are shown in Figs. 6-17 and 6-18. Flat pieces of irregular shape are dimensioned in a similar way (Figs. 6-19 and 6-20). The rule for dimensioning prisms is as follows:

For any flat piece, give the thickness in the edge view and all other dimensions in the outline view.

The *outline view* is the one that shows the shape of the flat surface or surfaces. The front views in Figs. 6-19 and 6-20 are the outline views.

Other prisms are shown in Fig. 6-21. A square prism needs two dimensions. A hexagonal or an octagonal prism may also use two dimensions. A triangular prism may use three dimensions, and so on for other regular or irregular prisms.

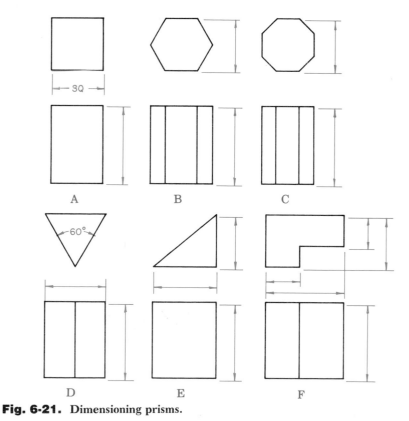

Fig. 6-21. Dimensioning prisms.

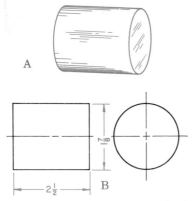

A

B

Fig. 6-22. Dimensioning a cylinder: the second shape.

The second shape is the *cylinder* (tube shaped). The cylinder needs two dimensions: the diameter and the length (Fig. 6-22). Three cylin-

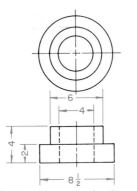

Fig. 6-23. The second rule applied.

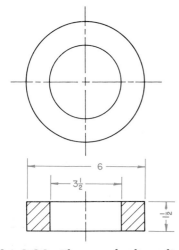

Fig. 6-24. The second rule applied.

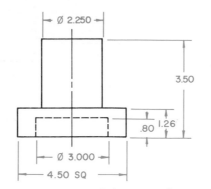

Fig. 6-25. Use of Ø on a single view.

ders are dimensioned in Fig. 6-23. One of these is the hole. A washer or other hollow cylinder can be thought of as two cylinders of the same length (Fig. 6-24). The second rule applies to the second shape:

For cylindrical pieces, give the diameter and the length on the same view.

When the circular view of a cylinder is not shown, the symbol Ø is placed with the diameter dimension (Fig. 6-25). The abbreviation DIA will be found on older drawings instead of the symbol.

Notes are generally used to give the sizes of holes. Such a note is usually placed on the outline view, especially when the method of forming the hole is also given (Fig. 6-26 at A). These notes show what operations are needed to form or finish the hole. For example, drilling, punching, reaming, lapping, tapping, countersinking, spot facing, and so forth, may be specified (Fig. 6-47). Either a dimension or a note can be used when a hole is to be formed by boring. When a hole in a casting is to be formed by a core, the word *core* is used in a note or with the dimension. However, the trend in industry is away from specifying machining operations. Sizes are given and the machining methods are determined by the manufacturer or machinist. For example, the note .500 DRILL would be Ø.500 in many industries.

When parts of cylinders occur, such as fillets (Fig. 6-26 at B), rounds (Fig. 6-26 at C), and rounded corners, they are dimensioned in the views where the curves show. The

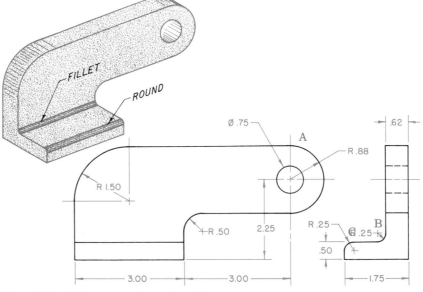

Fig. 6-26. Fillets, rounds, and radii.

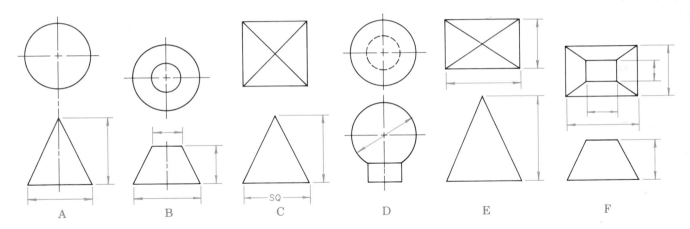

Fig. 6-27. Dimensioning some elementary shapes.

radius dimension is given and is preceded by the abbreviation R.

Some of the other shapes are the cone, the pyramid, and the sphere. The cone, the frustum, the square pyramid, and the sphere can be dimensioned in one view (Fig. 6-27). To dimension rectangular or other pyramids and parts of pyramids, two views are needed.

■ LOCATION DIMENSIONS

Location dimensions are used to show the relative positions of the basic shapes. They are also used to locate holes, surfaces, and other features. In general, location dimensions are needed in three mutually perpendicular directions: up and down, crossways, and forward and backward.

Finished surfaces and centerlines, or axes, are important for fixing the positions of parts by location dimensions. In fact, finished surfaces and axes are used to define positions. Prisms are located by surfaces, surfaces and axes, or axes. Cylinders are located by axes and bases. Three location dimensions are needed for both forms.

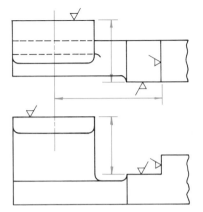

Fig. 6-28. Locating dimensions for prisms.

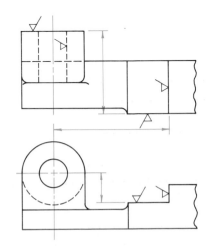

Fig. 6-29. Locating dimensions for prisms and cylinders.

All the surfaces and axes must be studied together so that the parts will go together as accurately as necessary. That means that in order to include the right location dimensions and notes on a drawing, drafters must be familiar with the engineering practices needed for the manufacture, assembly, and use of a product.

There are two general rules for showing location dimensions:

Prism forms are located by the axes and the surfaces (Fig. 6-28). Three dimensions are needed.

Cylinder forms are located by the axis and the base (Fig. 6-29). Three dimensions are needed.

Combinations of prisms and cylinders are shown in Figs. 6-30 and 6-31. Dimensions marked *L* (Fig. 6-31) are location dimensions.

■ DATUM DIMENSIONS

Refer to Fig. 6-32. *Datums* are points, lines, and surfaces that are assumed to be exact. Such datums are used to compute or locate other dimensions. Location dimensions

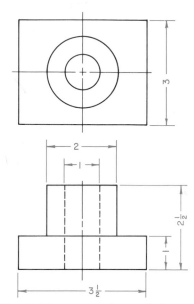

Fig. 6-30. First and second shapes.

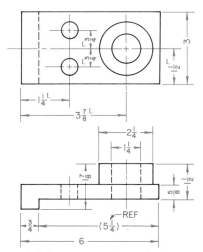

Fig. 6-31. First and second shapes.

are given from them. When positions are located from datums, the different features of a part are all located from the same datum.

Two surfaces, two centerlines, or a surface and a centerline are typical datums. In Fig. 6-32 at A, two surface datums are used. At B, two centerlines are used. At C, a surface and a centerline are used. A datum must be clear and visible while the part is

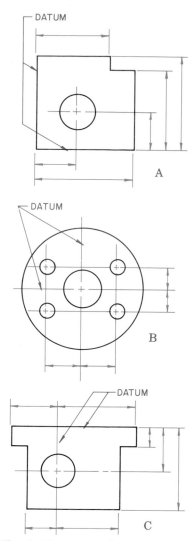

Fig. 6-32. Datum dimensioning.

being made. Mating parts should have the same datums, because they fit together.

■ GENERAL RULES

The following are general rules for dimensioning:

1. Dimension lines should be spaced about ¼ (.25) in. (6 mm) apart and about ⅜ (.38) in. (10 mm) from the view outline (see Fig. 6-3).

2. If the aligned system is used, dimensions must read in line with the dimension line and from the lower or right-hand side of the sheet (Fig. 6-12).

3. If the unidirectional system is used, all dimensions must read from the bottom of the sheet (see Fig. 6-14).

4. On machine drawings, dimensions should be given in decimal inches or millimeters, even if the drawings are of large objects such as airplanes or automobiles. In customary dimensioning, values are given to two digits, except when greater accuracy is required. Whole numbers use a decimal marker followed by two zeros to the right of the decimal point, for example, 3.00. In metric dimensioning, values are given to one digit, except when greater accuracy is required. Whole numbers do not need a decimal point or zeros. A millimeter value of less than 1 is shown with a zero to the left of the decimal point, for example, 0.2.

5. When all the dimensions are in inches or millimeters, the symbol is generally omitted. A note can be added to the drawing such as ALL DIMENSIONS ARE IN MILLIMETERS or the metric symbol can be used.

6. Very large areas on architectural and structural drawings may be dimensioned in meters. In that case, whole numbers stand for millimeters and decimalized dimensions stand for meters. However, this is true only on architectural and structural drawings. Customary measure is also used. Any measurement over 12 in. is given in feet and inches.

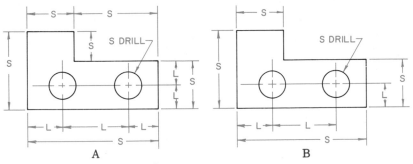

Fig. 6-33. Omit unnecessary dimensions.

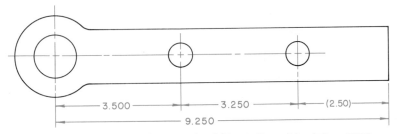

Fig. 6-34. A dimension for reference should be indicated by (□) or REF.

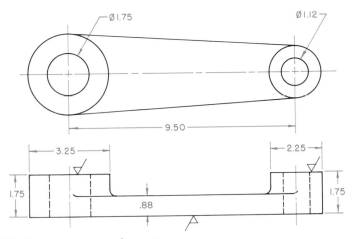

Fig. 6-35. Center-to-center dimensions.

12. Overall dimensions should be placed outside the smaller dimensions (Figs. 6-33 and 6-34). When the overall dimension is given, one of the smaller distances should not be dimensioned (Fig. 6-33 at B) unless it is needed for reference. Then REF or (□) should be added, as in Fig. 6-34.

13. On circular end parts, the center-to-center dimension is given instead of an overall dimension (Fig. 6-35).

14. When a dimension must be placed within a sectioned area, leave a clear space for the number (Fig. 6-36).

15. American National Standard practice is to avoid placing dimensions in the shaded area, as in Fig. 6-37, when the aligned system is used.

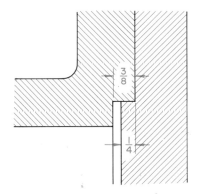

Fig. 6-36. Dimensions within a sectioned area.

7. Sheet-metal drawings are usually dimensioned in millimeters or in inches.

8. Furniture and cabinet drawings are usually dimensioned in millimeters or in inches.

9. When customary measure is used, feet and inches are shown as 7′-3. Where the dimension is in even feet, it is written 7′-0.

10. Dimensions should be positioned clearly. The same dimension is not repeated on different views.

11. Dimensions that are not needed should not be given. This is especially important for interchangeable manufacture where limits are used. Figure 6-33 at A shows unnecessary dimensions. They have been omitted in Fig. 6-33 at B.

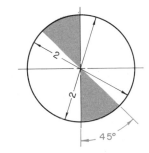

Fig. 6-37. Avoid shaded area with aligned dimensions.

16. Dimensions should be given from centerlines, finished surfaces, or datums where needed.

17. Do not use a centerline or a line of the drawing as a dimension line.

18. Do not have a dimension line that extends from a line of a view.

19. Do not cross a dimension line with another line.

20. Always give the diameter of a circle, not the radius. The symbol Ø is used before the dimension, unless it is obviously a diameter.

21. The radius of an arc should always be given with the abbreviation R placed before the dimension.

22. In general, dimensions should not be placed inside the view outlines.

23. Extension lines should not cross each other or cross dimension lines if this can be done without making the drawing more complicated.

24. Avoid dimensioning to hidden lines if possible.

25. Remember that there are no hard-and-fast rules or practices that are not subject to change under the special conditions or needs of a particular industry. However, when there is a variation of any rule, there must be a reason to justify it.

STANDARD DETAILS

The shape, the methods of manufacture, and the use of a part generally tell us which dimensions must be given and how accurate they must be. A knowledge of manufacturing methods, patternmaking, foundry and machine-shop procedures, forging, welding, and so on, is very useful when you are choosing and placing dimensions. The quantity of parts to be made must also be considered. If many copies are to be made, quantity-production methods have to be used. In addition, there are purchased parts, identified by name or brand, that call for few, if any, dimensions. Some companies have their own standard parts for use in different machines or constructions. The dimensioning of these parts depends on how they are used and produced.

There are, however, certain more-or-less standard details or conditions. For these, there are suggested ways of dimensioning.

ANGLES AND CHAMFERS

Angles are usually dimensioned in degrees (°), minutes ('), and seconds (") (Fig. 6-38 at A). However, decimalized angles are preferred. The abbreviation DEG may be used instead of the symbol ° when only degrees are given. Angular tolerance is usually bilateral (shown on both sides of the angles as plus or minus); for example, the tolerance is ± .08. If decimalized angles are not being used, the dimensions might be written ± 5' for minutes (Fig. 6-38 at B). Angular tolerance is stated either on the drawing or in the title block. Angular measurements on structural drawings are given by run and rise (Fig. 6-39 at A). A similar method is used for slopes, as at B and C. Here one side of the triangle is made equal to 1.

Two usual methods of dimensioning chamfers are shown in Fig. 6-40 at A and B.

TAPERS

Tapers can be dimensioned giving the length, one diameter, and the taper as a ratio, as in Fig. 6-41 at A. Another method is shown at B. In this method, one diameter or width, the length, and either the American National Standard or another standard taper number are given. For a close fit, the taper is dimensioned as at C. The diameter is given at a located gage line. At D, one diameter and the angle are given.

DIMENSIONING CURVES

A curve made up of arcs of circles is dimensioned by the radii that have centers located by points of tangency

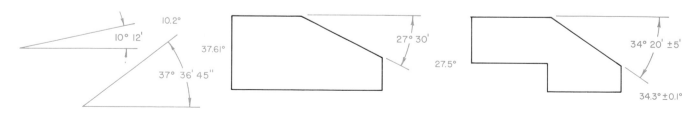

Fig. 6-38. Dimensioning an angle.

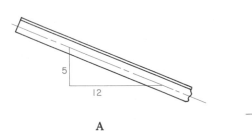
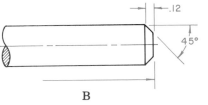
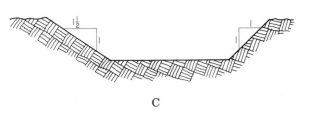

Fig. 6-39. Dimensioning angles of slopes.

Fig. 6-40. Dimensioning chamfers.

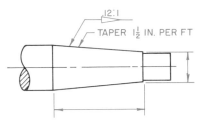
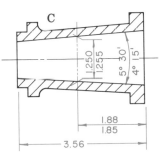

Fig. 6-41. Dimensioning tapers.

(Fig. 6-42 at A and B). Noncircular or irregular curves (Fig. 6-43) can be dimensioned as at A. They can also be dimensioned from datum lines, as at B. A regular curve can be described and dimensioned by showing the construction or naming the curve, as at C. The basic dimensions must also be given.

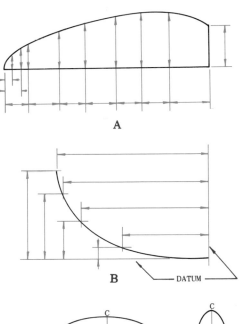

DATUM

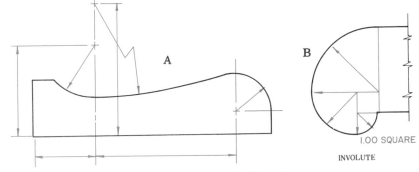

Fig. 6-42. Dimensioning curves composed of circular arcs.

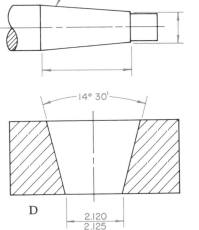

1.00 SQUARE

INVOLUTE

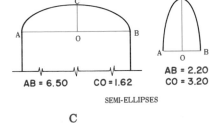

AB = 6.50   CO = 1.62   AB = 2.20
CO = 3.20

SEMI-ELLIPSES

Fig. 6-43. Dimensioning noncircular curves.

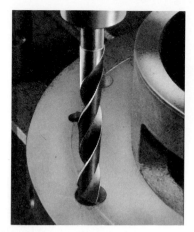

Fig. 6-44. Drilling a hole.

Fig. 6-45. Reaming a hole.

Fig. 6-46. Counterboring to a specified depth. Spot facing is generally used to provide smooth spots.

◼ DIMENSIONING A DETAIL DRAWING

A drawing for a single part that includes all the dimensions, notes, and information needed to make that part is called a *detail drawing.* The dimensioning should be done in the following order: (1) All the views of a drawing should be done before adding any dimensions or notes. (2) Think about the actual shape of the part and its characteristic views. With this in mind, draw all of the extension lines including the lengthening of any centerlines that may be needed. (3) Think about the size dimensions and the related location dimensions. Put on the dimension lines, leaders, and arrowheads. (4) After considering any changes, put in the dimensions and add any notes that may be needed.

◼ DIMENSIONING AN ASSEMBLY DRAWING

When the parts of a machine are shown together in their relative positions, the drawing is called an *assembly drawing.* If an assembly drawing needs a complete description of size,

the rules and methods of dimensioning apply.

Drawings of complete machines, constructions, and so on, are made for different uses. The dimensioning must show the information that a drawing is designed to supply.

1. If the drawing is only to show the appearance or arrangement of parts, the dimensions can be left off.
2. If a drawing is needed to tell the space a product requires, give overall dimensions.
3. If parts have to be located in relation to each other without giving all the detail dimensions, center-to-center distances are usually given. Dimensions needed for putting the machine together or erecting it in position may also be given. *Photodrawings* (photographs of products with dimensions, notes, and other details drawn on them), instead of regular drawings, can be made of a machine for the uses in paragraphs 1, 2, and 3.
4. In some industries, assembly drawings are completely dimensioned (Chap. 11). These *composite drawings* are used as both detail and assembly drawings.

For furniture and cabinetwork, sometimes only the major dimensions are given. For example, length, height, and sizes of stock may be given. The details of joints are left to the cabinetmaker or to the standard practice of the company. This is especially true if construction details are standard.

◼ NOTES FOR DIMENSIONS

Notes supply the information needed for some operations. Among these are the operations for making drilled holes (Fig. 6-44), making reamed holes (Fig. 6-45), and counterboring or spot facing (Fig. 6-46). Notes that specify these and other dimensions and operations are shown in Fig. 6-47.

Sometimes a hole is to be made in a piece after assembly with its mating piece. In that case, the words AT ASSEMBLY should be added to the note. Because such a hole is located when it is made, no dimensions are needed for its location. Some other dimensions with machining operations are suggested in Fig. 6-48.

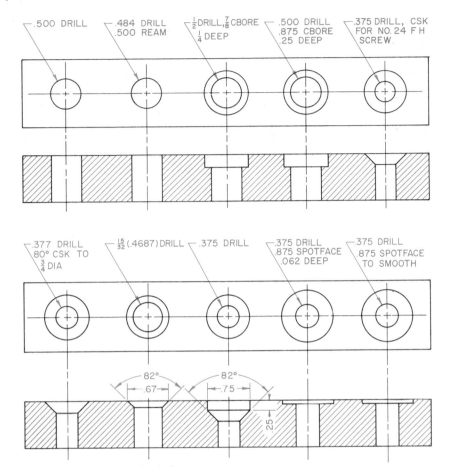

Fig. 6-47. Dimensions for holes.

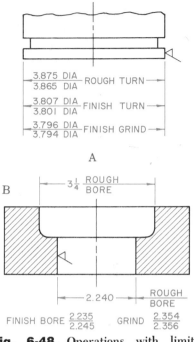

Fig. 6-48. Operations with limits specified.

The trend, however, in industry is to avoid the use of operational names in dimensions and notes. Terms such as turn, bore, grind, drill, ream, and other similar operational terms are no longer used in dimensioning in many industries. If the drawing is adequately dimensioned, it remains a shop problem to meet the drawing specifications. Table 6-1 shows a comparison between the traditional method and the newer method for dimensioning machined features. While this method is growing in popularity, the traditional method continues to be the most commonly used.

ABBREVIATIONS

You already know many of the abbreviations used in dimensioning. A few of the more familiar examples from American National Standards are listed here:

TABLE 6-1. NOTES FOR MACHINING OPERATIONS	
Traditional Method	**New Method**
½ Drill Or .500 Drill	Ø.500
.484 Drill .500 Ream	Ø.500
½ Drill, ⅞ Cbore ¼ Deep	Ø.500 ⊔.875 ⊤.25
.375 Drill 82° CSK To .750 DIA	Ø.375 Ø.750 × √82°
.375 Drill .875 Spotface .06 Deep	Ø.375 ⊔.875 ⊤.06

Allowance	ALLOW	Screw	SCR
Alloy	ALY	Spherical	SPHER
Aluminum	AL	Spotface	SF or ⊔
Babbitt	BAB	Square	SQ
Bevel	BEV	Stock	STK
Cast iron	CI	Surface	SUR
Centerline	CL or ℄	Tabulate	TAB
Chamfer	CHAM	Thread	THD
Cold-rolled steel	CRS	Tolerance	TOL
Counterbore	CBORE or ⊔	United States	
Countersink	CSK or ∨	Gage	USG
Deep	⊤	United States	
Degree	(°) DEG	Standard	USS
Diameter	DIA or Ø	Wrought Iron	WI
Dimension	DIM.		
Inch	IN.		
Key	K		
Keyseat	KST		
Keyway	KWY		
Left	L		
Left hand	LH		
Limit	LIM		
Material	MATL		
Maximum	MAX		
Millimeter	mm		
National	NATL		
Not to scale	NTS		
Outside diameter	OD		
Pattern	PATT		
Radial	RAD		
Radius	R		
Reference	REF		
Require	REQ		
Revise	REV		
Right hand	RH		

Other abbreviations may be found in the latest edition of *American National Standard Abbreviations for Use on Drawings*, ANSI Y1.1.

■ INTERCHANGEABLE MANUFACTURING

When one part is to be assembled with other parts, it must be made to fit into place without any more machining or handwork. These parts are called *mating parts*. For mating parts to fit together, specified size allowances are necessary. For example, suppose two mating parts are a rod or shaft and the hole in which it turns. For these parts to fit together, the diameter of the rod and the hole

would be limited. If the rod is too large in diameter, it will not turn. If it is too small, the rod will be too loose and will not work properly.

■ LIMIT DIMENSIONING

Absolute accuracy cannot be expected. Instead, workers have to keep within a fixed limit of accuracy. They are given a number of tenths, hundredths, thousandths, or ten-thousandths of an inch or a millimeter that the part is allowed to vary from the absolute measurements. This permitted variation is called the *tolerance*. The tolerance may be stated in a note on the drawing or written in a space in the title block. An example would be DIMENSION TOLERANCE ± .01 UNLESS OTHERWISE SPECIFIED. Limiting dimensions, or limits that give the maximum and minimum dimensions allowed, are also used to show the needed degree of accuracy. This is illustrated at A in Fig. 6-49. Note that the maximum limiting dimension is placed above the dimension line for both the shaft (external dimension) and the hole in the ring (internal dimension). The minimum limiting dimension is placed below the dimension line.

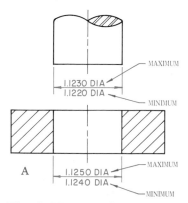

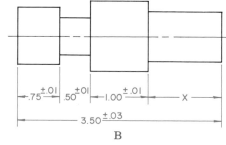

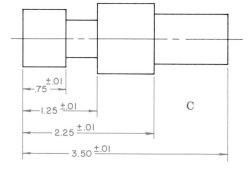

Fig. 6-49. Limit dimensions.

At B and C in Fig. 6-49, the basic sizes are given, and the plus-or-minus tolerance is shown. *Consecutive* dimensions (one after the other) are shown at B. In this case, the dimension *X* could have some variation. This dimension would not be given unless it was needed for reference. If it is given, it would be followed by the abbreviation REF or would be enclosed in parentheses. *Progressive* dimensions (each starting at the same place) are shown at C. Here they are all given from a single surface. This kind of dimensioning is sometimes called *baseline dimensioning*.

Very accurate or limiting dimensions should not be called for unless they are truly needed. They greatly increase the cost of making a part. The detail drawing in Fig. 6-50 has limits for only two dimensions. All the others are *nominal* dimensions. The amount of variation in these parts depends on their use. In this case, the general note calls for a tolerance of ± .01 in.

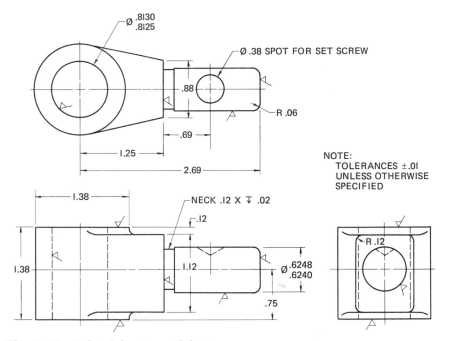

Fig. 6-50. A detail drawing with limits.

■ PRECISION OR EXACTNESS

The latest edition of ANSI Y14.5 gives precise information on accurate measurement and position dimensioning. The following paragraphs are adapted from *American National Standard Drafting Practices* with the permission of the publisher, The American Society of Mechanical Engineers, 345 East 47th St., New York, NY 10017.

■ EXPRESSING SIZE AND POSITION

Definitions Relating to Size. Size is a designation of magnitude. When a value is given to a dimen-

sion, it is called the size of that dimension. NOTE: The words *dimensions* and *size* are both used to convey the meaning of magnitude.

Nominal Size. The nominal size is used for general identification. Example: ½-in. (13-mm) pipe.

Basic Size. The basic size is the size to which allowances and tolerances are added to get the limits of size.

Design Size. The design size is the size to which tolerances are added to get the limits of size. When there is no allowance, the design size equals the basic size.

Actual Size. An actual size is a measured size.

Limits of Size. The limits of size (usually called *limits*) are the maximum and minimum sizes.

Position. Dimensions that fix position usually call for more analysis

than size dimensions. Linear and angular sizes locate features in relation to one another (point-to-point) or from a datum. Point-to-point distances may be enough for describing simple parts. If a part with more than one critical dimension must mate with another part, dimensions from a datum may be needed.

Locating Round Holes. Figures 6-51 through 6-56 show how to position round holes by giving distances, or distances and directions, to the hole centers. These methods can also be used to locate round pins and

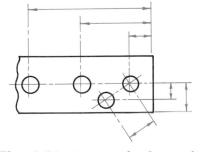

Fig. 6-51. Locating by linear distances.

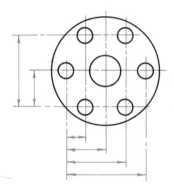

Fig. 6-52. Locating holes by rectangular coordinates. (ANSI.)

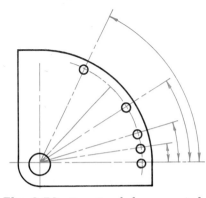

Fig. 6-53. Locating holes on a circle by polar coordinates. (ANSI.)

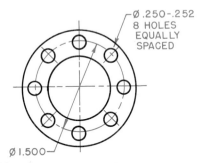

Fig. 6-54. Locating holes on a circle by radius or diameter and "equally spaced." (ANSI.)

other features. Allowable variations for the positioning dimensions are shown by a tolerance with each distance or angle. Variations may also be shown by stating limits of dimensions or angles, or by true position expressions.

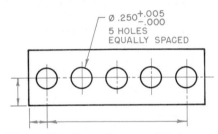

Fig. 6-55. "Equally spaced" holes in a line. (ANSI.)

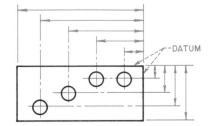

Fig. 6-56. Dimensions for datum lines. (ANSI.)

TOLERANCE

A *tolerance* is the total amount that a given dimension may vary. A tolerance should be expressed in the same form as its dimension. The tolerance of a decimal dimension should be expressed by a decimal to the same number of places. The tolerance of a dimension written as a common fraction should be expressed as a fraction. An exception is a close tolerance on an angle that can be expressed by a decimal representing a linear distance.

In a "chain" of dimensions with tolerances, the last dimension may have a tolerance equal to the sum of the tolerances between it and the first dimension. In other words, tolerances accumulate; that is, they are added together. The datum dimensioning method of Fig. 6-56 avoids overall accumulations. The tolerance on the distance between two features (first and second hole, for example) is equal to the tolerances on two di-

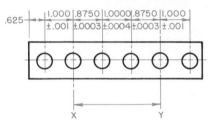

Fig. 6-57. Point-to-point, or chain, dimensioning. (ANSI.)

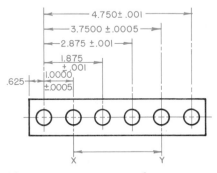

Fig. 6-58. Datum dimensioning. (ANSI.)

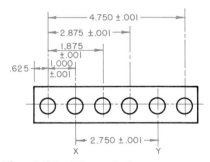

Fig. 6-59. Dimensioning to prevent tolerance accumulation between *X* and *Y*. (ANSI.)

mensions from the datum added together. Where the distance between two points must be controlled closely, the distance between the two points should be dimensioned directly, with a tolerance. Figure 6-57 illustrates a series of chain dimensions where tolerances accumulate between points *X* and *Y*. Datum dimensions in Fig. 6-58 show the same accumulation with larger tolerances. Figure 6-59 shows how to avoid the

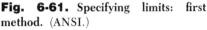

Fig. 6-60. Giving a tolerance by a plus figure and a minus figure. (ANSI.)

accumulation without the use of extremely small tolerances.

Unilateral Tolerance System. Unilateral tolerances allow variations in only one direction from a design size. This way of stating a tolerance is often helpful where a critical size is approached as material is removed during manufacture (see Fig. 6-60). For example, close-fitting holes and shafts are often given unilateral tolerances.

Bilateral Tolerance System. Bilateral tolerances allow variations in both directions from a design size. Bilateral variations are usually given with locating dimensions. They are also used with any dimensions that can be allowed to vary in either direction. See Fig. 6-60.

■ LIMIT SYSTEM

A limit system shows only the largest and smallest dimensions allowed. See Figs. 6-61 and 6-62. The tolerance is the difference between the limits.

Expressing Allowable Variations. The amount of variation permitted when dimensioning a drawing can be given in several ways. The ways recommended in this book are as follows:

1. If the plus tolerance is different from the minus tolerance, two tolerance numbers are used, one plus and one minus. See Fig. 6-60. NOTE: Two tolerances in

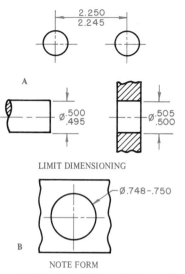

LIMIT DIMENSIONING

NOTE FORM

Fig. 6-61. Specifying limits: first method. (ANSI.)

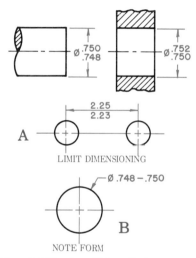

LIMIT DIMENSIONING

NOTE FORM

Fig. 6-62. Specifying limits: second method. (ANSI.)

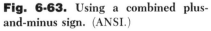

Fig. 6-63. Using a combined plus-and-minus sign. (ANSI.)

the same direction should never be called for.

2. When the plus tolerance is equal to the minus tolerance, use the combined plus-and-minus sign followed by a single tolerance number. See Fig. 6-63.

3. The maximum and minimum limits of size are shown. The numerals should be placed in one of two ways. Do not use both on the same drawing.

   a. The high limit is always above the low limit where dimensions are given directly. The low limit always comes before the high limit where dimensions are given in note form. See Fig. 6-61.

   b. For location dimensions given directly (not by note), the high-limit number (maximum dimension) is placed above. The low-limit number (minimum dimension) is placed below. For size dimensions given directly, the number representing the maximum material condition is placed above. The number representing the minimum material condition is placed below. Where the limits are given in note form, the minimum number is first and the maximum number is second. See Fig. 6-62 for an example.

4. You do not always have to give both limits.

   a. A unilateral tolerance is sometimes given without stating that the tolerance in the other direction is zero. See Fig. 6-64 at A.

   b. MIN or MAX is often placed after a number when the other limit is not important. Depths of holes, lengths of threads, chamfers, etc., are often lim-

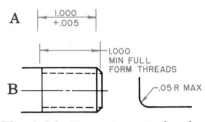

Fig. 6-64. Expressing a single tolerance or limit. (ANSI.)

ited in this way. See Fig. 6-64 at B.

5. The above recommendations are for *linear* (in-line) tolerances, but the same forms are also used for angular tolerances. Angular tolerances may be given in degrees, minutes and seconds, or decimals.

PLACING TOLERANCE AND LIMIT NUMERALS

You should place a tolerance numeral to the right of the dimension numeral and in line with it. You may also place the tolerance numeral below the dimension numeral with the dimension line between them. Figure 6-65 shows both arrangements.

DIMENSIONING FOR FITS

The tolerances on the dimensions of interchangeable parts must allow these parts to fit together at assembly. See Fig. 6-66.

Figure 6-67 shows a way to dimension mating parts that do not need to be interchangeable. The size of one part does not need to be held to a close tolerance. It is to be made the proper size at assembly for the desired fit.

"Hole Basis" and "Shaft Basis."
You need to start calculating dimensions and tolerances of cylindrical

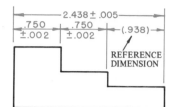

Fig. 6-65. Placing tolerance and limit numerals. Note the reference dimensions. (ANSI.)

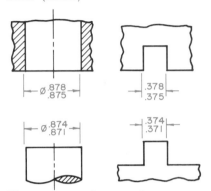

Fig. 6-66. Indicating dimensions or surfaces that are to fit closely. (ANSI.)

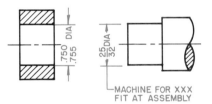

Fig. 6-67. Dimensioning noninterchangeable parts that are to fit closely. (ANSI.)

parts that must fit well together from the minimum hole size or the maximum shaft size.

A basic hole system is one in which the design size of the hole is the basic size and the allowance is applied to the shaft.

A basic shaft system is one in which the design size of the shaft is the basic size and the allowance is applied to the hole.

NOTE: For further information on limits and fits, see ANSI B4.1.

Basic Hole System.
To determine the limits for a fit in the basic

hole system, follow these steps: (1) Give the minimum hole size. (2) For a clearance fit, find the maximum shaft size by subtracting the desired allowance (minimum clearance) from the minimum hole size. For an interference fit, add the desired allowance (maximum interference). (3) Adjust the hole and shaft tolerances to get the desired maximum clearance or minimum interference. See Fig. 6-68. By using the basic hole system, tooling costs can often be kept down. This is possible because standard tools such as a reamer or broach can be used for machining.

Basic Shaft System.
To figure out the limits for a fit in the basic shaft system, follow these steps: (1) Give the maximum shaft size. (2) For a clearance fit, find the minimum hole size by adding the desired allowance (minimum clearance) to the maximum shaft size. Subtract for an interference fit. (3) Adjust the hole and shaft tolerances to get the desired maximum clearance or minimum interference. See Fig. 6-69. Use the basic shaft method only if there is a good reason for it, such as when a

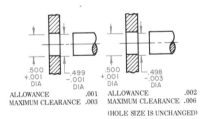

Fig. 6-68. "Basic hole" fits. (ANSI.)

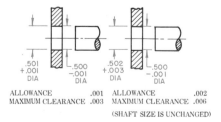

Fig. 6-69. "Basic shaft" fits. (ANSI.)

standard-size shaft is necessary or desired.

GEOMETRIC TOLERANCING

When you give geometric tolerances, you give the largest amount of variation in form (shape and size) and position (location) that a system of interchangeable parts can tolerate. Tolerances of position refer to the location of holes, slots, tabs, dovetails, and so on. Tolerances of form refer to such things as flatness, straightness, roundness, parallelism, perpendicularity, angularity, and so on.

SPECIFYING TOLERANCES OF POSITION AND FORM

Tolerances of position and form may be given in a note or written in symbol form (Fig. 6-70). Both methods are acceptable. However, most companies ask all drafters to use one or the other all the time.

GEOMETRIC CHARACTERISTIC SYMBOLS

The symbols that stand for geometric characteristics are shown in Fig. 6-71. These symbols are a drafter's shorthand way of writing form and position tolerances on drawings. These symbols can be used alone. They can also be used with other symbols and notes to give all the information needed for the shape or location of details.

DATUM

A *datum* is a reference that has been determined to be true and accurate.

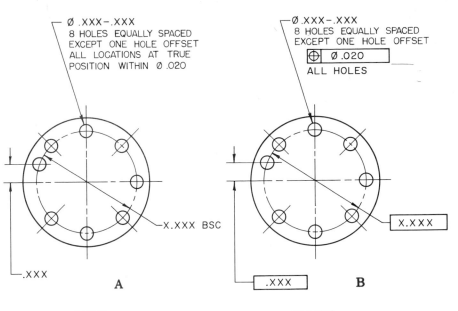

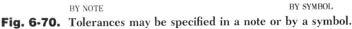

Fig. 6-70. Tolerances may be specified in a note or by a symbol.

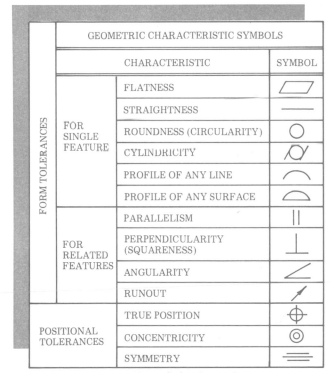

Fig. 6-71. Geometric characteristic symbols. (ANSI.)

It may be a point, a line, a plane, a cylinder, and so on. Because its dimensions are exact, measurements can be made from it. Therefore, a datum is used to figure out the location of other features.

DATUM REFERENCE LETTERS

Datum features that need to be identified are given a reference letter. Any letter of the alphabet may be used except I, O, and Q. These letters can be confusing and may be misread. Single letters are used unless more than 23 are needed. After that, combinations, such as AA through AZ, may be used. Figure 6-72 shows a reference letter in a datum symbol. The box around the letter is about .31 in. (8 mm) high. Its length may vary according to its application on the drawing.

SYMBOLS FOR MMC AND RFS

The symbols Ⓜ and Ⓢ stand for *maximum material condition* (MMC) and *regardless of feature size* (RFS). MMC exists when a feature has the maximum amount of material. That would be, for example, when a hole is at its maximum diameter. RFS means that tolerance of position or form must be met no matter where the feature lies within its size tolerance. The symbols are used only in feature control symbols.

FEATURE-CONTROL SYMBOLS

The *feature-control symbol* is a frame that contains the geometric characteristic symbol followed by the tolerance. In some cases, the symbol Ⓜ or Ⓢ is also included. A vertical line separates the symbol from the tolerance, as shown in Fig. 6-73. The datum-identifying symbol may also be added, as shown in Fig. 6-74. ANSI Y14.5M (ISO R129, R406), *Dimensioning and Tolerancing for Engineering Drawings*, describes terms

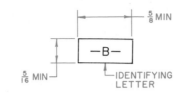

Fig. 6-72. Datum-identifying symbol.

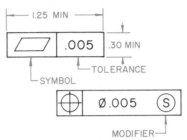

Fig. 6-73. Feature-control symbols.

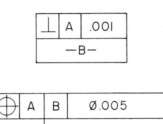

Fig. 6-74. Combined feature-control and datum-identifying symbols.

and symbols and their uses in detail.

SURFACE TEXTURE

There is no such thing as a perfectly smooth surface. All surfaces have irregularities. Sometimes a drafter has to tell how much roughness and waviness the surface of a material can have and the lay direction of both. There are standards for these characteristics. There are also symbols that represent them.

Surface texture is discussed fully in ANSI B46.1. Use that text for study and as a reference. The following paragraphs about surface texture

are adapted from *Surface Texture* (ANSI B46.1). They are included here with the permission of the publisher, The American Society of Mechanical Engineers, 345 East 47th St., New York, NY 10017.

Surfaces, in general, are very complex in character. This standard deals only with the height, width, and direction of the surface irregularities. These are of practical importance in specific applications.

CLASSIFICATION OF TERMS AND RATINGS RELATED TO SURFACES

The terms and ratings in this book have to do with surfaces made by various means. Among these are machining, abrading, extruding, casting, molding, forging, rolling, coating, plating, blasting, burnishing, and others.

Surface Texture. Surface texture includes roughness, waviness, lay, and flaws. It includes repetitive or random differences from the nominal surface that forms the pattern of the surface.

Profile. The profile is the contour (shape) of a surface in a plane perpendicular to the surface. Sometimes an angle other than a perpendicular one is specified.

Measured Profile. The measured profile is a representation of the profile obtained by instruments or other means. See Fig. 6-75.

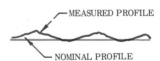

Fig. 6-75. An enlarged profile shows that a surface is not as it appears. (ANSI.)

TABLE 6-2. PREFERRED SERIES ROUGHNESS		
Roughness values		
		Grade
50	2000	12
25	1000	11
12.5	500	10
6.3	250	9
3.2	125	8
1.6	63	7
0.8	32	6
0.4	16	5
0.2	8	4
0.1	4	3
0.05	2	2
0.025	1	1

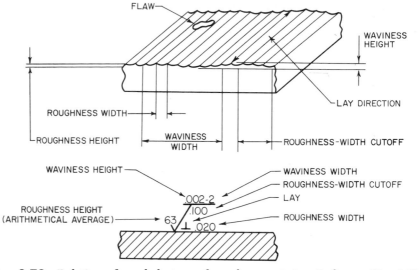

Fig. 6-76. Relation of symbols to surface characteristics. Refer to Fig. 6-78. (ANSI.)

Microinch. A microinch is one millionth of an inch (.000 001 in.). Microinches may be abbreviated μin.

Micrometer. A micrometer is one millionth of a meter (0.000 001 m). Micrometers may be abbreviated μm.

Roughness. Roughness is the finer irregularities in the surface texture. Roughness usually includes irregularities caused by the production process. Among these are traverse feed marks and other irregularities within the limits of the roughness-width cutoff. See Fig. 6-76.

Roughness Height. For the purpose of this book, roughness height is the arithmetical average deviation.

it is expressed in microinches or in micrometers measured normal to the centerline. The preferred series of roughness-height values is given in Table 6-2.

Roughness Width. Roughness width is the distance between two peaks or ridges that make up the pattern of the roughness. Roughness width is given in inches or in millimeters.

Roughness-Width Cutoff. This is the greatest spacing of repetitive surface irregularities to be included in the measurement of average roughness height. Roughness-width cutoff is rated in inches or in millimeters. Standard values are given in Table 6-3. Roughness-width cutoff must always be greater than the roughness width in order to obtain the total roughness-height rating.

Waviness. Waviness is covered by surface-texture standards. Geometric tolerancing now covers this surface condition under flatness. Flatness is a condition where all surface elements are in a single plane. Flatness tolerances are applied to surfaces to control variations in surface texture (see Fig. 6-76). Waviness was defined as the usually widely spaced component of surface texture. It generally was of wider spacing than the roughness-width cutoff. Waviness results from such factors as machine or work deflections, vibration, chatter, heat treatment, or warping strains. Roughness may be thought of as superimposed on a "wavy" surface.

Waviness Height. Waviness height is rated in inches as the peak to valley distance. The maximum waviness height is given in Table 6-3.

Waviness Width. Waviness width is rated in inches as the spacing of successive wave peaks or successive wave valleys. When specified, the values are the maximum amounts permissible.

TABLE 6-3. STANDARD ROUGHNESS-WIDTH CUTOFF VALUES*						
mm	0.075	0.250	0.750	2.500	7.500	25.000
in	0.003	0.010	0.030	0.100	0.300	1.000

*When no value is specified, the value .030 in. (0.750 mm) is assumed.

=	Lay parallel to the line representing the surface to which the symbol is applied.	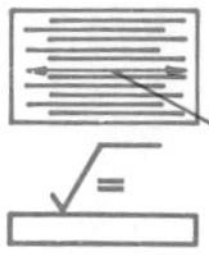 DIRECTION OF TOOL MARKS
⊥	Lay perpendicular to the line representing the surface to which the symbol is applied.	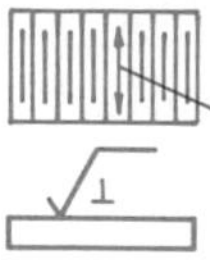 DIRECTION OF TOOL MARKS
X	Lay angular in both directions to line representing the surface to which symbol is applied.	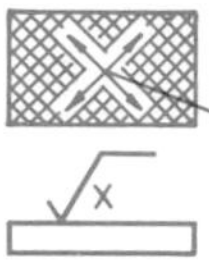 DIRECTION OF TOOL MARKS
M	Lay multidirectional.	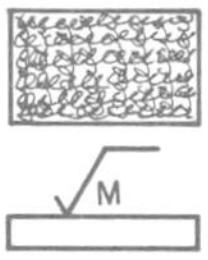
C	Lay approximately circular relative to the center of the surface to which the symbol is applied.	
R	Lay approximately radial relative to the center of the surface to which the symbol is applied.	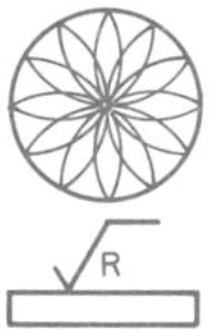

Fig. 6-77. Lay symbols. (ANSI.)

Lay. Lay is the direction of the predominant surface pattern. Ordinarily, it is determined by the production method used. Lay symbols are shown in Fig. 6-77.

Flaws. Flaws are irregularities that occur at one place or at relatively infrequent or widely varying intervals in a surface. Flaws include such defects as cracks, blowholes, checks, ridges, scratches, etc. The effect of flaws shall not be included in the roughness-height measurements, unless otherwise specified.

Contact Area. Contact area is the amount of area of the surface needed to be in contact with its mating surface. Contact area should be distributed over the surface with approximate uniformity. Contact is specified as shown in Fig. 6-78.

■ DESIGNATION OF SURFACE CHARACTERISTICS

Where no surface control is specified, you can assume that the surface produced by the operation will be satisfactory. If the surface is critical, the quality of surface needed should be indicated.

Surface Symbol. The symbol used to designate surface irregularities is the check mark with horizontal extension, as shown in Fig. 6-79. The point of the symbol must touch the line indicating which surface is meant. It may also touch the extension line, or a leader pointing to the surface. The long leg and extension is drawn to the right as the drawing is read. Where only roughness height is shown, the horizontal extension may be left off. Fig. 6-80 shows the typical use of the symbol on a drawing.

Where the symbol is used with a dimension, it affects all surfaces defined by the dimension. Areas of transition, such as chamfers and fillets, should usually be the same as the roughest finished area next to them.

Surface-roughness symbols always apply to the completed surface, unless otherwise indicated. Drawings or specifications for plated or coated parts must indicate whether the surface-roughness symbols apply before plating, after plating, or to both before and after plating.

Application of Symbols and Ratings. Fig. 6-78 shows the way roughness, waviness, and lay are called for on the surface symbol. Only those ratings needed to specify the desired surface needed to be shown in the symbol.

Symbols Indicating Direction of Lay. Symbols for lay are shown in Fig. 6-77.

Roughness ratings usually apply in a direction that gives the maximum reading. This is normally across the lay.

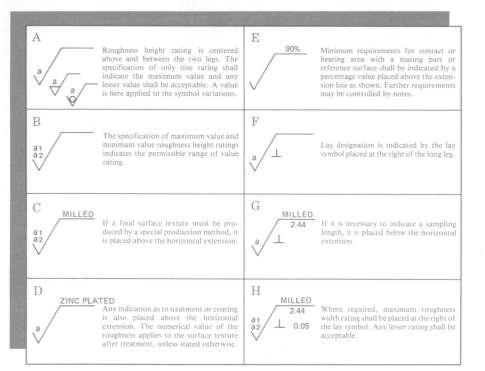

A	Roughness height rating is centered above and between the two legs. The specification of only one rating shall indicate the maximum value and any lesser value shall be acceptable. A value is here applied to the symbol variations.
B	The specification of maximum value and minimum value roughness height ratings indicates the permissible range of value rating.
C	If a final surface texture must be produced by a special production method, it is placed above the horizontal extension.
D	Any indication as to treatment or coating is also placed above the horizontal extension. The numerical value of the roughness applies to the surface texture after treatment, unless stated otherwise.
E	Minimum requirements for contact or bearing area with a mating part or reference surface shall be indicated by a percentage value placed above the extension line as shown. Further requirements may be controlled by notes.
F	Lay designation is indicated by the lay symbol placed at the right of the long leg.
G	If it is necessary to indicate a sampling length, it is placed below the horizontal extension.
H	Where required, maximum roughness width rating shall be placed at the right of the lay symbol. Any lesser rating shall be acceptable.

Fig. 6-78. Applications of symbols and ratings. (ANSI.)

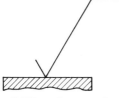

Fig. 6-79. The surface symbol. (ANSI.)

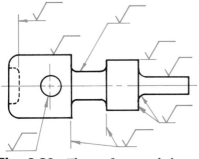

Fig. 6-80. The surface symbol on a drawing. (ANSI.)

This is the end of the material extracted and adjusted from ANSI *Surface Texture*, ANSI B46.1. For more information, use the complete ISO and ANSI standard.

LEARNING ACTIVITIES

1. Obtain several objects similar to the one shown in Fig. 6-15 and practice visualizing the basic geometric shapes from which they are made. Draw objects assigned and fully dimension.

2. Obtain technical drawings from industry. Mark size dimensions S and location dimensions L.

VOCABULARY

1. Dimension
2. Working drawing
3. Size description
4. Extension line
5. Dimension line
6. Leader
7. Finish mark
8. Unidirectional dimensioning
9. Aligned dimensioning
10. Dual dimensioning
11. Size dimension
12. Location dimension
13. Datum dimension
14. Reference dimension
15. Chamfer
16. Detail drawing
17. Assembly drawing
18. Composite drawing

CHAPTER 6

19. Counterbore
20. Countersink
21. Spotface
22. Limit dimension
23. Tolerance
24. Nominal size
25. Basic size
26. Design size
27. Actual size

28. Limits of size
29. Position
30. Unilateral tolerance
31. Bilateral tolerance
32. Allowance
33. Linear tolerance
34. Basic hole system
35. Basic shaft system
36. Geometric tolerancing

37. Datum
38. Surface texture
39. Surface profile
40. Waviness
41. Waviness height
42. Waviness width
43. Lay
44. Flaws

REVIEW

1. Name the line used to show where a measurement begins and ends.

2. Another name for size description is _____.

3. The dimensioning system in which the numerals are placed in line with the dimension lines is called the _____ system.

4. There are two basic kinds of dimensions: (1) size dimensions and (2) _____ dimensions.

5. Angles may be dimensioned by degrees, minutes, and seconds or by _____ angles.

6. Dimensions showing the distance between two features is called _____ dimensions.

7. Points, lines, and surfaces assumed to be exact are _____.

8. When all dimensions and notes read from the bottom of the sheet, the _____ dimensioning system is used.

9. Name the abbreviations for the following: diameter, inch, centerline, millimeter, radius, tolerance, outside diameter, and countersink.

10. The amount by which the accuracy of a part may vary from the absolute measurement is called _____.

Problems

Figures 6-81 through 6-99 offer a total of 60 dimensioning problems. Additional problems may be chosen from other chapters.

The problems in Figs. 6-81 through 6-84 are to be done as follows:

1. Take dimensions from the printed scale at the bottom of the page, using dividers.
2. Draw the complete views.
3. Add all necessary extension and dimension lines for size and location dimensions.

4. Add arrowheads, dimensions, and notes.

Figure 6-85 through 6-99 are more advanced. For these problems use the following procedure:

1. Determine the necessary views, and prepare a free-hand sketch.
2. Dimension the sketch.
3. Decide on a scale and draw views mechanically.
4. Add all necessary dimensions and notes.

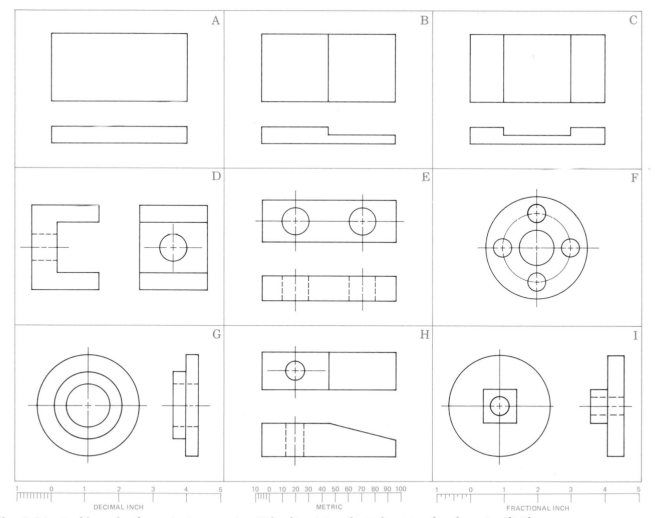

Fig. 6-81. Problems for dimensioning practice. Take dimensions from the printed scale, using dividers. Draw the views as shown and add all necessary size and location dimensions.

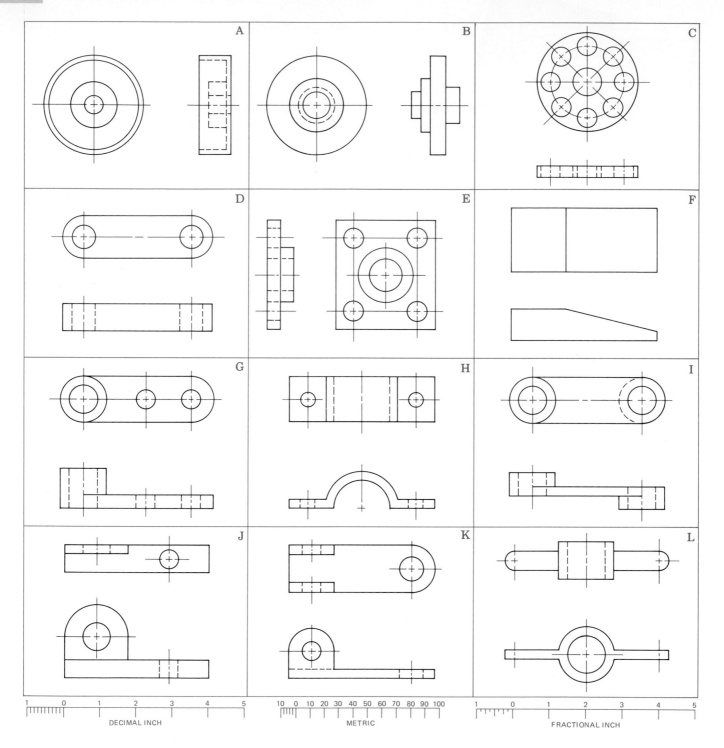

Fig. 6-82. Problems for dimensioning practice. Take dimensions from the printed scale, using dividers. Draw the views as shown and add all necessary size and location dimensions.

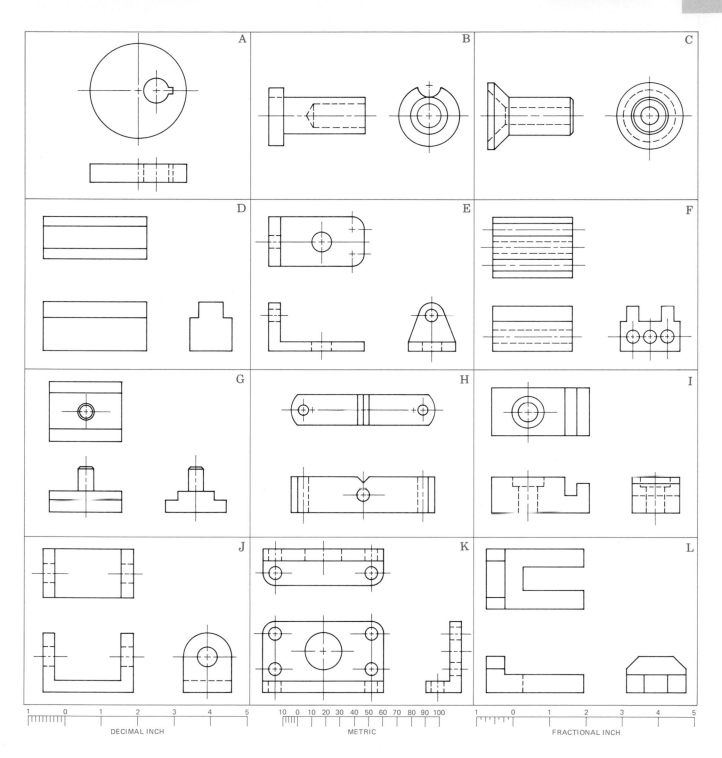

1 0 1 2 3 4 5
DECIMAL INCH

10 0 10 20 30 40 50 60 70 80 90 100
METRIC

1 0 1 2 3 4 5
FRACTIONAL INCH

Fig. 6-83. Problems for dimensioning practice. Take dimensions from the printed scale, using dividers. Draw the views as shown and add all necessary size and location dimensions.

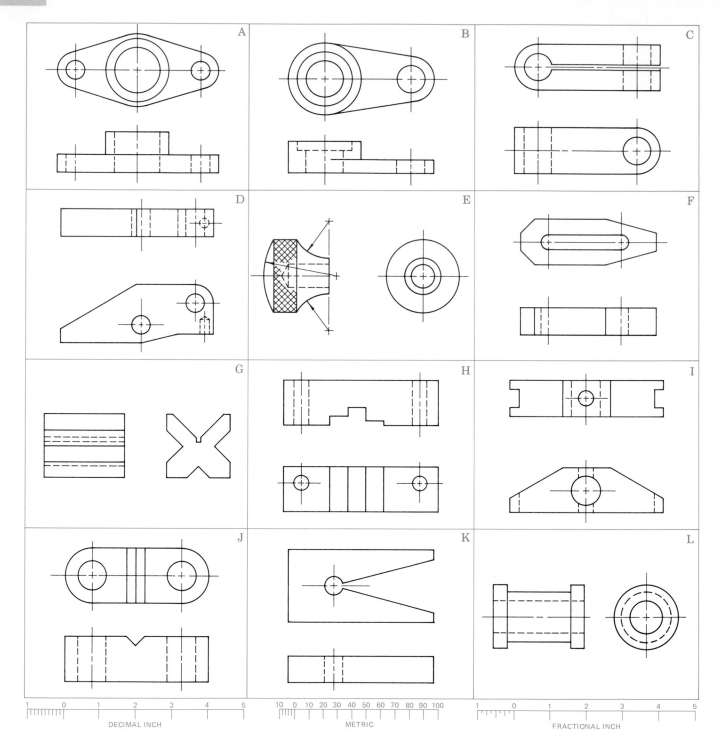

Fig. 6-84. Problems for dimensioning practice. Take dimensions from the printed scale, using dividers. Draw the views as shown and add all necessary size and location dimensions.

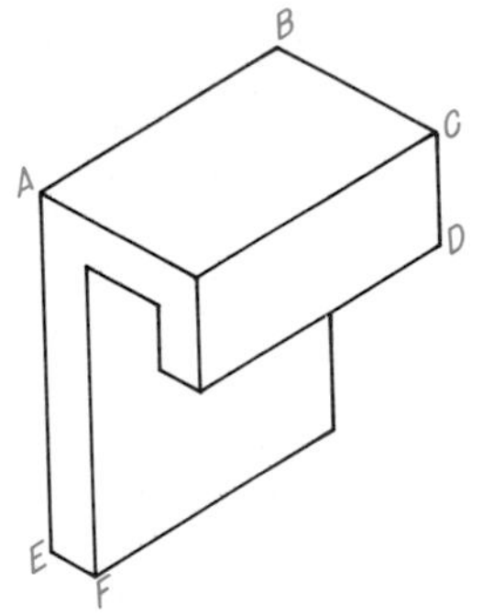

Fig. 6-85. Cutoff stop. Draw all necessary views and dimension. Scale: double size or as assigned. *AB* = 40 mm, *BC* = 26 mm, *CD* = 17 mm, *AE* = 53 mm, *EF* = 7.5 mm.

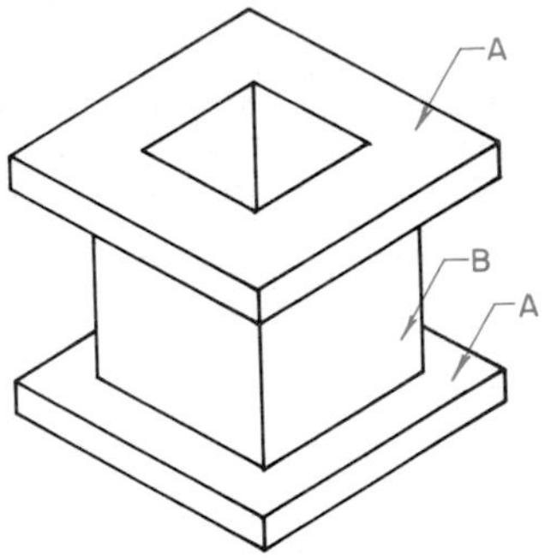

Fig. 6-86. Square guide. Draw all necessary views and dimension. Scale: full size or as assigned. *A* = 5 mm thick × 44 mm square, *B* = 30 mm square × 30 mm high, hole = 20 mm square.

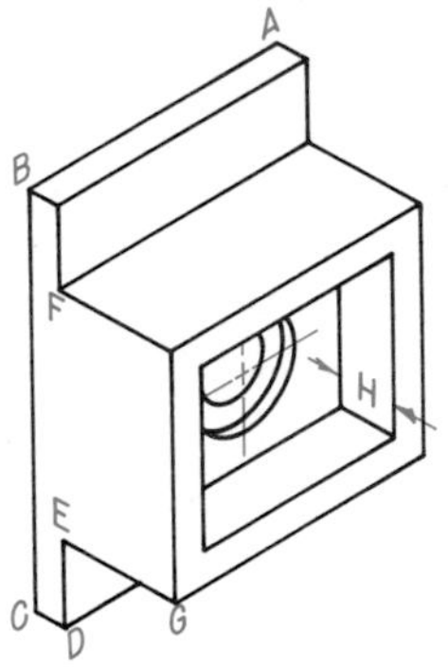

Fig. 6-87. Locator. Draw all necessary views and dimension. Scale: full size or as assigned. *AB* = 40 mm, *BC* = 60 mm, *CD* = 5 mm, *DE* = 12 mm, *EF* = 36 mm, *EG* = 18 mm, *H* = 8 mm, hole = 10 mm DIA through, 18 mm Cbore, 2 mm deep.

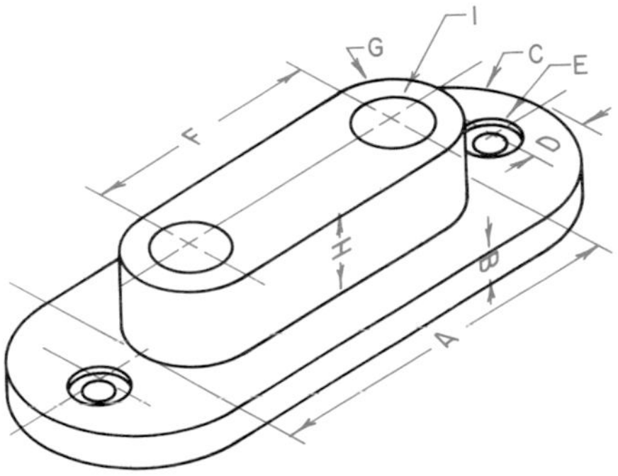

Fig. 6-88. Double-shaft support. Draw all necessary views and dimension. Scale: full size or as assigned. *A* = 67 mm, *B* = 7 mm, *C* = 21 mm R, *D* = 10 mm, *E* = 5 mm DIA through, 10 mm Cbore, 2 mm deep, 2 holes, *F* = 43 mm, G = 12 mm R, *H* = 14 mm.

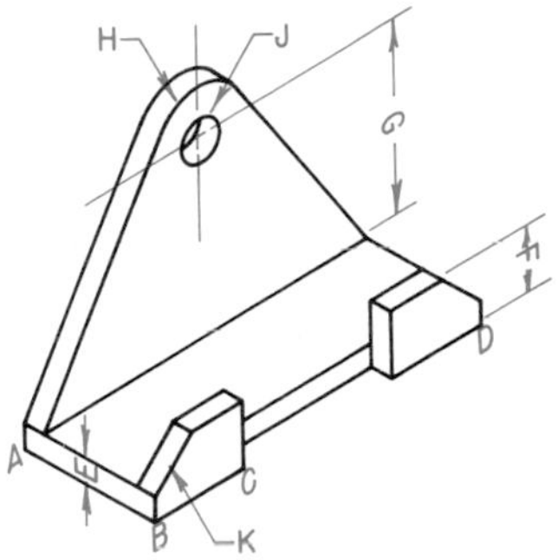

Fig. 6-89. Hanger. Draw all necessary views and dimension. Scale: three-quarter size or as assigned. *AB* = $1\frac{1}{4}$, *BC* = $\frac{13}{16}$, *BD* = 3, *E* = $\frac{3}{16}$, *F* = $\frac{1}{2}$, *G* = $1\frac{9}{16}$, *H* = $\frac{7}{16}$R, *J* = $\frac{3}{8}$ DIA, *K* = 45°.

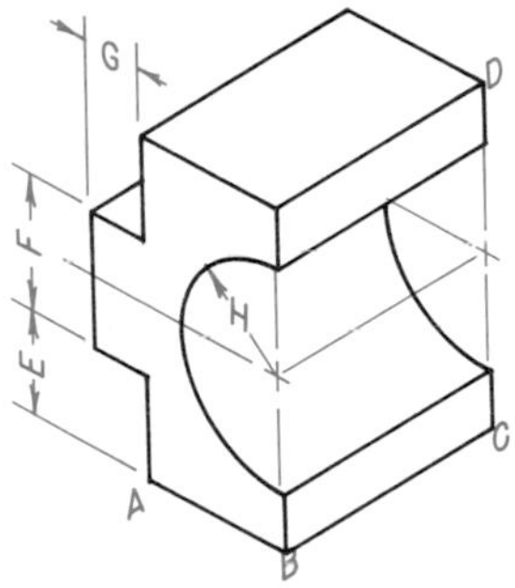

Fig. 6-90. Cradle slide. Draw all necessary views and dimension. Scale: full size or as assigned. *AB* = $2\frac{3}{8}$, *BC* = $3\frac{9}{16}$, *CD* = $5\frac{1}{8}$, *E* = $1\frac{1}{2}$, *F* = $2\frac{1}{8}$, *G* = $\frac{7}{8}$, *H* = $1\frac{5}{8}$R.

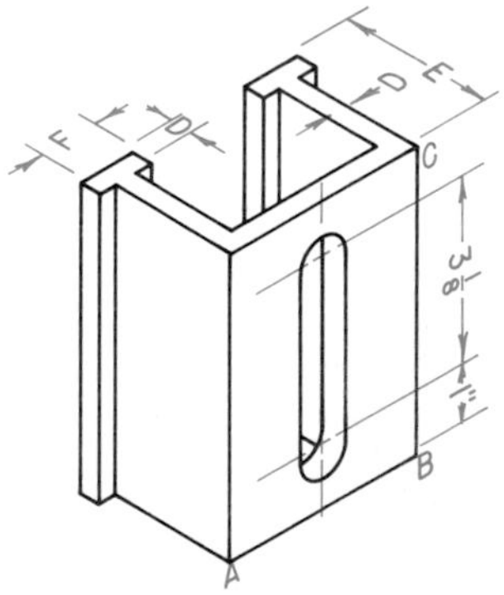

Fig. 6-91. Adjustable stop. Draw all necessary views and dimension. Scale: three-quarter size or as assigned. *AB* = $3\frac{5}{8}$, *BC* = $5\frac{1}{8}$, *D* = $\frac{5}{16}$, *E* = $2\frac{5}{8}$, *F* = 1, slot = $\frac{7}{8}$ wide.

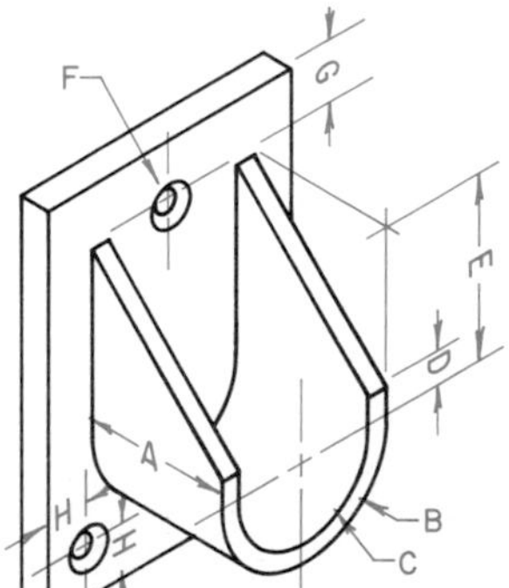

Fig. 6-92. Pipe support. Draw all necessary views and dimension. Scale: half size or as assigned. Base plate = $\frac{1}{2}$ thick × $4\frac{1}{2}$ wide × $6\frac{1}{2}$ long, *A* = $2\frac{3}{8}$, *B* = $1\frac{1}{2}$R, *C* = $1\frac{1}{8}$R, *D* = $\frac{1}{2}$, *E* = 3, *F* = $\frac{3}{8}$ DIA hole through, CSK to $\frac{3}{4}$ DIA, 3 holes, *G* = 1, *H* = $\frac{3}{4}$.

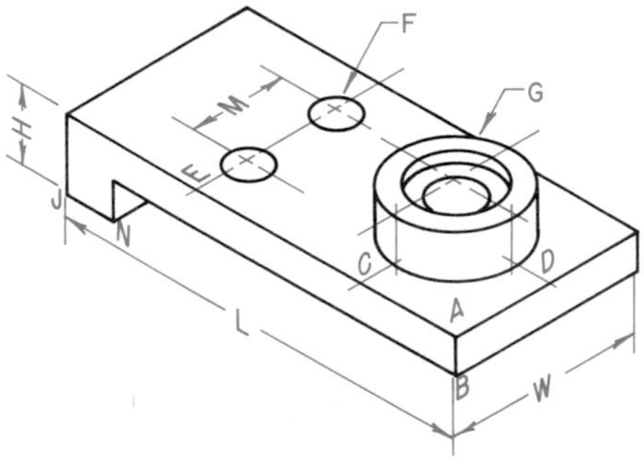

Fig. 6-93. Stop plate. Draw all necessary views and dimension. Scale: full size or as assigned. Overall sizes: *L* = $4\frac{1}{4}$, *W* = 2, *H* = $\frac{3}{4}$. *AB* = $\frac{3}{8}$, *AC* = 1, *AE* = $2\frac{3}{4}$, *AD* = 1, *JN* = $\frac{1}{2}$, *M* = 1, *F* = $\frac{7}{16}$ DIA, 2 holes, *G* = Boss: $1\frac{1}{4}$ DIA × $\frac{1}{2}$ high, $\frac{1}{2}$ DIA through, $\frac{7}{8}$ Cbore = $\frac{1}{8}$ deep.

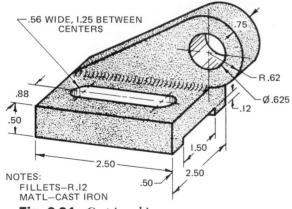

.56 WIDE, 1.25 BETWEEN
CENTERS

.75

R .62

Ø .625

.88

.50

.12

.50

1.50

2.50

2.50

NOTES:
FILLETS—R.12
MATL—CAST IRON

Fig. 6-94. Cast-iron hinge.

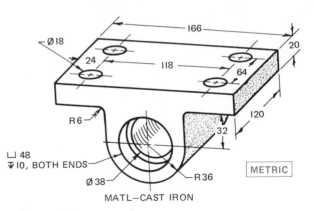

166

Ø 18

24

118

64

20

R 6

120

32

⌴ 48
⟱ 10, BOTH ENDS

Ø 38

R 36

METRIC

MATL—CAST IRON

Fig. 6-95. Bearing housing.

.30 THICK

Ø 4.00

.25

1.50

1.00 ID, 1.38 OD
1.50 LONG

NOTES:
FILLETS—R.12
MATL—CAST ALUMINUM

Fig. 6-96. Idler pulley.

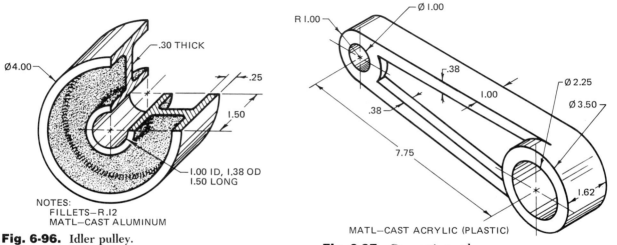

Ø 1.00

R 1.00

.38

1.00

Ø 2.25

Ø 3.50

.38

7.75

1.62

MATL—CAST ACRYLIC (PLASTIC)

Fig. 6-97. Connecting rod.

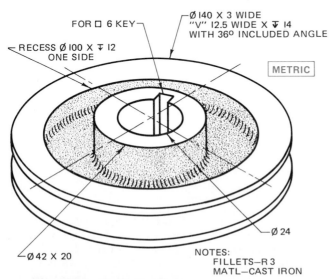

FOR □ 6 KEY

Ø 140 X 3 WIDE
"V" 12.5 WIDE X ⟱ 14
WITH 36° INCLUDED ANGLE

RECESS Ø 100 X ⟱ 12
ONE SIDE

METRIC

Ø 24

Ø 42 X 20

NOTES:
FILLETS—R 3
MATL—CAST IRON

Fig. 6-98. Single V-pulley.

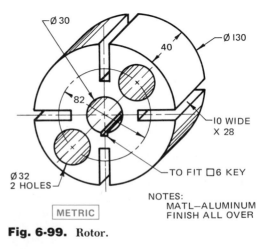

Ø 30

Ø 130

40

82

10 WIDE
X 28

Ø 32
2 HOLES

TO FIT □ 6 KEY

METRIC

NOTES:
MATL—ALUMINUM
FINISH ALL OVER

Fig. 6-99. Rotor.

7 Auxiliary Views and Revolutions

INTRODUCTION TO AUXILIARY VIEWS

In Chap. 5, Multiview Projections, you learned to describe an object with views on the three regular planes of projection. These are the top, or horizontal plane; the front, or vertical plane; and the side, or profile plane. With these planes, many graphic problems can be solved. However, to solve problems involving *inclined* (slanted) surfaces, you will need to draw views on *auxiliary* (additional) planes of projection. These are called *auxiliary views*. This chapter explains how to draw these views on planes that are parallel to the inclined surfaces (Fig. 7-1).

HOW TO LOOK AT OBJECTS

Imagine the many ways in which automated machines can form parts or assemble products. Stretch your imagination a little more and consider how robots can move or revolve to meet the demands of ma-

chine operations. The drafting technician must be able to see things form on the drawing board and to understand how they can be rotated into many positions. Therefore, it is necessary to study auxiliary views and revolutions in order to describe objects accurately as they are taking shape on the drawing board. High-technology drafting technicians need to draw "helper views" and to know how to look at objects.

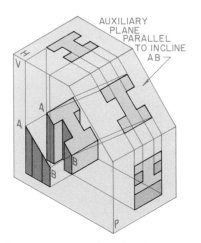

Fig. 7-1. Pictorial study of a primary auxiliary view.

AUXILIARY VIEWS ARE HELPER VIEWS

When an object has inclined surfaces, these do not show up in true shape in regular views. For example, in Fig. 7-2 at A, neither the front view, top view, or side view shows the true size and shape of the object's inclined surface. However, a view on a plane parallel to the inclined surface, as at B, does show its true size and shape. This is an auxiliary view. It and the other views at B describe the object better than the views at A.

An *auxiliary view* is a projection on an auxiliary plane that is parallel to an inclined (slanting) surface. It is a view looking directly at the inclined surface in a direction perpendicular to it.

An anchor with a slanting surface is pictured in Fig. 7-3 at A. At B are three views of the same anchor on the regular planes. These views are hard to draw and to understand. Also, they show three circular features of the anchor as ellipses. At C, by contrast, the anchor is described

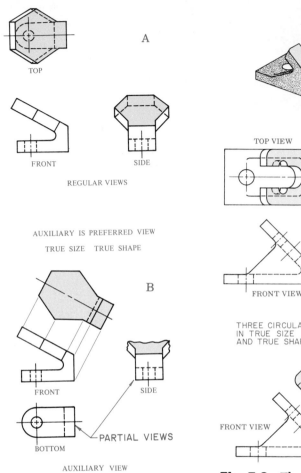

REGULAR VIEWS

AUXILIARY IS PREFERRED VIEW
TRUE SIZE   TRUE SHAPE

AUXILIARY VIEW

Fig. 7-2. Compare the regular views at A with the auxiliary views at B.

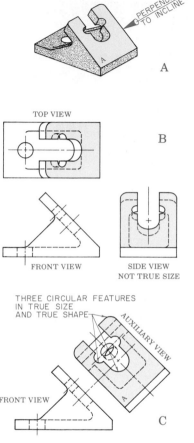

Fig. 7-3. The pictorial view at A and the three-view drawing at B are difficult to draw.

completely in two views, one of which is an auxiliary view.

Auxiliary projections are important for describing the true geometric shapes of inclined surfaces. They are also used for dimensioning these shapes.

THE RELATIONSHIP OF AUXILIARY VIEWS TO REGULAR VIEWS

In Fig. 7-4, at A, a simple inclined wedge block is shown in the regular views. In none of these views does the slanted surface, called A, appear

in its true shape. In the front view, all that shows is its edge line *MN*. In the side view, which is made by looking in the direction of arrow *Y*, surface A appears, but it is foreshortened. Surface A is also foreshortened in the top view. Line *MN* also appears in both views, but looking shorter than its true length, which shows only in the front view. To show surface A in its true shape, you need to imagine a plane parallel to it, as at B. This is called an *auxiliary plane*. A view of this plane from the direction of arrow *X*, which is perpendicular to it, will show the true size and shape of surface A at A^1. At

C this *auxiliary view* has been *revolved* (turned) to align with the plane of the paper. By following this method, you can show the true size and shape of any inclined surface.

The Auxiliary Plane in Relation to the Regular Planes

Figure 7-5, at A, shows a picture of a wedge along with the regular front, top, and side planes. The planes have been drawn to look "hinged" together. At B, the planes have been "unfolded, or revolved, to make a regular technical drawing. Auxiliary views are generally projected from one of these planes. At C, an auxiliary plane has been drawn hinged to the front and side planes. Note that the hinge line *XY* is parallel to the slanted surface of the wedge. At D, the auxiliary plane has been unfolded, or revolved, to align with the plane of the paper. The top view has been left out in this drawing, and the side view also could be, since the wedge is completely described by the front and auxiliary views.

KINDS OF AUXILIARY VIEWS

Auxiliary views are classified according to which of the three regular planes they are developed from. There are three primary auxiliary views (Fig. 7-6). Each is developed by projecting from one of the three regular views, using as a *primary reference* a dimension—height, width, or depth—obtained from another regular view.

The first primary auxiliary view is the front auxiliary view, shown at A. It is hinged on the front view, and its primary reference is the depth. The second auxiliary view is the top auxiliary view, shown at B. It is hinged on the top view and its primary ref-

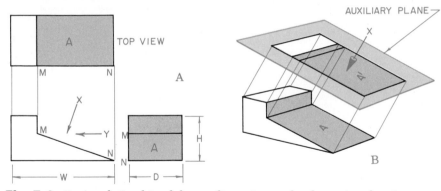

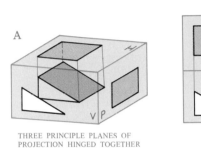

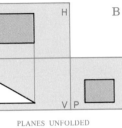

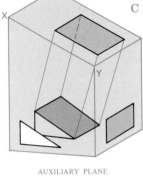

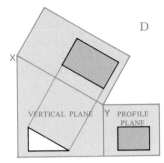

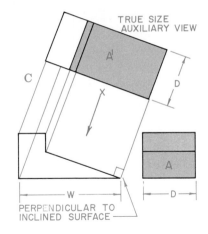

Fig. 7-4. Basic relationship of the auxiliary view to the three-view drawing.

TRUE SIZE
AUXILIARY VIEW

PERPENDICULAR TO
INCLINED SURFACE

AUXILIARY PLANE

TOP VIEW

THREE PRINCIPLE PLANES OF
PROJECTION HINGED TOGETHER

PLANES UNFOLDED

AUXILIARY PLANE

AUXILIARY PLANE REVOLVED

VERTICAL PLANE   PROFILE PLANE

Fig. 7-5. Basic relationship of the auxiliary plane to the regular planes.

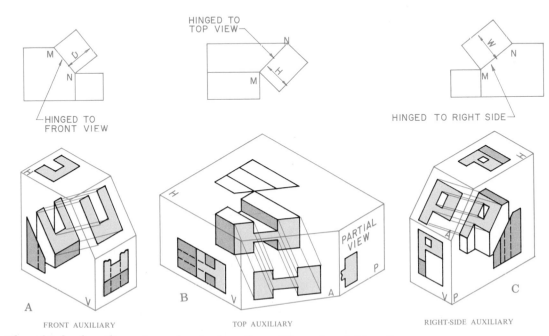

HINGED TO
TOP VIEW

HINGED TO
FRONT VIEW

HINGED TO RIGHT SIDE

PARTIAL
VIEW

FRONT AUXILIARY

TOP AUXILIARY

RIGHT-SIDE AUXILIARY

Fig. 7-6. Three kinds of auxiliary views, showing how the auxiliary plane is hinged.

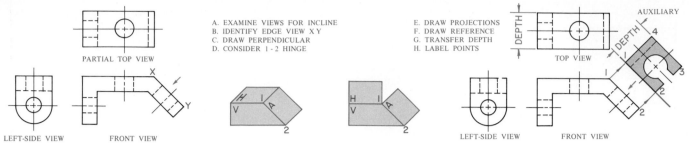

Fig. 7-7. Steps in constructing an auxiliary view. The eight steps apply to three kinds of auxiliary views.

erence is the height. The third auxiliary view is the right-side auxiliary view, shown at C. It is hinged to the side view and its primary reference is the width. The three principal planes always show the auxiliary plane as an inclined line *(MN)*. This line is considered the edge view of the auxiliary plane.

◼ CONSTRUCTING AN AUXILIARY VIEW

To construct any primary auxiliary view, use the following steps (Fig. 7-7). NOTE: The method shown is for a front auxiliary view.

A. Examine the views that are given for an inclined surface.
B. Find the line that is considered the edge view of the inclined plane.
C. In the front view, draw a light construction line at right angles to the inclined surface. This is the line of sight.
D. Think of the auxiliary plane as being attached by hinges to the front (vertical) plane from which it is developed.
E. From all points labeled on the front view, draw projection lines at right angles to the inclined surface (parallel to the line of sight).
F. Draw a reference line parallel to the edge view of the inclined surface and at a convenient distance from it.
G. Transfer the depth dimension, which in this case is the primary reference, to the reference line as shown.
H. Project the labeled points and connect them in sequence to form the auxiliary view. The points used to identify the shape

are for solving difficult problems (instructional purposes). You would not normally leave them on the final drawing.

Figure 7-8 shows how to make an auxiliary view of a symmetrical object. At A, the object is shown in a pictorial view. In this case, use a center plane as a reference plane, as at B (center plane construction). The edge view of this plane appears as a centerline, line *XY*, on the top view. Number the points on the top view. Then transfer these numbers to the edge view of the inclined surface on the front view, as shown. Parallel to this edge view and at a convenient distance from it, draw the line *X'Y'*, as at C. Now, in the top view, find the distances from the numbered points to the centerline. These are the depth measurement. Transfer them onto the corresponding cons-

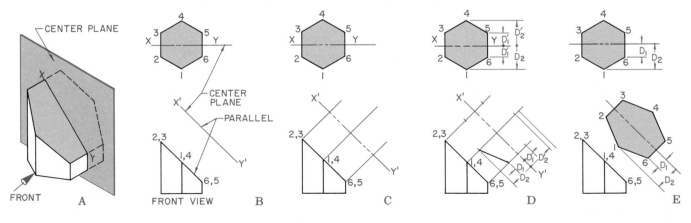

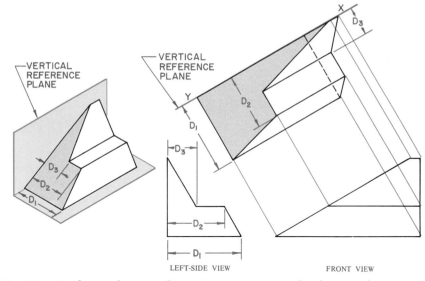

Fig. 7-9. To draw a front auxiliary view using a vertical reference plane.

tal reference plane. The object shown at A is a molding cut at a 30° angle. First, imagine a reference plane XY under the molding, as shown. Then find points 1 to 6 in the top and left-side views. In the top view, find the edge line of the slanted surface. Draw reference line $X'Y'$ parallel to it and at a convenient distance away. Then, from every point in the top view, project a line out to line $X'Y'$ and at right angles to it. Now, in the side view, find height measurements for the various numbered points by measuring up from XY. Lay off these same measurements up from $X'Y'$ along the lines leading to the corresponding points in the top view. Locate more points on the curve as needed in order to draw it more accurately. The result will be a top auxiliary view, with its base on line $X'Y'$.

ruction lines just drawn, measuring them off on either side of line $X'Y'$, as shown at D. The result will be a set of points on the construction lines. Connect and number these points, as at E, and you will have the front auxiliary view of the inclined surface. You could also, if desired, project the rest of the object from the center reference plane.

Figure 7-9 shows how to draw a front auxiliary view of a nonsymmetrical object by using a vertical reference plane. First, place the object on

reference planes, as shown. These planes are located strictly for convenience in taking reference measurements. The vertical plane can be in front or in back of the object. In this case it is in back. The construction is similar to that in Fig. 7-8, except that the depth measurements D_1, D_2, and D_3 are laid off in front of the vertical plane. The drawing shows the entire object projected onto the front auxiliary plane.

Figure 7-10 shows how to draw a top auxiliary view by using a horizon-

CURVES ON AUXILIARY VIEWS

You draw an auxiliary view of a curved line by locating a number of points along that line. A drawing of a simple curve is shown in Fig. 7-10 at B. Figure 7-11 shows how to make an auxiliary view of the curved cut sur-

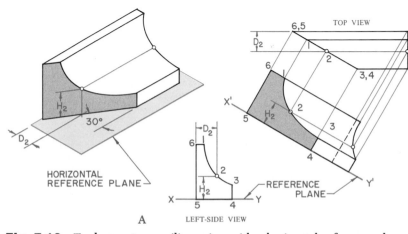

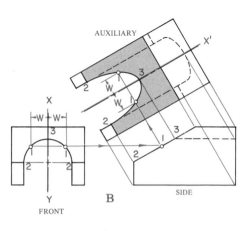

Fig. 7-10. To draw a top auxiliary view with a horizontal reference plane.

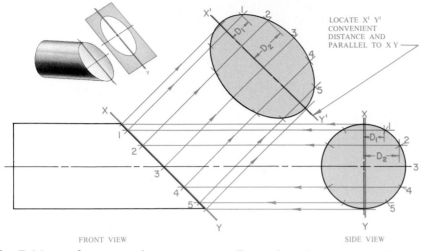

FRONT VIEW

SIDE VIEW

LOCATE X' Y' CONVENIENT DISTANCE AND PARALLEL TO X Y

Fig. 7-11. To draw an auxiliary-view curve (ellipse) about the centerline of the cut surface of a cylinder.

face of a cylinder. The cylinder is shown in a horizontal position. It has been cut at an angle, so the true shape of the slanting cut surface is an ellipse.

This auxiliary view is a front auxiliary view with the depth as its primary reference. To draw it, begin by locating the vertical centerline XY in the side view. This line will serve as a center reference plane. Next, locate a number of points along the rim of the side view. The more points you locate, the more accurate your

curve will be. Then project lines from these points over to the edge view of the cut surface in the front view. Now, parallel to this edge view and at a convenient dstance from it, draw the new centerline $X'Y'$. From the points you have located on the edge view, project lines out to line $X'Y'$ and perpendicular to it. Continue these lines beyond $X'Y'$. Finally, find your depth measurements in the side view by measuring off the distances D_1, D_2, etc., between the centerline XY and the points located

along the rim. Take these distances and measure them off on either side of $X'Y'$. Draw a smooth curve through the points marked to form the ellipse, as shown.

■ PARTIAL AUXILIARY VIEWS

Figure 7-12 at A shows a partial auxiliary view. If break lines and centerlines are used properly, you can leave out complex curves while still describing the object completely, as shown. A half view can be drawn if the object is symmetrical in a way that is simple and clearly understood.

■ AUXILIARY SECTIONS

Another practical use of auxiliary views is the auxiliary section (Fig. 7-12 at B). In this case, the slanted surface you are drawing is one made by a plane that cuts through the object. (See Chap. 9, Sectional Views.) Locate the surface in *cross section* (crosshatched) by using cutting-plane line *AA*.

■ SECONDARY AUXILIARY VIEWS

A view projected from a primary auxiliary view is called a *secondary auxiliary* (Fig. 7-13). Such a view must be used if you wish to show the true size and shape of a surface that is *oblique* (inclined to all three of the regular planes).

In Fig. 7-13, surface 1-2-3-4 is inclined to the three regular planes. At A, a first auxiliary view has been drawn. It is on a plane perpendicular to the inclined surface. Note that, in this view, points 1, 2, 3, and 4 appear as a line or edge view of the plane. At

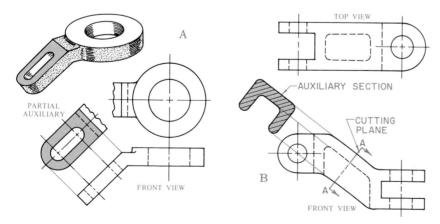

Fig. 7-12. Partial auxiliary views or sections are practical for understanding details.

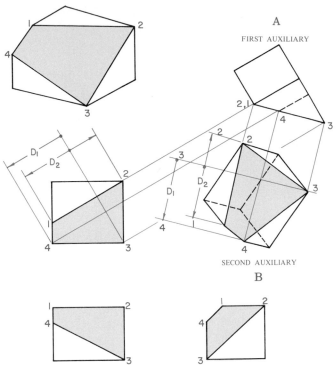

Fig. 7-13. Secondary auxiliary views assist in finding the true shape of a surface 1-2-3-4.

INTRODUCTION TO REVOLUTIONS

When the true size and shape of an inclined surface do not show in a drawing, one solution is, as we have seen, to make an auxiliary view. Another, however, is to keep using the regular reference planes while imagining that the object has been *revolved* (turned). (See Fig. 7-15.) Remember, in auxiliary views, you set up new reference planes to look at objects from new directions. Understanding *revolutions* (ways of revolving objects) should help you better understand auxiliary views. Both methods will be examined for spatial

B, a secondary auxiliary view has been drawn from the first. It is on a plane parallel to surface 1-2-3-4. This view shows the true shape of the surface.

Figure 7-14 shows another example. In this case, an octahedron (eight triangles making a regular solid) is shown in three views. Triangle surface 0-1-2 is inclined to all three. At A, a first auxiliary view has been drawn. It is on a plane perpendicular to triangle surface 0-1-2. Note that line 1-2 in the top view appears as point 1-2 in this auxiliary view and that the triangle now appears as an edge line 0'-1-2. At B, a secondary auxiliary view has been drawn. It is on a plane parallel to the edge view of triangle surface 0'-1-2 in the first auxiliary view. This secondary auxiliary view shows the true shape of triangle 0-1-2.

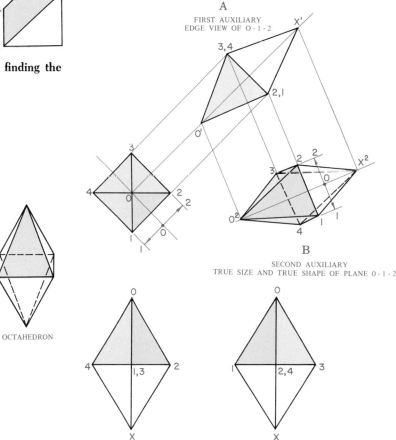

Fig. 7-14. True shape of a triangular surface on the octahedron is developed with a secondary auxiliary projection.

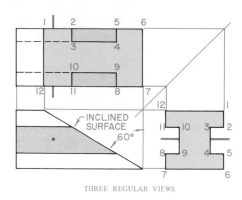

THREE REGULAR VIEWS

Fig. 7-15. The regular planes remain in order in revolution. The object is revolved behind the plane. Note that the front view is revolved 60° clockwise.

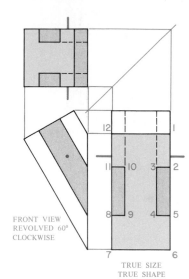

FRONT VIEW
REVOLVED 60°
CLOCKWISE

TRUE SIZE
TRUE SHAPE

problems in Chap. 8, Basic Descriptive Geometry.

THE AXIS OF REVOLUTION

An easy way to picture an object being revolved is to imagine that a shaft or an axis has been passed through it. Imagine, also, that this axis is perpendicular to one of the principal planes. In Fig. 7-16, the three principal planes are shown with an axis passing through each one and through the object beyond.

An object can be revolved to the right (clockwise) or to the left (counterclockwise) about an axis perpendicular to either the vertical or the horizontal plane. The object can be revolved forward (counterclockwise) or backward (clockwise) about an axis perpendicular to the profile plane.

SINGLE REVOLUTION

As we have seen, an axis of revolution can be perpendicular to the vertical, horizontal, or profile plane. In

Fig. 7-17 at A, the usual front and top views of an object are shown in space 1. Space 2 shows the same views of the object after it has been revolved 45° counterclockwise about an axis perpendicular to the vertical plane. Notice that the front view is the same in size and shape as in space 1, except that it has a new position. The new top view has been made by projecting up from the new front view and across from the old top view in space 1. Note that the depth remains the same from one top view to the other.

In Fig. 7-17 at B, a second object is shown in space 1 in the usual top and front views. Space 2 shows the same views of the object after it has been revolved 60° clockwise about an axis perpendicular to the horizontal plane. The new top view is the same in size and shape as in space 1. The new front view has been made by projecting down from the new top view and across from the old front view of space 1. Note that the height remains the same from the original front view to the revolved front view.

In Fig. 7-17 at C, a third object is shown in space 1 in the usual front and side views. Space 2 shows the same views of the object after it has been revolved forward (counterclockwise) 30° about an axis perpendicular to the profile plane. The new front view has been made by projecting across from the new side view and down from the old front view in space 1. Note that the width remains the same from one front view to the other. Revolution can be clockwise, as at Fig. 7-17 at B, or it can be counterclockwise, as at A and C.

THE RULE OF REVOLUTION

The rule of revolution has two parts.

1. *The view that is perpendicular to the axis of revolution stays the same except in position.* (This is true because the axis is perpendicular to the plane on which it is projected.)

2. *Distances parallel to the axis of revolution stay the same.* (This is true because they are parallel to the plane or planes on which they are projected.)

Figure 7-18 illustrates the two parts of the rule of revolution.

REVOLUTION ABOUT AN AXIS PERPENDICULAR TO THE VERTICAL PLANE

Figure 7-19 shows how to draw a primary revolution. At A, an imaginary axis *AX* is passed horizontally through a truncated right octagonal prism. In the front view, it shows as a dot. In the top view, and later in the

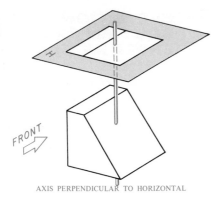
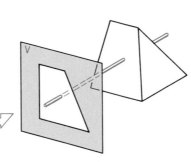
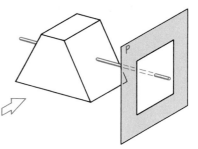

AXIS PERPENDICULAR TO HORIZONTAL   AXIS PERPENDICULAR TO VERTICAL   AXIS PERPENDICULAR TO PROFILE

Fig. 7-16. Three positions for the axis of revolution. The axis is perpendicular to the principal planes.

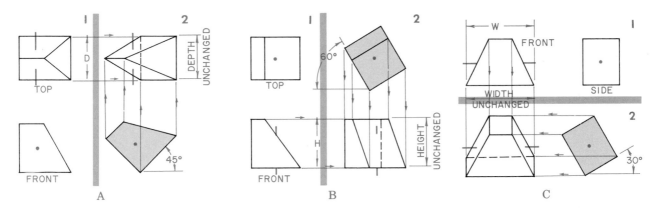

Fig. 7-17. Single revolution about the three axes.

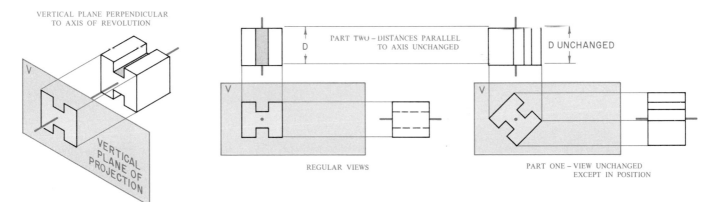

Fig. 7-18. The rule of revolution. Note the H-shape in the front view has changed only position, not shape.

side view, it shows as a line. At B, the prism has been revolved clockwise about the axis into a new position. You can see that the new front view has the same size and shape as the old. Just its position has changed. However, the side view now shows the true size and shape of the truncated surface. This side view is made by projecting across from the new front view and by transferring the depth from the top view.

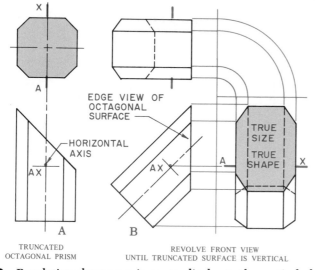

TRUNCATED
OCTAGONAL PRISM

REVOLVE FRONT VIEW
UNTIL TRUNCATED SURFACE IS VERTICAL

Fig. 7-19. Revolution about an axis perpendicular to the vertical plane.

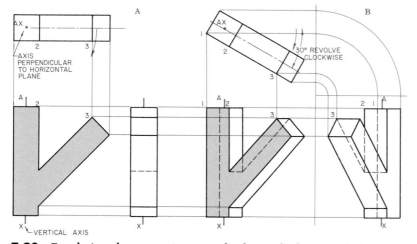

Fig. 7-20. Revolution about an axis perpendicular to the horizontal plane (clockwise).

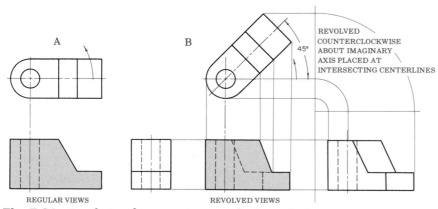

REGULAR VIEWS   REVOLVED VIEWS

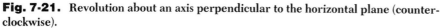

Fig. 7-21. Revolution about an axis perpendicular to the horizontal plane (counterclockwise).

REVOLUTION ABOUT AN AXIS PERPENDICULAR TO THE HORIZONTAL PLANE

Figure 7-20 shows how to draw an object that is revolved clockwise about an imaginary vertical axis *AX*. At A, the three regular views are given. At B, the top view has been revolved 30° clockwise about the axis *AX*. Since this is a vertical axis, revolution does not change the height of the object. Therefore, points from the old vertical plane at A can be projected to make the new one at B. The new side view is made from the front and top views in the usual way. Figure 7-21 shows how to draw an object revolved counterclockwise through 45°.

PRACTICAL REVOLVED VIEWS

In a working drawing, you can show an inclined surface by drawing one of the views, or part of a view with the object in a revolved position. In Fig. 7-22 at A, the top view shows the angle of a V-shaped part. In the front view, the part is revolved to show its true shape. In Fig. 7-22 at B, the front view shows the angles at which surfaces of a part are inclined. The inclined surfaces are then revolved in the front view. Next, their dimensions are transferred to the top view. There, the surfaces appear in true size and shape.

SUCCESSIVE REVOLUTIONS

After an object is revolved about an axis perpendicular to one plane, it can be revolved again about an axis perpendicular to another plane. This process is shown in Fig. 7-23. At A,

Fig. 7-22. Practical use of revolutions.

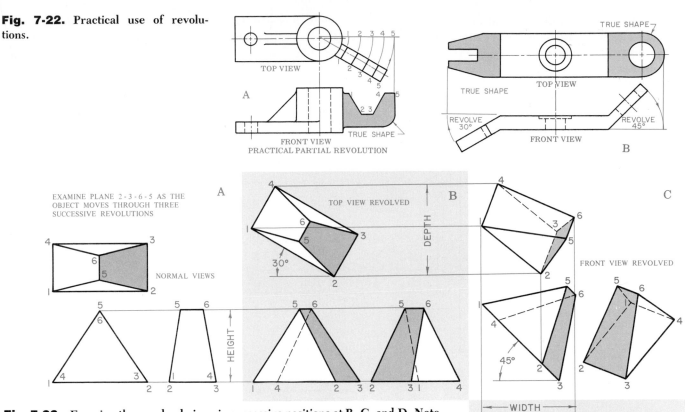

EXAMINE PLANE 2-3-6-5 AS THE OBJECT MOVES THROUGH THREE SUCCESSIVE REVOLUTIONS

NORMAL VIEWS

TOP VIEW REVOLVED

FRONT VIEW REVOLVED

SIDE VIEW REVOLVED

Fig. 7-23. Examine the revolved views in successive positions at B, C, and D. Note position of plane 2-3-5-6.

an object is shown in the normal views. At B, the object has been revolved 30° clockwise about an axis perpendicular to the horizontal plane. At C, the front view from B has been revolved 45° clockwise about an axis perpendicular to the vertical plane. At D, the side view from C has been revolved about an axis perpendicular to the profile plane until line 3-4 appears as a horizontal line. In all, then, three revolutions occurred. There has been

one in each of the three principal planes of projection.

AUXILIARY VIEWS AND REVOLVED VIEWS

You can show the true size of an inclined surface by either an auxiliary view (Fig. 7-24 at A) or a revolved view (Fig. 7-24 at B and C).

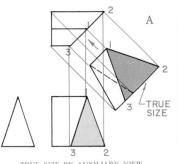
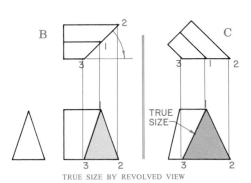

TRUE SIZE BY AUXILIARY VIEW

TRUE SIZE BY REVOLVED VIEW

Fig. 7-24. True size of a plane shown by an auxiliary view and compared to a revolved view.

In a revolved view, the inclined surface is turned until it is parallel to one of the principal planes. The revolved view at B and C is similar to the auxiliary at A.

In the auxiliary view, it is as if the observer has changed position to look at the object from a new direction. Conversely, in the revolved view, it is as if the object has changed

position. Both revolutions and auxiliaries help you visualize things better. They also work equally well in solving problems.

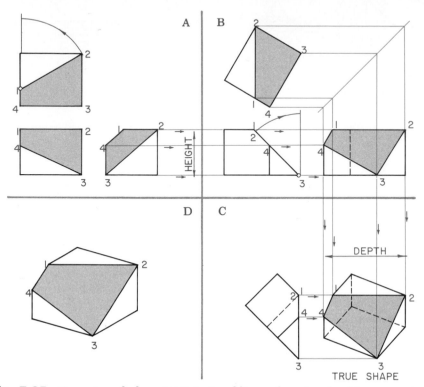

Fig. 7-25. True size of plane 1-2-3-4. An oblique plane is solved by successive revolution.

TRUE SHAPE OF AN OBLIQUE PLANE BY SUCCESSIVE REVOLUTIONS

A surface shows its true shape when it is parallel to a plane. In Fig. 7-25 at D, an object is pictured on which surface 1-2-3-4 is an oblique plane. It is oblique because it is inclined to all three of the normal planes. At A, the object is drawn in its normal position. At B, it has been revolved about an axis perpendicular to the horizontal plane until surface 1-2-3-4 is perpendicular to the vertical plane. Now, in the front view, all you see of this surface is its edge line. At C, the object has been revolved about an axis perpendicular to the front plane until surface 1-2-3-4 is parallel to the profile plane. There it shows its true shape.

TRUE LENGTH OF A LINE

Since an auxiliary view shows the true size and shape of an inclined surface, it can also be used to find the true length of a line. In Fig. 7-26 at A, the line OA does not show its true length in the top, front, or side view because it is inclined to all three of these planes of projection. At B, however, in the auxiliary plane, it does show its true length (TL). This is because the auxiliary plane is parallel to the surface OAB.

Figure 7-26 at C shows another way to show the true length (TL) of OA. This is to revolve the object about an axis perpendicular to the vertical plane until surface OAB is parallel to the profile plane. The side

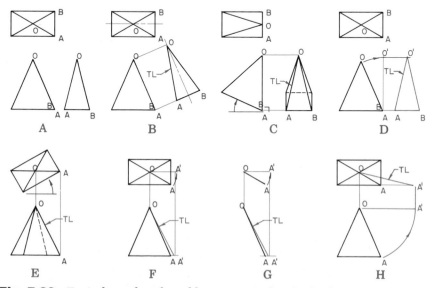

Fig. 7-26. Typical true-length problems examined and solved.

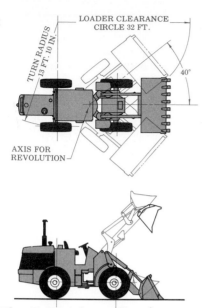

Fig. 7-27. The profile of a tractor shows several positions of the loading bucket. The plan view shows horizontal rotation.

Fig. 7-28. Note that the television profile has been revolved to show the screen in true shape.

Fig. 7-29. Civil engineers require prints and the knowledge of revolutions when using the automatic level for examining topography.

view will then show the true size of *OAB* and also the true length of *OA*. A shorter method of showing the true length of *OA* is to revolve only the surface *OAB*, as shown at D.

At E, the object is revolved in the top view until line *OA* in that view is horizontal. The front view now shows *OA* in its true length because this line is now parallel to the vertical plane.

At F and G, still another method is shown. In this case, instead of the whole object being revolved, just line *OA* is turned in the top view

until it is horizontal at *OA'*. The point *A'* then can be projected to the front view. There, *OA'* will be in true length.

You can revolve a line in any view to make it parallel to any one of the three principal planes. Projecting the line on the plane to which it is parallel will show its true length. In Fig. 7-26 at H, the line has been revolved parallel to the horizontal plane. The true length then shows the top view.

INDUSTRIAL APPLICATIONS

The following illustrations show how a drafter uses revolution. Figure 7-27 shows a tractor with various lift positions. Figure 7-28 shows a product designer's use of a revolved position. Figures 7-29 and 7-30 illustrate the transit and the axes of revolution used by the civil engineer.

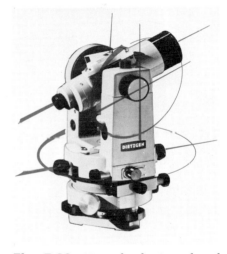

Fig. 7-30. Note the horizontal and vertical planes of revolution on the transit. (The Dietzgen Co.)

VOCABULARY

1. Revolution
2. Relationship
3. Primary
4. Secondary
5. Successive
6. Hinged
7. Reference
8. Symmetrical
9. Transit
10. Partial
11. Projection
12. Truncated
13. True length
14. Conversely
15. Visualize

REVIEW

1. Why do you need auxiliary views?

2. How do you place the auxiliary plane in relation to the inclined surface?

3. What are the two chief reasons for drawing auxiliary projections?

4. Name the three primary auxiliary projections.

5. List the steps in drawing an auxiliary view.

6. What is a reference-plane construction?

7. Can you plot curved lines on auxiliary views?

8. Is doing partial auxiliary views an accepted drawing practice in industry?

9. What auxiliary projection do you use to find the true size of an oblique surface?

10. What is the basic reason for revolving the view of an object?

11. What is the axis of revolution?

12. In your own words, describe the first rule of revolution.

13. Name the three basic single revolutions.

14. What is the chief use of successive revolutions?

15. Can you use both auxiliary views and revolved views to find the true lengths of inclined and oblique lines?

Problems

Figs. 7-31 through 7-50. In Figs. 7-32 through 7-50, only the top view is given. For each figure, draw the top and front views and either the complete auxiliary view or just the inclined surface, as directed by the instructor. Figure 7-31 has been worked as an example. Dimension the problems, as assigned by the instructor. The angle X may be 45° or 60°, as assigned. The total height of the front view is 3¾″ for Figs. 7-31 through 7-50.

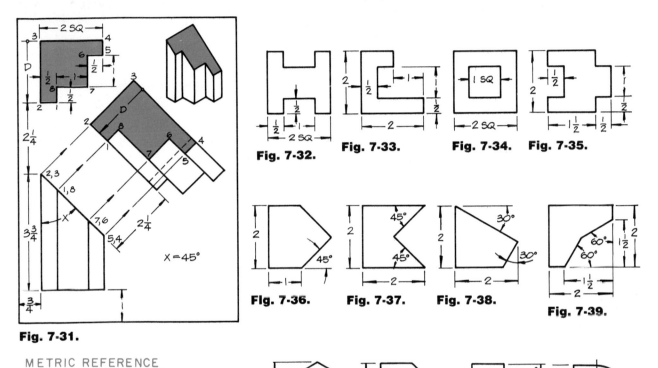

Fig. 7-31.

Fig. 7-32.

Fig. 7-33.

Fig. 7-34.

Fig. 7-35.

Flg. 7-36.

Fig. 7-37.

Fig. 7-38.

Fig. 7-39.

Fig. 7-40.

Fig. 7-41.

Fig. 7-42.

Fig. 7-43.

METRIC REFERENCE

in.	mm
1/2	13
3/4	19
1	25
1-1/2	38
1-3/4	44
2	50
3-3/4	95

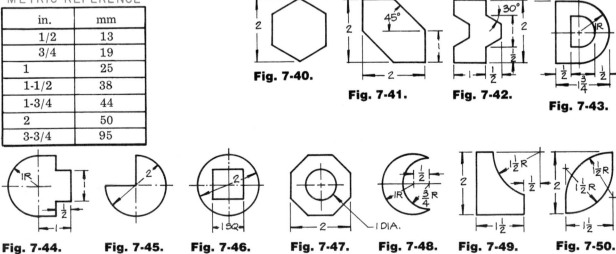

Fig. 7-44.

Fig. 7-45.

Fig. 7-46.

Fig. 7-47.

Fig. 7-48.

Fig. 7-49.

Fig. 7-50.

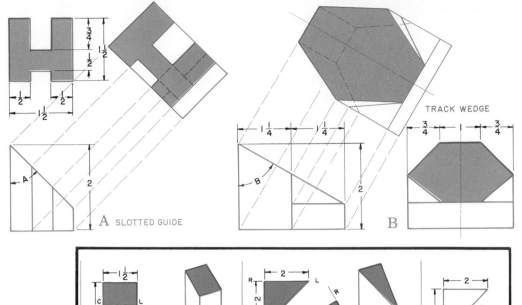

Fig. 7-51. Two parts. Problem 1: Draw the front, top, and side view of each figure, A and B. Complete the front auxiliary projection. Problem 2: Change the angles of the inclined surface at A to 30° and at B to 45°.

A SLOTTED GUIDE

TRACK WEDGE

B

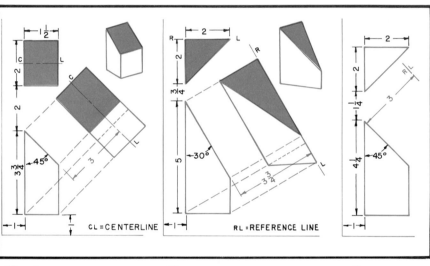

Figs. 7-52 through 7-54. Draw the front, top, and complete auxiliary views. The solutions are given for Figs. 7-52 and 7-53. Change the angle for Fig. 7-52 to 30° and the angle for Fig. 7-53 to 45°. Is there any difference in the solution?

CL = CENTERLINE

RL = REFERENCE LINE

Fig. 7-52.   **Fig. 7-53.**   **Fig. 7-54.**

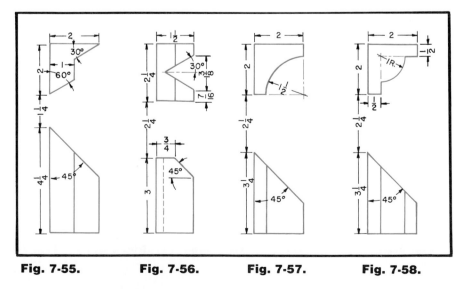

Figs. 7-55 through 7-58. Draw the front, top, and complete auxiliary views.

Fig. 7-55.   **Fig. 7-56.**   **Fig. 7-57.**   **Fig. 7-58.**

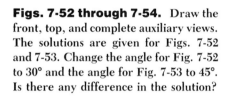

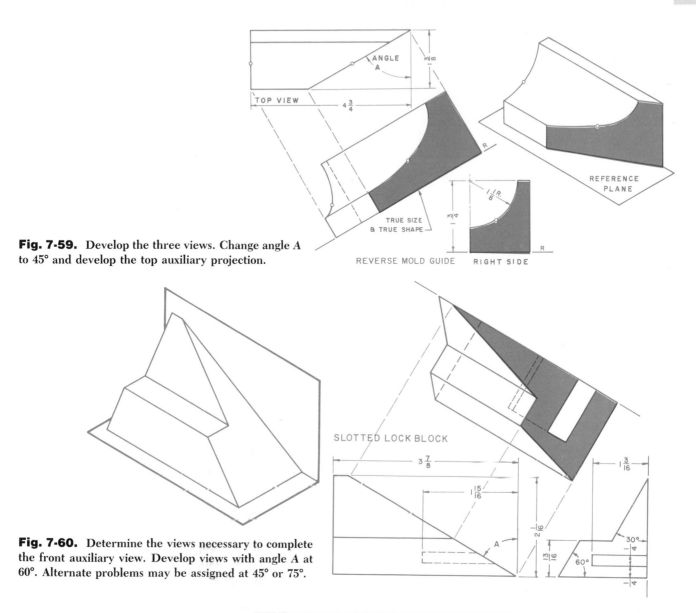

Fig. 7-59. Develop the three views. Change angle *A* to 45° and develop the top auxiliary projection.

Fig. 7-60. Determine the views necessary to complete the front auxiliary view. Develop views with angle *A* at 60°. Alternate problems may be assigned at 45° or 75°.

Fig. 7-61. A layout and pictorial for an angle plate are given. Draw the top view and the partial front view as shown. Draw a partial auxiliary view where indicated on the layout. Note that this is an auxiliary elevation. It is made on a plane perpendicular to the horizontal plane.

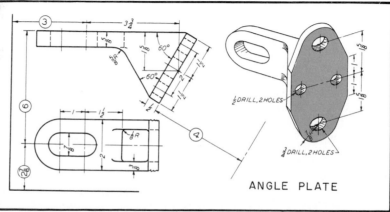

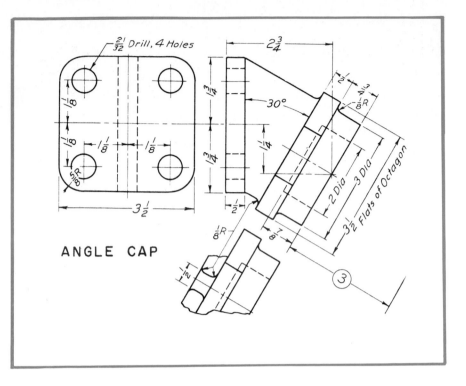

ANGLE CAP

Fig. 7-62. A part front view, a right-side view, and a part auxiliary view of the angle cap are shown on the layout. Draw the views given and another auxiliary view where indicated on the layout. This will be a rear auxiliary view. Dimensioning is required.

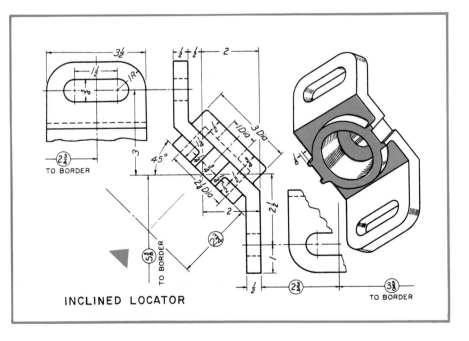

INCLINED LOCATOR

Fig. 7-63. A pictorial and layout for an inclined stop are shown. The complete view in the middle is the right-side view. Draw the complete view and the partial views as necessary. Draw an auxiliary view of the inclined middle, as indicated in layout.

Fig. 7-64. This figure shows the completed problem. It is given for comparison and is not to be copied. In space 1 is a three-view drawing of a block in its simplest position. In space 2 (upper right) the block is shown after being revolved from the position in space 1 through 45° about an axis perpendicular to the frontal plane. The front view was drawn first, copying the front view of space 1. The top view was obtained by projecting up from the front view and across from the top view of space 1.

In space 3 (lower left) the block has been revolved from position 1 through 30° about an axis perpendicular to the horizontal plane. The top view was drawn first, copied from the top of space 1. In space 4, the block has been tilted from position 2 about an axis perpendicular to the side plane 30°. The side view was drawn first, copied from the side view of space 2. The widths of the front and top views were projected from the front view of space 2.

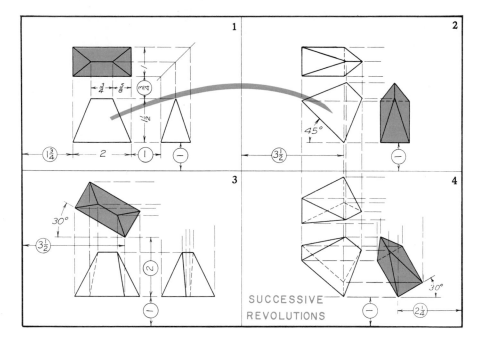

Fig. 7-65. Draw the revolved views of the wedge shown in space 1. In space 2, revolve 45° clockwise about an axis perpendicular to the frontal plane and draw three views. In space 3, revolve 30° forward from the position of space 2. In space 4, revolve from the position of space 1, 30° counterclockwise about an axis perpendicular to the horizontal plane.

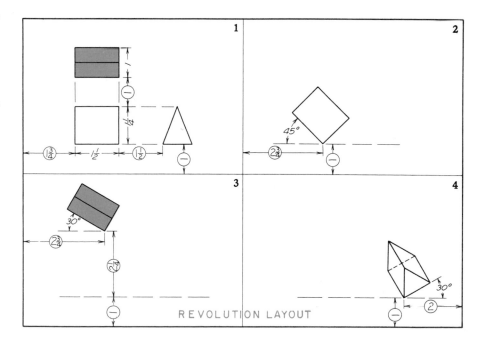

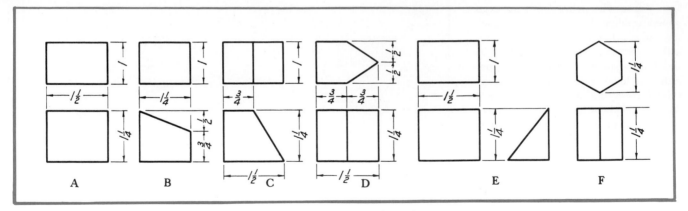

Fig. 7-66. Follow the directions for Fig. 7-65 for the objects in A through F, as assigned.

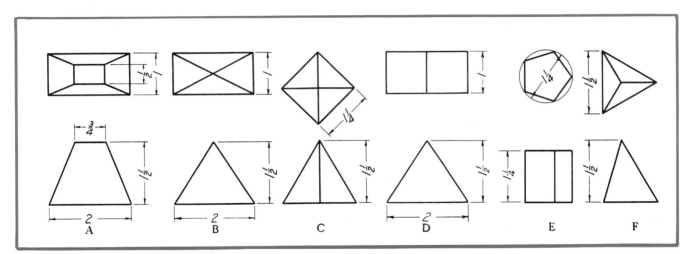

Fig. 7-67. Follow the directions for Fig. 7-65. Optional assignments may change revolutions to determine true sizes of inclined surfaces.

8 Basic Descriptive Geometry

■ GRAPHICS AND MATHEMATICS

The designer who works with an engineering team can help solve problems by producing drawings made up of geometric elements. *Geometric elements* are points, lines, and planes defined according to the rules of geometry. Every structure has a three-dimensional form made up of geometric elements (Fig. 8-1). In order to draw three-dimensional forms, you must understand how points, lines, and planes relate to each other in space to form a certain shape. Problems that you might think need mathematical solutions can often be solved through drawings that make manufacturing and construction possible. *Basic descriptive geometry* is one of the ways a designer thinks about and solves problems. In the eighteenth century a French mathematician, Gaspard Monge, developed a drawing system for solving *spacial* (space) problems related to military structures. Descriptive geometry was brought to the U.S. Military Academy at West

Fig. 8-1. Geometric space-frame structure. Franklin Park Mall, Toledo, Ohio. (Unistrut Corp.)

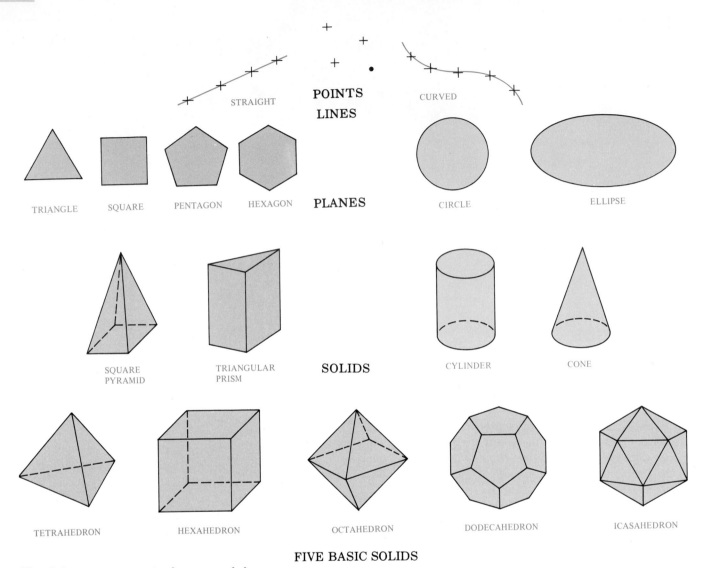

POINTS
LINES

STRAIGHT

CURVED

TRIANGLE   SQUARE   PENTAGON   HEXAGON   PLANES   CIRCLE   ELLIPSE

SQUARE
PYRAMID

TRIANGULAR
PRISM

SOLIDS

CYLINDER   CONE

TETRAHEDRON   HEXAHEDRON   OCTAHEDRON   DODECAHEDRON   ICASAHEDRON

FIVE BASIC SOLIDS

Fig. 8-2. Basic geometric elements and shapes.

Point by Claude Crozet in 1816. The Mongean method was changed over the years. However, its basic principles are still taught in engineering schools throughout the world. In studying descriptive geometry, you develop a reasoning ability to solve problems through drawing.

This chapter presents a way of drawing that lets you analyze all geometric elements. Learning to see geometric elements will make it possible to describe a structure of any shape. Most structures designed by people have been shaped like a rectangle. This is because it is easy to plan a structure with this shape. Figure 8-2 shows the basic geometric elements. It also shows some of the common geometric features found in engineering designs.

■ POINTS

A point can be thought of as having an actual physical existence. On a drawing, you can locate a point with a small dot or a small cross. Normally, a point is identified with two or more projections. In Fig. 8-3 at A, the small cross for point number 1 is shown in the front, top, and right-side views. At B in Fig. 8-3, the regular reference planes are shown in a pictorial view with point 1 projected to all three planes. The reference planes are shown again in Fig. 8-4. Notice that when the three planes are unfolded, a flat two-dimensional

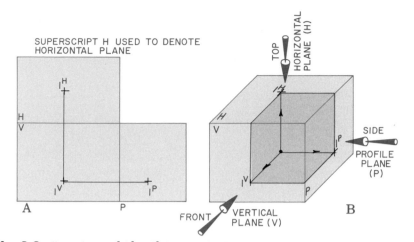

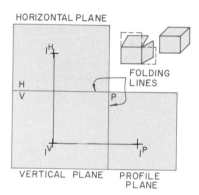

Fig. 8-3. Locating and identifying a point in space.

Fig. 8-4. Points identified on unfolded reference planes.

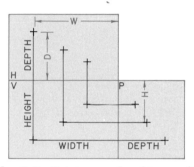

Fig. 8-5. Points are related to coordinated reference planes.

surface with fold lines is formed. The fold lines are labeled as shown to indicate that *V* stands for the vertical view, *H* stands for the horizontal, or top, view, and *P* stands for the profile, or right-side, view. Points may also stand for the intersection of two lines or the corners on an object.

FIXED POINTS

Figure 8-5 shows a group of points. Points are related to each other by distance and direction. These are measured on the coordinated reference planes. The vertical height relation can be seen in the front and side

views. The relative width dimensions can be seen in the front and top views. The relative depth dimensions can be seen in the top and side views. Note that the three basic dimensions are indicated by *H*, *W*, and *D* (Fig. 8-5).

LINES

If a point moves away from a fixed place, its path forms a line. A line has location, direction, and length. A straight line moves in only one direction. It is easy to draw circular and straight lines. However, plot irregular curves very carefully. A straight line can be determined by two points. Or it can be determined by one point and a fixed direction.

THE BASIC LINES

There are three kinds of straight lines. The kind of line depends on how it relates to the coordinating reference planes.

1. *Normal lines.* A normal line is one that is *perpendicular* (at right angles) to one of the three reference planes. It will project on that plane as a point (Fig. 8-6 at A, B, and C). If a normal line is parallel to the other two reference planes, as shown in Fig. 8-6 at D, E, and F, it is shown true length (TL), as noted.
2. *Inclined lines.* An inclined line is slanted in one of the three main reference planes but is parallel to one other. Inclined lines are shown in Fig. 8-7 at A, B, and C. An inclined line is *foreshortened* (not in true size) in two planes. It is shown true length (TL), as noted.
3. *Oblique lines.* An oblique line appears inclined in all three reference planes (Fig. 8-8). It makes an angle other than a right angle with all three planes. In other words, it is not perpendicular or parallel to any of the three planes. The true length (TL) is not shown in any of these views. Also, the angles of direction cannot be measured on the main reference planes.

AUXILIARY REFERENCE PLANE—TRUE LENGTH OF AN OBLIQUE LINE

A normal line and an inclined line each project parallel to at least one of the main planes of projection. Therefore, a line parallel to a plane of projection shows true length in that

A LINE HAS LOCATION, DIRECTION, AND LENGTH

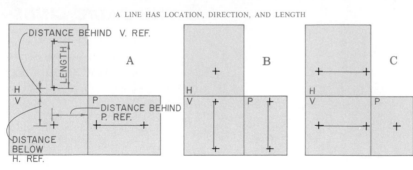

NORMAL LINES - PERPENDICULAR TO A PRINCIPAL REFERENCE PLANE

NORMAL LINES - PERPENDICULAR AND PARALLEL WILL ALWAYS
BE TRUE LENGTH (T. L.) AS SHOWN.

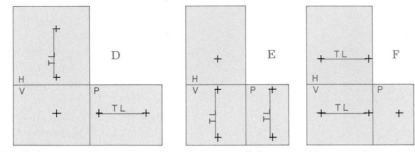

Fig. 8-6. Normal lines in true length are parallel to two reference planes.

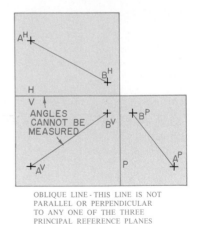

OBLIQUE LINE - THIS LINE IS NOT
PARALLEL OR PERPENDICULAR
TO ANY ONE OF THE THREE
PRINCIPAL REFERENCE PLANES

Fig. 8-8. Oblique lines appear inclined in all projections.

projection. An oblique line is not parallel to any of the three main reference planes. Thus, in order to show an oblique line true length, you must place an auxiliary reference plane parallel to the oblique line in any plane. This is shown in Fig. 8-9 at A, B, and C. The auxiliary and regular planes of projection will have the same relationship as any two main planes. First, they must always

be perpendicular to each other (see Fig. 8-9 at D). Second, they must be measured in relation to the previous plane on which they are related. The true length is obtained as noted.

However, they reflect real things. In that sense, they are always being used. The following terms related to lines are used in mining, geology, engineering, and navigation.

■ LINE TERMINOLOGY

It may seem that lines drawn on paper mean little and are worth little.

Slope

A line that makes an angle with the horizontal plane has a *slope* measured in degrees. In Fig. 8-10 at A, slope is shown in the front view when the line is in true length. At B, slope is found for an oblique line in true length in an auxiliary projection perpendicular to a horizontal reference.

Bearing

The angle a line makes in the top view with a north-south line is called its *bearing*. The north-south line is generally vertical. North is at the top. Therefore, right is east and left is west. The measurement should be made in the horizontal projection. It should be dimensioned in degrees, as shown in Fig. 8-11 at both A and B.

Azimuth

A measurement that defines the direction of a line off due north is the

TRUE LENGTH SHOWS ON INCLINED PROJECTION

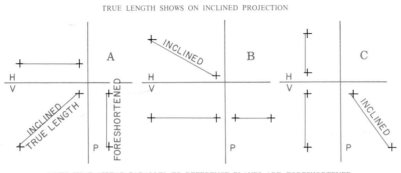

LINES THAT APPEAR PARALLEL TO REFERENCE PLANES ARE FORESHORTENED

Fig. 8-7. Inclined lines will be parallel to one reference plane.

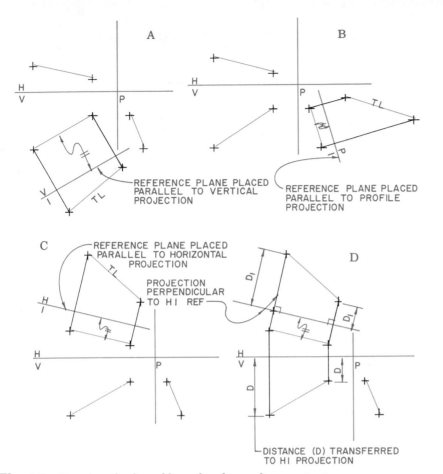

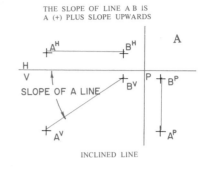

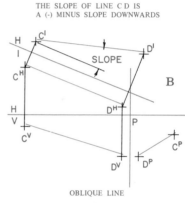

Fig. 8-9. True length of an oblique line by auxiliary projection.

Fig. 8-10. Slope of a line in the vertical (elevation) projection.

azimuth. The azimuth is always measured off the north-south line in the horizontal plane with clockwise dimensioning, as shown in Fig. 8-11 at C.

Grade

Incline, or *grade*, is measured as a percentage. Figure 8-12 shows the scale for constructing a highway with a +12 percent grade. The grade rises 12 ft (3.6 m) in every 100 ft (30 m) of horizontal distance.

◼ LINES IN SPACE

If two lines intersect, they will have at least one point in common. Figure 8-13 shows the alignment needed to

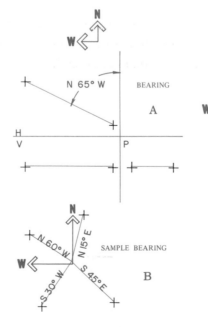

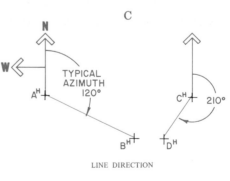

Fig. 8-11. (A) Bearing is identified. (B) Typical readings. (C) Azimuth readings related to due north.

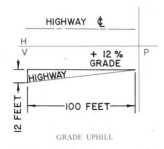

Fig. 8-12. Grade is measured in the vertical projection.

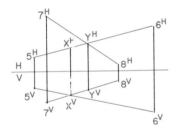

Fig. 8-13. Intersecting lines with aligned points.

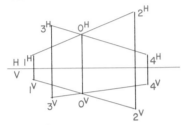

Fig. 8-14. Lines in space examined for intersection.

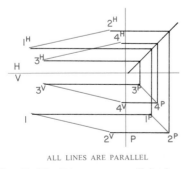

Fig. 8-15. Lines are parallel when all three of the reference projections are parallel.

check the point of intersection of two straight lines. In Fig. 8-14, the points of intersection in the H and V projection are not aligned. Thus, the intersection is incomplete. How would the intersection look if completed? (Note that the cross symbol for locating points will no longer be used in the figures.)

Figure 8-15 shows the relationship of parallel lines in a three-view study. All line projections are parallel if they appear parallel in all three reference planes. Note that the lines in Fig. 8-16 seem parallel in the front

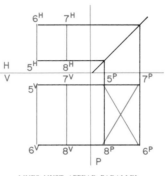

LINES MUST APPEAR PARALLEL IN ALL THREE VIEWS TO BE PARALLEL

Fig. 8-16. Lines in space examined for parallel relationship.

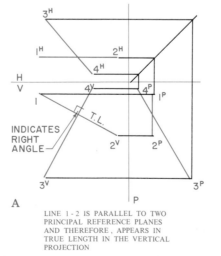

A

LINE 1-2 IS PARALLEL TO TWO PRINCIPAL REFERENCE PLANES AND THEREFORE, APPEARS IN TRUE LENGTH IN THE VERTICAL PROJECTION

and top views but are not parallel in the side view.

Perpendicular lines are examined in Fig. 8-17 at A and B. To find out if two lines are perpendicular, and thus have a right angle between them, first find the true length of one line. This lets you know if the angle between the lines is actually a right angle. Note that the true length at A indicates that one line is parallel to a main plane of projection. The oblique lines at B are examined in an auxiliary projection to find true length and the right angle.

PLANES

If a line moves away from a fixed place, its path forms a plane. In drawings, planes are thought of as having no thickness. They can also be extended as far as desired. A plane can be shown or determined by intersecting lines, two parallel lines, a line and a point, three points, or a triangle.

BASIC PLANES

There are three basic planes. What kind a plane is depends on how it

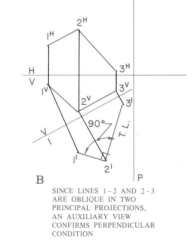

B

SINCE LINES 1-2 AND 2-3 ARE OBLIQUE IN TWO PRINCIPAL PROJECTIONS, AN AUXILIARY VIEW CONFIRMS PERPENDICULAR CONDITION

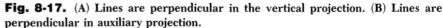

Fig. 8-17. (A) Lines are perpendicular in the vertical projection. (B) Lines are perpendicular in auxiliary projection.

relates to the three main reference planes. Plane 1 is perpendicular to two of the reference planes and parallel to the third. Plane 2 is perpendicular to one reference plane and inclined to the other two. Plane 3 is inclined to all three reference planes.

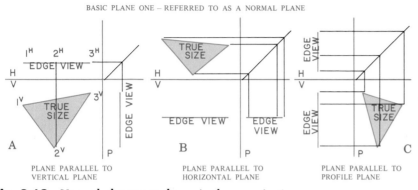

BASIC PLANE ONE – REFERRED TO AS A NORMAL PLANE

Fig. 8-18. Normal planes are shown in three projections.

BASIC PLANE TWO – INCLINED PLANE

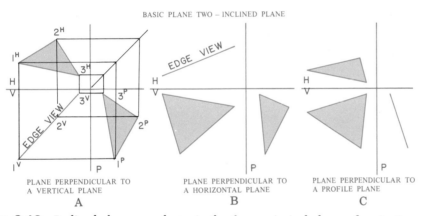

Fig. 8-19. Inclined planes are shown in the three principal planes of projection.

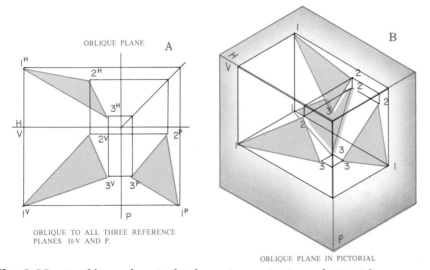

Fig. 8-20. An oblique plane in the three-view projection and pictorial.

Basic Plane 1

Sometimes this plane is called a *normal plane*. Three similar examples are shown in Fig. 8-18 at A, B, and C. Two of the main reference planes in each example show the edgewise view (a line) of the plane. At A, the plane is parallel to the vertical reference plane and perpendicular to the horizontal and profile planes. At B, the plane is parallel to the horizontal reference plane and perpendicular to the vertical and profile planes. At C, the plane is parallel to the profile reference plane and perpendicular to the vertical and horizontal planes.

Basic Plane 2

This type of plane is also called the *inclined plane*. Three similar examples are shown in Fig. 8-19 at A, B, and C. In one of the main reference planes, the plane shows as a line (edge view). Thus, it is perpendicular to that plane. The other two reference planes show the plane as a foreshortened surface. At A, the inclined plane is perpendicular to the vertical reference plane. It is inclined to the horizontal and profile planes, where it is foreshortened. At B, the inclined plane is perpendicular to the horizontal reference plane, where it shows as a line. The other two reference planes show the plane foreshortened. At C, the inclined plane is perpendicular to the profile reference plane, where it shows as a line. The plane shows as a foreshortened surface in the other two reference planes.

Basic Plane 3

This third type of plane is also called an *oblique plane*. An example is shown in Fig. 8-20 at A. The oblique plane is not perpendicular to any of the three main reference planes. It therefore cannot be parallel to any one of the three planes. Thus, it shows as a foreshortened plane in each of the three regular views. Figure 8-20 at B shows the same oblique plane in a pictorial rendering.

■ A POINT ON A LINE

In Fig. 8-21 at A, the line AB on the vertical plane has a point X. To place the point on the line in the other two reference planes, project construction lines perpendicular to the folding lines, as at B. The construction lines are projected across to $A^H B^H$ and $A^P B^P$ to locate point X in the horizontal and profile projections. Straight lines can be extended to new points on either end, as needed, to solve problems (Fig. 8-22). A point may seem to be on a line in one view. However, another view may show that it is really in front, on top, or in back of the line, as in Fig. 8-23 at A, B, and C.

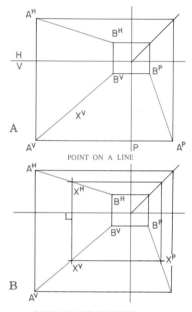

A POINT ON A LINE

B POINT ON LINE PROJECTED PERPENDICULAR TO FOLD LINES

Fig. 8-21. A point located on a line.

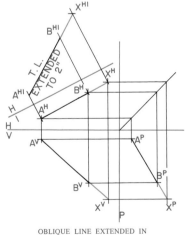

OBLIQUE LINE EXTENDED IN HORIZONTAL AUXILIARY

Fig. 8-22. Straight lines may be extended.

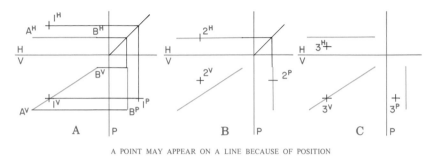

A POINT MAY APPEAR ON A LINE BECAUSE OF POSITION

Fig. 8-23. Defining a point-line relationship.

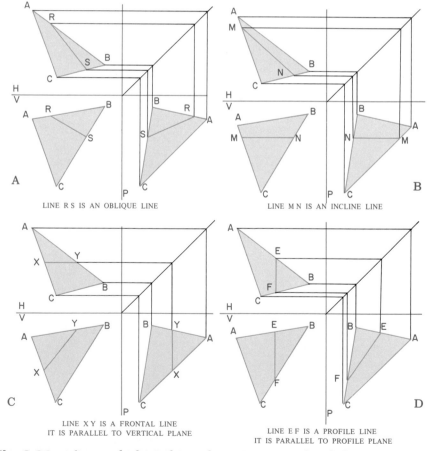

A LINE R S IS AN OBLIQUE LINE

B LINE M N IS AN INCLINE LINE

C LINE X Y IS A FRONTAL LINE IT IS PARALLEL TO VERTICAL PLANE

D LINE E F IS A PROFILE LINE IT IS PARALLEL TO PROFILE PLANE

Fig. 8-24. A line can be located in a plane. Four types identified.

A LINE IN A PLANE

A line lies in a plane if it intersects two lines of a plane. It also lies in a plane if it intersects one line and is parallel to another line of that plane. Figure 8-24 at A, B, C, and D shows how lines can be added to planes. At A, the line *RS* must be a part of the plane *ABC*, since *R* is on line *AB* and *S* is on *BC*. You know that line *RS* is an oblique line because it is not parallel to a main plane of projection and is clearly not perpendicular to the reference planes.

At B, a horizontal line *MN* is constructed in the vertical projection of plane *ABC*. This line is called a *level line*. Projecting *MN* to the right lines in the horizontal projection shows that it is an inclined line. The top view will show the true length.

At C, a line *XY* is constructed parallel to the HV reference line in the horizontal reference plane. Projected into the vertical plane, it shows as an inclined line. It will be in true length. This line is called a *frontal line*, since it is parallel to the vertical plane.

At D, a vertical line *EF* is constructed within plane *ABC*. It is parallel to the profile reference plane. The line *EF* projected to the profile reference shows in true length. The line *EF* is called a *profile line*.

LOCATING A POINT IN A PLANE

A point can be located in a plane by adding a line containing the point to the plane. Figure 8-25 at A shows that a point *O* appears within the plane *ABC*. At Fig. 8-25 at B, the line *AX* containing point *O* is projected. The line *AX* at Fig. 8-25 at C is projected to *ABC* in the horizontal reference plane. Point *O* is located on the line by drawing a ver-

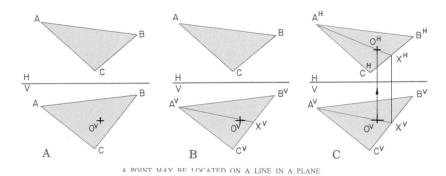

A POINT MAY BE LOCATED ON A LINE IN A PLANE

Fig. 8-25. Locating a point in a plane.

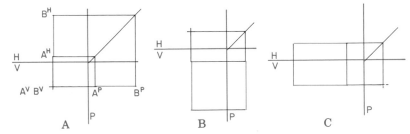

A NORMAL LINE HAS A POINT PROJECTION WHEN PERPENDICULAR TO REFERENCE PLANE

Fig. 8-26. A point view of a normal line in three positions.

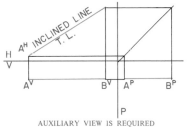

AUXILIARY VIEW IS REQUIRED TO FIND A POINT PROJECTION

Fig. 8-27. A point projection of an inclined line can be found in an auxiliary projection.

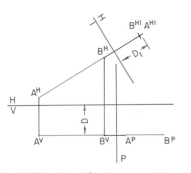

Fig. 8-28. Transferring a point projection of an inclined line.

tical projection to line *AX* in the horizontal reference plane.

A POINT VIEW OF A LINE

If a line is perpendicular to a reference plane, it will project as a point on that plane. In Fig. 8-26 at A, the line *AB* is parallel to two main reference planes. It therefore shows as a point in the third vertical reference plane. At B and C, the same conditions exist. The line projects as a point in the horizontal plane, at B, and the profile plane, at C.

When a line is parallel to only one main reference plane and inclined to the other two, as in Fig. 8-27, the point is projected by auxiliary projection. As shown in Fig. 8-28, a reference plane is placed perpendicular to the inclined line at a chosen distance. It is labeled H/1. The distance *D* is transferred as shown for a verti-

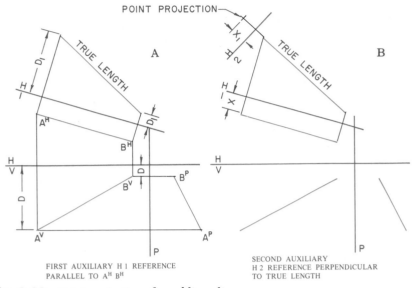

FIRST AUXILIARY H 1 REFERENCE
PARALLEL TO A^H B^H

SECOND AUXILIARY
H 2 REFERENCE PERPENDICULAR
TO TRUE LENGTH

Fig. 8-29. Point projection of an oblique line.

cal or a horizontal auxiliary projection.

If a line seems inclined in all three reference planes (an oblique line), the point can be projected by using two auxiliary projections. As shown in Fig. 8-29 at A, set up the first auxiliary reference plane parallel to the oblique inclined line. Then find the true length. The second auxiliary reference plane is placed perpendicular to the true-length line of the first auxiliary. In Fig. 8-29 at B, the point

projection is located by transferring the distance X.

◼ PARALLEL LINES

Point projection is one way to show the true distance between two parallel lines. In Fig. 8-30, the parallel lines MN and RS are considered oblique. Two auxiliary projections are needed to find the point projections. The first auxiliary reference plane V/1 is parallel to MN and RS. In it, lines MN and RS are shown true length. The second auxiliary reference plane V/2 is perpendicular to the true-length lines in the first auxiliary. The distance between the point projections of the lines is a true distance.

A second way of finding the shortest distance between two parallel lines is shown in Fig. 8-31. Lines AB and CD are thought of as parts of a plane. Connect the points A, B, C, and D to form a plane. Draw a horizontal line in the top view DX. Project the point X into the vertical view. Draw the line DX in the verti-

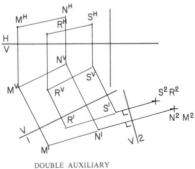

DOUBLE AUXILIARY
POINT-PROJECTION OF PARALLEL
OBLIQUE LINES TO FIND SHORTEST
DISTANCE BETWEEN LINES

Fig. 8-30. Distance between parallel lines through point projection.

cal plane. Draw the first reference plane V/1 perpendicular to DX in the vertical view. The edge view of the plane ABCD is found by transferring distances 1, 2, 3, and 4, as shown. In the second auxiliary V/2, the true lengths of AB and CD show. The plane formed is in true size. The true distance between the lines is measured perpendicularly from AB to CD, as shown.

◼ POINT-LINE RELATIONS

To find the shortest distance from a point to a line, project the line as a point. Point A and oblique line CD in Fig. 8-32 are projected into the first auxiliary projection H/1. In H/1, the true length of CD is labeled. The second auxiliary reference plane H/2 is placed perpendicular to line CD. Line CD is projected as a point in this plane. As shown, the distance between points in this projection is true length.

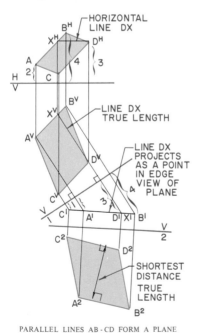

PARALLEL LINES AB-CD FORM A PLANE

Fig. 8-31. Distance between lines forming a plane.

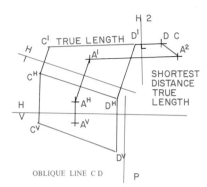

Fig. 8-32. Distance from a point to a line.

SHORTEST DISTANCE BETWEEN SKEW LINES

Skew lines *AB* and *CD* in Fig. 8-33 are not parallel and do not intersect. They are basically oblique. They show inclined in all main views. The shortest distance is a perpendicular line between the point view of one of the lines and the other line, as shown in Fig. 8-33 at B. First, find the true length of one of the lines, *CD*, in the first auxiliary. Do this by placing a

V/1 reference line parallel to line *CD*. The second auxiliary reference 1/2 is placed perpendicular to the true length of line *CD*. The point projection of line *CD* is found as shown. Extend line *AB* as shown. Construct a perpendicular line from the point projection of *CD* to line *AB*. The perpendicular line intersects *AB* extended at point *X*. You can then transfer the intersecting projection back to the first auxiliary, as shown on the extension of line *AB*.

THE TRUE SIZE OF AN INCLINED PLANE

In Fig. 8-34, the plane *ABC* shows as an edge view in the top view. The auxiliary reference plane H/1 is placed conveniently parallel to the edge view. Perpendicular projections are made. The distances *XY* and *Z* are transferred as shown. This lets you find the true size of the plane in the first auxiliary projection.

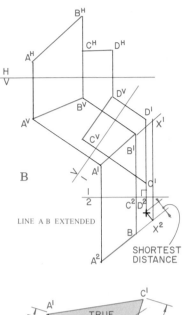

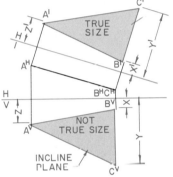

Fig. 8-34. True size of an inclined plane.

THE TRUE SIZE OF AN OBLIQUE PLANE

If the plane *ABC* in Fig. 8-35 at A is projected on a plane perpendicular to any line in the figure, it will show an edge view in the first auxiliary. In the top view, a line *BX* is drawn parallel to the reference plane. Reference line V/1 is placed perpendicular to the front view of *BX*. The front view of *BX* is now projected into a point projection in the first auxiliary. The point projection is in the edge view of plane *ABC*, as shown. In Fig.

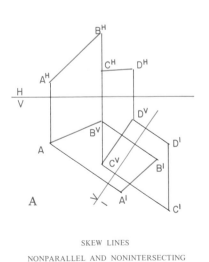

SKEW LINES
NONPARALLEL AND NONINTERSECTING

Fig. 8-33. Distance between skew lines.

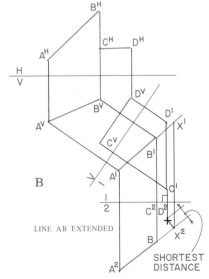

LINE AB EXTENDED

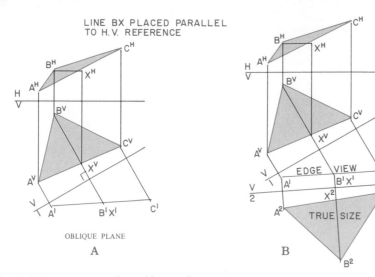

Fig. 8-35. True size of an oblique plane.

8-35 at B, the second auxiliary reference line V/2 is placed parallel to the edge view. The projection of plane *ABC* in the secondary auxiliary shows the true size.

TRUE ANGLES BETWEEN LINES

If two lines show in true length, the angle between them will appear in its true value. In Fig. 8-36 at A, the two lines show as an inclined plane. This is because the vertical view shows that lines *AB* and *AC* coincide (lie in a single line). The V/1 auxiliary

reference is placed parallel to the two lines in the vertical view. The two lines are drawn true length in the auxiliary view. The first auxiliary shows the true angle between the lines.

The oblique condition of lines *MN* and *NS* does not show in an edgewise view. Figure 8-36 at B shows how to solve the problem with two auxiliary planes. The first reference plane is perpendicular to the plane formed by the lines *NA*. The second reference plane is parallel to the first auxiliary view. That is, it is parallel to the edge view of the lines *MN* and *NS*. The second auxiliary view shows

MN and *NS* in true length. It also shows the true angle between them.

■ PIERCING POINTS

If a line is not in, or parallel to, a plane, it must intersect the plane. The point of intersection, which is common to the plane and the line, is called a *piercing point*. The line can be thought of as piercing the plane.

■ EDGE-VIEW SYSTEM

The edge view of a plane contains all the points in the plane. Therefore, a line crossing the edge view will show the point where the line pierces the plane. If a line lies in a plane or is parallel to it, it cannot intersect the plane. In Fig. 8-37, the straight line is neither in the plane nor parallel to the plane. It intersects the plane at a point common to both. The edge view of plane *ABC* is shown in the vertical plane. The line *RS* in the horizontal plane pierces the plane at point *P* when projected to the vertical plane. If you look at line *RS* closely in the vertical projection, you

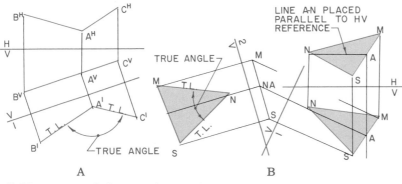

Fig. 8-36. True angle between lines with two auxiliary projections.

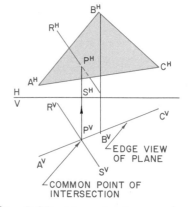

Fig. 8-37. Line piercing a plane. Edge-view system.

will see that element *A* of the triangle is lower than point *R* of the piercing line. Therefore, the dashed portion of line *RS* is invisible.

You can solve the problem of the piercing point of a line *MN* intersecting an oblique plane *ABC* with the edge-view system, as shown in Fig. 8-38. The first auxiliary will determine the edge view of plane *ABC*. The piercing point *P* of line *MN* can then be carried back to the vertical and horizontal with the projections shown by arrows.

CUTTING-PLANE SYSTEM

When a line *RS* intersects an oblique plane *ABC*, a cutting plane containing the line will determine other working points, as shown in Fig. 8-39. The cutting plane seems to intersect the triangle in the front view at points 3 and 4. Points 3 and 4 are projected to the top view on lines *AC* and *BC*. When the points are connected across the plane, the piercing point *P* is found. Point *P* is projected to the front view. Since you now know what parts of line *RS* are not visible, you can add hidden lines. Figure 8-40 shows the cutting plane

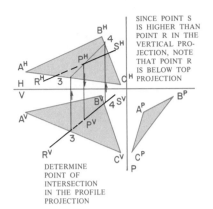

Fig. 8-39. Piercing-pointcutting-plane method.

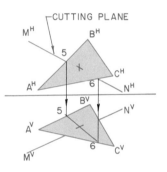

Fig. 8-40. Piercing point developed.

in the horizontal projection intersecting line *AB* at point 5 and line *AC* at point 6. Can the visibility of line *MN* be determined by inspection?

ANGLE BETWEEN INTERSECTING PLANES

A *dihedral angle* is formed when planes intersect. Two planes that intersect have a straight line in common. The dihedral angle formed can only be measured perpendicular to the line of intersection. When a point view of the line of intersection is found, the planes are shown as edges. The angle between the planes will be true in size. In Fig. 8-41 at A, the planes *ABC* and *ACD* have a common line of intersection, *AC*. The first auxiliary projection H/1 in Fig. 8-41, at B, is taken in the horizontal plane of the line *AC*. The H/1 auxiliary lets *AC* be drawn in true length (TL). In the second auxiliary, the point projection of the true-length line *AC* also shows the two given planes as edges. The true angle is measured in the second auxiliary, as shown in Fig. 8-41 at B.

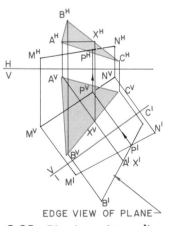

Fig. 8-38. Piercing point—a line with an oblique plane.

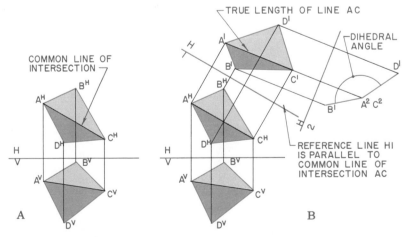

Fig. 8-41. Angle formed between intersecting planes.

ANGLE BETWEEN A LINE AND A PLANE

A view that shows a plane in the edge view along with a true-length line will also show the true angle in between.

Plane Method

In Fig. 8-42, oblique line *XY* intersects the oblique plane *ABC*. First find the edge view of the oblique plane *ABC*. The first auxiliary is found by placing H/1 perpendicular to a true-length line in the horizontal projection. Note that line *XY* is not shown true length in this first auxiliary view. After projecting the edge view in H/1, place the reference plane H/2 parallel to plane *ABC*. In the second auxiliary, the true size of plane *ABC* is plotted. In the third auxiliary, 2/3, the reference line is placed parallel to line *XY*. The new edge view of the plane and the true-length line form the true angle.

Line Method

In Fig. 8-43, oblique plane *ABC* and the oblique line *XY* intersect in the top view. The first auxiliary V/1 is placed parallel to the line *XY*. The

true length of *XY* is thus found. In the second auxiliary, a reference V/2 is placed perpendicular to the true length of line *XY*. The point projection of *XY* is thus found. Finally, in the third auxiliary, the 2/3 reference line is set perpendicular to the true length of B^2O^2. Plane *ABC* is thereby shown as an edge view. The intersection of the line and plane shows the true size of the angle.

REVOLUTION

We read about the rule of revolution in Chapter 7. Basic problems can often be solved by changing the position of an object so that the new view can show needed information. Descriptive geometry problems can be solved by revolving an object.

TRUE SIZE OF AN OBLIQUE PLANE BY REVOLUTION

In Fig. 8-44, it is clear that the plane *ABC* is inclined to all main planes. A line *AX* is placed in the plane *ABC* parallel to the horizontal reference line. Line *AX* projected to the top view appears as an inclined line in

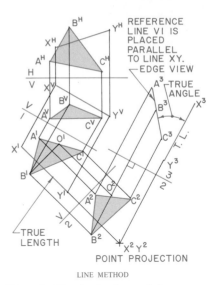
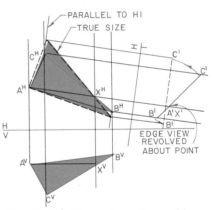

Fig. 8-43. Angle formed between a line and a plane (line method).

Fig. 8-44. True size of an oblique plane by revolution.

true length. The first auxiliary projected in the horizontal has H/1 reference perpendicular to line *AX*. The point view of *AX* is found in the first auxiliary. Line *ABC* appears as an edge view, with *X* within. At point *AX*, the edge view B'A'C' is revolved (dashed line) so as to appear parallel to reference line H/1. Projecting points B and C to the horizontal projection allows the true size of plane *ABC* to be drawn in the top view, as shown.

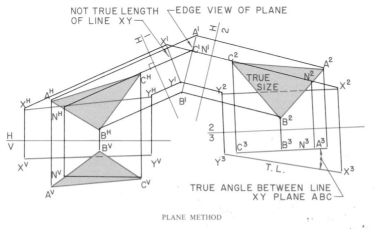

Fig. 8-42. Angle formed between a line and a plane (plane method).

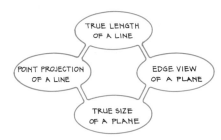

Fig. 8-45. Four basic approaches for solving problems.

Almost all problems in descriptive geometry can be worked out by using auxiliary planes. You can solve problems (Fig. 8-45) by knowing how to find

1. The true length of a line
2. The point projection of a line
3. The edge view of a plane
4. The true size of a plane figure

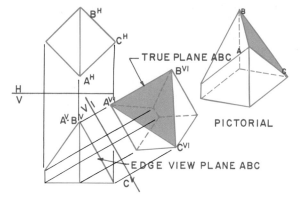

Fig. 8-46. Basic problems can be solved with descriptive geometry.

■ CONCLUSION

In Fig. 8-46, the basics of descriptive geometry can be applied to one face of a geometric solid. It should be clear that Chaps. 7 and 8 are closely related. Problems are easily resolved through analysis of point-line and plane relationships.

VOCABULARY

1. Analyze
2. Reference plane
3. Irregular
4. Plot
5. Normal line
6. Inclined line
7. Oblique line
8. Slope
9. Bearing
10. Azimuth
11. Grade
12. Foreshortened
13. Oblique plane
14. Level line
15. Frontal line
16. Skew lines
17. Transferred
18. Coincide
19. Dihedral angle
20. Edge view

REVIEW

1. Any given shape may be formed by which three basic geometric elements?

2. What is the basic shape of most structures designed by people? Why is this form common?

3. Do you believe that basic descriptive geometry is one of the designer's ways of thinking and of solving problems? Why?

4. How is a point located in a drawing?

5. What are the three characteristics of a line?

6. Name the three basic lines used in descriptive geometry.

7. How does a normal line relate to the three main planes of projection?

8. How does an inclined line relate to the three main planes of projection?

9. In how many of the main planes

of projection will an inclined line show in true length?

10. How does an oblique line relate to the three main planes of projection?

11. How many auxiliary projections are needed to find the point projection of an oblique line?

12. Name the three basic planes.

13. What are the major characteristics of a plane?

14. How can you find whether two lines really intersect?

15. What is the difference between an inclined and an oblique plane?

16. How can a line be added to a plane in space?

17. How can you determine the distance between two parallel inclined lines?

18. How is the auxiliary reference plane placed in finding the true size of an inclined line?

19. How can you find the true angle between intersecting lines?

20. Why is Chap. 7 closely related to Chap. 8?

Problems

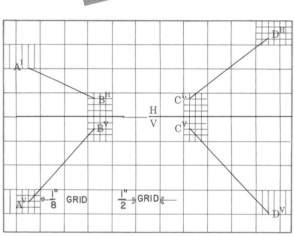

Fig. 8-47. Find the true length of line *AB* and the true length and slope of line *CD*.

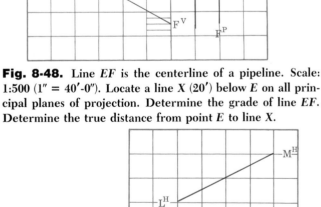

Fig. 8-48. Line *EF* is the centerline of a pipeline. Scale: 1:500 (1″ = 40′-0″). Locate a line *X* (20′) below *E* on all principal planes of projection. Determine the grade of line *EF*. Determine the true distance from point *E* to line *X*.

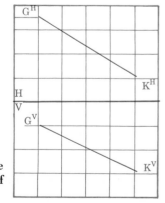

Fig. 8-49. Determine the point projection of line *GK*.

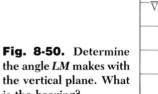
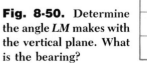

Fig. 8-50. Determine the angle *LM* makes with the vertical plane. What is the bearing?

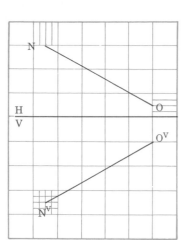

Fig. 8-51. Determine the slope of line *NO*. Extend *NO* to measure 2¼″ (56-mm) long. Draw all three views.

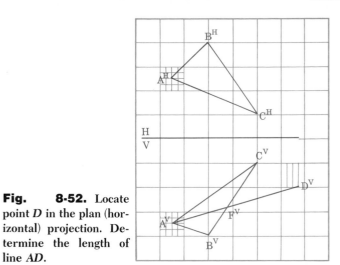

Fig. 8-52. Locate point *D* in the plan (horizontal) projection. Determine the length of line *AD*.

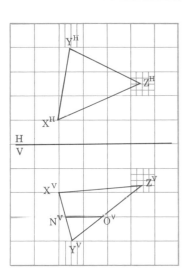

Fig. 8-53. What is the bearing of line *NO* located on plane *XYZ*? Determine the true size of plane *XYZ*.

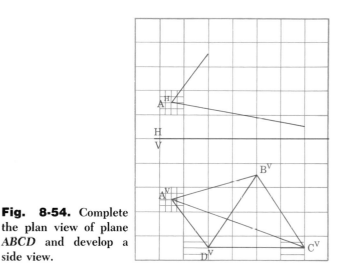

Fig. 8-54. Complete the plan view of plane *ABCD* and develop a side view.

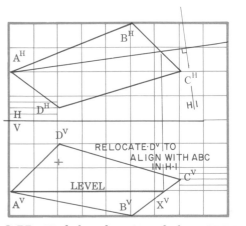

Fig. 8-55. Find the edge view of plane *ABCD* and determine the angle it makes with the horizontal plane.

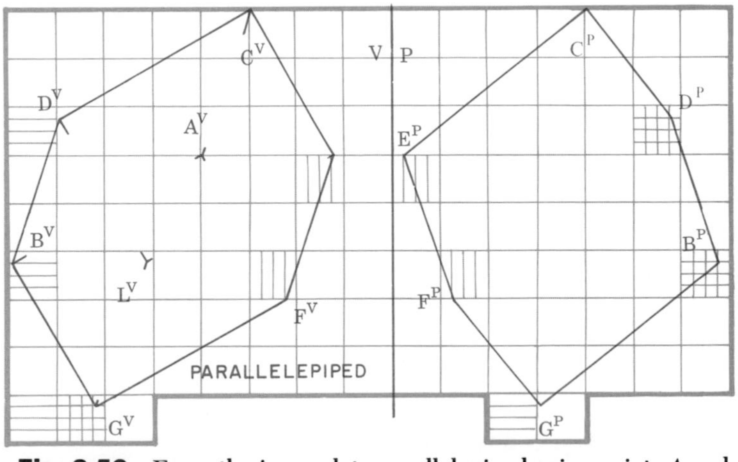

Fig. 8-56. Form the incomplete parallelepiped using points *A* and *L*. Determine the proper visibility in all three views.

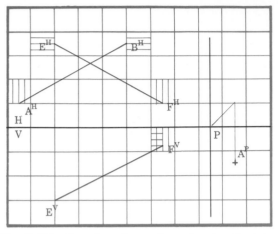

Fig. 8-57. Complete three views showing the intersection of *AB* and *EF*.

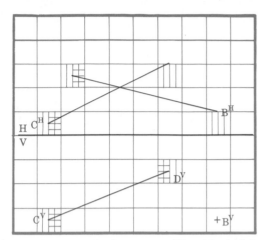

Fig. 8-58. Draw the front view of line *AB* which intersects line *CD*. What is the distance from *C* to *A*?

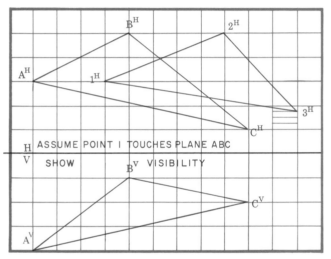

Fig. 8-59. Create a location for plane 1-2-3 in the vertical plane.

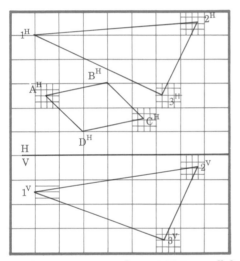

Fig. 8-60. Construct a plane *ABCD* parallel to plane 1-2-3.

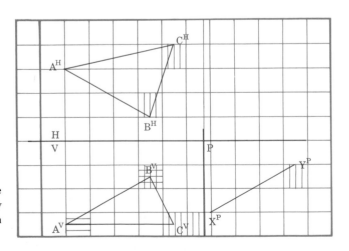

Fig. 8-61. Determine the true size of oblique plane *ABC*. Draw line *XY* parallel to plane *ABC* in the plan view.

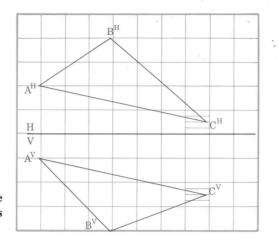

Fig. 8-62. Determine the true size of plane *ABC* and label its slope.

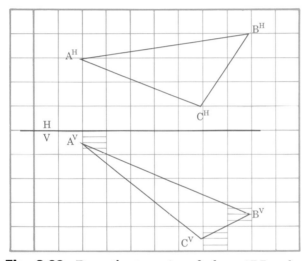

Fig. 8-63. Draw the true size of plane *ABC* and dimension the three angles of the plane.

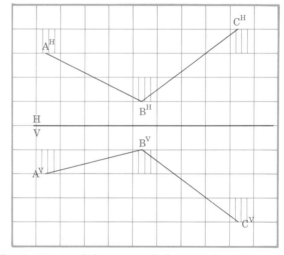

Fig. 8-64. Find the true angle between lines *AB* and *BC*.

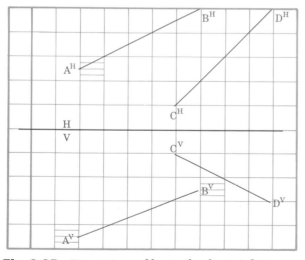

Fig. 8-65. Determine and locate the shortest distance between skew lines *AB* and *CD*.

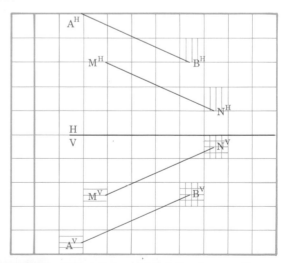

Fig. 8-66. Determine and label the shortest distance between parallel lines *AB* and *MN*.

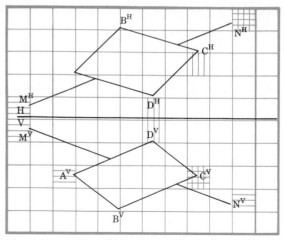

Fig. 8-67. Draw three views of the line and plane and determine if line *MN* pierces the plane.

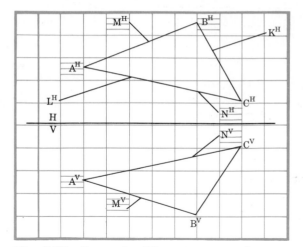

Fig. 8-68. Determine if line *MN* pierces plane *ABC*. Locate line *KL* so that it pierces the center of plane *ABC*.

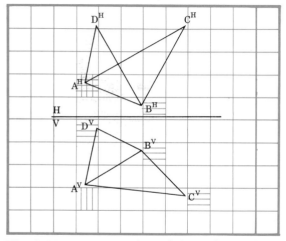

Fig. 8-69. Determine the visibility and angle formed between planes *ABC* and *ABD*.

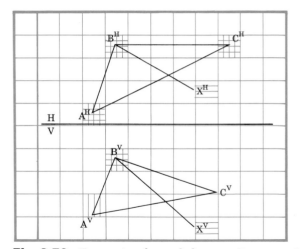

Fig. 8-70. Determine the angle between line *BX* and plane *ABC*.

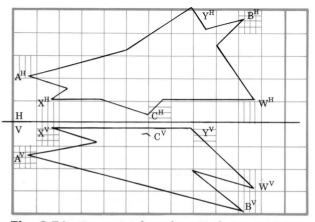

Fig. 8-71. Determine how the two planes intersect and show proper visibility.

Fig. 8-72. After examining Figs. 8-37 through 8-40 on piercing points, create a problem of your own design in which an oblique plane is pierced by one inclined line and one oblique line. Determine your own dimensions or use a grid paper while showing construction using the cutting plane.

Fig. 8-73. Study Figs. 8-41 through 8-43 carefully. Create two intersecting oblique planes of your own design, and determine the angle formed. Develop your own sizes through experimentation.

9 Sectional Views and Conventions

SECTIONAL VIEWS

Technical drawings must show all parts of an object, including the insides and other parts not easily seen. These hidden details can be drawn with *hidden lines* made with short dashes. But this method works well only if the hidden part has a fairly simple shape. If the shape is complicated, dashed lines may show it poorly. They can also be confusing, as shown in Fig. 9-1. Instead of dashed lines, a special view called a *section* or *sectional view* can be drawn. A sectional view shows an object as if part of it were cut away to expose the insides (Fig. 9-2).

To draw a sectional view, imagine that a wide-blade knife has cut through the object. Call this knife a *cutting plane*. Then imagine that everything in front of the plane has been taken away, so that the cut surface and whatever is inside can be seen (Fig. 9-2). On a normal view, you can show where a cutting plane will pass by drawing a special line, the *cutting-plane line* (Fig. 9-3). On a sectional view, show the cut sur-

face by marking it with evenly spaced thin lines. This is called *section lining* or *crosshatching*. The basic section lining pattern described above is not the only pattern. However, it is the one that is used in most cases. You can use it for objects made of any material, so it is called the

general-purpose symbol. It is used especially when more than one kind of material need not be shown, such as in a drawing of a single part. However, special section lining patterns (also called *symbols*) can be used to show what materials are to be used. The American National Standards

Fig. 9-1. Pictorial view of object and the three normal views.

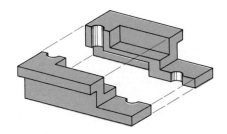

Fig. 9-2. Object cut to show inside details. The front of the object has been removed.

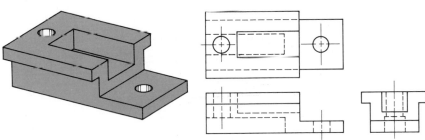

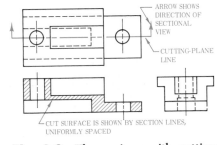

ARROW SHOWS DIRECTION OF SECTIONAL VIEW

CUTTING-PLANE LINE

CUT SURFACE IS SHOWN BY SECTION LINES, UNIFORMLY SPACED

Fig. 9-3. Three views with cutting-plane line and section lining.

provide for many symbols to stand for different materials (Fig. 9-4). Under this system, the general-purpose symbol can also mean that an object is made of cast iron. These special symbols are most useful in a drawing showing several objects made of different materials. These would be used, for example, in an *assembly drawing* (a drawing showing how different parts fit together). However, materials to be used should not be shown with just these symbols. The exact materials needed are *specified* in a note or in a list of materials.

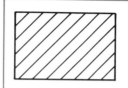

CAST IRON AND MALLE-ABLE IRON. ALSO FOR GENERAL USE FOR ALL MATERIALS

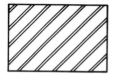

STEEL

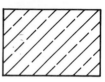

BRONZE, BRASS, COPPER, AND COMPOSITIONS

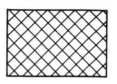

WHITE METAL, ZINC, LEAD, BABBITT, AND ALLOYS

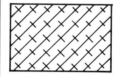

MAGNESIUM, ALUMINUM, AND ALUMINUM ALLOYS

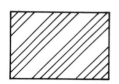

RUBBER, PLASTIC ELECTRICAL INSULA-TION

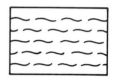

CORK, FELT, FABRIC, LEATHER, FIBER

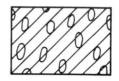

SOUND INSULATION

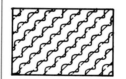

THERMAL INSULATION

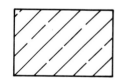

FIREBRICK AND REFRACTORY MATERIAL

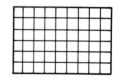

ELECTRIC WINDINGS, ELECTROMAGNETS, RESISTANCE, ETC.

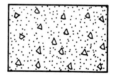

CONCRETE

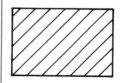

BRICK AND STONE MASONRY

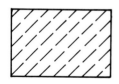

MARBLE, SLATE, GLASS, PORCELAIN, ETC.

EARTH

ROCK

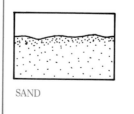

SAND

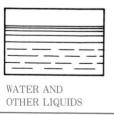

WATER AND OTHER LIQUIDS

WOOD ACROSS GRAIN WOOD WITH GRAIN

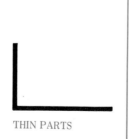

THIN PARTS

Fig. 9-4. American National Standard symbols for section lining. (ANSI Y14.2.)

SPACING OF SECTION LINES

Section lines are drawn together or far apart, depending upon how much space must be filled (Fig. 9-5). According to the American National Standards, section lines can be spaced from about ⅟₃₂ (.03) in. (1.0

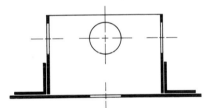

Fig. 9-5. Section lines are spaced by eye. Their distance apart varies according to the size of the space to be sectioned.

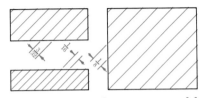

Fig. 9-6. Thin section, blacked in.

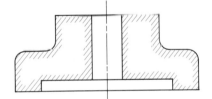

Fig. 9-7. Outline sectioning.

Fig. 9-8. Cut surface may by grayed.

Fig. 9-9. Cut surface may have grayed outline.

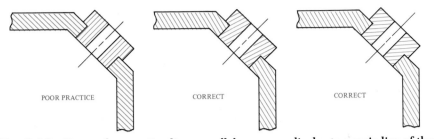

POOR PRACTICE   CORRECT   CORRECT

Fig. 9-10. Do not draw section lines parallel or perpendicular to a main line of the view.

mm) to ⅛ (.12) in. (3.0 mm) apart. However, they must be evenly spaced and they are usually slanted at a 45° angle. The drawing will be neater if you *do not* space section lines extremely close together. This will also save time. In most cases, the lines will look best spaced about ³⁄₃₂ (.09) in. (2 mm) apart. The distance between section lines need not be measured; they may be spaced by eye. If the area to cover is large, space the lines farther apart, up to ⅛ (.12) in. (3 mm) or more. If the area is small, space the lines closer together, down to ⅟₁₆ (.06) in. (1.5 mm) or less apart. If the area is very small, as for thin plates, sheets, and structural shapes, *blacked-in* (solid black) sections may be used. These are shown in Fig. 9-6. Note the white space between the parts.

When you are drawing a large sectioned area, one way to save time is to use outline sectioning. This method is shown in Fig. 9-7. Drafters who use it often draw the section lines freehand and spaced widely apart. You can also gray the sectioned area (Fig. 9-8), gray only along its outline (Fig. 9-9), or rub pencil dust over it. Apply a *fixative* (sealing substance) to prevent smudging.

Never draw section lines parallel to or at right angles to an important visible line (Fig. 9-10). However,

they may be drawn at any other suitable angle and spaced at any width. Section lines may be drawn at different angles and spaced to identify different sectioned parts.

THE CUTTING-PLANE LINE

The cutting-plane line represents the cutting plane as viewed from an edge (Fig. 9-11). This line may be drawn in either of two ways approved by the American National Standards (Fig. 9-12). The first form is more commonly used. The second shows up well on complicated drawings. At each end of the line, draw a short arrow to show the direction for looking at the section. Make the arrows at right angles to the line. Place bold capital letters at the corners as shown, if needed for reference to the section.

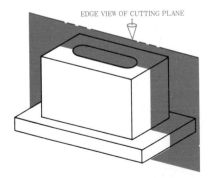

EDGE VIEW OF CUTTING PLANE

Fig. 9-11. The cutting-plane line represents the edge view of the cutting plane.

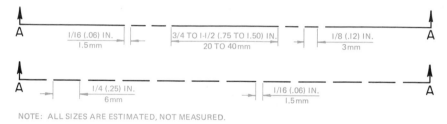

NOTE: ALL SIZES ARE ESTIMATED, NOT MEASURED.

Fig. 9-12. Cutting-plane lines.

A cutting-plane line is not needed when it is clear that the section is taken along an object's main centerline or at some other obvious place (Fig. 9-13).

SECTIONS THROUGH ASSEMBLED PIECES

If a drawing shows more than one piece in section, draw the section lines in a different direction on each piece (Figs. 9-14 and 9-15). Remember, however, that any piece can, in turn, show several cut surfaces. Make sure that all the cut surfaces of any one piece have section lines in the same direction, as in Fig. 9-15.

FULL SECTIONS

A *full-sectional view* shows an object as if it were cut completely apart from one end or side to another, as in Fig. 9-16. Such views are usually just called *sections*. The two most common types of full sections are *vertical* and *profile* (Figs. 9-17 and 9-18).

OFFSET SECTIONS

In sections, the cutting plane is usually taken straight through the object. But it can also be *offset* (shifted) at one or more places in order to show some detail or to miss some part. An offset section is shown in Fig. 9-19. Here a cutting plane is offset to pass through the two bolt holes. If the plane were not offset, the bolt holes would not show in the sectional view. Indicate an offset section by drawing it on the cutting-

plane line. No indication is given on the sectional view. If reference letters are needed on the cutting-plane line, place them at the ends, opposite the arrowheads. Use capital letters.

HALF SECTIONS

A *half section* is one half of a full-sectional view. Remember, a full-sectional view makes an object look as if half of it has been cut away. A half-sectional view looks as if one quarter of the original object has been cut away. Imagine that two cutting planes at right angles to each other slice through the object to cut away one quarter of it, as in Fig. 9-20. The half-sectional view shows one half of the front view in section (Fig. 9-20E). Figure 9-20D shows the *exterior* (outside) of the object. Half sections are useful when you are drawing a *symmetrical* object (one whose halves are mirror images of

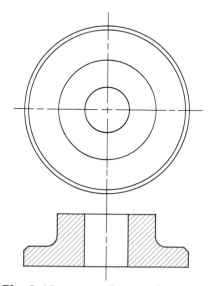

Fig. 9-13. A centerline may be used to represent a cutting-plane line.

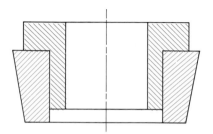

Fig. 9-14. Two pieces. Section lines in two directions.

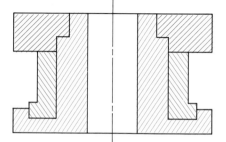

Fig. 9-15. Three pieces. Section lines in three directions.

A

PHOTOGRAPH OF BRACKET

B

BRACKET WITH CUTTING PLANE

C

FRONT OF BRACKET
MOVED AWAY TO
EXPOSE CUT SURFACE

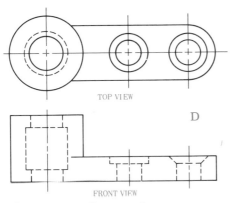

TOP VIEW

D

FRONT VIEW

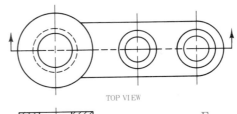

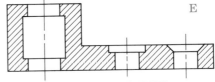

TOP VIEW

E

FRONT FULL-SECTIONAL VIEW

Fig. 9-16. Full-sectional view.

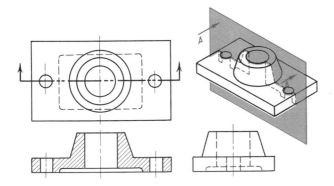

Fig. 9-17. Vertical section.

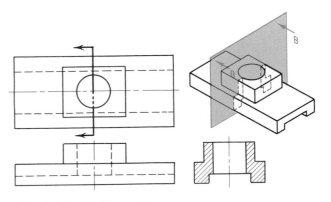

Fig. 9-18. Profile section.

PHOTOGRAPH OF BEARING FLANGE

FLANGE WITH CUTTING PLANE

FRONT OF FLANGE
MOVED AWAY TO EXPOSE
CUT SURFACE

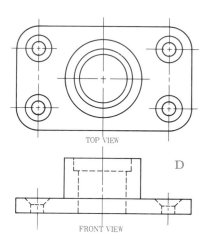

Fig. 9-19. An offset section.

each other). Both the inside and the outside can be shown in one view. Use a centerline where the exterior and half-sectional views meet, since the object is not actually cut. In the top view, show the complete object, since no part is actually removed. If the direction of viewing is needed, use only one arrow, as at E. In the top view, at E, the cutting-plane line could have been left out, for there is no doubt where the section is taken.

HIDDEN AND VISIBLE LINES ON SECTIONAL VIEWS

Do not draw hidden lines on sectional views, unless they are needed for dimensioning or for clearly describing the shape. In Fig. 9-21 at A,

a hub is clearly described using no hidden lines. Compare it with B.

On *sectional assembly drawings* (sectional views of how parts fit together), most hidden lines are generally omitted. This keeps the drawing from becoming cluttered and hard to read (Fig. 9-22). Sometimes a good way to avoid using hidden lines is to draw a half section or part section.

Normally, in a sectional view, include all the lines that would be visible on or beyond the plane of the section. In Fig. 9-23, for example, the section drawing at A correctly includes the numbered lines, which match the lines on the drawing at B. A drawing without these lines, as at C, would have little value. *Never* draw sectional views in this manner. Figure 9-24 is another example of

how wrong it is to leave out lines visible beyond the plane of the section.

BROKEN-OUT SECTIONS

A view with a *broken-out section* shows an object as it would look if a portion of it were cut partly away from the rest by a cutting plane and then "broken off" to reveal the cut surface and insides (Fig. 9-25). This view shows some inside detail without drawing a full or half section.

Note that a broken-out section is bounded by a *short break line* drawn freehand the same thickness as visible lines. Figure 9-26 shows two more examples of broken-out sections.

A

PHOTOGRAPH OF PACKING GLAND

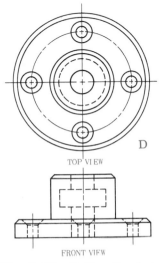

B

PACKING GLAND WITH CUTTING PLANE

C

ONE QUARTER MOVED AWAY
TO EXPOSE CUT SURFACE

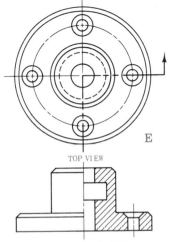

D

TOP VIEW

FRONT VIEW

E

TOP VIEW

HALF-SECTIONAL VIEW

Fig. 9-20. Half-sectional view.

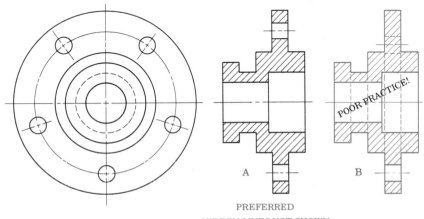

A

B   POOR PRACTICE!

PREFERRED

HIDDEN LINES NOT SHOWN

Fig. 9-21. Omit hidden lines when not needed for clearness
or dimensioning.

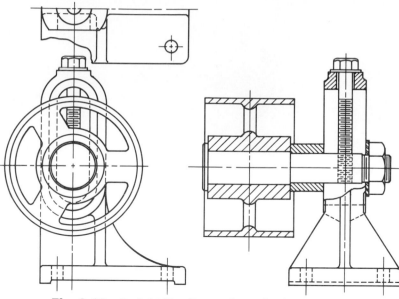

Fig. 9-22. Omit hidden lines to keep the drawing from becoming confusing.

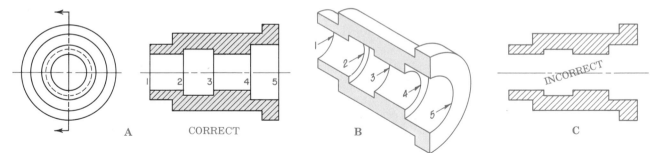

A   CORRECT   B   C

Fig. 9-23. Show all visible lines beyond the sectioned surface.

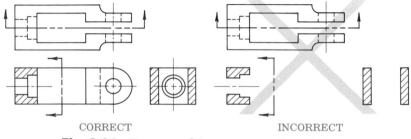

CORRECT   INCORRECT

Fig. 9-24. Correct and incorrect uses of visible lines beyond the plane of the section.

A

PHOTOGRAPH OF DOUBLE PACKING GLAND

B

GLAND WITH CUTTING PLANE

C

PART OF GLAND MOVED
AWAY TO EXPOSE CUT SURFACE

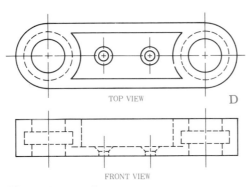

TOP VIEW D

FRONT VIEW

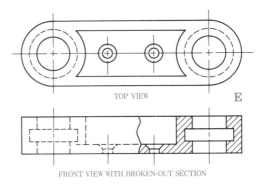

TOP VIEW E

FRONT VIEW WITH BROKEN-OUT SECTION

Fig. 9-25. Broken-out section.

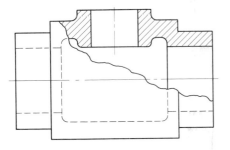

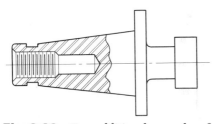

Fig. 9-26. Two additional examples of broken-out sections.

■ *REVOLVED SECTIONS*

Think of a cutting plane passing through a part of an object, as in Fig. 9-27. Now think of that cut surface as *revolved* (turned) 90°, so that its shape can be seen clearly (Fig. 9-28). The result is a *revolved* or *rotated section*.

If the part is long and thin and its shape in cross section is the same throughout, a revolved section may be used (Fig. 9-29). In such cases, the view may be shortened but the full length of the part must be given by a dimension. This lets you draw a large part with a revolved-sectional view in a short space.

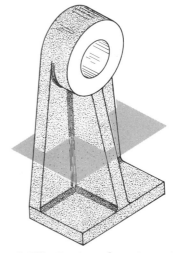

Fig. 9-27. Cutting plane in position for revolved section.

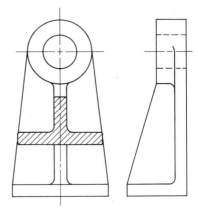

MOST COMMON METHOD

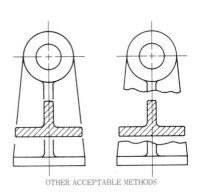

OTHER ACCEPTABLE METHODS

Fig. 9-28. Revolved section.

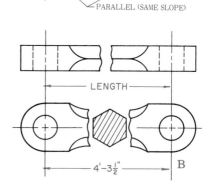

Fig. 9-29. Revolved sections in long parts.

■ *REMOVED SECTIONS*

When a sectional view is taken from its normal place on the view and moved somewhere else on the drawing sheet, the result is a *removed-sectional view*. It can also be called a *removed section*. Remember, however, the removed section will be easier to understand if it is positioned to look just as it would if it were in its normal place on the view. In other words, do not rotate it in just any direction. Fig. 9-30 shows right and wrong ways to position removed sections. Use bold letters to identify a removed section and its corresponding cutting plane on the regular view (Fig. 9-30).

A removed section can be a *sliced section* (the same as a revolved section). Or it can show some additional detail visible beyond the cutting

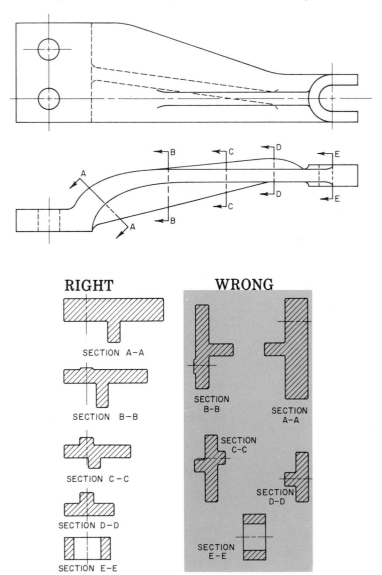

Fig. 9-30. Correct and incorrect positions of removed sections.

plane. It can be drawn to a larger scale to show details clearly or to fit in dimensions.

Besides removed-*sectional* views, removed *views* of the *exterior* (outside) of an object can be drawn. These can be made to the same scale or to a larger one. They can also be complete views or partial views.

AUXILIARY SECTIONS

When a cutting plane is passed through an object at an angle, as in Fig. 9-31 at A, the resulting sectional view, taken at the angle of that plane, is called an *auxiliary section*. It is drawn like any other auxiliary view (see Chap. 7).

Usually, on working drawings, only the auxiliary section is shown on the cut surface. However, if needed, any or all parts beyond the auxiliary cutting plane may be shown. In Fig. 9-31 at B, the auxiliary section contains one hidden line. There are also three incomplete views.

RIBS AND WEBS IN SECTION

Ribs and *webs* are thin, flat parts of an object used to brace or strengthen another part of the object. When a cutting plane passes through a rib or web *parallel to* (in line with) the flat side, section lining should not be drawn for that part, as in Fig. 9-32 at B. Instead, think of the plane passing just in front of the rib. A true section, as at A, would give the idea of a very heavy, solid piece. This would not appear to be a true description of the part.

If a cutting plane passes through a rib, a web, or any other thin, flat part at right angles to the flat side, draw in the section lining for that part. Figure 9-33 is an example.

ALTERNATE SECTION LINING

Alternate (or wide) section lining is a sectioning pattern made by leaving out every other section line. It can

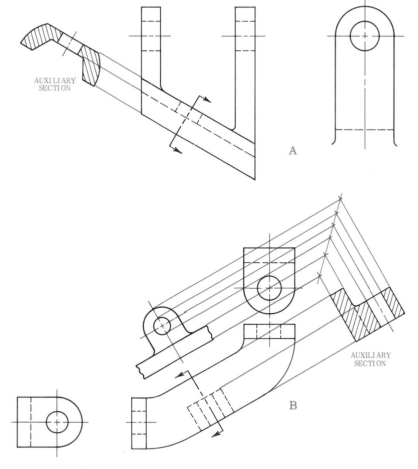

Fig. 9-31. Auxiliary section.

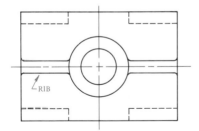

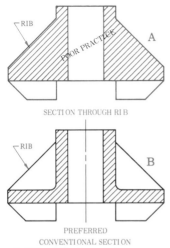

Fig. 9-32. Ribs in section.

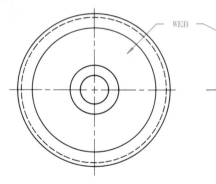

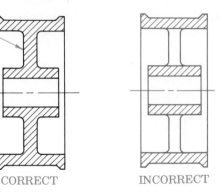

CORRECT · INCORRECT

Fig. 9-33. Web in section.

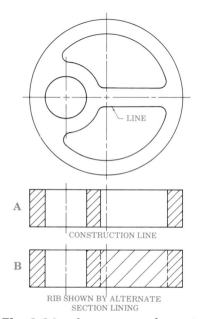

Fig. 9-34. Alternate, or wide, section lining.

out a rib would look exactly the same. The problem is solved, however, at B. Here, alternate section lining is used with hidden lines to show the extent of the rib.

Alternate section lines are useful to show ribs and other thin, flat pieces in one-view drawings of parts or in assembly drawings.

OTHER PARTS USUALLY NOT SECTIONED

Do not draw section lining on spokes and gear teeth when the cutting plane passes through them. Leave them as shown in Fig. 9-35. Do not draw section lining either on shafts, bolts, pins, rivets, or similar items when the cutting plane passes through them *lengthwise* (through the axis) as shown in Fig. 9-36. These objects are not sectioned because they have no inside details. Also, sectioning might give a wrong idea of the part. A drawing showing them in full is easier to read. It also takes less time to draw. However, when such parts are cut *across* the axis, they should be sectioned (Fig. 9-37). See

be used to indicate a rib or another flat part in a sectional view when that part otherwise would not show up clearly. In Fig. 9-34 at A, an eccentric piece (two or more circular shapes not using the same centerlines) is drawn in section. A rib is visible in the top view, but in the sectional view it is not shown by any section lining. As described above, this is standard practice with a flat part like a rib. But there are no visible lines to represent the rib either, because its top and bottom are both even with the surfaces they join. In fact, without the top view, you might not know that the rib was there. A drawing of an eccentric piece with-

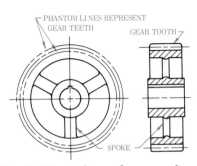

Fig. 9-35. Spokes and gear teeth not sectioned.

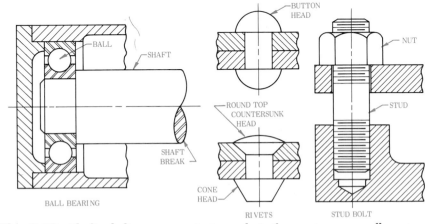

Fig. 9-36. Shafts, bolts, screws, rivets, and similar parts are usually not sectioned.

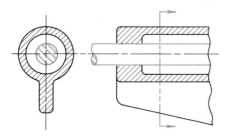

Fig. 9-37. A cross section.

the sectional assembly in Fig. 9-38 for names and drawings of a number of other items that are not sectioned.

PHANTOM (HIDDEN) SECTIONS

Use a *phantom*, or *hidden*, section to show in one view both the inside and the outside of an object that is not completely symmetrical. Figure 9-39 shows an object with a circular boss on one side. Since the object is not symmetrical, the inside cannot be shown with a half section. A phantom section is used instead. A partial phantom section can sometimes be better than a broken-out section to show interior detail on an exterior view.

ROTATED FEATURES IN SECTION

A section or an *elevation* (side, front, or rear view) of a symmetrical piece can sometimes be hard to read if drawn in true projection. It can also be hard to draw. When drawing such a view, follow the example in Fig. 9-40. In this example, a symmetrical

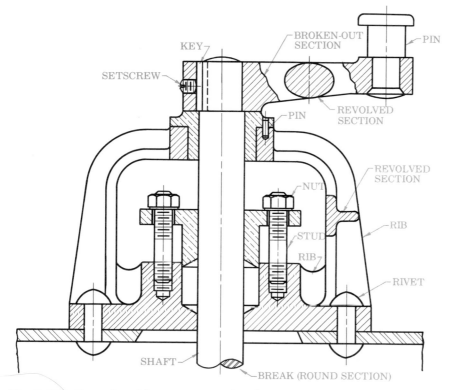

Fig. 9-38. Examples of features not sectioned.

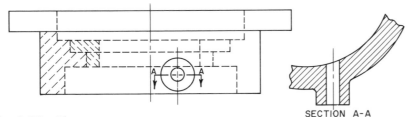

Fig. 9-39. Phantom section.

SECTION A-A

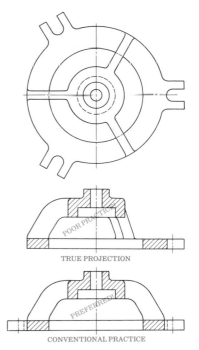

Fig. 9-40. Some features should be rotated to show true shape.

piece with ribs and lugs is drawn twice in section. The first section drawing is a true projection. But it does not show the true shape of the ribs and lugs. In the second section, the ribs and lugs have been *rotated* (turned) on the vertical axis until they appear as mirror images of each other on either side of the centerline. Their true shape can now be

shown. This is the right way to draw this kind of object. Note that only the parts that extend all the way around the vertical axis are drawn with section lining. In Fig. 9-41, the lugs are rotated to show true shape. Note that they are not drawn with section lining.

When a section passes through spokes, section lining is not drawn on the spokes. They are left as in the section drawing in Fig. 9-42, at A. Compare this drawing with the section drawing for a solid web (Fig. 9-42 at B). It is the section lining that

shows that the web is made solid rather than made with spokes.

When drawing a section or elevation of a part with holes arranged in a circle, follow the *good practice* examples in Fig. 9-43. In these examples, the holes have been rotated for the section drawing until two of them lie squarely on the cutting plane. These views then show the true distance of the holes from the center, when a true projection would not.

Rotating features in drawings are very useful when you want to show

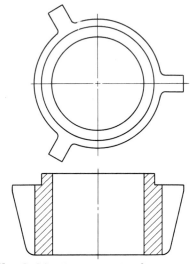

Fig. 9-41. Do not section lugs.

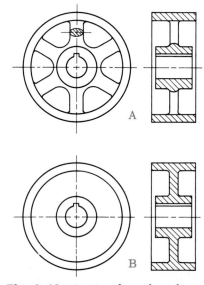

Fig. 9-42. Section through spokes.

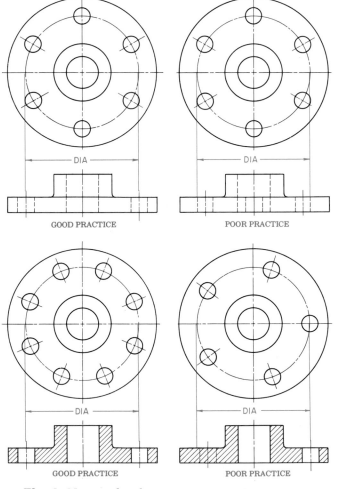

Fig. 9-43. Good and poor practice for showing holes.

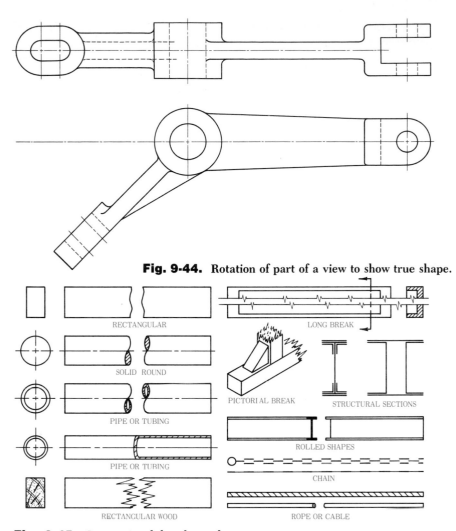

Fig. 9-44. Rotation of part of a view to show true shape.

RECTANGULAR

LONG BREAK

SOLID ROUND

PICTORIAL BREAK

STRUCTURAL SECTIONS

PIPE OR TUBING

ROLLED SHAPES

PIPE OR TUBING

CHAIN

RECTANGULAR WOOD

ROPE OR CABLE

Fig. 9-45. Conventional breaks and symbols.

true conditions or distances that would not show in a true projection. Moreover, for some objects, only part of the view is rotated, as shown by the bent lever in Fig. 9-44.

■ CONVENTIONAL BREAKS AND SYMBOLS

Conventional breaks and symbols are used to make some details in a drawing easier to draw and easier to understand. Figure 9-45 shows the methods used to draw long, evenly shaped parts and to *break out* (shorten) the drawing of parts. Using a break lets you draw a view to a larger scale. Since the break shows how the part looks in cross section, an end view usually need not be drawn. Give the length by a dimension. The symbols for conventional breaks are usually drawn freehand. However, on larger drawings, conventional breaks are often drawn with instruments to give a neat appearance. Figure 9-46 shows how to draw the break for cylinders and pipes.

$\frac{1}{3}$ R

R

30°

SOLID ROUND

$\frac{1}{2}$ R

R

30°

45°

45°

30°   30°

PIPE OR TUBING

Fig. 9-46. Drawing the break symbols for cylinders and pipes.

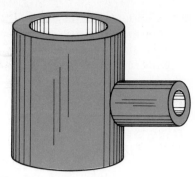

Fig. 9-47. Intersecting parts.

INTERSECTIONS IN SECTION

An *intersection* is a point where two parts join together (Fig. 9-47). Drawing a true projection of an intersection is difficult and takes too much time. Also, such accuracy of detail is of little or no use to a print reader. Therefore, approximated and preferred sections, such as those shown in Fig. 9-48, are generally drawn.

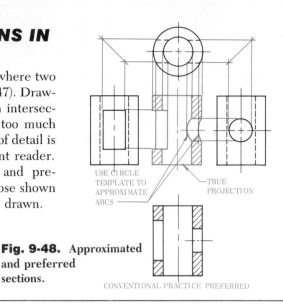

Fig. 9-48. Approximated and preferred sections.

LEARNING ACTIVITIES

1. Obtain sectional-view drawings from industry. Study them carefully to determine if the appropriate sections were selected and drawn properly. See if any alternate methods would have worked as well or perhaps better.

2. Obtain several objects that have interior detail, such as those shown throughout Chap. 9. Decide which type of section would be most practical and make complete working drawings with sectional views and dimensions.

VOCABULARY

1. Cutting plane
2. Cutting-plane line
3. Section lining
4. Crosshatching
5. Full section
6. Half section
7. Offset section
8. Broken-out section
9. Revolved section
10. Removed section
11. Auxiliary section
12. Ribs
13. Webs
14. Phantom section
15. Conventional break

REVIEW

1. General-purpose section lining may also be used to designate what material?

2. What type of line is used to show where the section is to be taken?

3. When a cutting-plane line passes through an entire view, a _____-sectional view results.

4. How are thin sections usually shown on a drawing?

5. Sketch two types of cutting-plane lines.

6. When one quarter of the object is cut away, a _____-sectional view results.

7. A revolved-sectional view is rotated _____°.

8. Name four items that are generally not sectioned even when the cutting-plane line passes through their axis.

9. A _____ section is used to show both inside and outside details of a nonsymmetrical object.

Problems

In Fig. 9-49, problems A and B show examples of full- and half-sectional-view drawings. In the half-sectional view, the hidden line is optional. Study these examples carefully before attempting any of the drawing assignments for this chapter.

For problems C through L in Fig. 9-49, use dividers to take dimensions from the printed scales below. Draw both views using the scale assigned. Make a full- or half-sectional view as assigned. Add dimensions if required. Estimate the sizes of fillets and rounds (small radii).

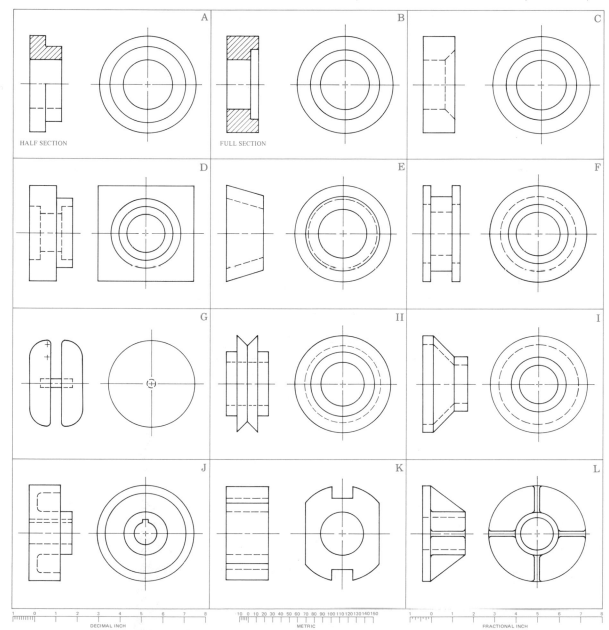

Fig. 9-49. Problems for practice in sectioning.

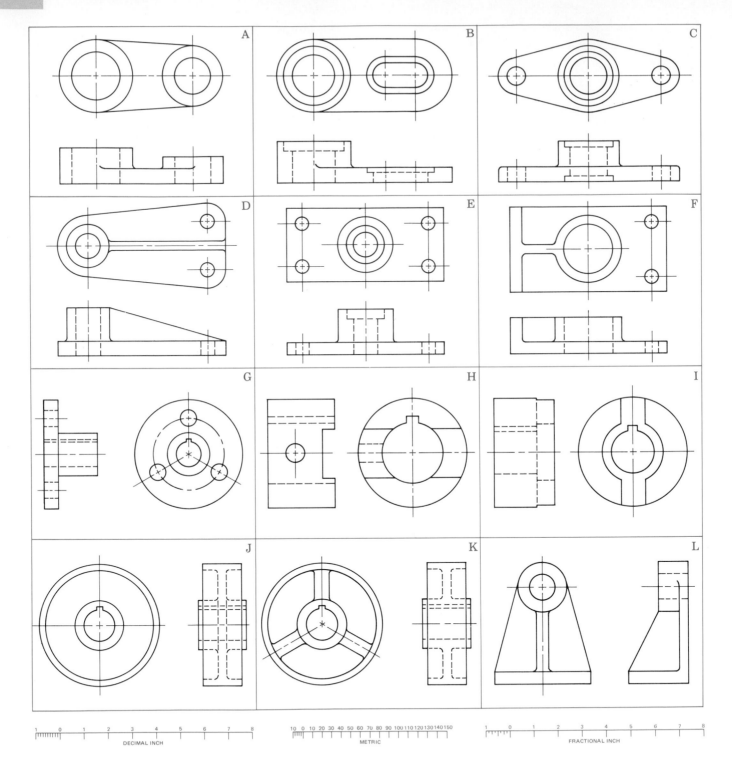

DECIMAL INCH

METRIC

FRACTIONAL INCH

Fig. 9-50. Problems for practice in sectioning. Take dimensions from the printed scale, using dividers. Draw both views. Make a sectional view as assigned. Add dimensions if required. Estimate sizes of fillets and rounds (small radii).

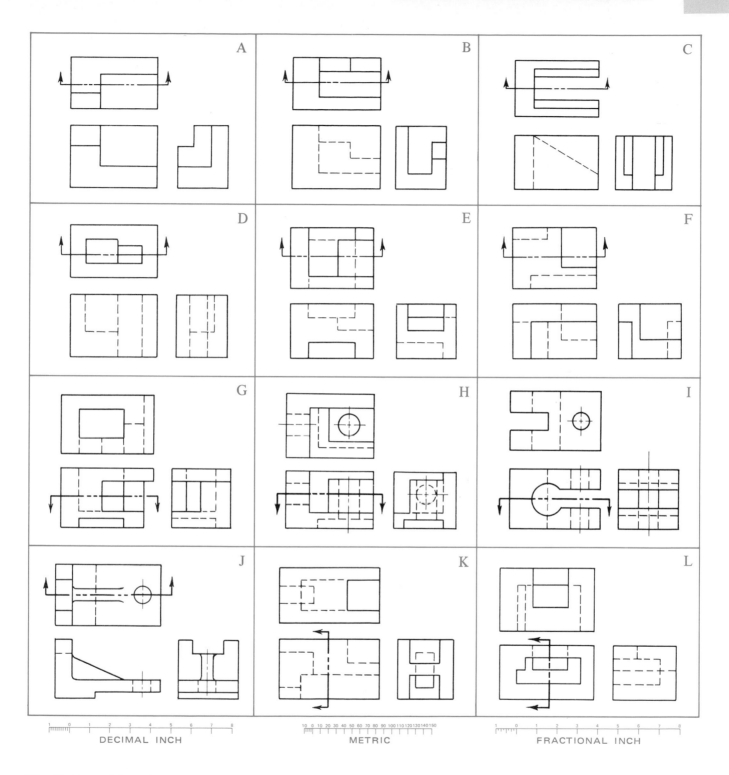

Fig. 9-51. Problems for practice in sectioning. Take dimensions from the assigned scale, using dividers. Draw all views and section the view indicated by the cutting-plane line. Add dimensions if required.

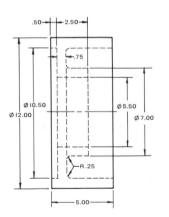

Fig. 9-52. Make a two-view drawing of the collar, showing a full or half section as assigned.

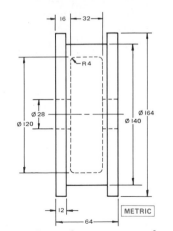

METRIC

Fig. 9-53. Make a two-view drawing of the steam piston, showing a full or half section as assigned.

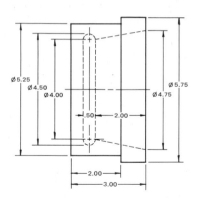

Fig. 9-54. Make a two-view drawing of the shaft cap, showing a full or half section as assigned.

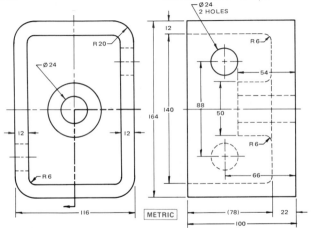

METRIC

Fig. 9-55. Make a two-view drawing of the protected bearing, showing the right-hand view as a half section.

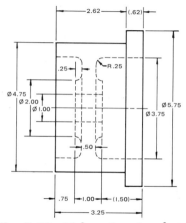

Fig. 9-56. Make a two-view drawing of the water-piston body, showing a full or half section as assigned.

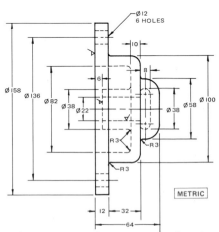

METRIC

Fig. 9-57. Make a two-view drawing of the cylinder head. Show a full or half section.

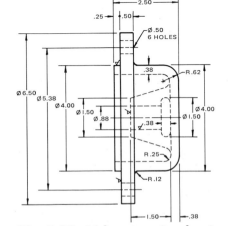

Fig. 9-58. Make a two-view drawing of the cylinder cap. Show a full or half section.

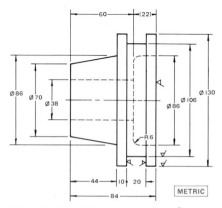

METRIC

Fig. 9-59. Make a two-view drawing of the cone spacer, showing a full or half section.

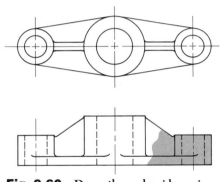

Fig. 9-60. Draw the rod guide, using the assigned scale at the bottom of the page. Make top and front views. Show broken-out section as indicated by the colored screen.

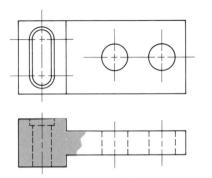

Fig. 9-61. Draw the chisel, using the assigned scale at the bottom of the page. Make revolved or removed sections on colored centerlines. A is a ¼″ (.25″) × 3″ (3.00″) (6.3 × 76 mm) rectangle, B is a 1¼″ (1.25″) (32 mm) (across flats) octagon, C and D are circular cross sections. The chisel may be drawn full size or to a reduced scale.

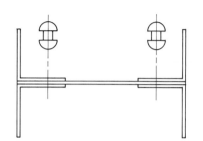

Fig. 9-62. Draw the structural joint, using the assigned scale at the bottom of the page. Make a full-sectional view of the joint with rivets moved into their proper position on the centerlines.

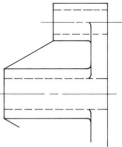

Fig. 9-63. Draw the adjusting plate, using the assigned scale at the bottom of the page. Draw front and top views. Make the broken-out section as indicated by the colored screen.

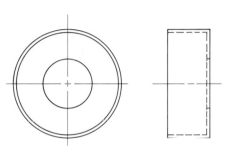

Fig. 9-64. Draw the grease cap using the assigned scale at the bottom of the page. Make front and right full- or half-sectional views as assigned.

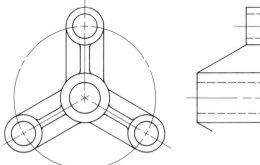

Fig. 9-65. Draw the rotator using the assigned scale at the bottom of the page. Complete the right-side view and make full- or half-sectional view.

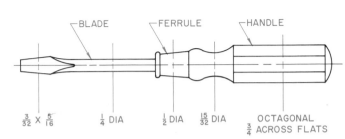

BLADE   FERRULE   HANDLE

$\frac{3}{32}$ × $\frac{5}{16}$ ¼ DIA   ½ DIA $\frac{15}{32}$ DIA   OCTAGONAL ¾ ACROSS FLATS

Fig. 9-66. Draw the screwdriver twice the size shown. Add removed or revolved sections on the colored centerlines.

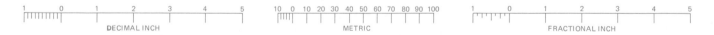

1 0 1 2 3 4 5
DECIMAL INCH

10 0 10 20 30 40 50 60 70 80 90 100
METRIC

1 0 1 2 3 4 5
FRACTIONAL INCH

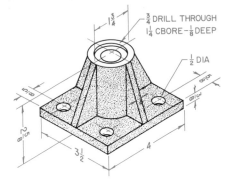

Fig. 9-67. Base plate. Scale: full size or as assigned. Material: cast iron.

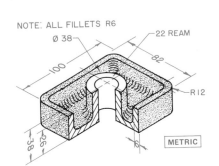

Fig. 9-68. Shaft base. Scale: 1:1 or as assigned. Material: cast iron.

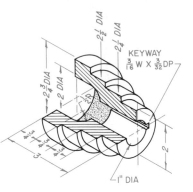

Fig. 9-69. Step pulley. Scale: full size or as assigned. Material: cast iron.

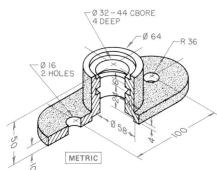

Fig. 9-70. Lever bracket. Scale: 1:1 or as assigned. Material: cast iron.

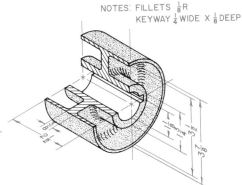

Fig. 9-71. Idler pulley. Scale: full size or as assigned. Material: cast iron.

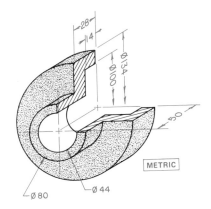

Fig. 9-72. Retainer. Scale: 1:1 or as assigned. Material: cast aluminum.

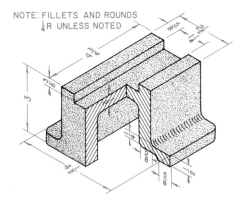

Fig. 9-73. Rest. Scale: three-quarter size or as assigned. Material: cast aluminum.

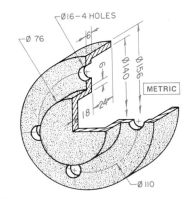

Fig. 9-74. End cap. Scale: 1:1 or as assigned. Material: cast iron.

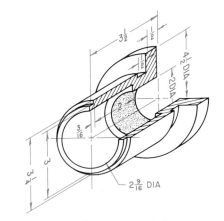

Fig. 9-75. Flange. Scale: full size or as assigned. Material: cast aluminum.

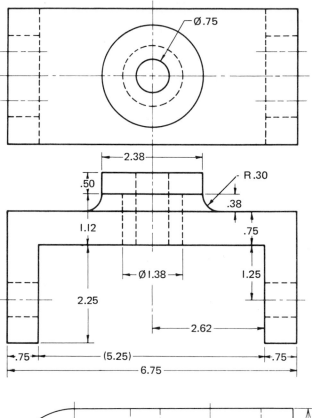

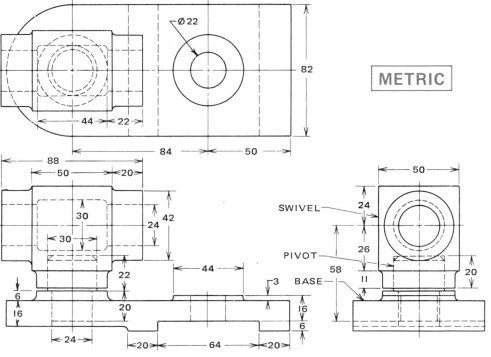

Fig. 9-76. Draw three views of the yoke with the front view in section. There are two pieces: the yoke and the bushing. Do not copy the picture.

METRIC

Fig. 9-77. Draw three views of the swivel base with the front view in section.

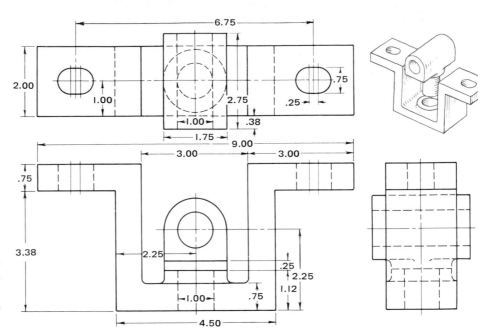

Fig. 9-78. Draw three views of the swivel hanger with the right-side view in section. There are two pieces: the hanger and the bearing.

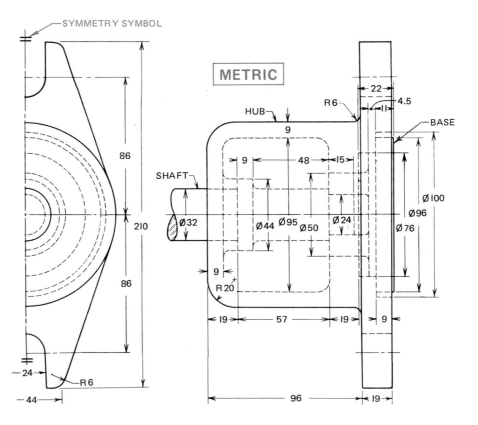

Fig. 9-79. Draw three views of the thrust bearing with the right-hand view in section. There are three parts: the shaft, the hub, and the base.

10 Fasteners

◼ THE FUNCTION OF FASTENERS

A *fastener* is any kind of device or method for holding parts together. Screws, bolts and nuts, rivets, welding, brazing, soldering, adhesives, collars, clutches, and keys are all fasteners. Each of these fastens in a different way. Each can also fasten parts permanently or so that they can later be taken apart again or adjusted.

◼ SCREWS AND SCREW THREADS

The principle of the screw thread has been known for so long that no one knows who discovered it. Archimedes (287–212 B.C.), a Greek mathematician, put the screw to practical use. He used it in designing a screw conveyor to raise water. Similar devices are still used today to move flour and sugar in commercial bakeries, to raise wheat in grain elevators, to move coal in stokers, and for many other purposes.

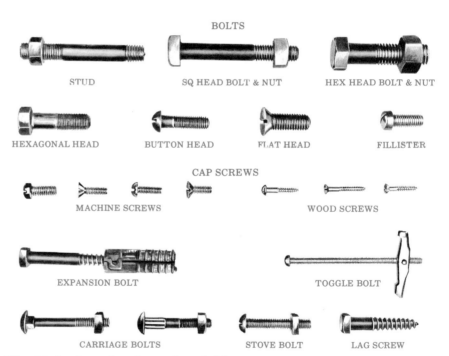

BOLTS

STUD — SQ HEAD BOLT & NUT — HEX HEAD BOLT & NUT

HEXAGONAL HEAD — BUTTON HEAD — FLAT HEAD — FILLISTER

CAP SCREWS

MACHINE SCREWS — WOOD SCREWS

EXPANSION BOLT — TOGGLE BOLT

CARRIAGE BOLTS — STOVE BOLT — LAG SCREW

Fig. 10-1. Examples of some threaded fasteners.

Screws and other fasteners have so many uses and have become so important that engineers, drafters, and technicians must become familiar with their different forms (Fig. 10-1). They must also be able to draw and *specify* (call for) each type correctly.

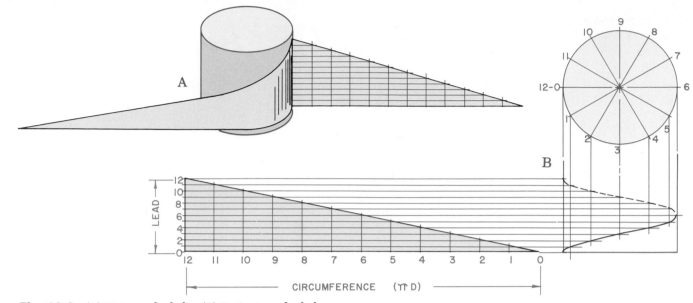

Fig. 10-2. (A) Picture of a helix. (B) Projection of a helix.

■ THE TRUE SHAPE OF A SCREW THREAD

All screw threads are shaped basically as a *helix*, or *helical curve*. Technically, a *helix* is the curving path that a point would follow if it were to travel in an even spiral around a cylinder and *parallel* to (in line with) the axis of that cylinder. In simpler terms, if a wire is wrapped around a cylinder in evenly spaced coils, it forms helical curves. A coil spring is another example of helical curves.

Another way to visualize the shape of screw threads is to cut out a right triangle in paper and wrap it around a cylinder, as shown in Fig. 10-2 at A. If the triangle's base is the same length as the cylinder's circumference, its hypotenuse, wrapped around the cylinder, will form one turn of a helix. The triangle's altitude will be the pitch of the helix. A right triangle and the projections of the

corresponding helix are shown in Fig. 10-2 at B.

To draw the projections of a helix, follow the method shown in Fig. 10-3 in spaces A and B. First, draw two projections of a cylinder, as in space A. Lay off the *pitch* (the distance from a point on the thread form to the corresponding point on the next form). Divide the circumference into a number of equal parts. Divide the pitch into the same number of equal parts. From each division point on the circumference, draw lines parallel to the axis. From each division point on the pitch, draw lines at right angles to the axis. Then draw a smooth curve through the points where these lines meet, as in B. This will give the projection of the helix.

The application of the helix is shown in space C, the actual projection of a square thread. However, such drawings are seldom made, since they take too much time. Also, they are no more useful than conventional representations.

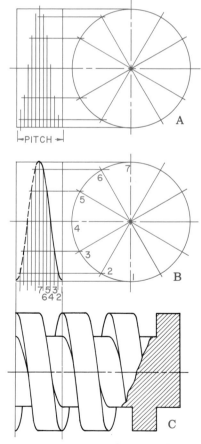

Fig. 10-3. Helix and square thread.

SCREW-THREAD STANDARDS

The first screws were made for one purpose. They were made without any thought of how anyone else might make one of the same diameter. Later, as industry developed, goods were produced in quantity by using *interchangeable* parts (parts in standard sizes that can be substituted for each other). The need then arose for standards for screws and screw threads.

Screw-thread standards in the United States were developed from a system that William Sellers presented to the Franklin Institute in Philadelphia in 1864. Screw-thread standards in England came from a paper presented to the Institution of Civil Engineers in 1841 by Sir Joseph Whitworth. These two standards were not interchangeable.

More and better screw-thread standards have been drawn up as industrial production has grown more complex. In 1948, standardization committees of Canada, Great Britain, and the United States agreed on the Unified Thread Standards. These standards are now the basic American National Standards. They are listed in the *American National Standard Unified Screw Threads for Screws, Bolts, Nuts and Other Threaded Parts* (ANSI B1.1) and in Handbook H–28, *Federal Screw Thread Specifications*.

Since 1948, the different thread systems have been brought increasingly into line with each other. In 1968, the International Organization for Standardization (ISO) adopted the Unified Standard as its inch screw-thread system standard. The two systems are alike in some ways.

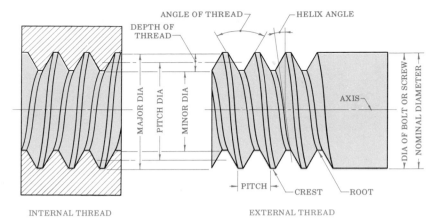

Fig. 10-4. Screw-thread terms.

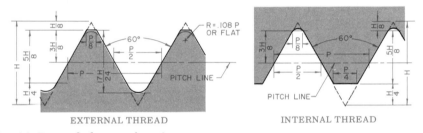

Fig. 10-5. Unified screw-thread terms.

These similarities are very important as the metric system of measurement comes into world wide use. To help bring other screw-thread standards into line with this system, the American National Standards Institute has published the *American National Standard Unified Screw Threads—Metric Translation* (ANSI B1.1a).

SCREW-THREAD TERMS

Figure 10-4 shows the main terms used to describe screw threads. The Unified and American (National) screw-thread profile shown in Fig. 10-5 is the form used for fastening in general. Other forms of threads are used to meet special fastening needs.

Some of these threads are shown in Fig. 10-6. The sharp V is seldom used today. The square thread and similar forms (worm thread and acme thread) are made especially to transmit motion or power along the line of the screw's axis. The knuckle thread is the thread used in most electric-light sockets. It may also be a "cast" thread. The Dardelet thread automatically locks a screw in place. It was designed by a French military officer but is seldom used today. The former British Standard (Whitworth) has rounded crests and roots. Its profile forms 55° angles. The former United States Standard had flat crests and roots and 60° angles. The buttress thread takes pressure in one direction only: against the surface at an angle of 7°.

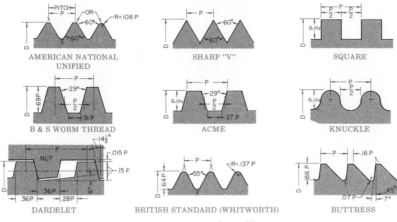

Fig. 10-6. Some of the various screw-thread profiles.

DEFINITION OF A SCREW THREAD

A *screw thread* is a helical ridge on the external or internal surface of a cylinder. A screw thread is also made in the form of a conical spiral on the external or internal surface of a cone or frustum of a cone.

The *pitch* of a thread is the distance from one point on the thread form to the corresponding point on the next form. It may be defined as a formula as follows:

$$\text{Pitch} = \frac{1}{\text{no. of threads per inch}}$$

Pitch is measured parallel to the thread's axis (Fig. 10-7). The *lead* (say "leed") *L* is the distance along this same axis that the threaded part moves against a fixed mating part when given one full turn. It is the distance a screw enters a threaded hole in one turn.

SINGLE AND MULTIPLE THREADS

Most screws have single threads (Fig. 10-7 at A). A screw has a single thread unless it is marked otherwise. A single thread is a single ridge in the form of a helix. If you give a single-thread screw one full turn, the distance it will advance into the nut (lead) will equal the pitch of the thread.

A double thread (Fig. 10-7 at B) is two helical ridges side by side. The lead, in this case, is twice the pitch. A triple thread (Fig. 10-7 at C) is three ridges side by side. The lead for this thread is 3 times the pitch.

Multiple threads are used where parts must screw together quickly. For example, technical pen caps and toothpaste tube caps have multiple threads.

RIGHT- AND LEFT-HAND THREADS

A *right-hand thread* screws in when turned clockwise as you view it from the outside end (Fig. 10-8 at A). A *left-hand thread* screws in when turned counterclockwise (Fig. 10-8 at B). Threads are always right hand unless marked with the initials *LH*, meaning a left-hand thread. Some devices, such as the turnbuckle (Fig. 10-9), have both right- and left-hand threads. Others, such as bicycle pedals, can have either. A left-hand pedal has left-hand threads; a right-hand pedal, right-hand threads.

REPRESENTATIONS OF SCREW THREADS ON DRAWINGS

When drawing screw-threads, you use special symbols. These are the same whether the threads are coarse

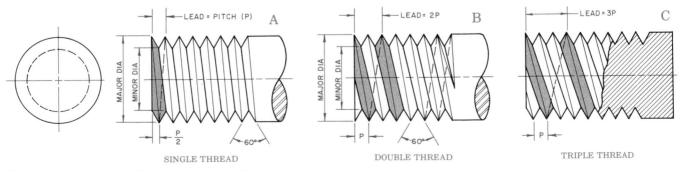

Fig. 10-7. Single, double, and triple threads.

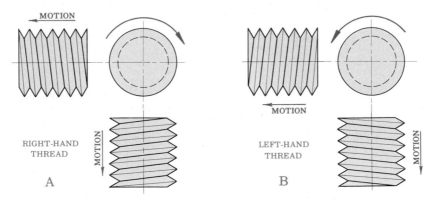

Fig. 10-8. (A) A right-hand screw thread. (B) A left-hand screw thread.

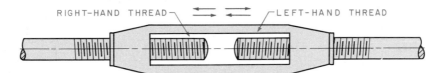

Fig. 10-9. A turnbuckle uses right- and left-hand screw threads.

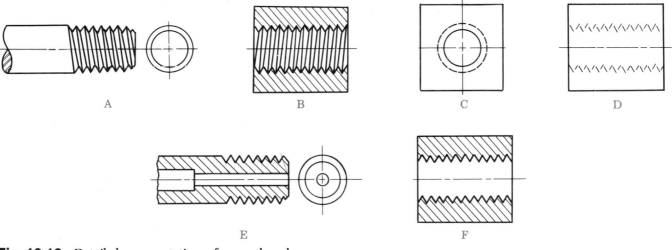

Fig. 10-10. Detailed representations of screw threads.

or fine, right hand or left hand. You use notes to give the necessary information.

Under ANSI rules, screw-threads can be drawn in three ways:

1. A *detailed representation* approximates the real look of threads (Fig. 10-10). For this kind of drawing, it is not necessary to draw the pitch exactly to scale. Instead, estimate the right size for the drawing. Draw the helixes as straight lines. Draw the threads as sharp Vs. In general, detailed representation is not used in working drawings (see Chap. 11), except where needed for clearness. It is also not usually used if the screw is less than 1(1.00) in. (25 mm) in diameter.

2. A *schematic representation* shows the threads with symbols, rather than as they really look. For this kind of drawing, leave out the Vs (Fig. 10-11). Also the pitch need not be drawn to scale. Make it about the right size for the drawing. Then draw the crest and root lines accordingly. Space

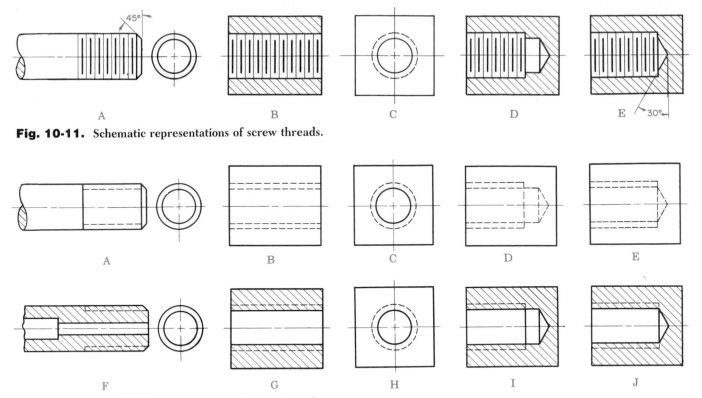

Fig. 10-11. Schematic representations of screw threads.

Fig. 10-12. Simplified representations of screw threads.

them by eye to make them look good. These lines can be at right angles to the axis or slanted to show the helix angle (see Fig. 10-17). The American National Standards calls for crest lines to be thin and root lines to be thick. However, the lines may all be made the same thickness to save time. This is especially so on pencil drawings.

3. *A simplified representation* is much like a schematic representation. In this case, however, draw the crest and root lines as dashed lines, except where either of them would normally show as a visible solid line (Fig. 10-12 at A, C, F, G, H, I, and J). The drafter saves time by using the simplified representation because it leaves out useless details. Therefore, simplified representation has be-

come the most commonly used in industry.

■ TO DRAW THE DETAILED REPRESENTATION OF SCREW THREADS

For a detailed representation, draw the screw threads with the sharp-V profile. Use straight lines to represent the helixes of the crest and root lines. To draw the V-form thread, follow the steps shown in Fig. 10-13. First, lay off the pitch *P*, as at A. The pitch need not be drawn to scale. Make it about the right size for the drawing. Next, lay off the half pitch *P*/2 at the end of the thread, as shown. Construct a right triangle with a base of *P*/2 and an altitude equal to the outside diameter of the

screw. Adjust your triangle, or the ruling arm of your drafting machine, to the slope of this right triangle. Now draw all the crest lines with this slope. Next, use the 30°-60° triangle (or the drafting machine ruling arm set at a 30° angle) to draw one side of the V for the threads, as at B. Then reverse the triangle, or ruling arm, and complete the Vs. Now set the triangle, or the ruling arm of the drafting machine, to the slope of the root lines. Draw them as shown at C. Notice that the root lines do not *parallel* (have the same slope as) the crest lines. This is because the minor (root) diameter is less than the major (crest) diameter. Finally, draw 45° chamfer lines using the construction shown at D.

A hole with an internal right-hand thread is shown as a sectional view in Fig. 10-14. Notice that the thread-

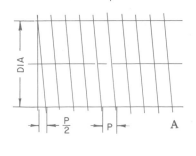

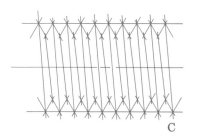

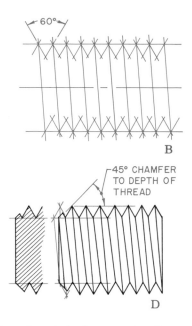

Fig. 10-13. To draw the detailed representation of screw threads.

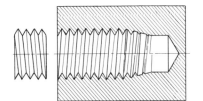

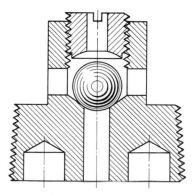

Fig. 10-14. Internal threads in section (threaded hole).

Fig. 10-15. Threads in section on assembled pieces.

clearer where two or more threaded pieces are shown in section (Fig. 10-15).

TO DRAW THE SCHEMATIC REPRESENTATION OF SCREW THREADS

Follow the steps in Fig. 10-16. First, lay off the outside diameter of the screw thread, as at A. Then follow the steps shown at B to lay off the thread depth and the chamfer. Next, draw thin crest lines at right angles to the axis, as at C. Then draw thick root lines parallel to the crest lines, as at D.

Crest and root lines may be drawn at a slope (Fig. 10-17 at A). To do so, give each thread a slope of half the pitch. Also, crest and root lines may be made the same width on pencil drawings (Fig. 10-17 at B). Finally, the pitch need not be made to scale. Just space it to look good.

line slope is in the opposite direction from the *external* right-hand thread lines on a mating screw. This is because the internal thread lines must match the far side of the screw.

The realistic, or V-form, thread representation can make a drawing

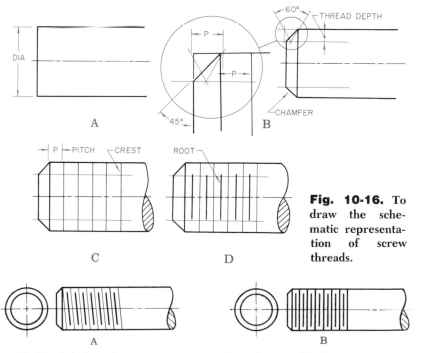

Fig. 10-16. To draw the schematic representation of screw threads.

Fig. 10-17. (A) Slope-line representation. (B) Uniform-width lines.

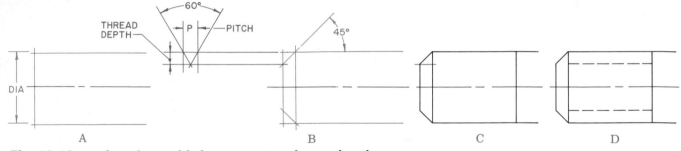

Fig. 10-18. To draw the simplified representation of screw threads.

TO DRAW THE SIMPLIFIED REPRESENTATION OF SCREW THREADS

Follow the steps in Fig. 10-18. First, lay off the outside diameter of the screw, as at A. Next, follow the steps shown at B to lay off the screw-thread depth and the chamfer. Then draw the chamfer and a line to show the length of the thread, as at C. Finally, draw dashed lines for the threads, as at D, to complete the drawing.

TO DRAW SQUARE SCREW THREADS

The square thread has a depth that is one half its pitch. To draw square threads, follow the steps in Fig. 10-19. First, lay off the diameter, the pitch *P*, one-half-pitch spaces, and the depth of the thread, as at A. Next, draw the crest lines, as at B. Then draw the root lines, as at C. At D, an internal square thread is drawn in section.

TO DRAW ACME SCREW THREADS

The acme thread has a depth that is one half its pitch. To draw acme threads, follow the method shown in Fig. 10-20 at A. First, lay off the out-

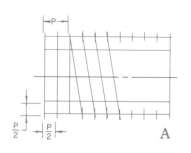

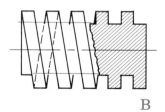

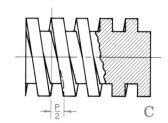

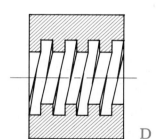

Fig. 10-19. To draw square threads.

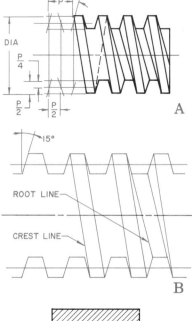

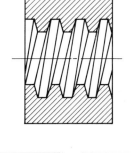

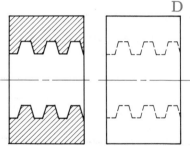

Fig. 10-20. To draw acme threads.

side diameter, the pitch, and the depth of the thread. Midway between the outside diameter and the depth of the thread is the *pitch diameter*. Along it, draw the *pitch line*. On the pitch line, lay off one-half-pitch spaces. Use these to draw the thread profile. Now draw the crest lines and root lines to complete the view. See space B for an enlarged view.

Internal acme threads in section are drawn as shown at C. They can also be drawn in other ways, including with dashes for the hidden lines or with the outline in section, as shown at D.

■ THREAD SERIES FOR UNIFIED AND AMERICAN NATIONAL STANDARD SCREW THREADS

Screws of the same diameter are made with different pitches (number of threads per inch) for different uses. The various combinations of diameter and pitch have been grouped in *screw-thread series*. These series are listed in ANSI B1.1. Each is denoted by letter symbols, as follows:

Coarse-Thread Series (UNC or NC). In this series the pitch for each diameter is relatively large. This series is used for engineering in general.

Fine-Thread Series (UNF or NF). In this series, the pitch for each diameter is smaller (there are more threads per inch) than in the coarse-thread series. This series is used where a finer thread is needed, as in making automobiles and airplanes.

Extra-Fine-Thread Series (UNEF or NEF). In this series, the pitch is even smaller than in the fine-thread

series. This series is used where the thread depth must be very small, as on aircraft parts or thin-walled tubes.

Constant-Pitch-Thread Series (UN). There are also several constant-pitch-thread series. These series have 4, 6, 8, 12, 16, 20, 28, or 32 threads per inch. They offer a variety of pitch-diameter combinations that can be used where the coarse, fine, and extra-fine series are not suitable. However, when selecting a constant-pitch series, the first choices should be the 8-, 12-, or 16-thread series. Constant-pitch threads are generally used as a continuation of coarse-, fine-, and extra-fine-thread series in larger diameters.

Eight-Thread Series (8UN or 8N). This series uses 8 threads per inch for all diameters.

Twelve-Thread Series (12UN or 12N). This series uses 12 threads per inch for all diameters.

Sixteen-Thread Series (16 UN or 16N). This series uses 16 threads per inch for all diameters.

Special Threads (UNS, UN, or NS). These are nonstandard, or special, combinations of diameter and pitch.

American National Standard Thread Series. These series have been largely replaced by the Unified Standard. However, you should still know what they and their letter symbols are. These series and their symbols are listed below.

Coarse-Thread Series (NC). This series is used for screws, bolts, and nuts produced in quantity, and also for fastening in general.

Fine-Thread Series (NF). This series has a smaller pitch for each di-

ameter than NC. It is used where the coarse series is not suitable.

Extra-Fine-Thread Series (NEF). This series is used when an even smaller pitch than NF is needed, as on thin-walled tubes.

The symbol NS denotes special threads.

Eight-Thread Series (8N). This series is a constant-pitch series for large diameters.

Twelve-Thread Series (12N). This series is a constant-pitch series that has a medium-fine pitch for large diameters.

Sixteen-Thread Series (16N). This series is a constant-pitch series that has an even smaller pitch per diameter.

Notice that the American system has only three constant-pitch series, while the Unified system has eight, including all three from the American system. The three in the American system are used most.

■ CLASSES OF FITS FOR UNIFIED AND AMERICAN NATIONAL SCREW THREADS

Screw threads are also divided into *screw-thread classes* based on their *tolerances* (amount of size difference from exact size) and *allowances* (how loosely or tightly they fit their mating parts). The exact screw thread needed can be obtained by choosing both a series and a class. In brief, the classes for Unified threads are Classes 1A, 2A, and 3A for external threads only and Classes 1B, 2B, and 3B for internal threads only.

Classes 1A and 1B. These have a large allowance (loose fit). They are used on parts that must be put together quickly and easily.

Classes 2A and 2B. These are the thread standards most used for general purposes, such as for bolts, screws, nuts, and similar threaded items.

Classes 3A and 3B. These are stricter standards for fit and tolerance than the others. They are used where thread size must be more exact.

Classes 2 and 3. These are American National Standard. They are described, with tables of dimensions, in Appendix 1 of ANSI B1.1.

■ SCREW-THREAD SPECIFICATIONS

A screw thread is specified (called for) by telling its diameter (nominal, or major, diameter), number of threads per inch, length of thread, initial letters of the series, class of fit, and external (A) or internal (B), as shown in Fig. 10-21. Any thread you specify will be assumed to be both single and right hand unless stated otherwise. If you mean the thread to be left hand, include the letters LH after the class symbol. If they are to be double or triple, include DOUBLE or TRIPLE.

Some examples using fractional sizes follow:

- 1¼—7UNC—1A (1¼-in. diameter, 7 threads per inch, Unified threads, coarse threads, Class 1, external)
- .750—10UNC—2A (.750-in. diameter, 10 threads per inch, Unified threads, coarse threads, Class 2, external) (See Fig. 10-21.)
- ⅞—14UNF—2B (⅞-in. diameter, 14 threads per inch, Unified threads, fine threads, Class 2, internal)

- 1⅝—18UNEF—3B—LH (1⅝-in. diameter, 18 threads per inch, Unified threads, extra-fine threads, Class 3, internal, left hand)

Specify tapped (threaded) holes by a note giving the diameter of the tap drill (²⁷⁄₆₄″), depth of hole (1⅜″), thread information (½ diameter, American National threads, Class 2), and length of thread (1″), as:
²⁷⁄₆₄ DRILL × 1⅜ DEEP
½—13NC—2 × 1 DEEP

■ THREAD DESIGNATION: METRIC THREADS

An ISO metric screw thread is *designated* (specified) by giving its nominal size (basic major diameter) and pitch, both expressed in millimeters. Include an M to denote that the thread is an ISO metric screw thread. Place the M before the nominal size. Use X to separate the nominal size from the pitch. For the coarse thread series only, do not show the pitch unless you are also giving the length of the thread. If the length is given, separate it from the

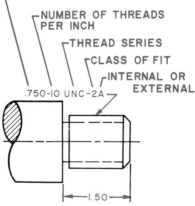

CUSTOMARY INCH THREAD CALLOUT

Fig. 10-21. Customary thread specifications.

other designations with an X. For external threads, the length of thread may be given as a dimension.

For example, designate a 10-mm diameter, 1.25 pitch, fine-thread series as M 10 X 1.25. A 10-mm diameter, 1.5 pitch, coarse-thread series would, however, be designated as M 10. Remember that for coarse threads, the pitch is not given unless the length of the thread is given. If, in this last example, the length were 25 mm, the thread would be designated as M 10 X 1.5 X 25.

In addition to the basic designation, the designation for the thread's tolerance class must be given. This designation is separated from the basic designation by a dash. It consists of two sets of symbols. The first denotes the pitch diameter tolerance. The second, which follows immediately, denotes the crest diameter tolerance. Each of these two sets of symbols is composed of a number denoting the grade tolerance, followed by a letter (capital for internal threads and lowercase for external threads) denoting the tolerance position. If the symbols for the pitch diameter tolerance are the same as those for the crest diameter tolerances, one set may be left out (see Fig. 10-22).

Two classes of fits are generally used in the ISO metric system. The first is a general-purpose class and is specified as 6H/6g. The second is for fits that require a closer tolerance and is specified as 6H/5g6g. More detailed information may be obtained from ISO and ANSI standards. However, unless a very specific tolerance class is necessary, many industries have adopted a simplified system for designating thread fits. If the general-purpose fit is acceptable, no tolerance class is speci-

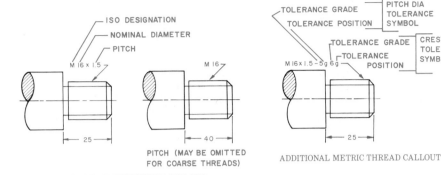

Fig. 10-22. ISO metric thread specifications.

fied. If a close fit is necessary, the letter C is placed after the pitch in the thread note. An example would be M10 X 1.25C.

THREADED FASTENINGS

Fasteners are made in many forms for different uses. The following sections describe how to identify and draw the threaded fasteners most commonly used on machines, constructions, and engineering projects. These include square- and hexagonal-head bolts, square and hexagonal nuts, studs, machine screws, capscrews and setscrews, etc. You will find dimensions for drawing bolts, nuts, and some of the other generally used threaded fasteners in Appendix A.

Certain bolt and nut dimensions have been designated as Unified Standard for use in the United States, Great Britain, and Canada. For complete metric information, consult the latest American National Standard or ISO standard.

STANDARD SQUARE AND HEXAGON BOLTS AND NUTS

These are so important that you should know the main terms used (Fig. 10-23). You should also be able to draw the necessary views.

In general, bolts and nuts are either regular or heavy. They are also either square or hexagon. Regular bolts and nuts are used for the general run of work. Heavy bolts and nuts are somewhat larger. They are used where the bearing surface or the hole in the part being held must be larger. The types of bolts and nuts made in regular sizes include square bolts and nuts, hexagon bolts and nuts, and semifinished hexagon bolts and nuts. Types made in heavy sizes include hexagon bolts and nuts, semifinished hexagon bolts and nuts, finished hexagon bolts, and square nuts.

Regular bolts and nuts are not finished on any surface. Semifinished bolts and nuts are processed to have

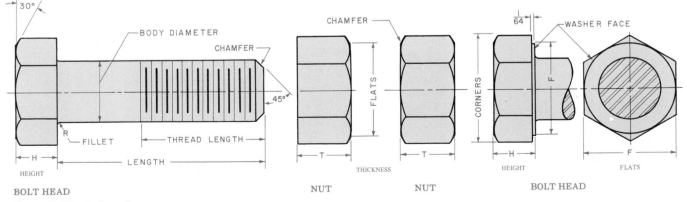

Fig. 10-23. Bolt and nut terms.

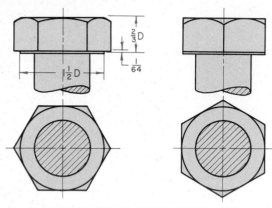

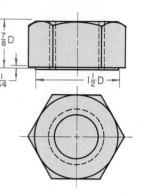

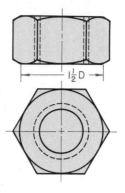

HEX. BOLT HEAD WITH WASHER FACE

HEX. NUT WITH
WASHER FACE

HEX. NUT WITH
CHAMFER FACE

Fig. 10-24. Semifinished boltheads and nuts.

a flat bearing surface. "Finished bolts and nuts" are so called only because of the quality of their manufacture and the closeness of their tolerance. They do not have completely machined surfaces. Semifinished boltheads and nuts (Fig. 10-24) have on the bearing surface either a washer face or a face with chamfered corners. Each face has a diameter equal to the distance *across the flats* (between opposite sides of the bolthead or nut). The thickness of the washer face is about 1/64 (.02) in. (0.4 mm).

REGULAR BOLTHEADS AND NUTS

When drawing regular boltheads and nuts, figure the dimensions from the proportions given in Figs. 10-25 and 10-26 or from Appendix A. Draw the chamfer angle at 30° for either the hexagon or the square forms. (The standard for the square form is actually 25°.) Find the radii for the chamfer arcs by trial. Note that you can find one half the distance across corners, *ab*, by the construction shown.

HEAVY HEXAGON BOLTHEADS AND NUTS

When drawing heavy hexagon boltheads and nuts, figure the dimen-

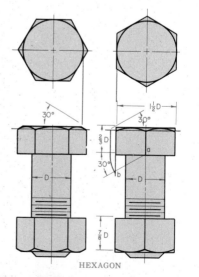

HEXAGON

Fig. 10-25. Regular hexagon bolthead and nut.

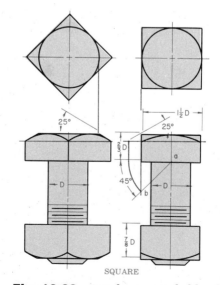

SQUARE

Fig. 10-26. Regular square bolthead and nut.

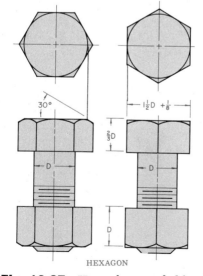

HEXAGON

Fig. 10-27. Heavy hexagon bolthead and nut.

sions from the proportions given in Fig. 10-27 or from Appendix A. See also ANSI B18.2 and B18.2.1.

A new standard covers hexagon structural bolts. These bolts are made of high-strength steel. They are used for structural steel joints. The dimensions for drawing these high-strength steel bolts are listed in Appendix A.

TO DRAW A REGULAR SQUARE BOLTHEAD ACROSS FLATS

(Square boltheads are not made in heavy sizes.) To draw a regular square bolthead across the flats, use the following proportions. If D is the major diameter of the bolt and W is the width of the bolthead across the flats, then $W = 1.5D$. The height of the head $H = 0.67D$. Now, follow the method shown in Fig. 10-28. Draw centerlines, then start the top view, as in space A. Draw a chamfer circle with a diameter equal to the distance across the flats. Then draw a square about this circle. Below the square, start the front view, as in space B. Draw a horizontal line representing the bearing surface of the head. Lay off the height of the head H. Then draw the top line of the head. To get the sides of the head, project lines down from the top view. Finally, to draw the chamfer, begin by drawing line ox, as in space A. Revolve the line to make line oy, as shown. Now, run a line from point y straight down to the front view (space B). From point a in the front view, draw a 30° chamfer line ab out to meet the line from point y. This forms the chamfer depth. From point b, extend a line horizontally across to establish points c and d, as shown. Now draw the chamfer arc through points c, e, and d, using

radius R. Find the length of R by trial, or make it equal to W, the width across the flats. Complete the view, as in space D.

TO DRAW A REGULAR SQUARE BOLTHEAD ACROSS CORNERS

(Square boltheads are not made in heavy sizes.) To draw a regular square bolthead across the corners,

use the following proportions. If D is the major diameter of the bolt and W is the width across the flats, then $W = 1.5D$. The height of the head $H = 0.67D$. Begin your drawing, as in space A of Fig. 10-29, by drawing centerlines. Start the top view by drawing a chamfer circle with a diameter equal to W, the distance across the flats. Then, using a 45° triangle, draw a square about this circle, as shown. Next, start the side

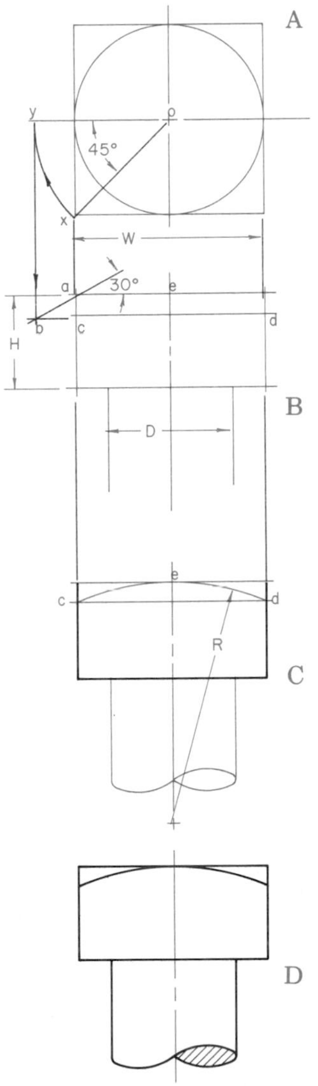

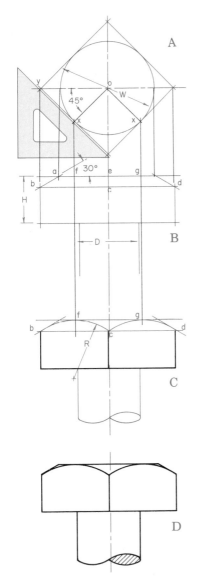

Fig. 10-28. To draw a regular square bolthead across flats.

Fig. 10-29. To draw a regular square bolthead across corners.

view, as in space B. Draw a horizontal line to represent the bearing surface of the head H. Then draw the top line of the head. To get the two sides of the head, drop lines down from the corners of the square in the top view. Next, project lines down from the diameter of the chamfer circle in the top view to meet the top line in the front view. From this point, draw 30° chamfer lines to meet the lines at the sides of the head.

Now, find the two tangent points x shown in space A. Project these points to the top view to get points f and g, as shown in spaces B and C. Also draw line bcd. Then draw the chamfer arcs bfc and cgd using radius R. Find radius R by trial, or figure that $R = yo = \frac{1}{2}$ distance across corners. Now, complete the view as in space D. Use the same method to draw a nut except that $T = 0.88D$ for regular nuts, and $T = D$ for heavy nuts.

Boltheads and nuts are usually drawn across corners on all views of design drawings, no matter what the projection. This is done to show the largest *clearance* (space needed for turning) that the bolt or nut must have. It is also done to keep hexagon heads and nuts from being confused with square heads and nuts.

■ TO DRAW A HEXAGON BOLTHEAD ACROSS CORNERS

Use the same proportions as for the preceding drawings. Start the top view as in Fig. 10-30, space A. Draw a chamfer circle with a diameter of W, the distance across the flats. About this circle, draw a hexagon as shown by the lines 1, 2, 3, 4, 5, 6. Begin the front view as in space B. Draw a horizontal line representing the bearing surface or undersurface of the head. Lay off the height of the head. Now draw the top line. Then project the edges from the corners of the top view. Also project the chamfer points. With these guides, draw in the chamfer line.

Draw line $abcd$, as in space C, to locate the chamfer intersections.

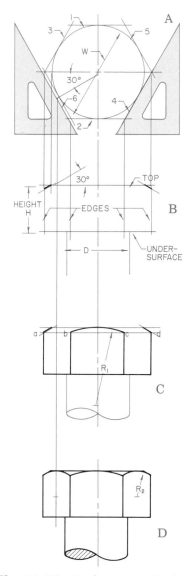

Fig. 10-30. To draw a regular hexagon bolthead across corners.

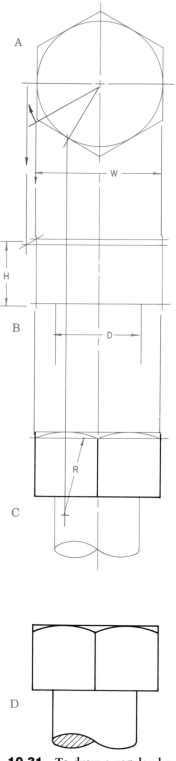

Fig. 10-31. To draw a regular hexagon bolthead across flats.

Now, draw the arc *bc*, using radius R_1. Find this radius by trial. Complete the view as in space D by drawing arcs *ab* and *cd*, using radius R_2. Find this radius, too, by trial.

To draw a hexagon bolthead across flats, proceed as shown in Fig. 10-31. Draw hexagon nuts in the same way, but note the difference between the height of the head and the thickness of the nut.

BOLTHEADS AND NUTS

Boltheads and nuts come in standard sizes. As a result, they are generally not dimensioned on drawings. In-stead, give the needed information in a note, as in Fig. 10-32 at A. The note specifies a 1-in. diameter, 8 threads per inch, Unified coarse-thread series, Class 2A fit, 2¾ in. long, regular hex-head bolt.

A bolt may hold one part to another by passing through both of them and using a nut, as at B. A bolt may also pass through the first part and screw into a threaded hole in the second, as at C.

A stud or stud bolt (Fig. 10-32 at D) has threads on both ends. It is used where a bolt is not suitable and for parts that must be removed often. Dimension the length of thread from each end as shown. A tapped (threaded) hole is dimensioned as at E. One way to use a stud is to screw it into a threaded hole permanently. The hole in another part is put over the stud and a nut is screwed onto the end, as at F. In certain cases, a stud can be passed through two parts and a nut placed on each end.

LOCK NUTS

Lock nuts and various devices are used to keep nuts, or bolts and screws, from working loose. Many special devices are available.

Some forms of lock nuts are shown in Fig. 10-33. There are many spe-

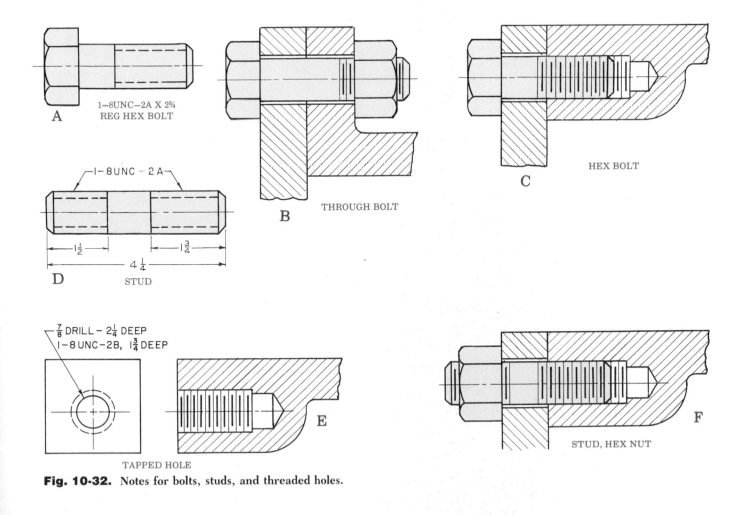

Fig. 10-32. Notes for bolts, studs, and threaded holes.

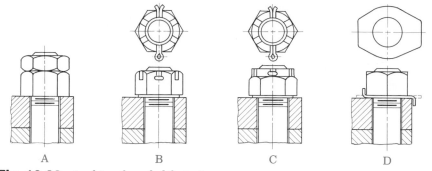

Fig. 10-33. Locking threaded fastenings.

Fig. 10-34. Chemical methods of locking a fastener.

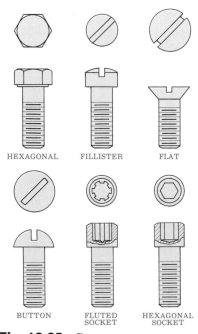

HEXAGONAL   FILLISTER   FLAT

BUTTON   FLUTED SOCKET   HEXAGONAL SOCKET

Fig. 10-35. Capscrews.

cial forms made to provide positive locking.

A self-locking fastener, the Jay Lock (Fig. 10-34), consists of an epoxy chemical locking agent that combines with a hardening agent. When a bolt or screw is screwed into it, it provides a strong, vibration-proof bond.

CAPSCREWS

A capscrew fastens two parts together by passing through a clearance hole in one and screwing into a tapped hole in the other (Fig. 10-35). In most cases, clearance holes are not shown on drawings. Capscrews have a naturally bright finish. This is in keeping with the machined parts with which they are used. Coarse, fine, or 8 threads may be used on capscrews. Socket-head capscrews may have Class 3A threads. Class 2A threads will be found on other head styles. Dimensions for drawing various sizes of American National Standard capscrews are given in Appendix A.

MACHINE SCREWS

Machine screws are used where the fastener must have a small diameter (Fig. 10-36). Sizes below ¼ (.25) in. (6 mm) in diameter have been specified by number. Machine screws may screw into a tapped hole or extend through a clearance hole and into a nut. The nuts used are square nuts. The finish on machine screws is bright. The ends are flat, as shown. Machine screws up to 2 (2.00) in. (50 mm) long are threaded full length. Coarse or fine threads and Class 2 threads may be used on machine screws.

SETSCREWS

Setscrews are used to hold two parts together in a desired position. They do so by screwing through a threaded hole in one part and *bearing* (pushing) against the other (Fig. 10-37). There are two general types: square head and headless. Square-head setscrews can cause accidents when used on *rotating* (turning) parts. They may violate safety codes. Headless setscrews can have either a slot or a socket. Any of the points shown can be used on any setscrew.

WOOD SCREWS

Wood screws are made of steel, brass, or aluminum. They are finished in various ways (Fig. 10-38). Steel screws can be bright (natural finish), blued, galvanized, or copper

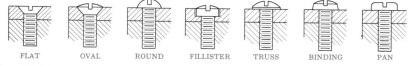

FLAT   OVAL   ROUND   FILLISTER   TRUSS   BINDING   PAN

Fig. 10-36. Machine screws.

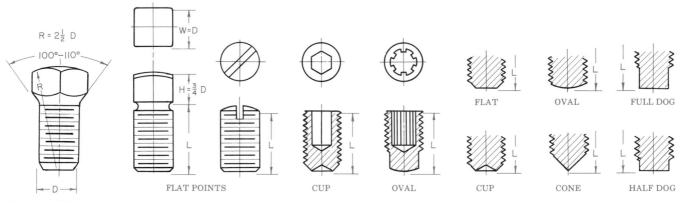

Fig. 10-37. Setscrews.

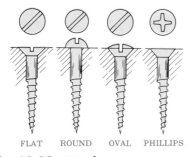

Fig. 10-38. Wood screws.

plated. Both steel and brass screws are sometimes nickel plated. Round-head screws are set with the head above the wood. Flat-head screws are set flush, or countersunk. Wood screws may be drawn as shown. Specify them by number, length, style of head, and finish. For flat-head screws, the length is measured overall. For round-head screws, it is measured from under the head to the point. For oval-head screws, it is measured from the largest diameter of the countersink to the point. You will find sizes and dimensions listed in Appendix A.

■ SOME MISCELLANEOUS THREADED FASTENINGS

These are shown in Fig. 10-39. The names denote the ways in which they are used. Screw hooks and screw eyes are specified by diameter and overall length.

A lag screw, or lag bolt, is used to fasten machinery to wood supports. It is also used in heavy wood constructions when a regular bolt cannot be used. It is similar to a regular bolt but has wood-screw threads. Specify a lag bolt by its diameter and the length from under the head to the point. The head of the lag bolt has the same proportions as a regular bolthead.

■ MATERIALS FOR THREADED FASTENERS

Threaded fasteners are usually made from steel, brass, bronze, aluminum, cast iron, wood, and nylon. Nylon screws and bolts are made in various bright colors, such as red, blue-green, yellow, and white.

■ KEYS

Keys are used to secure pulleys, gears, cranks, and similar parts to a shaft (Fig. 10-40). Keys are made in different forms for different uses. They range from the saddle key, for light duty, to special forms, such as two square keys, for heavy duty. The

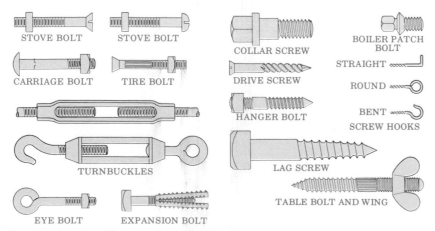

Fig. 10-39. Miscellaneous threaded fastenings.

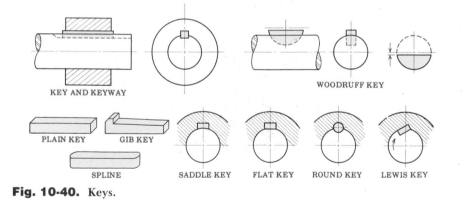

Fig. 10-40. Keys.

common sunk key can have a breadth of about one fourth the shaft diameter. Its thickness can vary from five eighths the breadth to the full breadth. The Woodruff key is often used in machine-tool work. It is made in standard sizes. Specify it by number (see Appendix A). Special forms of pins have been developed to replace keys for some uses. These pins need only a drilled hole instead of the machining that is needed to make keyseats and keyways.

■ RIVETS

Rivets are rods of metal with a pre-formed head on one end. They are used to permanently fasten sheet-metal plates, structual steel shapes, boilers, tanks, and many other items. The rivet is first heated red hot. Then it is placed through the parts to be joined. It is held there in place while a head is formed on the projecting end. The rivet is then said to have been "driven."

Large rivets (Fig. 10-41) have nominal diameters ranging in size from ½ (.50) to 1¾ (1.75) in. (12 mm to 45 mm). Small rivets (Fig. 10-42) range from ¹⁄₁₆ (.06) or ³⁄₃₂ (.09) to ⁷⁄₁₆ (.44) in. (2 or 3 to 11 mm) in diameter.

Some rivets are made especially for use where one side of the plates cannot be reached or where the space is too small to use a regular rivet. These are called *blind* rivets. One type is the du Point explosive rivet (Fig. 10-43). It has a small explosive charge in a cavity. After the rivet is inserted, the charge is exploded, forming a head. This rivet is thus excellent for blind riveting, since the head can be formed inside places that are closed or impossible to reach.

Sometimes plates that are riveted together need to clear surfaces. This requires flush riveting (Fig. 10-44) on one or both sides. These rivets are used on airplanes, automobiles, spacecraft, etc.

Riveted joints (Fig. 10-45) are used for joining plates. They may

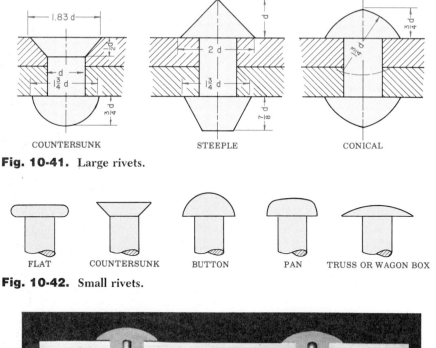

Fig. 10-41. Large rivets.

Fig. 10-42. Small rivets.

Fig. 10-43. Explosive rivet. (E. I. Du Pont de Nemours & Co.)

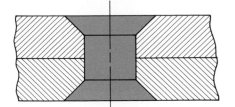

FLUSH BOTH SIDES
THICK PLATES

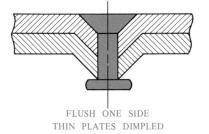

FLUSH ONE SIDE
THIN PLATES DIMPLED

Fig. 10-44. Flush rivets.

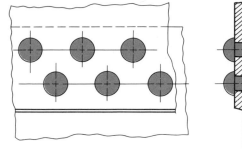

SINGLE-RIVETED LAP JOINT

DOUBLE-RIVETED LAP JOINT
STAGGERED RIVETING

Fig. 10-45. Riveted joints.

have lap or butt joints. They may also have single or multiple riveting. (See Appendix A for American National Standard rivet dimensions.)

For some uses, as in tanks, steel buildings, etc., high-strength structural bolts are used. Welding is also in wide use.

LEARNING ACTIVITIES

1. Collect as many different types of fasteners as possible. Identify each by name. Determine size and thread specifications of each if applicable.

2. Obtain technical drawings from industry. Locate threads and fasteners on the drawings, and study thread symbols and methods of specifying threads and fasteners.

Convert some of the threads and specifications to ISO metric equivalents.

VOCABULARY

1. Helix
2. Helical curve
3. Internal thread
4. External thread
5. Pitch
6. Lead
7. Axis
8. Major diameter
9. Minor diameter
10. Crest
11. Root
12. Single thread
13. Double thread
14. Triple thread
15. Multiple threads
16. Right-hand thread
17. Left-hand thread
18. Thread representation
19. Detailed representation
20. Schematic representation
21. Simplified representation
22. Chamfer
23. Pitch diameter
24. Acme thread
25. Screw-thread series
26. Tolerance
27. Allowance
28. ISO
29. Capscrew
30. Stud bolt
31. Machine screw
32. Setscrew
33. Rivet

REVIEW

1. A device for holding parts together is called a _____.

2. The true shape of a screw thread is based on a _____ curve.

3. The distance from a point on one thread form to a corresponding point on the next thread form is the _____.

4. The distance a threaded part advances with one complete revolution (turn) is called _____.

5. Name the three types of thread representation.

6. Double and triple threads may also be called _____ threads.

7. The three most common Unified thread series are coarse, _____, and _____.

8. A fastener with threads on both ends is called a _____.

9. What is the difference in designating metric fine- and coarse-thread series?

10. If a close fit is necessary for a metric screw thread, the letter _____ may be used in the note.

Problems

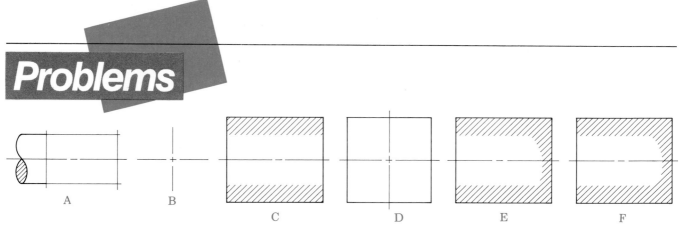

A   B   C   D   E   F

Fig. 10-46. Use dividers to take dimensions from the assigned scale below. Draw the views as follows: (A) schematic representation of 1.00—8UNC—2A threads; (B) end view of A; (C) schematic representation of section through 1.00—8UNC—2B (internal) threads; (D) right-side view of C; (E) schematic representation of section through Ø.88 × 1.50 deep, 1.00—8UNC—2B × 1.12 deep; (F) schematic representation of section through Ø.88 × 1.50 deep, 1.00—8UNC—2B × 1.50 deep.

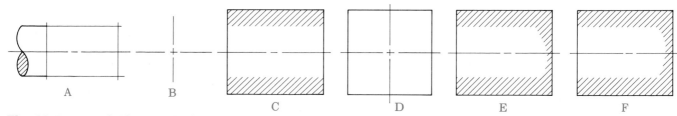

A   B   C   D   E   F

Fig. 10-47. Use dividers to take dimensions from the assigned scale below. Draw the views as follows: (A) simplification representation of 1.00—8UNC—2A threads; (B) end view of A; (C) simplified representation of section through 1.00—8UNC—2B (internal) threads; (D) right-side view of C; (E) simplified representation of section through Ø.88 × 1.50 deep, 1.00—8UNC—2B × 1.12 deep; (F) simplified representation of section through Ø.88 × 1.50 deep, 1.00—8UNC—2B × 1.50 deep.

DECIMAL INCH   FRACTIONAL INCH

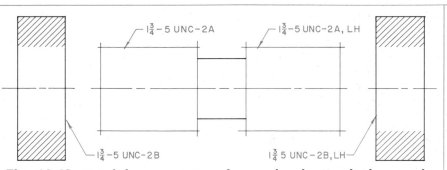

Fig. 10-48. Detailed representation of screw threads. Use dividers to take dimensions from the printed scale assigned at the bottom of the page. Draw the views as shown and complete each according to the specifications noted on each. Use detailed thread representation.

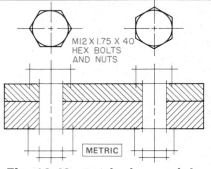

Fig. 10-49. Regular hexagon bolt and nut. Draw the views and complete the bolts and nuts in the sectional view. See Appendix A for bolt and nut detail sizes.

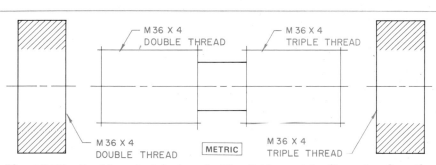

Fig. 10-50. Double and triple threads. Use dividers to take dimensions from the printed scale assigned at the bottom of the page. Draw the views as shown and complete each according to the specifications noted on each. Use detailed thread representation.

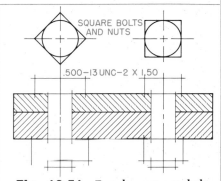

Fig. 10-51. Regular square bolt and nut. Draw the views and complete the bolts and nuts in the sectional view. See Appendix A for bolt and nut detail sizes.

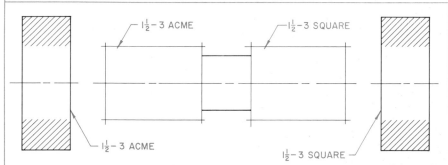

Fig. 10-52. Acme and square threads. Use dividers to take dimensions from the printed scale assigned at the bottom of the page. Draw the views as shown and complete each according to the specifications noted on each. Use detailed thread representation.

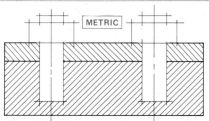

Fig. 10-53. Studs. Draw the view as shown and complete the Ø12 × 44 studs and hexagon nuts. Check Appendix A for specific nut sizes. Other dimensions may be taken from the printed scale assigned at the bottom of the page. Use schematic thread representation.

DECIMAL INCH
FRACTIONAL INCH
1 0 1 2 3 4 5

10 0 10 20 30 40 50 60 70 80 90 100 110 120 130 140
METRIC

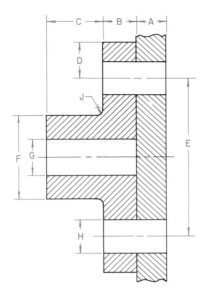

Fig.	Bolt Ø	A	B	C	D	E	F	G	H	J
U.S. Customary (in.)										
10-54	¼ (.250)	.25	.32	.62	.50	2.00	.75	.375	.281	.10
10-55	⅜ (.375)	.38	.44	.62	.62	2.25	.75	.375	.406	.12
10-56	½ (.500)	.50	.56	.88	.62	2.75	1.25	.625	.562	.12
10-57	⅝ (.625)	.62	.70	1.12	.75	3.25	1.75	.750	.688	.12
10-58	¾ (.750)	.75	.80	1.25	.88	3.50	2.00	.875	.812	.20
10-59	1 (1.000)	1.00	1.12	1.50	1.12	4.00	2.50	1.125	1.125	.25
Metric (mm)										
10-60	6	6	8	16	12	50	20	9.6	6.3	3
10-61	10	10	12	20	20	60	20	9.6	10.3	4
10-62	16	16	18	24	24	80	40	16.5	16.5	4
10-63	24	24	26	32	26	100	60	29.0	25.0	6

Figs. 10-54 through 10-63. Take all dimensions from the table for the problem assigned and draw the flange and head plate as shown. On the colored centerlines, draw American National Standard hex or square bolts and nuts or metric hex bolts and nuts as assigned. Place bolthead at the left and show bolthead across flats; nut across corners.

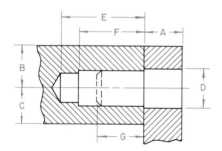

Fig.	Stud Ø	Nut	A	B	C	D	E	F	G
10-64	.750	Hex	.80	.88	.75	.812	1.75	1.38	1.00
10-64	.875	Sq	.94	1.25	.88	.938	2.00	1.56	1.12
10-66	1.000	Hex	1.12	1.12	1.00	1.125	2.25	1.75	1.50
10-67	1.125	Sq	1.25	1.50	1.12	1.250	2.75	2.12	1.50

Figs. 10-64 through 10-67. On the centerline shown, draw a stud with hexagon or square nut, across flats or corners, as directed by the instructor. Take dimensions from the table.

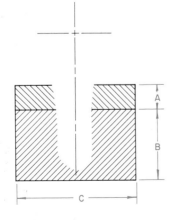

Fig.	Bolt Ø	Head style	A	B	C
U.S. customary (in.)					
10-68	.500	Button (round)	.75	1.75	3.00
10-69	.500	Flat	.75	1.75	3.00
10-70	.500	Fillister	.75	1.75	3.00
Metric (mm)					
10-71	10	Flat	12	40	64
10-72	12	Fillister	12	40	64
10-73	14	Round	12	40	64

Figs. 10-68 through 10-73. Take dimensions from the table for the problem assigned and draw the figure shown at the left. Refer to Appendix A for sizes and draw the assigned style of head and size of capscrew. Also, draw a top view of the screw head.

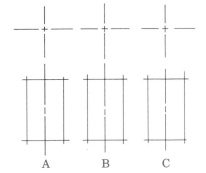

Fig. 10-74. Setscrews. Draw three setscrews: (A) Ø.75 × 1.25 long, square head, flat point; (B) Ø.75 × 1.25 long, slotted head, oval point; (C) Ø.75 × 1.25 long, socket head, cup point. Use schematic thread representation. Do not section.

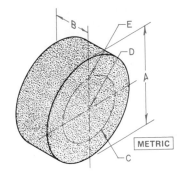

Fig. 10-75. Spacer. Draw two views of the spacer. Use schematic representation to show the threaded holes. A = Ø200, B = 36, C = Ø60, D = M 24 (through), E = M 10 (through). Add notes and all necessary dimensions.

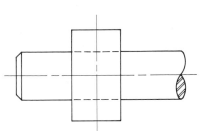

Fig. 10-76. Draw the Ø1.00 × 3.50 long shaft and Ø1.88 × 1.00 collar as shown. Draw the collar in full section. Add a No. 4 × 2.00 American National Standard taper pin on the colored centerline. Estimate sizes not given. Materials: shaft—steel; collar—cast iron.

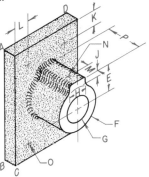

Fig. 10-77. Shaft support. Draw necessary views. At N, draw a Ø.31 setscrew (square head, flat point). At O (four places), draw Ø.38 coarse threads (simplified representation). All fillets and rounds = .12R. AB = 4.50, BC = .50, AD = 3.00, E = 1.50, F = .88R, G = Ø1.00, H = .75, J = .25, K = .62, L = .62, M = .50, P = 1.50.

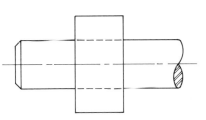

Fig. 10-78. Draw the Ø1.00 × 3.50 long shaft and Ø2.00 × 1.00 collar as shown. Draw the collar in half section and add an ANSI No. 608 Woodruff key at the top of the shaft. Estimate sizes not given. Materials: shaft—steel; collar—cast aluminum.

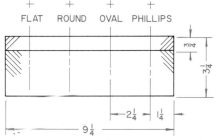

Fig. 10-79. Wood screws. Draw the view shown above. Add 2½" #12 wood screws on the colored centerlines. Show the four head types as specified and draw a top view of each on the center mark above the view.

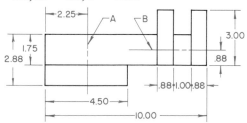

Fig. 10-80. Draw the view shown. On centerline A, draw Ø.50 × 2.50 fillister-head capscrew (head at top). At B, draw a Ø.38 × 4.00 flat-head capscrew. Show view in section. Material: steel. Use simplified or schematic thread representation as assigned.

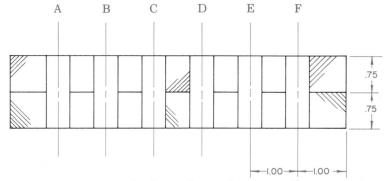

Fig. 10-81. Rivets. Draw the figure shown above. Overall sizes = 1.50 × 7.00. Refer to Chap. 10 and Appendix A and draw the heads (top and bottom) for Ø.50 rivets. Do not dimension. (A) Button head, (B) high button head, (C) cone head, (D) flat-top countersunk head, (E) round-top countersunk head, (F) pan head.

11 Working Drawings

DEFINITIONS

A *working drawing* tells all that needs to be known for making a single part or a complete machine or structure. It tells precisely what the shape and size should be. It also tells exactly what kinds of material should be used, how the finishing should be done, and what degree of accuracy is needed. A photograph and a working drawing of a simple machine part are shown in Fig. 11-1. A working drawing can be a *detail drawing*, an *assembly drawing*, or an *assembly working drawing*.

WORKING DRAWINGS

Working drawings are usually multiview drawings with complete dimen-

sions and notes added. They must provide all necessary information for the manufacture of parts and for assembly. Nothing must be left to guess.

A good working drawing must follow the style and practices of the office or industry where it is made. Most industries follow the standard recommended by the American Na-

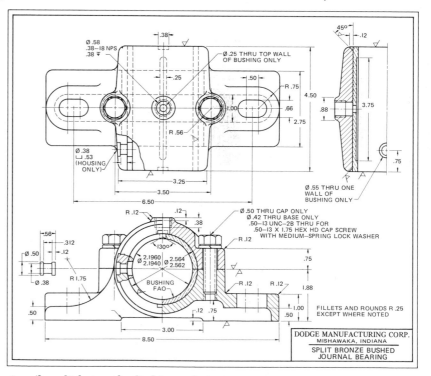

Fig. 11-1. A photograph and a working drawing of a split bronze-bushed journal bearing. (Dodge Manufacturing Corp.)

tional Standards Institute (ANSI). That way, plans can be easily read and understood from one industry to another. However, in some cases, shortcuts are used to save time.

No matter what special styles and practices are used in working drawings, there are certain rules that must be followed. For example, proper line technique must be used to make the contrast sharp. Also, dimensions and notes must be easy to read and accurate. The terms and abbreviations must be the standard ones. When the drawing is finished, it must be carefully checked to avoid mistakes and ensure accuracy. This should be done before submitting the drawing to the supervisor or checker for approval.

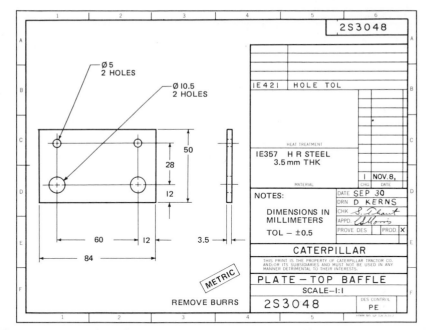

Fig. 11-2. A working drawing of a simple part. (Caterpillar Tractor Co.)

MAKING DRAWINGS FOR INDUSTRIAL USE

In the language of drawing, an object is described giving its shape and its size. All drawings follow this principle, whether they are for steam or gas engines, machines, buildings, airplanes, automobiles, missiles, or satellites.

A student's drawing can look rough and unfinished next to one made by a drafter or engineer. The drafter's skill in giving a finished look comes from a thorough knowledge of drafting and a great deal of practice. In this chapter, you will learn the right order of going about the work and some of the procedures that drafters usually follow. You must become thoroughly familiar with these practices. Otherwise, your drawings will not have the style and good form that is demanded by today's industry.

Many drawings done for industry are done in pencil. However, pencil drawings must have good line technique. They must be neat, have few erasures, and the lines must be dense and sharp.

There is also, however, a growing need for high-quality drawings in ink. This is because many drawings are put through modern micro-reproduction, storage, and retrieval systems. Lines must therefore be thick and sharp, and lettering must be large and neat.

DETAIL DRAWINGS

A drawing of a single piece that gives all the information needed for making it is called a *detail drawing*. An example is the drawing of a simple part in Fig. 11-2. A detail working drawing must be a full and exact description of the piece. It should show carefully selected views and include well-placed dimensions (Fig. 11-3).

When a large number of machines are to be produced, it is usual to make a detail drawing of each part on a separate sheet. This is done especially when some of the parts may be used on other machines. In some industries, however, it is usual to detail several parts of a machine on the same sheet. Sometimes a separate detail drawing is made for each of several workers involved, such as the patternmaker, machinist, or welder. Such a drawing shows only the dimensions and information needed by the worker for whom it is made. Figure 11-4 shows an index-plunger operating handle made by the Hartford Special Machinery Company. The picture shows the piece as a forging and also after it has been machined. Figure 11-5 shows a working drawing of the same piece. Notice how the parts to be removed after all work is done are shown. Notice also the detailed list of machine operations. Figure 11-6 shows a *combination drawing* (two-part detail drawing) for an oil-pump drive gear. The right-hand half gives the dimensions for the forging. The left-hand half

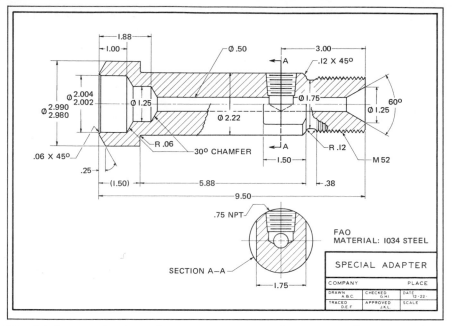

Fig. 11-3. A single view and an extra section provide a com̲ ̲ description of the special adapter.

Within the drawing (Fig. 11-3) the following text/labels appear:

1.88
1.00
Ø .50
3.00
.12 X 45°
A
Ø 2.004 / 2.002
Ø 2.990 / 2.980
Ø 1.25
Ø 1.75
Ø 1.25
60°
Ø 2.22
.06 X 45°
R .06
30° CHAMFER
A
R .12
M 52
.25
(1.50)
5.88
1.50
.38
9.50
.75 NPT
FAO
MATERIAL: 1034 STEEL
SECTION A—A
1.75

SPECIAL ADAPTER
COMPANY   PLACE
DRAWN A.B.C. / CHECKED G.H.I / DATE 12-22-
TRACED D.E.F / APPROVED J.K.L / SCALE

gives the machining dimensions and notes. A detail drawing can also contain calculated data.

Detail drawings are often made for standard parts that come in a range of sizes. Figure 11-7 is a *tabulated,* or *tabular,* drawing of a bushing. In this kind of drawing, the different

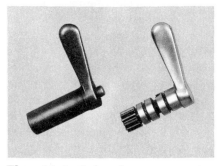

Fig. 11-4. Index-plunger operating handle—forging and finished part. (The Hartford Special Machinery Co.)

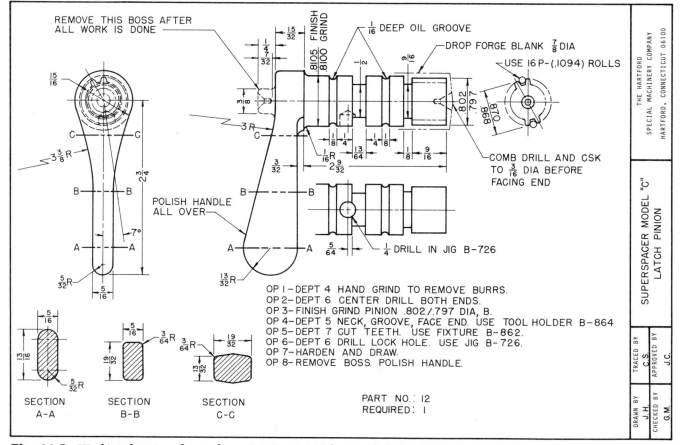

Within the working drawing (Fig. 11-5) the following text/labels appear:

REMOVE THIS BOSS AFTER ALL WORK IS DONE
15/32 FINISH GRIND
1/16 DEEP OIL GROOVE
1/4 7/32
.8105 / .8100
DROP FORGE BLANK 7/8 DIA
USE 16 P-(.1094) ROLLS
15/16
3/8
9/16
.802 / .797
.870 / .868
3 R
1/8 1/4
13/64
1/4 1/8
1/8
9/16
COMB DRILL AND CSK TO 3/16 DIA BEFORE FACING END
3 3/8 R
1/16 R
3/32
2 9/32
2 3/4
C C
B B
A A
POLISH HANDLE ALL OVER
7°
5/32 R
5/16
13/32 R
5/64
1/4 DRILL IN JIG B-726

5/16
13/16
5/32 R
SECTION A-A

5/16
19/32
3/64 R
SECTION B-B

3/64 R
19/32
13/32
SECTION C-C

OP 1—DEPT 4 HAND GRIND TO REMOVE BURRS.
OP 2—DEPT 6 CENTER DRILL BOTH ENDS.
OP 3—FINISH GRIND PINION .802/.797 DIA, B.
OP 4—DEPT 5 NECK, GROOVE, FACE END. USE TOOL HOLDER B-864
OP 5—DEPT 7 CUT TEETH. USE FIXTURE B-862.
OP 6—DEPT 6 DRILL LOCK HOLE. USE JIG B-726.
OP 7—HARDEN AND DRAW.
OP 8—REMOVE BOSS. POLISH HANDLE.

PART NO.: 12
REQUIRED: 1

THE HARTFORD SPECIAL MACHINERY COMPANY HARTFORD, CONNECTICUT 06100

SUPERSPACER MODEL "C" LATCH PINION

TRACED BY C.S.
APPROVED BY J.C.
DRAWN BY J.H.
CHECKED BY G.M.

Fig. 11-5. Working drawing of part shown in Fig. 11-4. (The Hartford Special Machinery Co.)

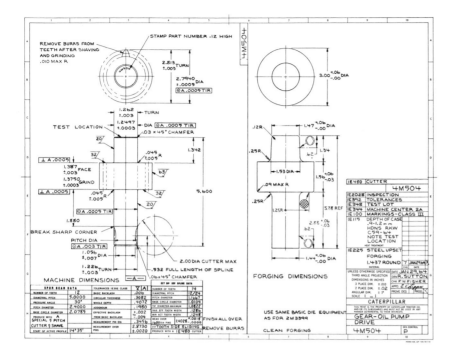

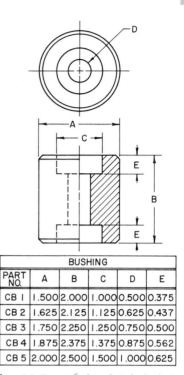

Fig. 11·6. A two-part detail drawing showing separate information for forging and machining. (Caterpillar Tractor Co.)

Fig. 11·7. Tabulated (tabular) drawing.

BUSHING					
PART NO.	A	B	C	D	E
CB 1	1.500	2.000	1.000	0.500	0.375
CB 2	1.625	2.125	1.125	0.625	0.437
CB 3	1.750	2.250	1.250	0.750	0.500
CB 4	1.875	2.375	1.375	0.875	0.562
CB 5	2.000	2.500	1.500	1.000	0.625

dimensions are identified by letters. A table placed on the drawing tells what each dimension is for different sizes of the part. Either all or some of the dimensions can be given in this way. A similar kind of drawing is one in which the views are drawn with blank spaces for dimensions and notes (Fig. 11-8 at A). These are then filled in as needed with the required

information (as at B). The views will not be to scale, except perhaps for one size.

■ ASSEMBLY DRAWINGS

A drawing of a fully assembled construction is called an *assembly drawing*. Such drawings vary greatly in

how completely they show detail and dimensioning. Their special value is that they show how the parts fit together, the look of the construction as a whole, the dimensions needed for installation, the space the construction needs, the foundation, the electrical or water connections, and so forth. When an assembly drawing gives complete information, it can be

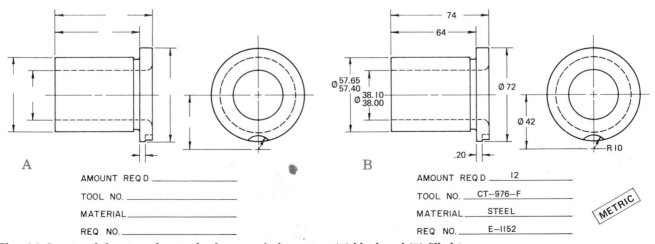

Fig. 11·8. Detail drawing of a standard part with dimensions (A) blank and (B) filled in.

used as a working drawing. It is then called an *assembly working drawing*. This kind of drawing can be made only when there is little or no complex detail. An example is shown in Fig. 11-9. You can often show furniture and other wood construction in assembly working drawings by adding enlarged details or partial views as needed (Fig. 11-10).

Assembly drawings of machines are generally made to a small scale. Dimensions are chosen to tell overall distances, important center-to-center distances, and local dimensions. All, or almost all, hidden lines may be omitted. Also, if the drawing is made to a very small scale, unnecessary detail may be omitted. This has

been done, for example, in Fig. 11-11. Either exterior or sectional views may be used. When the main aim is just to show the general look of the construction, only one or two views need to be drawn. Because some assembled constructions are so large, different views may need to be drawn on separate sheets. However, the same scale must be used on all sheets.

A special assembly drawing is the *reference assembly drawing* (Fig. 11-12). It is made for reference to identify parts to be assembled. Note the tabular list in the upper right-hand corner. Note the dimensions shown.

Many other kinds of assembly drawings are made for special pur-

poses. These include part assemblies for groups of parts, drawings for use in assembling or erecting a machine, drawings to give directions for upkeep and use, and so forth. A most important kind of assembly drawing is the *design layout*. It is from this kind of drawing that the detail drawings are made.

■ CHOICE OF VIEWS

Your drawings will be easier to use if the views are chosen properly. To completely describe an object, two views are generally needed. Although a drawing is not a picture, the views chosen should always be those that are easiest to read. Each view

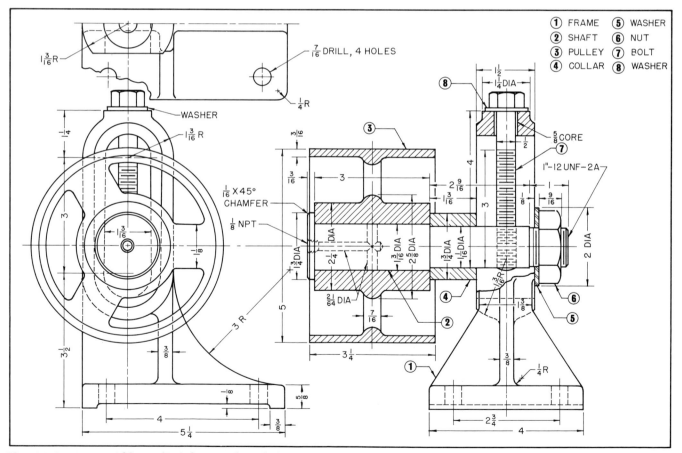

Fig. 11-9. An assembly working drawing for a belt tightener.

PICTORIAL RENDERING
OF STEREO CABINET

Fig. 11-10. An assembly working drawing with enlarged details and partial views.

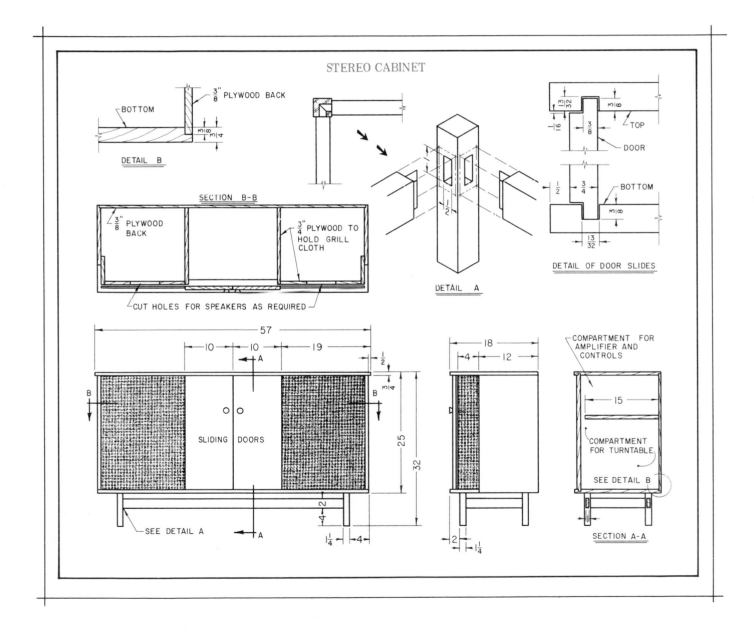

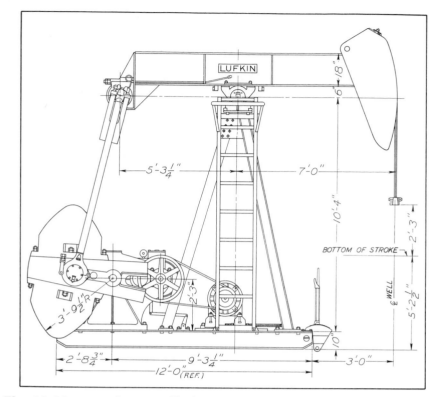

Fig. 11-11. An outline assembly drawing. (Lufkin Foundry and Machine Co.)

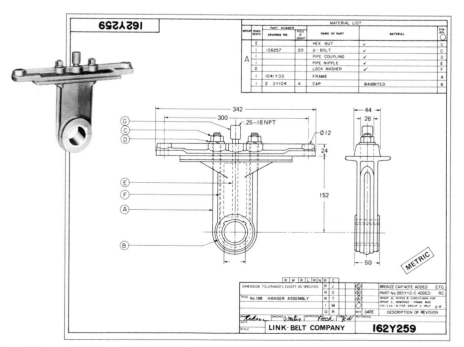

Fig. 11-12. A reference assembly drawing. (Link-Belt Co.)

must have a part in the description. Otherwise, it is not needed and it should not be drawn. In some cases, only one view is needed, as long as the shape and size are clearly understood. Also, information may be given in a note rather than in a drawing of a second view. For complex pieces, more than three views may be needed. Some of these may be partial views, auxiliary views, or sectional views. If there is any question of which views to draw, think of why you are making the drawing. A drawing is judged on how clearly and precisely it gives the information needed for its purpose.

■ CHOICE OF SCALE

The scale for a detail drawing is chosen according to three factors:

1. How big the drawing must be to show details clearly
2. How big it must be to include all dimensions without crowding
3. How big the paper is

In most cases the best choice is a full-size scale. Other scales commonly used are half, quarter, and eighth. Avoid such scales as 1:6 (2″ = 1′), 1:3 (4″ = 1′), and 1:1.33 (9″ = 1′). If a part is very small, you can sometimes draw it to an enlarged scale, perhaps twice full size or more.

When drawing a number of details on one sheet, make them all to the same scale if possible. If different scales are used, note them near each drawing. It is often useful to draw a detail, or part detail, to a larger scale on the main drawing. This will save both time and work in making separate detail drawings. For general assembly drawings, choose a scale that will show the details you want and look good on the size of paper being used. Sheet-metal pattern drawings are generally made full size for direct transfer to the sheet material in the shop.

For complete assemblies, a small scale is generally used. The scale is often fixed by the size of the paper the company has chosen for assemblies. For part assemblies, choose a scale to suit the purpose of the drawing. This might be to show how parts fit together, to identify the parts, to explain an operation, or to give other information.

TITLE BLOCKS

Every sketch or drawing must have some kind of title block. However, the form, completeness, and location of the title block can vary. On working drawings, it can be placed in a box in the lower right-hand corner (Fig. 11-13 at A) or it can be included in a record strip running across the bottom or end of the sheet (Fig. 11-13 at B) as far as needed.

The title block gives as much as is needed of the following information:

1. The name of the construction, machine, or project.
2. The name of the part or parts shown, or simple details.
3. Manufacturer, company, or firm name and address.
4. The date, usually the date of completion.
5. The scale or scales.
6. Heat treatment, working tolerances, and so forth.
7. Numbers of the drawing: of the shop order or of the customer's order, according to the system used.
8. Drafting-room record: names or initials, with dates, of drafter, tracer, and checker, and approval of chief drafter, engineer, and so forth.
9. A revision block or space for recording the changes, when needed, should be placed above or at the left of the title block.

A form for a basic layout of a title block is shown in Fig. 11-14. However, the arrangement, size, and content may vary.

In large drafting rooms, the title block is generally printed on the paper, cloth, or film, leaving spaces to be filled in. However, many firms use separate printed adhesive title blocks.

LIST OF MATERIALS OR PARTS LIST

Most working drawings include a list of parts, the materials of which they are made, identification numbers, or other important information. This information is especially important on assembly drawings of various kinds. It is also needed on detail drawings where a number of parts are shown on the same sheet.

The names of parts, material, number required, part numbers, and so forth are often given in notes near the views of each part. It is better, however, to place just the part numbers near the views, link them to the views with leaders, and then collect all the other information in tabulated lists. Such a list is called a *list of material, bill of material,* or *parts list* (Fig. 11-15) and is often placed above the title block. However, it may also be placed in the upper right corner of the sheet. It can also be written or typed on a separate sheet. If this method is used, it would be titled "Parts List for Drawing No. 00" to identify it. The recommended form is shown in Fig. 11-16. However, the column widths and names may vary as needed.

GROUPING AND PLACING PARTS

When a number of details are used for a single machine, they are often grouped on a single sheet or set of sheets. A convenient method is to group the forging details together, the casting details together, the brass details together, and so on for other materials. In general, parts should be shown in the position they will occupy in the assembled machine. That way, related parts will

RIVETS EXC. NOTED
HOLES EXC. NOTED
PAINT:
BLUE PRINT RECORD
NO. DATE ISSUED FOR
CENTRAL TEXAS IRON WORKS
WACO TEXAS ABILENE
BUILDING
LOCATION
CUSTOMER
ARCHITECT
MADE BY CHK. BY SHEET NO.
DATE REVISIONS: ORDER NO.
A

B
UNIT NAME OF PIECE
DR. DATE SYMBOL OF MACHINES USED ON SUPERSEDES DRAW. STOCK CASTING DROP FORGING
DR.
TR. THE LODGE & SHIPLEY MACHINE TOOL CO. SUPERSEDED BY DR. MATERIAL PIECE NO.
TR. CH. FORM 795 CINCINNATI, OHIO. U. S. A.

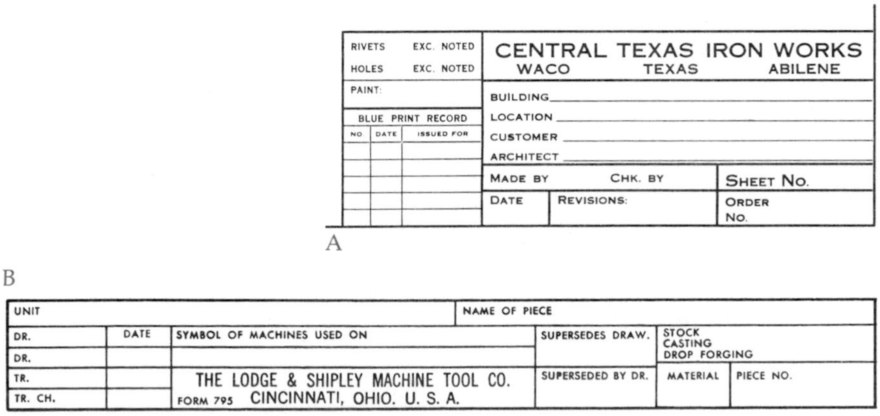

Fig. 11-13. Title blocks. (A) A boxed title. (B) A strip title.

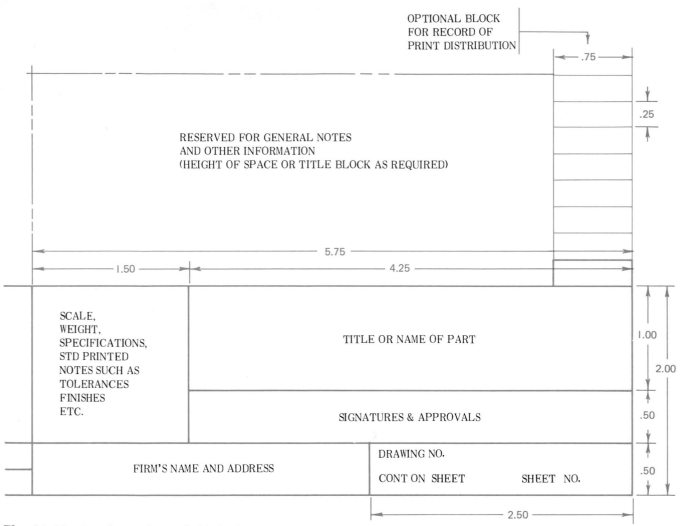

OPTIONAL BLOCK
FOR RECORD OF
PRINT DISTRIBUTION

.75

.25

RESERVED FOR GENERAL NOTES
AND OTHER INFORMATION
(HEIGHT OF SPACE OR TITLE BLOCK AS REQUIRED)

5.75

1.50   4.25

SCALE,
WEIGHT,
SPECIFICATIONS,
STD PRINTED
NOTES SUCH AS
TOLERANCES
FINISHES
ETC.

TITLE OR NAME OF PART

1.00

2.00

SIGNATURES & APPROVALS

.50

FIRM'S NAME AND ADDRESS

DRAWING NO.

CONT ON SHEET   SHEET NO.

.50

2.50

Fig. 11-14. Basic layout for a title block. (ANSI Y14.1)

BILL OF MATERIAL FOR IDLER PULLEY			
NAME	REQD	MATL	NOTES
IDLER PULLEY	I	CI	
IDLER PULLEY FRAME	I	CI	
IDLER PULLEY BUSHING	I	BRO	
IDLER PULLEY SHAFT	I	CRS	
Ø .62 HEX NUT	I		PURCHASED
WOODRUFF KEY # 405	I		PURCHASED
.I2 OILER	I		PURCHASED

Fig. 11-15. A bill of material.

appear near each other. Long pieces, however, such as shafts and bolts, should not be drawn this way. They are drawn with their long dimensions parallel to the long dimension of the sheet (Fig. 11-17).

■ NOTES AND SPECIFICATIONS

Information that you cannot make clear in a drawing must be given in lettered notes and symbols. For ex-

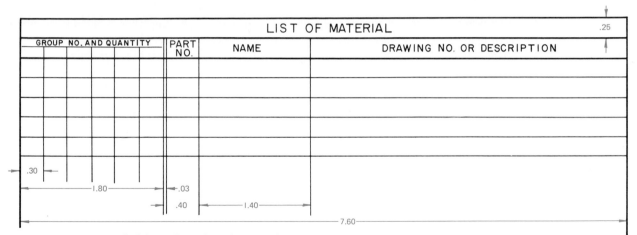

LIST OF MATERIAL					
GROUP NO. AND QUANTITY			PART NO.	NAME	DRAWING NO. OR DESCRIPTION

Fig. 11-16. Recommended form for a list of materials. (ANSI Y14.1)

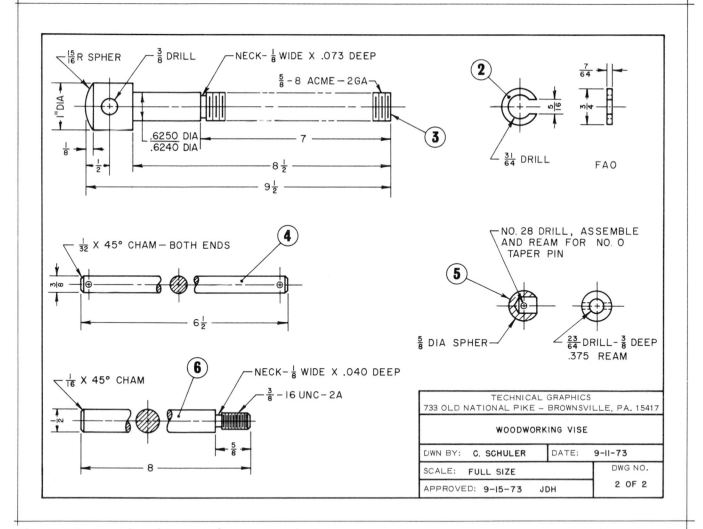

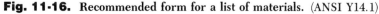

TECHNICAL GRAPHICS 733 OLD NATIONAL PIKE – BROWNSVILLE, PA. 15417		
WOODWORKING VISE		
DWN BY: C. SCHULER	DATE: 9-11-73	
SCALE: FULL SIZE	DWG NO. 2 OF 2	
APPROVED: 9-15-73 JDH		

Fig. 11-17. Several machine parts drawn on one sheet.

ample, trade information that is generally understood by those on the job is often given in this way. Notes such as these are used for the following items: the number (quantity) needed, the material, the kind of finish, the kind of fit, the method of machining, the kinds of screw threads, the kinds of bolts and nuts, the sizes of wire, the thickness of sheet metal, and other similar information.

The materials in general use are wood, plastic, cast iron, wrought iron, steel, brass, aluminum, and various alloys. All parts to go together must be of the proper size so that they will fit. Pieces may be left rough, partly finished, or completely finished. Wood used in furniture is shaped with woodworking tools and machines. Many metals, such as cast iron, brass, aluminum, and so forth, are made into shapes by molding, casting, and machining. First, a wooden pattern of the required shape and size is made. Then this pattern is placed in sand to make an impression, or mold, into which the molten metal is poured. Wrought iron and steel are made into shapes by rolling or forging in the rolling mill or blacksmith shop. Some kinds of steel may be cast.

There are many interesting ways of forming metals for special purposes. There are also many special alloys that this book does not have room to describe. However, you will learn much by observing the different shapes of machine parts and the many materials of which they are made.

After a part is cast or forged, it must be machined on all surfaces that are to fit with other surfaces.

Round surfaces are generally formed on a lathe. Flat surfaces are finished or smoothed on a planer, milling machine, or shaper. Holes are made with drill presses, boring mills, or lathes. Extra metal is allowed for surfaces that are to be finished. To specify such surfaces, place a $\sqrt{}$ symbol on the lines that represent their edges. If the entire piece is to be finished, write a note such as FINISH ALL OVER or FAO. No other mark is needed.

The kinds of machining, finish, or other treatment are specified in notes. Such notes would read, for example, SPOT FACE, GRIND, POLISH, KNURL, CORE, DRILL, REAM, COUNTERSINK, COUNTERBORE, HARDEN, CASE-HARDEN, BLUE, or TEMPER. Often other notes are added for special directions, to explain the assembly, for example, or the order of doing work.

■ CHECKING A DRAWING

After a drawing is finished, it must be looked over very carefully before it is used. This is called *checking the drawing*. It is very important work. A drawing you have made should be checked by someone other than you. A person who has not worked on the drawing will be better able to spot errors.

To make a thorough check, there is a set order of procedure that must be followed. These checks should be made:

1. See that the views completely describe the shape of each piece.

2. See that there are no unnecessary views.
3. See that the scale used was large enough to show all detail clearly.
4. See that all views are to scale and that the right dimensions are given.
5. See that there will be no parts that will interfere with each other during assembly or operation and that necessary clearance space is provided around all parts that need it.
6. See that enough dimensions are given to define the sizes of all parts completely, and that no unnecessary or duplicate dimensions are given.
7. See that all necessary location or positioning dimensions are given with necessary precision.
8. See that necessary tolerances, limits and fits, and other precision information is given.
9. See that the kind of material and the quantity needed of each part are specified.
10. See that the kind of finish is specified, that all finished surfaces are marked, and that a finished surface is not called for where one is not needed.
11. See that standard parts and stock items, such as bolts, screws, pins, keys or other fastenings, handles, and catches, are used where suitable.
12. See that all necessary explanatory notes are given and that they are properly placed on the drawing.

Each drafter must inspect his or her own work for errors or omissions before having it further checked.

LEARNING ACTIVITIES

1. Visit industrial shops and study various methods of forming and machining materials such as metal, wood, and plastic.

2. Obtain different types of working drawings from industry. Study each to determine which are detail drawings, assembly drawings, or assembly working drawings.

VOCABULARY

1. Working drawing
2. Detail drawing
3. Assembly drawing
4. Assembly working drawing
5. Reference assembly drawing
6. Bill of material
7. Title block
8. FAO
9. ANSI
10. Combination drawing

REVIEW

1. A drawing showing all that needs to be known for manufacturing a part or complete product is a _____ drawing.

2. If an entire part is to be finished (machined), what general note is used?

3. Is Fig. 11-1 a detail drawing, an assembly drawing, or an assembly working drawing?

4. A drawing of a single part with all information needed for making it is called a _____ drawing.

5. What is another name for a list of material or parts list?

6. What is the outside diameter of Part No. CB 4 in Fig. 11-7?

7. What three factors help determine the scale to be used on a drawing?

8. What is another way of specifying the scale used for the drawing in Fig. 11-17?

9. What symbol is used to call for a finished surface?

10. Is Fig. 11-5 a detail drawing, an assembly drawing, or an assembly working drawing?

Problems

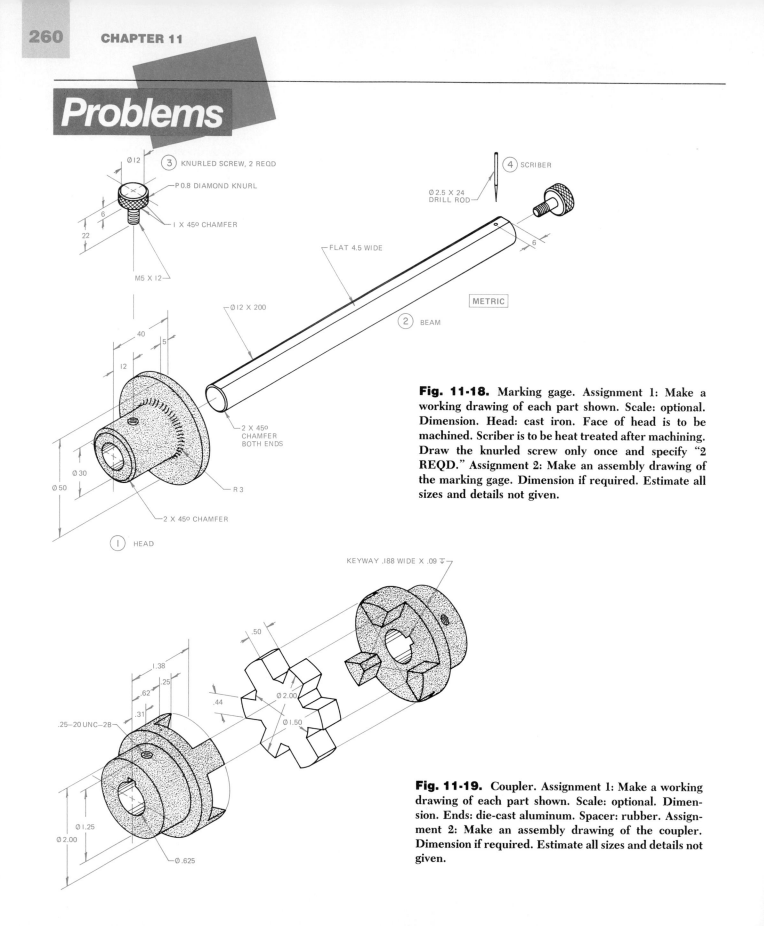

Ø12
(3) KNURLED SCREW, 2 REQD
P 0.8 DIAMOND KNURL
6
1 X 45° CHAMFER
22
M5 X 12

(4) SCRIBER
Ø 2.5 X 24 DRILL ROD
6

FLAT 4.5 WIDE

METRIC

Ø12 X 200
(2) BEAM

40
5
12
Ø 30
Ø 50
2 X 45° CHAMFER BOTH ENDS
R 3
2 X 45° CHAMFER

(1) HEAD

Fig. 11-18. Marking gage. Assignment 1: Make a working drawing of each part shown. Scale: optional. Dimension. Head: cast iron. Face of head is to be machined. Scriber is to be heat treated after machining. Draw the knurled screw only once and specify "2 REQD." Assignment 2: Make an assembly drawing of the marking gage. Dimension if required. Estimate all sizes and details not given.

KEYWAY .188 WIDE X .09

.50
1.38
.62   .25
.44
.31
Ø 2.00
Ø 1.50
.25—20 UNC—2B
Ø 1.25
Ø 2.00
Ø .625

Fig. 11-19. Coupler. Assignment 1: Make a working drawing of each part shown. Scale: optional. Dimension. Ends: die-cast aluminum. Spacer: rubber. Assignment 2: Make an assembly drawing of the coupler. Dimension if required. Estimate all sizes and details not given.

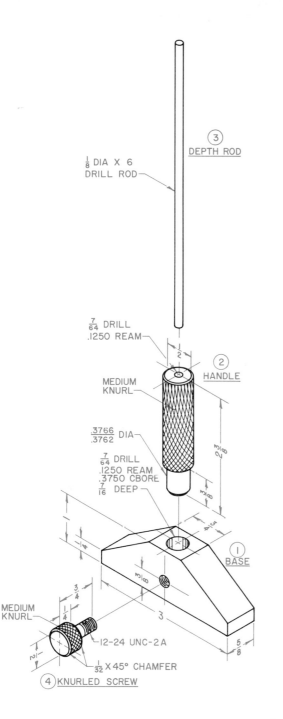

$\frac{1}{8}$ DIA X 6
DRILL ROD

③ DEPTH ROD

$\frac{7}{64}$ DRILL
.1250 REAM

② HANDLE

MEDIUM
KNURL

$\frac{.3766}{.3762}$ DIA

$\frac{7}{64}$ DRILL
.1250 REAM
.3750 CBORE
$\frac{7}{16}$ DEEP

① BASE

MEDIUM
KNURL

12-24 UNC-2A
$\frac{1}{32}$ X 45° CHAMFER

④ KNURLED SCREW

Fig. 11-20. Depth gage. Assignment 1: Make a working drawing of each part shown. Scale: optional. Dimension. All parts: cold-rolled steel. Assignment 2: Make an assembly drawing of the depth gage. Dimension if required. Estimate all sizes and details not given.

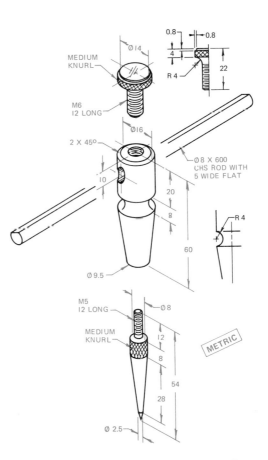

0.8   0.8
MEDIUM
KNURL
Ø14
4   22
R 4

M6
12 LONG

Ø16

2 X 45°
Ø 8 X 600
CRS ROD WITH
5 WIDE FLAT

10

20

8

R 4

60

Ø 9.5

M5
12 LONG   Ø 8
MEDIUM
KNURL

12

8

METRIC

54

28

Ø 2.5

Fig. 11-21. Trammel. Assignment 1: Make a working drawing of each part shown. Scale: optional. Dimension. Specify "2 REQD" for the point, body, and knurled screw. The point is to be heat treated after machining. Assignment 2: Make an assembly drawing of the trammel. Estimate all sizes and details not given.

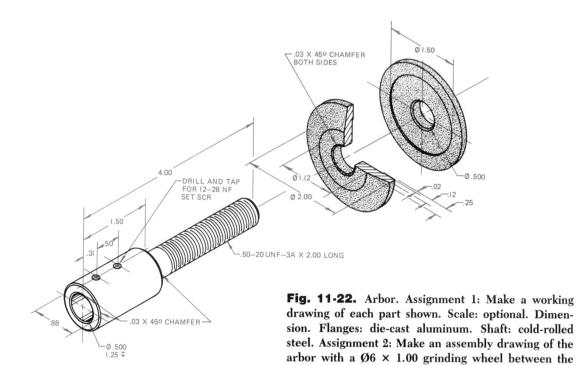

Fig. 11-22. Arbor. Assignment 1: Make a working drawing of each part shown. Scale: optional. Dimension. Flanges: die-cast aluminum. Shaft: cold-rolled steel. Assignment 2: Make an assembly drawing of the arbor with a Ø6 × 1.00 grinding wheel between the flanges. Show sectional views where necessary. Draw all fasteners. Estimate all sizes and details not given.

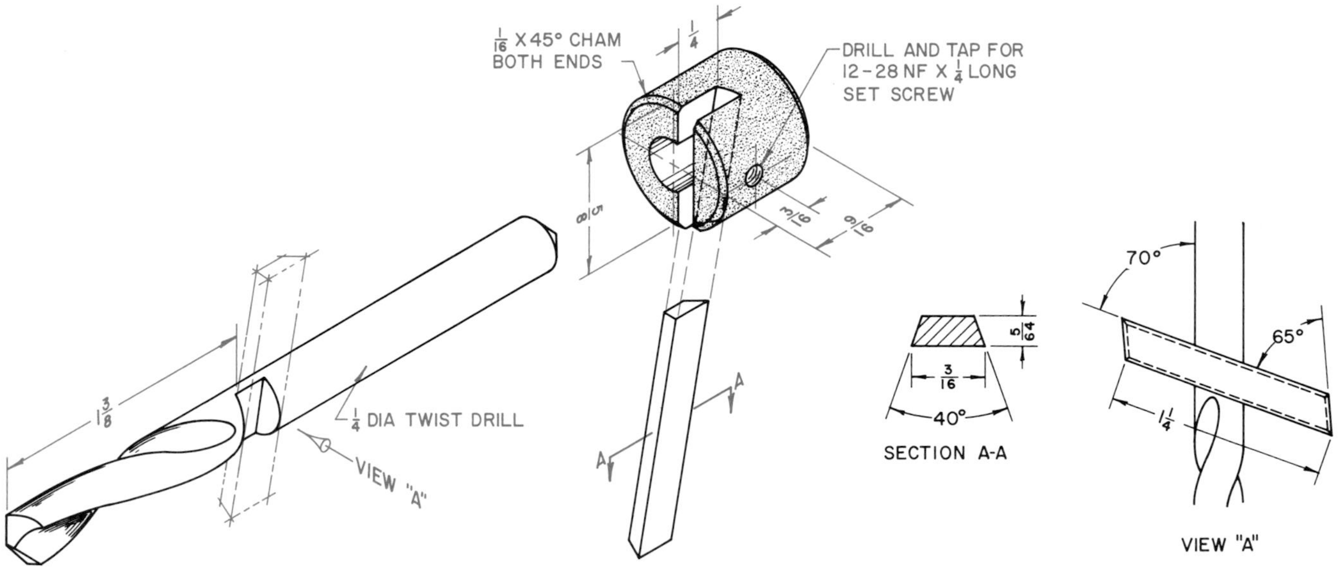

Fig. 11-23. Power expansion bit. Assignment 1: Make a working drawing of each part shown. Scale: optional. Dimension. Cutter: tool steel. Body: cast iron. Use sectional views where necessary. Assignment 2: Make an assembly drawing of the power expansion bit. Dimension if required. Estimate all sizes and details not given.

Fig. 11-24. Level. Assignment 1: Make a working drawing of each part shown except the level glass. Body: die-cast aluminum. Top plate: cold-rolled steel. Dimension. Use sectional views where necessary. Fillets = ⅛ R. Assignment 2: Make an assembly drawing of the level. Dimension if required. Redesign the level to include vertical and 45° angle level glasses if desired. Estimate all sizes and details not given.

DRILL AND
CSK FOR 6-32 UNC X ¼
FLAT HEAD MACHINE SCREW

¼ DIA X 1¾ LEVEL GLASS TO
BE ATTACHED WITH LATEX
CAULK. STRIPES TO BE
ADDED AFTER ASSEMBLY

NOTE: ALL CHAMFERS
1/32 X 45°

¼ - 20 UNC-3B FOR SET SCR

½ REAM

¼ SQ X 2¼
CUTTER TO BE
SHARPENED AS
DESIRED

4 ¾

SLOT 5/32 WIDE X 1/32 DP

¼ DIA X 1¾ LONG
TWIST DRILL

¼ SQUARE HOLE
BROACHED THROUGH

¼ - 20 UNC-3B FOR
SET SCREW

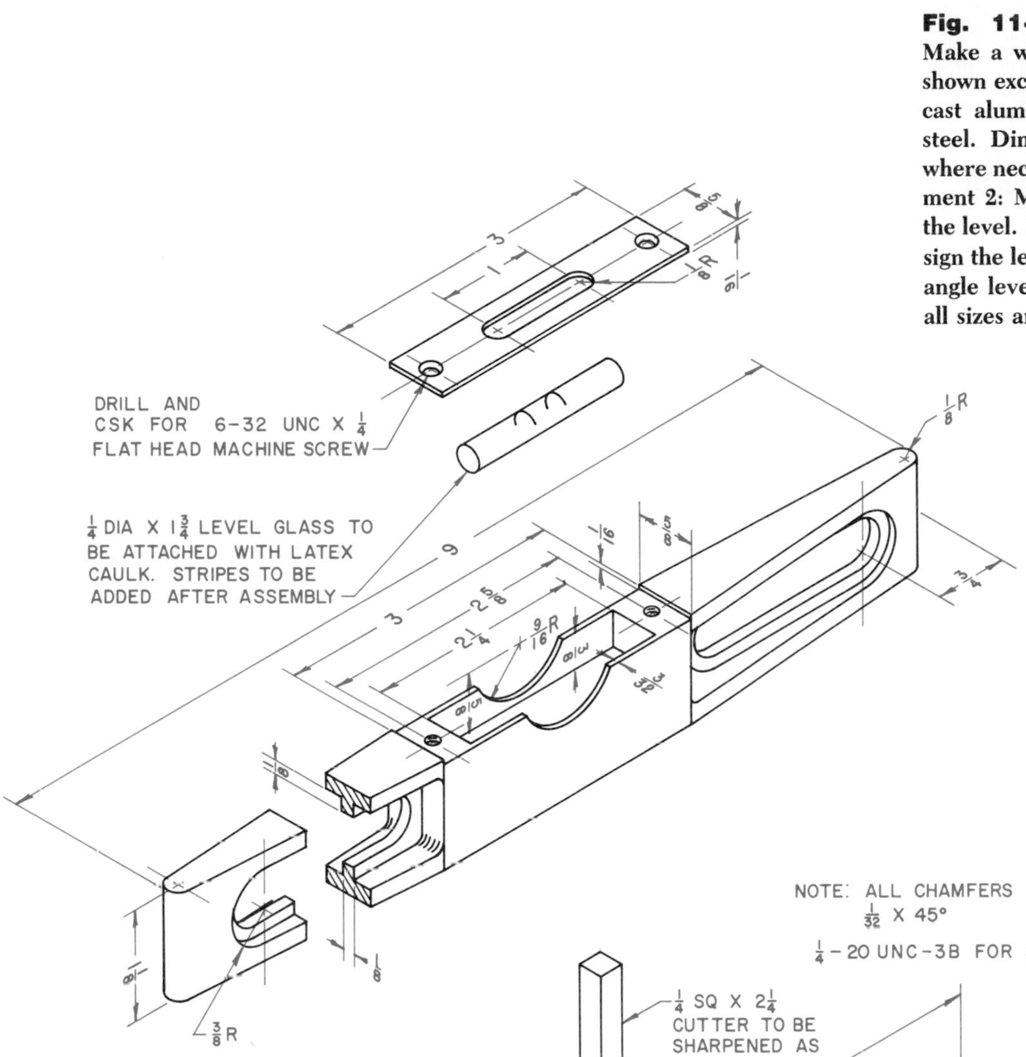

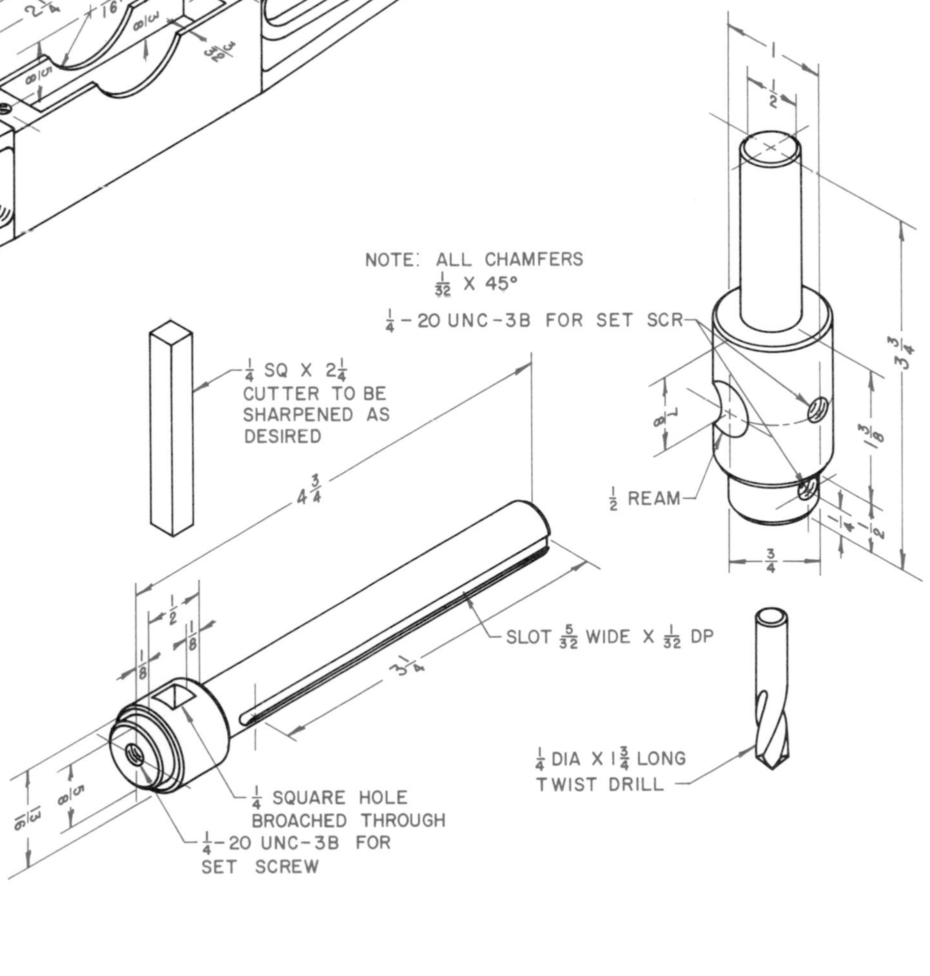

Fig. 11-25. Circle cutter. Assignment 1: Make a working drawing of each part shown. Scale: optional. Dimension. Cutter: tool steel. Body and tool holder: cold-rolled steel. Assignment 2: Make an assembly drawing of the circle cutter. Show sectional views where necessary. Draw all fasteners. Estimate all sizes and details not given.

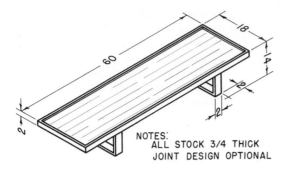

Fig. 11-26. Coffee table. Make an assembly working drawing with all necessary details and partial views. Scale: optional. Dimension. Wood and types of joints optional (see. Fig. 11-10). Make a complete bill of material. Estimate all sizes not given.

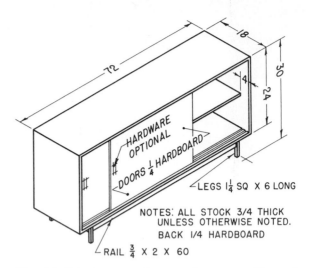

Fig. 11-27. Storage cabinet. Make an assembly working drawing with all necessary details and partial views. Scale: optional. Dimension. Wood and types of joints optional (see Fig. 11-10). Make a complete bill of material. Estimate all sizes not given.

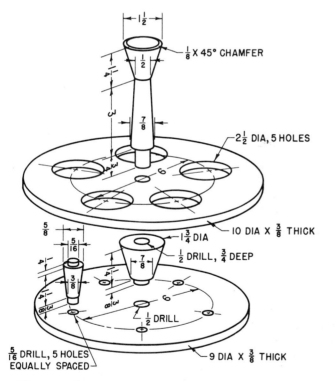

Fig. 11-28. Beverage server. Assignment 1: Make detail working drawings of each part shown. Scale: optional. Dimension. Material: hardwood. Assignment 2: Make an assembly drawing of the beverage server. Dimension if required. Assignment 3: Make a complete bill of material.

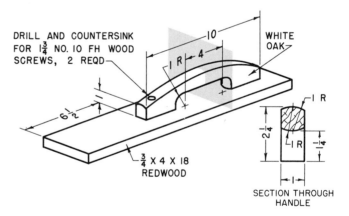

Fig. 11-29. Cement float. Make an assembly working drawing of the cement float. Scale: optional. Dimension. Materials: as noted. Use a 1″ grid for curved outline of handle. Add a bill of material on the same drawing sheet.

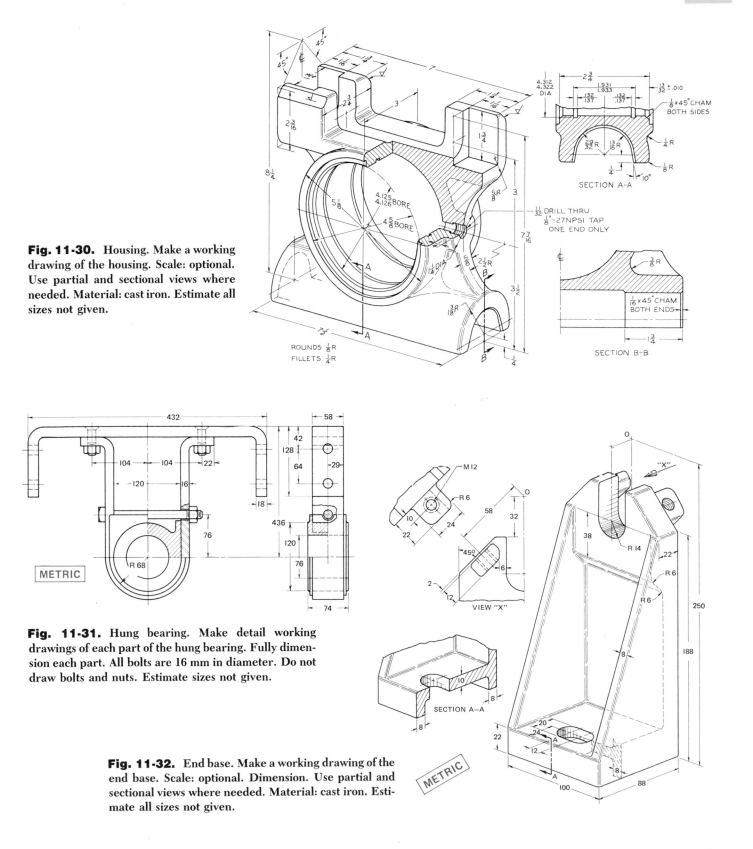

Fig. 11-30. Housing. Make a working drawing of the housing. Scale: optional. Use partial and sectional views where needed. Material: cast iron. Estimate all sizes not given.

Fig. 11-31. Hung bearing. Make detail working drawings of each part of the hung bearing. Fully dimension each part. All bolts are 16 mm in diameter. Do not draw bolts and nuts. Estimate sizes not given.

Fig. 11-32. End base. Make a working drawing of the end base. Scale: optional. Dimension. Use partial and sectional views where needed. Material: cast iron. Estimate all sizes not given.

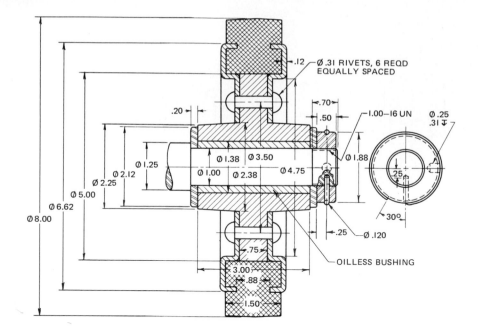

Fig. 11-33. Cushion wheel. Assignment 1: Make a front view and section of the cushion wheel, full size. This type of wheel is used on warehouse or platform trucks to reduce noise and vibration. Assignment 2: Make a complete set of detail drawings, full size, with a bill of material, for the cushion wheel. Three sheets will be needed. Rivets are purchased and, therefore, need not be detailed but would be listed in the bill of material.

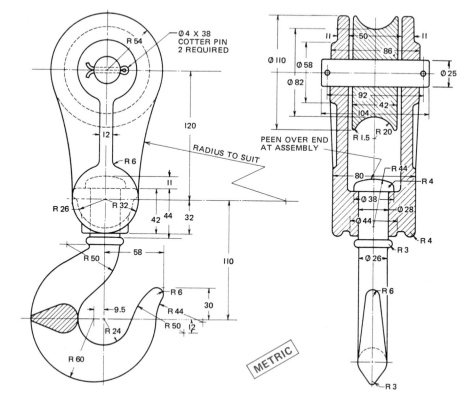

Fig. 11-34. Crane hook. Make detail drawings of the crane hook parts.

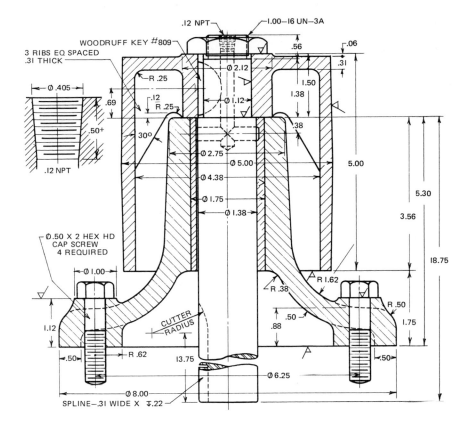

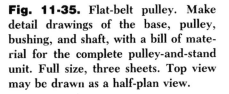

Fig. 11-35. Flat-belt pulley. Make detail drawings of the base, pulley, bushing, and shaft, with a bill of material for the complete pulley-and-stand unit. Full size, three sheets. Top view may be drawn as a half-plan view.

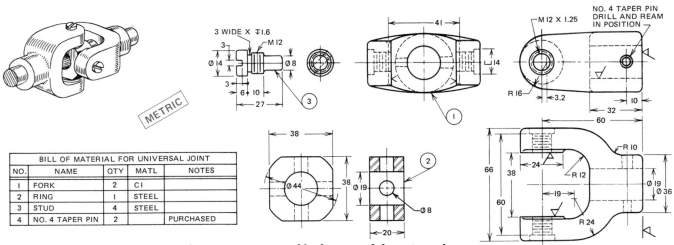

BILL OF MATERIAL FOR UNIVERSAL JOINT				
NO.	NAME	QTY	MATL	NOTES
1	FORK	2	CI	
2	RING	1	STEEL	
3	STUD	4	STEEL	
4	NO. 4 TAPER PIN	2		PURCHASED

Fig. 11-36. Universal joint. Make a two-view assembly drawing of the universal joint in section.

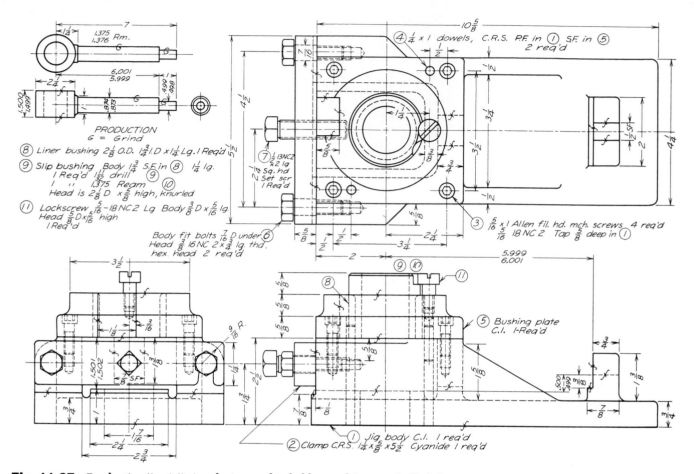

Fig. 11-37. Production jig. A jig is a device used to hold a machine part (called the *work* or *production*) while it is being machined, or produced, so that all the parts will be alike within specified limits of accuracy. Notice the work, or production, in the upper left corner of the drawing. Notice also that the drawing shown does not follow present ANSI standards. Be sure to make necessary changes to update each drawing in the following assignments. Assignment 1: Make a detail working drawing of the jig body. Assignment 2: Make a complete set of detail drawings for the jig, with a bill of material. Use as many sheets as needed. Assignment 3: Make a complete assembly drawing of the jig, three views. Give only such dimensions as are needed for putting the parts together and using the jig.

Fig. 11-38. Refer to Fig. 11-1, journal bearing. Prepare a detail drawing of the journal base. Scale: full size.

Fig. 11-39. Refer to Fig. 11-9, belt tightener. Prepare detail drawings of the frame, shaft, and pulley. Scale: full size.

Fig. 11-40. Refer to Fig. 11-17, machine parts. Make detail drawings of the parts shown. Scale: full size. Convert all dimensions to metric sizes (mm). Use simplified thread representation. Be sure to use current ANSI dimensioning standards.

12 Pictorial Drawing—Technical Illustration

▪ USES OF PICTORIAL DRAWING

Pictorial drawing is an essential part of the graphic language. It is very important in engineering, architecture, science, electronics, technical illustration, and in many other professions. It appears in all kinds of technical literature as well as in catalogs and assembly, service, and operating manuals. Architects use pictorial drawing to show what the finished building will look like. Advertising agencies use pictorial drawings to display a new product.

Pictorial drawing is often used in exploded views on production and assembly drawings (Fig. 12-1). These views are made to illustrate parts lists (Fig. 12-2), to explain the operation of machines, apparatus, and equipment, and for many other commercial and technical purposes. In addition, most people use some form of pictorial sketches to help convey ideas that are hard to describe in words.

▪ TYPES OF PICTORIAL VIEWS

Figure 12-3 shows different kinds of pictorial drawings. The first is a perspective view. It shows an object (in this case a stereo cabinet) as it actually looks to the eye. Drafters who are familiar with technical subjects often make perspective sketches. The other views are made with either of two pictorial projection methods: isometric or oblique. The isometric and oblique views are easier to draw than the perspective view, but they do not look as good. All three drawing methods allow the drafter to show three sides of an object in one view.

Cabinet drawing is a special kind of oblique drawing. In a cabinet drawing, distances on the receding axes are reduced by one half. In other words, an oblique cabinet drawing is drawn only half as deep as a normal oblique view.

Isometric Drawing

A multiview drawing of a filler block is shown in Fig. 12-4 at A. To make an isometric drawing of the block, begin with the three lines shown at B. These lines are the *isometric axes*. They will represent three edges of the block. Draw them to form three equal angles, each of 120° (3 × 120° = 360°). Draw axis line OA vertically. Draw axes OB and OC with the 30°-60° triangle. The point at which the lines meet represents the upper front corner O of the block as shown at C.

Measure off the width W, the depth D, and the height H of the block on the three axes. Then draw lines parallel to the axes to make the isometric drawing of each block. To locate the rectangular hole shown at D, lay off 1 in. along OC to c. Then

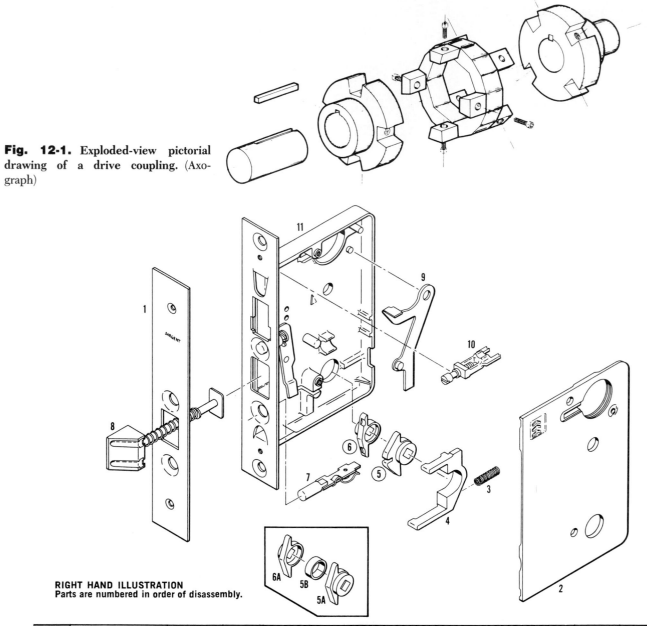

Fig. 12-1. Exploded-view pictorial drawing of a drive coupling. (Axograph)

RIGHT HAND ILLUSTRATION
Parts are numbered in order of disassembly.

Ident. Number	Part	For Lock	Specify Finish	Order Number	Price Each	Ident. Number	Part	For Lock	Specify Finish	Order Number	Price Each
1	Outside Front	7714 7715	x	77-0106	$2.20	6	Swivel Hub	7714 7746		77-0267	$0.60
		7746	x	77-0111	2.20	7	Long Stop Assembly	7714		77-2165	1.00
2	Cap	7715		77-0274	1.90			7746	x	77-2166	1.00
		7714 7746		77-2222	1.90						

Fig. 12-2. An exploded assembly drawing may be used to illustrate a parts list. (Sargent & Co.)

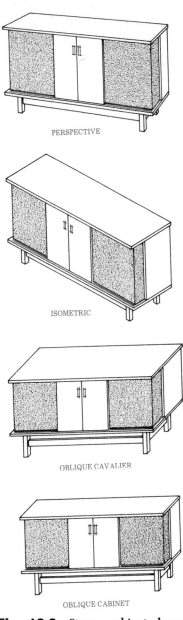

PERSPECTIVE

ISOMETRIC

OBLIQUE CAVALIER

OBLIQUE CABINET

Fig. 12-3. Stereo cabinet drawn in various types of pictorial views.

from *c* lay off 2 in. to *c*^{1}. Through *c* and *c*^{1} draw lines parallel to *OB*. In like manner, locate *b* and *b*^{1} on axis *OB* and draw lines parallel to *OC*. Draw a vertical line from corner 3. NOTE: *The dimensions, letters, and numerals are for instructional purposes. You would not normally put*

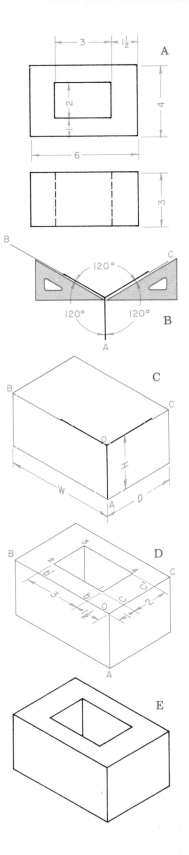

Fig. 12-4. Steps in making an isometric drawing.

them on your drawing. Darken all necessary lines to complete the drawing (Fig. 12-4E).

Pictorial drawings, in general, are made to show how something looks. Hidden edges (lines) are not "part of the picture"; therefore, they are normally left out. However, they might be included in some particular case when you want to show a certain feature for explanation.

Position for the Isometric Axes

Isometric axes can be arranged in different ways, provided they remain at 120° angles from each other. Several positions are shown and identified in Fig. 12-5. You will see how to apply them later in this chapter.

The arrangement of the axes in Fig. 12-4 is called the first position. Here the axes represent the three edges of the cube that meet at the upper front corner. Often it is more convenient to place the axes in the second position. There they represent the edges that meet at the lower front corner (Fig. 12-6).

Any line of an object parallel to one of the edges of a cube is drawn parallel to an isometric axis. Such a line is called an *isometric line*. An important rule of isometric drawing is:

Measurements can be made only on isometric lines.

Nonisometric Lines

Lines that are not parallel to any of the isometric axes are called *noniso-*

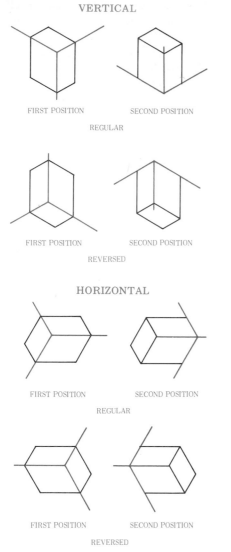

VERTICAL

FIRST POSITION   SECOND POSITION

REGULAR

FIRST POSITION   SECOND POSITION

REVERSED

HORIZONTAL

FIRST POSITION   SECOND POSITION

REGULAR

FIRST POSITION   SECOND POSITION

REVERSED

Fig. 12-5. Positions for isometric axes.

metric lines (Fig. 12-7). Such lines do not show in their true length and cannot be measured. To draw them, first locate their two ends and then connect the points. Angles on isometric drawings also do not show in their true size. Therefore, they cannot be measured in degrees.

Figure 12-8 shows how to locate and draw nonisometric lines in an isometric drawing. The method is called the *box method*. The lines in

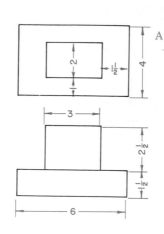

A

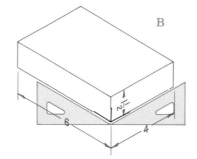

B

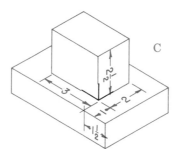

C

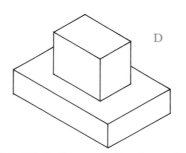

D

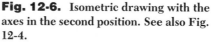

Fig. 12-6. Isometric drawing with the axes in the second position. See also Fig. 12-4.

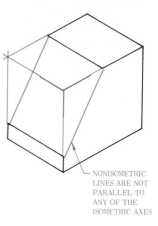

NONISOMETRIC LINES ARE NOT PARALLEL TO ANY OF THE ISOMETRIC AXES

Fig. 12-7. Nonisometric lines.

question are the slanted sides of the packing block shown in a multiview drawing at A in Fig. 12-8. NOTE: *The colored lines are for instructional purposes only. You would not normally put them on your finished drawing.* To make an isometric drawing of the block, use the following procedure:

1. Block in the overall sizes of the packing block to make the isometric box figure as shown at B.
2. Use dividers or a scale to transfer the distances *AG* and *HB* from the multiview drawing to the isometric figure. Lay these distances off along line *AB* to locate points *G* and *H*. You can then draw the lines connecting points *D* with *G* and *C* with *H*. This is shown in Fig. 12-8C.
3. Complete the layout by drawing *GJ* and *HI* and by connecting points *E* and *J* to form a third nonisometric line. Erase the construction lines to complete the drawing (Fig. 12-8E).

Angles in Isometric

Angles in some isometric drawings can be measured with a protractor. To draw an angle such as the 40°

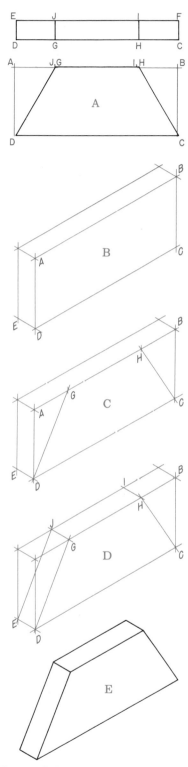

Fig. 12-8. Drawing nonisometric lines.

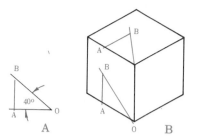

Fig. 12-9. Constructing angles in isometric drawing.

angle shown in Fig. 12-9A, use the following procedure:

1. Make *AO* and *AB* any convenient length. Draw *AB* perpendicular to *AO* at any convenient place.
2. Transfer *AO* and *AB* to the isometric cube (Fig. 12-9B). Lay off *AO* along the base of the cube. Draw *AB* parallel to the vertical axis.
3. Connect points *O* and *B* to complete the isometric angle. If this

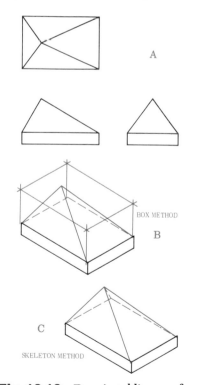

Fig. 12-10. Drawing oblique surfaces in isometric.

angle is checked with a protractor, it will not measure 40°.

Follow the same steps to construct the angle on the top of the isometric cube. This method can be used to lay out any angle on any isometric plane.

Figure 12-10 is a multiview drawing of an object with four oblique surfaces. An isometric view of this object can be made by either the box or the skeleton method, as shown at B and C.

Isometric Circles

In isometric drawing, circles appear as ellipses. Since it takes a long time to plot a true ellipse, a four-centered approximation is generally drawn. The way to do this is shown and explained in Fig. 12-11. Figure 12-12 shows isometric circles drawn on three surfaces of a cube.

Figure 12-13 shows how to make an isometric drawing of the cylinder shown as a multiview drawing at A. First, draw an ellipse of the 3-in. circle following the procedure given in Fig. 12-11. Next, drop centers *A*, *C*, and *D* a distance equal to the height of the cylinder (in this case, 4 in.) as shown at B. Draw lines *A'C'* and *A'D'*. Complete the drawing as shown at C. A line through *C'D'* will locate the points of tangency. Draw the arcs using the same radii as in the ellipse at the top. Finally, draw the vertical lines to complete the cylinder. Notice that the radii for the arcs at the bottom match those at the top.

Quarter rounds are drawn in isometric the same way you draw quarters of a circle. This is illustrated in Fig. 12-14. Notice that in each case the radius is measured along the tangent lines from the corner. Then the actual perpendiculars are drawn to locate the centers for the isometric

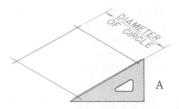

Draw an isometric square with the sides equal to the diameter of the circle.

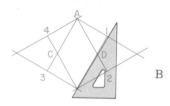

Use a 30°-60° triangle to locate points *A, B, C, D,* and 1, 2, 3, 4.

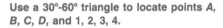
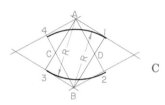

With *A* and *B* as centers and a radius equal to *A2*, draw arcs as shown.

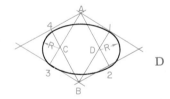

With *C* and *D* as centers and a radius equal to *C4*, draw arcs to complete the isometric circle (ellipse).

Fig. 12-11. Steps in drawing an isometric circle.

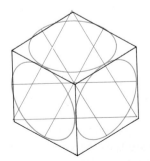

Fig. 12-12. Isometric circles on a cube.

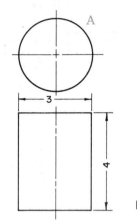
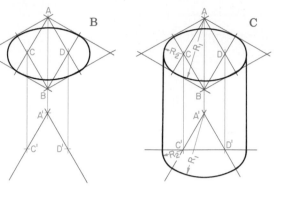

Fig. 12-13. Steps in drawing an isometric cylinder.

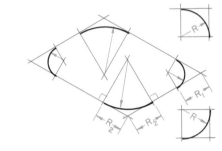

☐ = RIGHT ANGLE (90°)

Fig. 12-14. Drawing quarter rounds in isometric.

arcs. Observe that r_1 and r_2 are found in the same way as the radii of an isometric circle.

When an arc is more or less than a quarter circle, it can sometimes be plotted by drawing all or part of a complete isometric circle and using as much of the circle as needed.

Figure 12-15 shows how to draw outside and inside corner arcs. Note

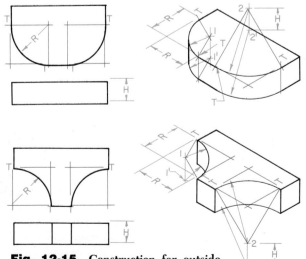

Fig. 12-15. Construction for outside and inside arcs.

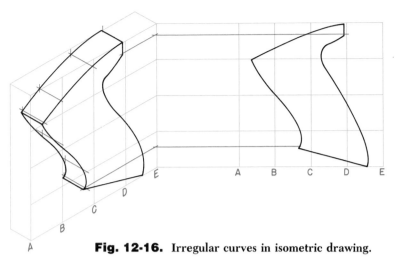

Fig. 12-16. Irregular curves in isometric drawing.

the tangent points *T* and centers 1 and 1′ and 2 and 2′.

Irregular Curves in Isometric

Irregular curves in isometric cannot be drawn using the four-center method. You must first plot points and then connect them using a French curve, as shown in Fig. 12-16.

Isometric Templates

Isometric templates are made in a variety of forms. They are convenient and will save you time when you have to make many isometric drawings. Many of them have openings for drawing ellipses, as well as 60° and 90° guiding edges. Simple homemade guides (Fig. 12-17) are convenient for straightline work in isometric. Ellipse templates are very convenient for drawing true ellipses (Fig. 12-18). If you use these templates, your drawings will look better

and you will not have to spend time plotting approximate ellipses. See Chap. 3 for information on how to use and care for templates.

Making an Isometric Drawing

Figure 12-19 shows how to make an isometric drawing of a guide. The guide is shown in a multiview drawing at A. Study the size, shape, and relationship of views before you proceed.

1. Draw the axes *AB*, *AC*, and *AD* in the second position (Fig. 12-19B). If you do not recall this position, refer to Fig. 12-5, which shows the positions of the isometric axes. Next, lay off the length, width, and thickness measurements given at A; that is, measure from A the length 3 in. on *AB*.

Measure from *A* the width 2 in. on *AC*. Measure from *A* the thickness 5/8 in. on *AD*. Through the points found, draw isometric lines parallel to the axes. This "blocking in" will produce an isometric view of the base.

2. Block in the upright part in the same way, using the measurements of 2 in. and 3/4 in. given in the top view at A.

3. Find the center of the hole and draw centerlines as shown. Block in a 3/4-in. isometric square and draw the hole as an approximate ellipse. To make the two quarter rounds, measure the 1/2-in. radius along the tangent lines from both upper corners (Fig. 12-19C). Draw real perpendiculars to find the centers of the quarter circles. See Fig. 12-14 for more information on drawing isometric quarter rounds.

4. Darken all necessary lines. Erase all construction lines to complete the isometric drawing as in Fig. 12-19D.

Isometric Sections

Isometric drawings are generally outside views. Sometimes, however, a sectional view is needed. If so, take a section on an *isometric plane* (a plane parallel to one of the faces of the cube). Figure 12-20 shows isometric full sections taken on a different plane for each of three objects. Note the construction lines showing

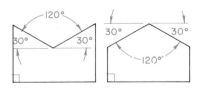

Fig. 12-17. Simple isometric templates.

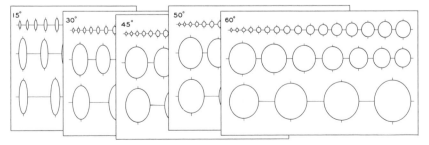

Fig. 12-18. Ellipse templates.

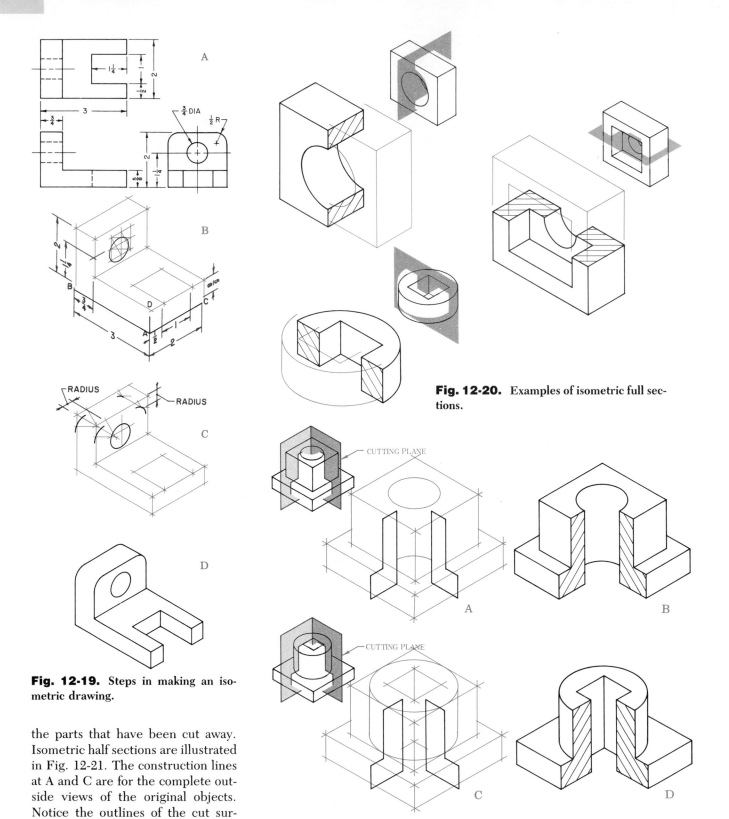

Fig. 12-19. Steps in making an isometric drawing.

the parts that have been cut away. Isometric half sections are illustrated in Fig. 12-21. The construction lines at A and C are for the complete outside views of the original objects. Notice the outlines of the cut surfaces in both views. There are two

Fig. 12-20. Examples of isometric full sections.

CUTTING PLANE

CUTTING PLANE

Fig. 12-21. Examples of isometric half sections.

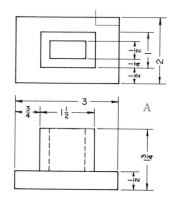

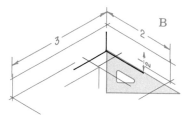

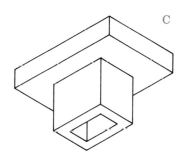

Fig. 12-22. Steps in making an isometric drawing with reversed axes.

ways to make the sectional views shown at B and D. In one, the cut method, the complete outside view plus the isometric cutting plane is drawn. Then erase that part of the view that the cutting plane has cut away. In the other method, first draw the section on the isometric cutting plane. Then work from it to complete the view.

Reversed Axes

Sometimes you will want to draw an object as if it were being viewed from below. This is done in isometric drawing by reversing the position of

the axes. Follow the example in Fig. 12-22. Consider how an object appears in a regular multiview drawing, as at A. Then begin the isometric view, as at B, by drawing the axes in reversed position. Complete the view, as at C, with dimensions taken from the multiview drawing. Darken the lines to finish the drawing.

Long Axis Horizontal

When long pieces are drawn in isometric, make the long axis horizontal. A drawing of this kind is illustrated in Fig. 12-23. At A, a long object is shown in a multiview drawing. At B is the start of an isometric view, with the axes shown by thick black lines. At C, the view is completed with dimensions taken from the drawing at A. Remember, in isometric drawing, draw circles first as isometric squares; then complete them by the four-center method or with an ellipse template.

Dimensioning Isometric Drawings

There are two general ways to put dimensions on isometric drawings. The older method is to place them in the isometric planes, or extensions of them, and to adjust the letters, numerals, and arrowheads to isometric shapes, as shown in Fig. 12-24 at A. The newer unidirectional system, shown at B, is simpler. In this system, numerals and lettering are read from the bottom of the sheet. However, since isometric drawings are usually not used as working drawings, they are seldom dimensioned at all.

■ AXONOMETRIC PROJECTION

Isometric projection is one form of *axonometric* projection. Other forms are *dimetric* and *trimetric* projec-

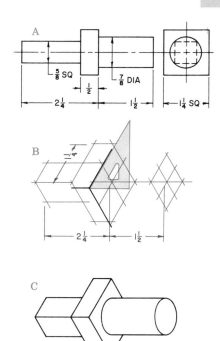

Fig. 12-23. Steps in making an isometric drawing with the long axis horizontal.

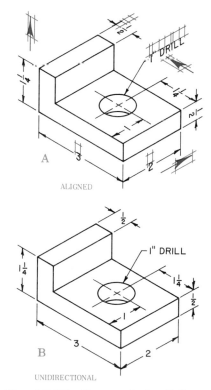

Fig. 12-24. Two methods of dimensioning isometric views.

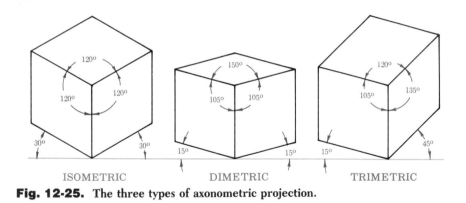

Fig. 12-25. The three types of axonometric projection.

tion. All three are made according to the same theory; the difference is the angle of projection (Fig. 12-25). In isometric projection, the axes form three equal angles of 120° on the plane of projection. Only one scale is needed for measurements along each axis. Isometric drawings are the easiest type of axonometric drawing to make. In dimetric projection, only two of the angles are equal, and two special foreshortened scales are needed to make measurements. In trimetric projection, all three angles are different, and three special foreshortened scales are needed.

Dimetric and trimetric drawings are complicated to draw. Therefore, drafters use them less often than other types of pictorial drawing.

ISOMETRIC PROJECTION AND ISOMETRIC DRAWING

Isometric projection and isometric drawing are not the same thing, even though many people think they are. In this section, we will learn what the differences are. You probably will not have to make pictorial drawings using isometric projection, but it is a good idea to understand the theory behind it.

One way to make an isometric projection is by a method called *revolution* (drawing an object as if it were revolved or turned). Figure 12-26 shows a cube in the three normal views of a multiview drawing. In Fig. 12-26B, each of the three views has been revolved 45°. Notice that the front and side views now show as two equal rectangles. On the side view, a diagonal is drawn from point *O* to point *B*. This is called the *body diagonal*. The body diagonal is the longest straight line that can be drawn in a cube.

In Fig. 12-26C, the cube is revolved upward until the body diagonal is horizontal, as shown in the side view. Notice that the cube was revolved 35°16′ to achieve this. The front view now forms an isometric projection. Its lower edges form an angle of 30° to the horizontal.

Since the front view (isometric projection) of the revolved cube is made by projection, its lines are foreshortened. The actual difference is 0.8165 to 1 in. In other words, 1 in. on the cube in Fig. 12-26A has been reduced to 0.8165 in. on the isometric projection in Fig. 12-26C.

Figure 12-26D shows an isometric drawing and an isometric projection of the same cube. In the isometric drawing, all edges are drawn their true length instead of the shortened

length. The drawing shows the shape of the cube just as well as the projection. Its advantage is that it is easier to draw, because all its measurements can be made with a regular scale. It can also be drawn without projecting from other views or without using a special scale.

OBLIQUE PROJECTION AND OBLIQUE DRAWING

Oblique projection (Fig. 12-27), like isometric, is a way of showing depth. A main difference, however, is that whereas isometric shows an object as if viewed from on edge, oblique shows it as if viewed face on. That is, one side of the object is seen squarely, with no distortion, because it is parallel to the *picture plane* (the plane on which the view is drawn). Depth is shown, as in isometric, by *projectors* (lines representing receding edges of the object). These lines are drawn at an angle other than 90° from the picture plane, to make the receding planes visible in the front view. And, as in isometric, lines on these receding planes that are actually parallel to each other are drawn parallel. Figure 12-27 shows how an oblique projection is developed. You probably will never have to develop an oblique projection in this way, but it is a good idea to understand the theory behind it.

Usually, no distinction is made between oblique projection and oblique drawing. This is another difference from isometric. In oblique, if the receding lines are drawn full length, the projection is called *cavalier*. If they are drawn one-half size, the projection is called *cabinet*. Some drafters use three-quarter size. This is sometimes called *normal*, or *general*, *oblique*. Generally,

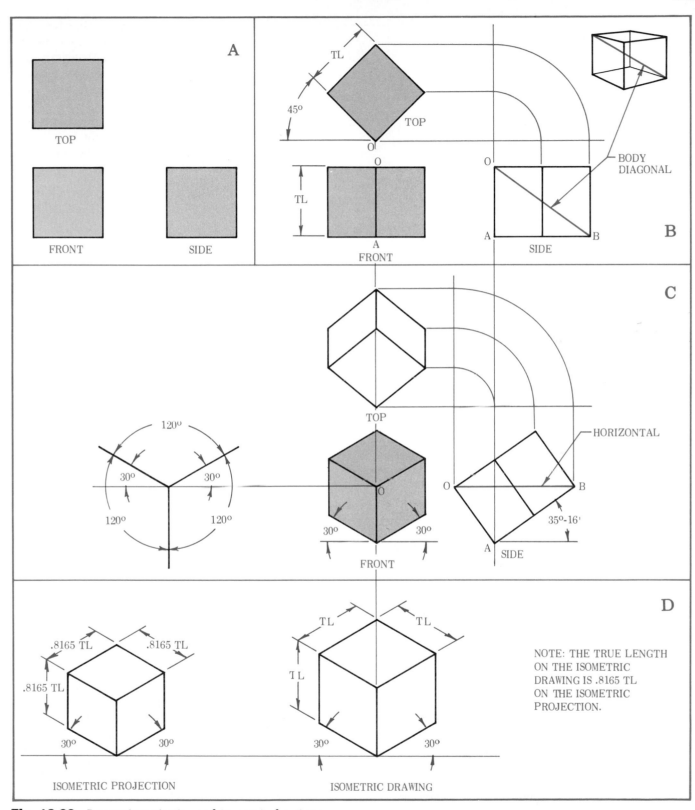

Fig. 12-26. Isometric projection and isometric drawing.

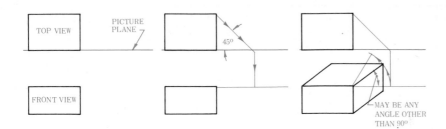

Fig. 12-27. Oblique projection.

however, all three types are simply called oblique drawings.

Because oblique drawing can show one face of an object without distortion, it has a distinct advantage over isometric. It is especially useful for showing objects with irregular outlines.

Oblique Drawing

Oblique drawings are plotted in the same way as isometric drawings, that

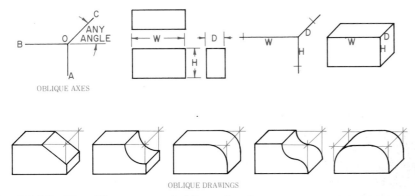

Fig. 12-28. The oblique axes and oblique drawings.

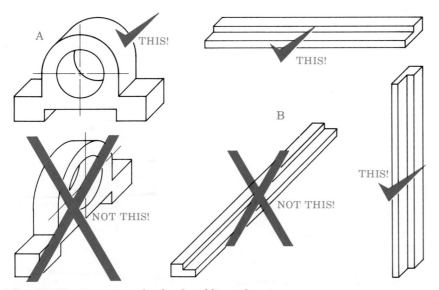

Fig. 12-29. Two general rules for oblique drawings.

is, on three axes. However, in oblique, two axes are parallel to the picture plane, rather than just one, as in isometric. These two axes always make right angles with each other (Fig. 12-28).

The methods and rules of isometric drawing also apply to oblique drawing. Oblique also has some special rules. The first is:

Place the object so that the irregular outline or contour faces the front (Fig. 12-29A).

The second rule is:

Place the object so that the longest dimension is parallel to the picture plane (Fig. 12-29B).

Positions for the Oblique Axes

Figure 12-30 shows several positions for oblique axes. In all cases, two of the axes, *AO* and *OB*, are drawn at right angles. The oblique axis *OC* can be at any angle to the right, left, up, or down, as illustrated. The best way to draw an object is usually at the angle from which it would normally be viewed.

Angles and Inclined Surfaces on Oblique Drawings

Angles that are parallel to the picture plane show in their true size. Other angles can be laid off by locating both ends of the slanting line.

Figure 12-31 shows a plate with the corners cut off at angles. At B, C, and D, the plate is shown in oblique drawings. At B, the angles are parallel to the picture plane. At C, they are parallel to the profile plane. In each case, the angle is laid off by measurements parallel to one of the oblique axes. These measurements are shown by the construction lines.

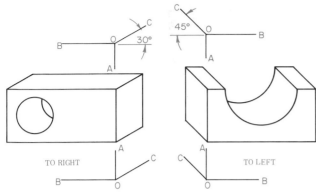

Fig. 12-30. Positions for oblique axes.

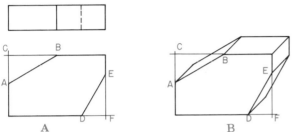

Fig. 12-31. Angles on oblique drawings.

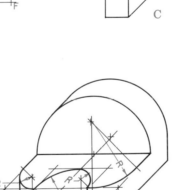
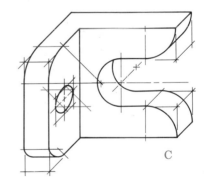

Fig. 12-32. Circles parallel to the picture plane are true circles; on other planes, ellipses.

Oblique Circles

On the front face, circles and curves show in their true shape (Fig. 12-32). On other faces, they show as ellipses. These ellipses can be drawn by the four-center method. In Fig. 12-32A, a circle is shown as it would be drawn on a front plane, a side plane, and a top plane.

Figure 12-32B is an oblique drawing with some arcs in a horizontal plane. At C is an oblique drawing with some arcs in a profile plane.

When oblique circles are drawn by the four-center method, the results will be satisfactory for some purposes, but not pleasing. Ellipse templates, when available, give

much better results. If you use a template, first block in the oblique circle as an oblique square. This will show where to place the ellipse. Blocking in the circle first also helps you choose the proper size and shape of the ellipse. If you do not have a template, plot the ellipse as shown in Fig. 12-33.

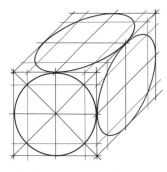

Fig. 12-33. To plot oblique circles.

To Make an Oblique Drawing

Figure 12-34 shows the steps in making an oblique drawing. Notice that the drawing can show everything but the two small circles in true shape.

Oblique Sections

Oblique drawings are generally outside views. Sometimes, however, you need to draw a sectional view.

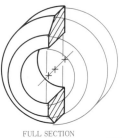

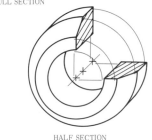

FULL SECTION

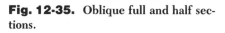

HALF SECTION

Fig. 12-35. Oblique full and half sections.

To do so, take a section on a plane parallel to one of the faces of an oblique cube. Figure 12-35 shows an oblique full section and an oblique half section. Note the construction lines indicating the parts that have been cut away.

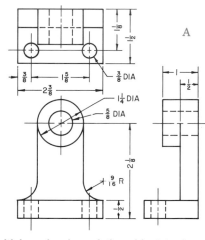

Multiview drawing of the object to be drawn in oblique.

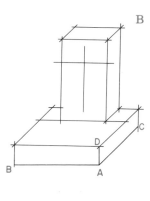

Draw the axes *AB, AC,* and *AD* for the base in second position and on them measure the length, width, and thickness of the base. Draw the base. On it, block in the upright, omitting the projecting boss as shown.

Cabinet Drawings

A *cabinet drawing* is an oblique drawing in which distances parallel to the oblique (receding) axis are drawn one-half size. Figure 12-36 shows a bookcase drawn in cavalier, normal oblique, and cabinet drawings. Cabinet drawings are so named because they are often used in the furniture industry.

■ PERSPECTIVE DRAWING

A *perspective drawing* (Fig. 12-37) is a three-dimensional representation of an object as it looks to the eye from a particular point. Of all pictorial drawings, perspective drawings look the most like photographs. The dis-

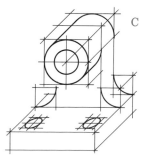

Block in the boss and find the centers of all circles and arcs. Draw the circles and arcs.

Fig. 12-34. Steps in making an oblique drawing.

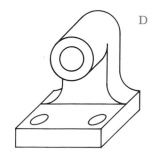

Darken all necessary lines, and erase construction lines to complete the drawing.

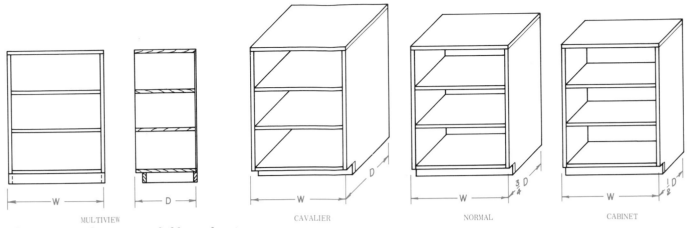

MULTIVIEW   CAVALIER   NORMAL   CABINET

Fig. 12-36. Three types of oblique drawings.

tinctive feature of perspective drawing is that in perspective, lines on
the receding planes that are actually
parallel are not drawn parallel, as
they are in isometric and oblique
drawing. Instead, they are drawn as
if they are *converging* (coming together).

Definition of Terms

Figure 12-38 illustrates perspective
terms. A card appears on the plane at
the right. The sight lines leading
from points on this card and converging at the eye of the observer are
called *visual rays*. The *picture plane*
(PP) is the plane on which the card at
right is drawn. The *station point* (SP)
is the point from which the observer
is looking at the card. A horizontal
plane passes through the observer's
eye. Where it meets the picture
plane, it forms the *horizon line* (HL).

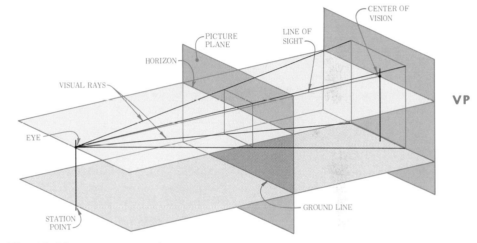

Fig. 12-38. Some perspective terms.

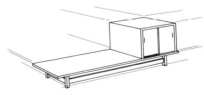

Fig. 12-37. Perspective drawing of a
music center.

Where the ground plane on which
the observer stands meets the picture plane, it forms the *ground line*
(GL). The *center of vision* (CV) is the
point at which the *line of sight* (LS)
(visual ray from the eye perpendicular to the picture plane) pierces the
picture plane.

Figure 12-39 shows how, in perspective drawing, the *projectors* (receding axes) converge. The point at
which they meet is called the *vanishing point* (VP).

Figure 12-39 also shows how the
observer's eye level affects the per-

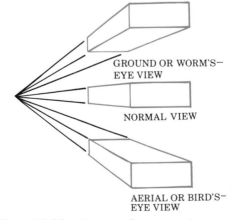

Fig. 12-39. Types of perspective
views.

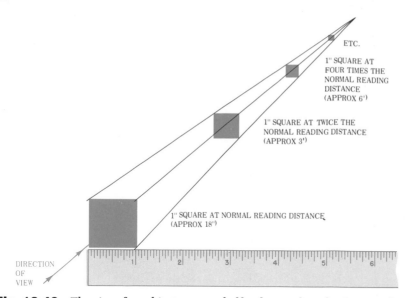

Fig. 12-40. The size of an object appears half as large when the distance from the observer is doubled.

How Distance Affects the View

The size of an object seems to change as you move toward or away from it. The farther away from the object you go, the smaller it looks. The closer you get, the larger it seems to grow. Figure 12-40 shows a graphic explanation of this effect of distance. An object is placed against a scale at a normal reading distance from the viewer. In that position, it looks to be the size indicated by the scale. However, if it is moved back from the scale to a point twice as far away from the viewer, it looks only half as large. Notice that each time the distance is doubled, the object looks only half as large as before.

How Position Affects the View

The shape of an object also seems to change when seen from different positions (angles). This is illustrated in Fig. 12-41. If you look at a square face on, the top and bottom edges are parallel. But if the square is rotated so that you see it at an angle, these edges seem to converge. The

spective view. This eye level can be anywhere on, above, or below the ground. If the object is seen from above, the view is an *aerial*, or *bird's-eye view*. If the object is seen from underneath, the view is a *ground*, or *worm's-eye view*. If the object is seen face on, so that the line of sight is directly on it rather than

above or below, the view is a *normal view*. The view in Fig. 12-38 is this kind of view.

Factors That Affect Appearance

Two factors affect how an object looks in perspective: its *distance* from the viewer and its *position* (angle) in relation to the viewer.

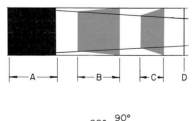

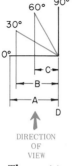

Fig. 12-41. The position of the object in relationship to the observer affects its appearance.

Fig. 12-42. The lines of the porch rail, deck, roof, and building's side appear to converge at a distance. (Designed by Howard A. Friedman & Associates.)

square also appears to grow narrower. This foreshortening occurs because one side of the square is now farther away from you (see Fig. 12-40 for an illustration of how distance affects the view).

One-Point Perspective

One-point perspective (also called *parallel perspective*) means *one vanishing point*. Figure 12-42 shows how one-point perspective looks. Notice that if the lines of the porch rail, deck, roof, and siding were extended, they would converge at a single point near the center of the small square at the far end of the deck near the center of the picture. Figure 12-43 shows an object in multiview and isometric drawings. Figure 12-44 shows how to draw the same object in a one-point bird's-eye-view perspective. In Figs. 12-45 and 12-46, the object is drawn in one-point perspective in the other positions. Notice that in all three cases, one face of the object is placed on the picture plane (thus the name parallel perspective). Therefore, this face appears in true size and shape. True-scale measurements can be made on it.

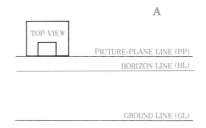

A

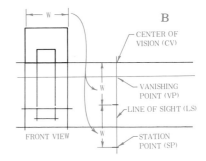

B

Decide on the scale to be used and draw the top view near the top of the drawing sheet. A more interesting view is obtained if the top view is drawn slightly to the right or to the left of center. Draw an edge (top) view of the picture plane (PP) through the front edge of the top view. Draw the horizon line (HL). The location will depend upon whether you want the object to be viewed from above, on, or below eye level. Draw the ground line. Its location in relation to the horizon line will determine *how far* above or below eye level the object will be viewed.

Locate the station point (SP). (a) Draw a vertical line (line of sight) from the picture plane toward the bottom of the sheet. Draw the line slightly to the right or to the left of the top view. (b) Set your dividers at a distance equal to the width (*W*) of the top view. (c) Begin at the center of vision on the picture plane and step off 2 to 3 times the width (*W*) of the top view, along the line of sight, to locate the station point (SP). Project downward from the top view to establish the width of the front view on the ground line. Complete the front view.

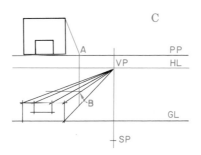

C

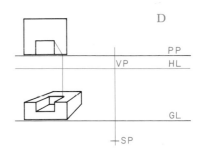

D

The vanishing point (VP) is the intersection of the line of sight (LS) and the horizon line (HL). Project lines from points on the front view to the vanishing point. Establish depth dimensions in the following way: (a) Project a line from the back corner of the top view to the station point. (b) At point *A* on the PP, drop a vertical line to the perspective view to establish the back vertical edge. (c) Draw a horizontal line through point *B* to establish the back top edge.

Proceed as in the previous step to lay out the slot detail. Darken all necessary lines, and erase construction lines as desired to complete the drawing.

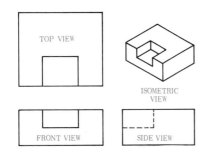

Fig. 12-43. Multiview and isometric drawings of an object to be drawn in single-point perspective.

Fig. 12-44. Procedure for making a single-point-, or parallel-, perspective drawing (bird's-eye view).

Two-Point Perspective

Two-point perspective means *two vanishing points*. It is also called *angular perspective*, since none of the faces are drawn parallel to the picture plane. Figure 12-47 shows how two-point perspective looks.

Figure 12-48 shows an object in multiview and isometric drawings.

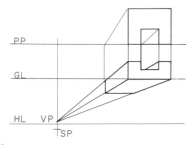

Fig. 12-45. Single-point perspective, worm's-eye view.

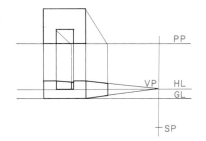
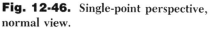

Fig. 12-46. Single-point perspective, normal view.

Fig. 12-47. When a building is viewed at an angle, two sides can be seen. The top and ground lines of each side appear to converge toward points. This is the effect in two-point, or angular, perspective. (Bruning.)

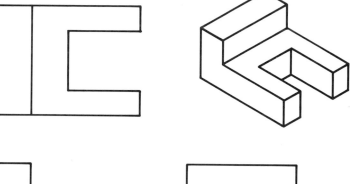

Fig. 12-48. Multiview and isometric drawings of an object to be drawn in two-point perspective.

Figure 12-49 shows how to draw this same object in two-point bird's-eye-view perspective. Figures 12-50 and 12-51 show the object drawn in two-point perspective in the other positions.

Inclined Surfaces

Inclined surfaces are plotted in perspective by finding the ends of inclined lines and connecting them. This method of drawing is shown in Fig. 12-52.

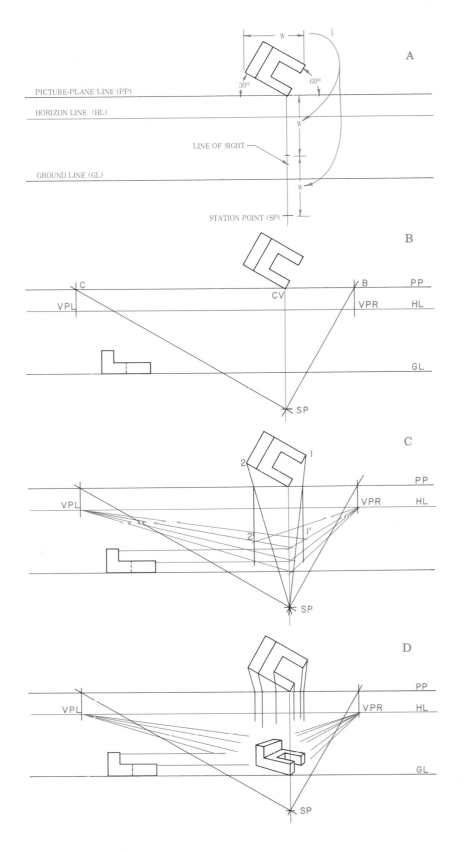

Draw an edge view of the picture plane (PP). Allow enough space at the top of the sheet for the top view. Draw the top view with one corner touching the PP. In this case, the front and side of the top view form angles of 30° and 60°, respectively. Other angles may be used, but 30° and 60° seem to give the best appearance on the finished perspective drawing. The side with the most detail is usually placed along the smaller angle for a better view. Draw the horizon line (HL) and the ground line (GL). (Follow the procedure given in Fig. 12-44.) Draw a vertical line (line of sight) from the center of vision (CV) toward the bottom of the sheet to locate the station point. (Follow the procedure given in Fig. 12-44.)

Draw line *SP-B* parallel to the end of the top view and line *SP-C* parallel to the front of the top view. (Use a 30°-60° triangle.) Drop vertical lines from the picture plane (PP) to the horizon line (HL) to locate vanishing point left (VPL) and vanishing point right (VPR). Draw the front or side view of the object on the ground line as shown.

Begin to block in the perspective view by projecting vertical dimensions from the front view to the line of sight (also called *measuring line*) and then to the vanishing points. Finish blocking in the view as follows: (a) Project lines from points 1 and 2 on the top view to the station point. (b) Where these lines cross the picture plane (PP), drop vertical lines to the perspective view to establish the length and width dimensions. (c) Project point 1¹ to VPL and 2¹ to VPR.

Add detail by following the procedure described in the previous two main steps. Darken all necessary lines and erase construction lines as desired.

Fig. 12-49. Procedure for making a two-point-perspective drawing (bird's-eye view).

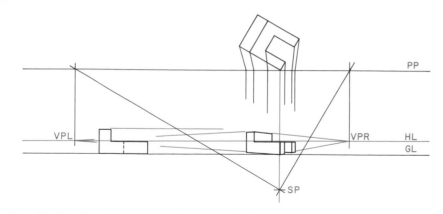

Fig. 12-50. Two-point perspective, normal view.

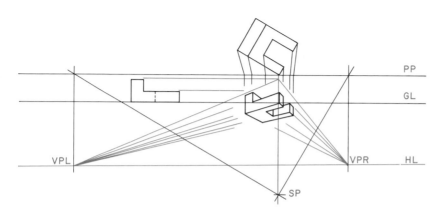

Fig. 12-51. Two-point perspective, worm's-eye view.

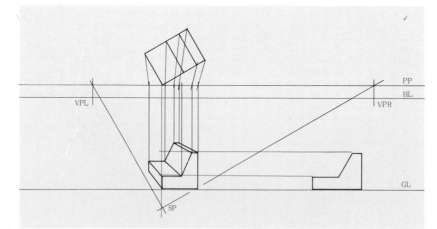

Fig. 12-52. Two-point perspective with an inclined surface.

Circles and Arcs in Perspective

Figure 12-53 shows how to make a perspective view of an object with a cylindrical surface. Notice that points are first located on the front and top views. They are then projected to the perspective view. Where the projection lines meet, a path is formed. Along the path, the drafter draws the perspective arc, using a French curve or an ellipse template.

Perspective Drawing Shortcuts

Perspective drawing can take a lot of time. This is because so much layout work is needed before you can start the actual perspective view. Also, a large drawing surface is often needed in order to locate distant points. However, you can offset these disadvantages by using various shortcuts. These are described below.

Perspective Grids

One of the simpler shortcuts is the perspective grid. Examples are shown in Fig. 12-54. There are many advantages in using grids. But there is one major disadvantage: a grid cannot show a variety of views. It is limited to one type of view based on one set of points and one view location. However, for the work done in some industrial drafting rooms, this may be all that is needed.

Perspective grids can be bought or the drafter can make them. This is only practical, however, if the drafter has a number of perspective drawings to make in a special style.

Perspective Drawing Boards

The Klok perspective board (Fig. 12-55) is another timesaving device for

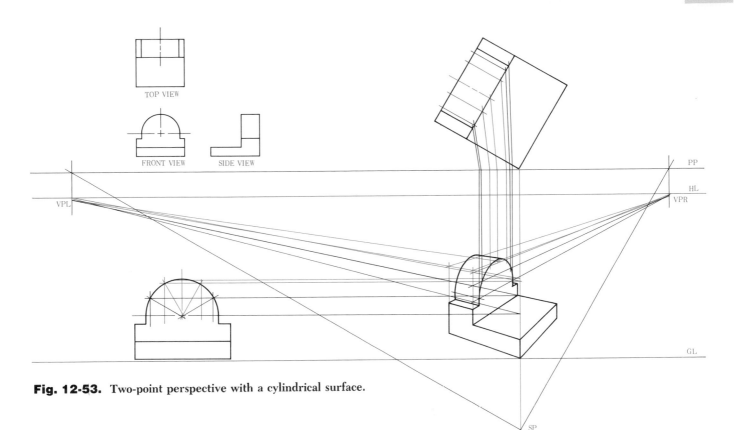

Fig. 12-53. Two-point perspective with a cylindrical surface.

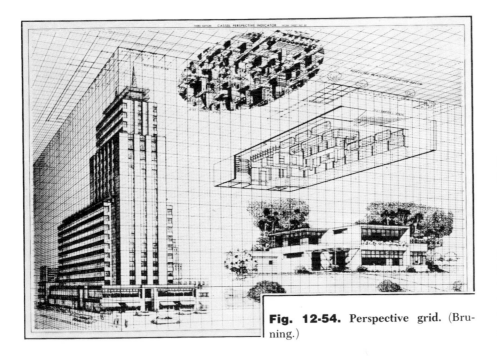

Fig. 12-54. Perspective grid. (Bruning.)

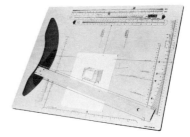

Fig. 12-55. Klok perspective drawing board. (The UTLEY Co.)

making accurate perspective drawings. It includes a special T-square with the top edge of the blade centered on the head for drawing converging lines. It also includes perspective scales. These eliminate the need for extensive projection or calculation.

You can make your own perspective drawing board by using a regular drawing board and some thin pieces of wood, cardboard, or hardboard and by making or modifying a T-square (Fig. 12-56). Like the perspective grid, this device is practical only if you have to make many drawings of about the same size and style, such as perspective drawings of houses.

The way to make a perspective board is as follows:

1. Follow the procedure in Fig. 12-49 for locating two vanishing points. Use the top view of an object similar in size to the ones that will be drawn on the finished board. Make your layout on a large sheet of paper in order to locate distant vanishing points.
2. Place a drawing board within the layout, as shown in Fig. 12-56. For guides, use heavy cardboard or thin wood. Fasten the guide material in place. Then strike the

arcs, using the vanishing points as centers and any convenient radius.
3. Cut the arcs with a sharp knife to form the guides. Draw the PP, HL, GL, LS, and SP on the board. You can now remove the board from the layout sheet.
4. Construct the T-square as shown. Notice that the top edge of the blade falls on the centerline of the heads. You must use thin material for the head as well as for the blade.

To use the perspective board:

1. Place a sheet of tracing paper in position between the guides (Fig. 12-56). Tracing paper will allow you to use the lines drawn on the board without redrawing them.
2. Draw the top view of the desired object in its proper position.
3. Proceed as in Fig. 12-49. Draw the lines that project toward VPR with the T-square head against the left guide. For those that project toward VPL, use the

right-hand guide. Draw the vertical lines with the T-square head against the bottom edge of the board.

TECHNICAL ILLUSTRATION

Technical illustration has an important place in all areas of engineering and science. Technical illustrations form a necessary part of the technical and service manuals for machine tools, automobiles, machines, and appliances. In technical illustration, pictorial drawings are used to describe parts and the methods for making them. Pictorial drawings show how the parts fit together. They also show the steps that need to be followed to complete the product on the assembly line. Technical illustrations may be used to set up the assembly line. They are useful for industrial, engineering, and scientific purposes.

Technical illustration drawings can range from simple sketches to rather detailed shaded drawings. They may be based on any of the pictorial methods: isometric, perspective, oblique, and so forth. The complete project or parts or groups of parts may be shown. The views may be exterior, interior, sectional, cutaway, or phantom. The purpose in all cases is to provide a clear and easily understood description.

Drawings for use within a company's plant can sometimes be made by a drafter with artistic talent. However, the special skills needed for such drawings call for the work of a professional technical illustrator. Technical illustrations have been used for many years in illustrated parts lists, operation and service manuals, and process manuals (Fig. 12-57). The aircraft industry in par-

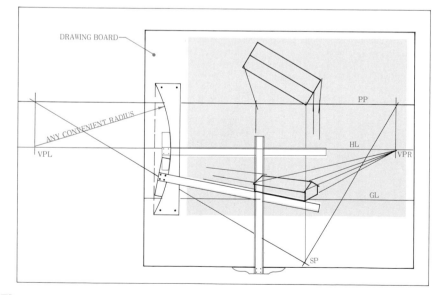

Fig. 12-56. A homemade perspective drawing board.

PART NO.	PART NAME	NO. REQD
1	BASE	1
2	MOVABLE JAW	1
3	MOVABLE JAW PLATE	1
4	MACHINE SCREW	1
5	LOCKING PIN	1
6	HANDLE STOP	2
7	HANDLE	1
8	CLAMP SCREW	1
9	JAW FACE	2
10	CAP SCREW	2

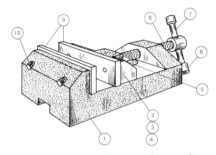

Fig. 12-57. An illustrated parts list.

ticular has found production illustration especially valuable. In aircraft construction, pictorial drawings are used when the plane is first designed. These drawings are also used throughout the many phases of a plane's production and completion on the assembly line. When the plane is delivered to the customer, the service, repair, and operation manuals are also illustrated with pictorial drawings.

■ DEFINITION

Generally, a technical illustration is a pictorial drawing that provides technical information by visual methods. This is usually done by turning a multiview drawing into a three-dimensional pictorial drawing. It must show shapes and relative positions in a clear and accurate way. Shading may be used to bring out the shape. A technical illustration, however, is not necessarily a work of art. Therefore, the shading must serve a practical purpose, not an artistic one.

In addition to pictorials, technical illustrations include graphic charts, schematics, flowcharts, diagrams, and sometimes circuit layouts. Dimensions are not a part of technical illustrations because they are not working drawings.

■ TOOLS AND TIPS

Technical illustrators use the tools and regular drafting equipment described in Chaps. 3 and 13. An H or a 2H pencil, with the point kept well sharpened, is the most useful tool. A few other useful items include felt-tip pens, technical pens, masking tape, Scotch tape, an X-Acto knife, paper stomps, two or three artist's brushes, and an airbrush.

Craftint, Zip-a-tone, Chart-Pak, and similar press-on section linings, screen tints, letters, and the like are also useful timesaving materials. You should know about the various methods of graphic reproduction. In addition, be aware of the effect of reduction when your drawing is to be used in a smaller size. Lines must be firm and black. Erasures must be clean. The part of a drawing not being worked on should be kept covered with paper or sheet plastic.

■ LETTERING

Lettering is an important element of technical illustration. In some cases, templates can be used. Good freehand lettering is a required skill for a technical illustrator.

■ PICTORIAL LINE DRAWINGS

Basically, all technical illustrations are pictorial line drawings. Therefore, you should have a complete understanding of the various types of

pictorial line drawings and their uses. The first half of this chapter describes the various types of pictorial drawings and the procedure for drawing each.

Usually, any type of pictorial drawing can be used as the basis for a technical illustration. However, some types are more suitable than others. This is especially true if the illustration is to be *rendered* (shaded). Figure 12-58 shows a V-block drawn in several types of pictorial drawing. Notice the difference in the appearance of each. Isometric is the least natural in appearance. Perspective is the most natural. This might suggest, then, that all technical illustrations should be drawn in perspective. This is not necessarily true. While perspective is more natural than isometric in appearance, it takes more time to do. It is also more difficult to draw. Thus, it is often a more costly method to use.

The shape of the object also helps to determine the type of pictorial drawing to use. Figure 12-59 shows a pipe bracket drawn in isometric and oblique. The shape of this object is most easily and quickly drawn in oblique. In many cases, it will look more natural than in isometric.

If an illustration is to be used only in-plant, the illustrator will usually make the pictorial drawing in isometric or oblique. These are quickest and the least costly to make. If the illustration is to be used in a publication such as a journal, operator's manual, or technical publication, dimetric, trimetric, or perspective may be used.

■ EXPLODED VIEWS

Perhaps the easiest way to understand an exploded view is this: Take an object and separate it into its indi-

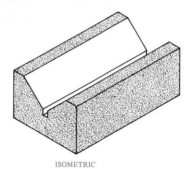

ISOMETRIC

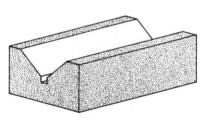

DIMETRIC

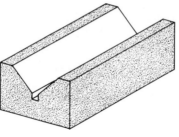

TRIMETRIC

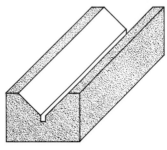

OBLIQUE   CAVALIER

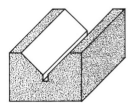

OBLIOUE   CABINET

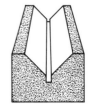

SINGLE - POINT
(PARALLEL) PERSPECTIVE

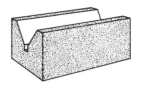

TWO - POINT (ANGULAR)
PERSPECTIVE

Fig. 12-58. A V-block in various types of pictorial drawing.

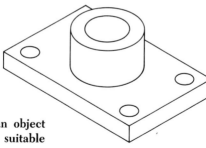

ISOMETRIC

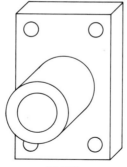

OBLIQUE

Fig. 12-59. The shape of an object helps to determine the most suitable type of pictorial drawing to use.

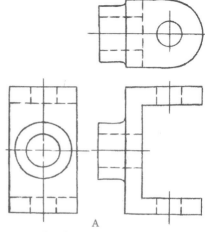

A

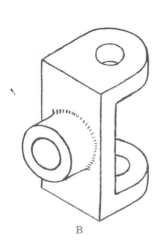

B

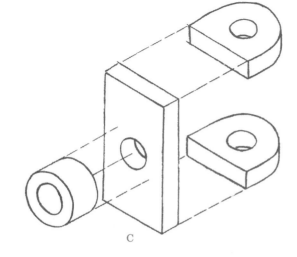

C

Fig. 12-60. How a view is exploded.

Fig. 12-61. Photograph of a high-pressure pump. (Industrial Division, Standard Precision, Inc.)

vidual parts, as in Fig. 12-60. Three views are shown at A, and a pictorial view is shown at B. At C, an "explosion" has projected the elementary parts away from each other. This il-

lustrates the principle of exploded views.

All such views are based upon the same principle: projecting the parts from the positions they occupy when put together. Simply, they are just pulled apart. The exterior of a high-pressure piston pump is shown in Fig. 12-61. An exploded illustration of the pump is shown in Fig. 12-62. Note that all parts are easily identifiable.

▪ IDENTIFICATION ILLUSTRATIONS

Pictorial drawings are very useful for identifying parts. They help save time when the parts are manufactured or assembled in place. They are useful for illustrating operating instruction manuals and spare-parts catalogs and are also used for many other purposes.

Identification illustrations are usually presented in exploded views. If there are only a few parts, they can be identified by names and pointing arrows. The identification illustration in Fig. 12-63 is an example showing numbers for the parts. A tabulation shows names and quantities.

▪ RENDERING

Surface shading or rendering of some kind may be used when shapes are difficult to read or for other purposes. For most industrial illustrations, accurate descriptions of shapes and positions are more important than fine artistic effects. Satisfactory results can often be obtained without any shading. In general, you should limit surface shading. Shade the least amount necessary to define the shapes that are being illustrated.

1. PUMP BODY (M - 10091)
2. CYLINDER HEADS (M - 10095)
3. PISTON (M - 10097)
4. VALVE ASSEMBLY (M - 10147)
5. PUMP SHAFT (10 - 10050)
6. OUTER BALL BEARING (10 - 10050)
7. WASHER (10 - 10050)
8. INNER ROLLER BEARING (10 - 10050)
9. GREASE ZERK (¼ - 28)
10. "O" RINGS (125)
11. "O" RINGS (132)
12. "O" RINGS (220)
13. BACK-UP RING (9)
14. HEAD BOLTS (¼ - 20 X 1 ¼)

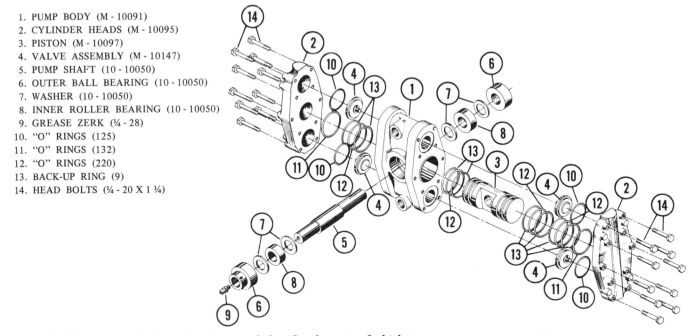

Fig. 12-62. An exploded view that shows and identifies the parts of a high-pressure piston pump. (Industrial Division, Standard Precision, Inc.)

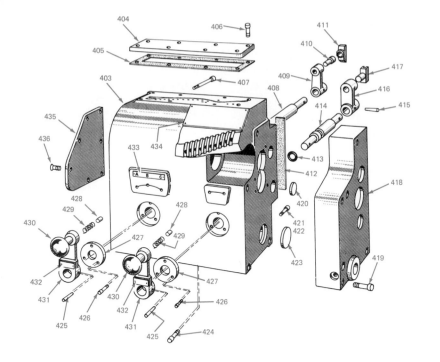

PART NO.	PART NAME	QTY.
403	QUICK CHANGE BOX	1
404	COVER, TOP	1
405	GASKET, COVER	1
406	SCREW, SOCKET HEAD CAP	8
407	SCREW	2
408	SHAFT, SHIFTER	1
409	LINK, SHIFTER	1
410	PIN	1
411	SHOE, SHIFTER	1
412	GASKET (MAKE IN PATTERN SHOP-BOX TO BED)	2
413	"O" RING	2
414	SHAFT, SHIFTER	1
415	PIN, TAPER	2
416	LINK, SHIFTER	1
417	SHOE, SHIFTER	1
418	COVER, SLIP GEAR	1
419	SCREW	4
420	PLUG	2
421	SCREW	3
422	SCREW	1
423	PLUG (NOT USED WITH SCREW REVERSE)	1
424	SCREW	3
425	PIN	2
426	SCREW	6
427	COLLAR	2
428	PLUNGER	2
429	SPRING	2
430	KNOB	2
431	LEVER	2
432	PLATE, FEED-THD.	1
433	PLATE, COMPOUND	1
434	PLATE, ENGLISH INDEX	1
435	COVER	1
436	SCREW	7

Fig. 12-63. An identification illustration. (The R. K. LeBlond Machine Tool Co.)

Line drawings are used most for both pictorials and schematics. In addition, there are halftone renderings, photographs, and isometric, oblique, and perspective pictorials. Different ways of rendering technical illustrations include the use of screen tints, pen and ink, wash, stipple, felt-tip pen and ink, smudge, edge emphasis, and other means. These various means of rendering can be seen in the technical illustra- tions of aircraft companies, automobile manufacturers, and machine-tool makers. They can even be found in the directions that come with your TV set.

■ OUTLINE SHADING

Outline shading may be done mechanically or freehand. Sometimes a combination of both methods is used. The light is generally consid- ered to come from in back of and above the left shoulder of the observer. It also comes across the diagonal of the object, as at A in Fig. 12-64. This is a convention, or a standard method, used by drafters and renderers. At B, the upper left and top edges are in the light. They are drawn with thin lines. The lower right and bottom edges should be shaded. They should be drawn with thick lines. At C, the edges meeting

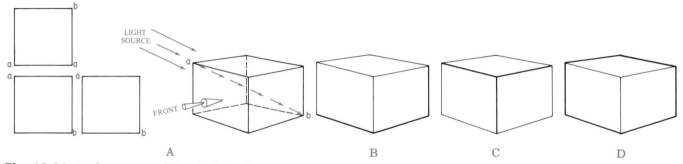

Fig. 12-64. Light source and line-shaded cubes.

in the center are made with thick lines to accent the shape. At D, the edges meeting at the center are made with thin lines. Thick lines are used on the other edges to bring out the shape.

An example of the use of a small amount of line shading is shown and described in Figs. 12-65 and 12-66.

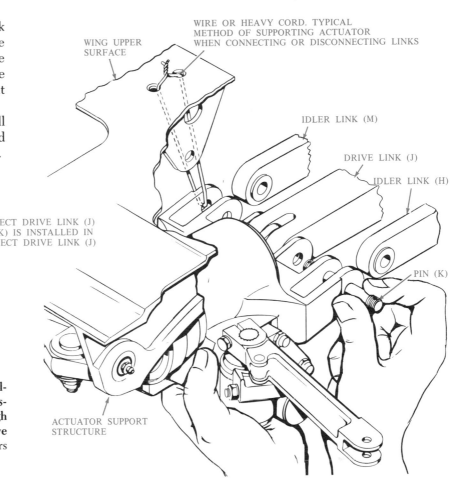

WING UPPER SURFACE

WIRE OR HEAVY CORD. TYPICAL METHOD OF SUPPORTING ACTUATOR WHEN CONNECTING OR DISCONNECTING LINKS

IDLER LINK (M)

DRIVE LINK (J)

IDLER LINK (H)

PIN (K)

REMOVE PIN (K) TO DISCONNECT DRIVE LINK (J) FROM THE ACTUATOR. PIN (K) IS INSTALLED IN THE SAME MANNER TO CONNECT DRIVE LINK (J) TO THE ACTUATOR

ACTUATOR SUPPORT STRUCTURE

Fig. 12-65. A maintenance manual illustration. Notice that only the necessary detail is shown and that just enough shading is used to emphasize and give form to the parts. (Technical Illustrators Association.)

Fig. 12-66. Outline emphasis by a thick black or white line is an effective method of making a shape stand out. (Rockford Clutch Division, Borg-Warner.)

SURFACE SHADING

With the light rays coming in the usual conventional direction, as at A in Fig. 12-67, the top and front surfaces should be lighted. The right-hand surface should then be shaded, as at B. The front surface can have light shading with heavy shading on the right-hand side, as at C. Solid black may be used on the right-hand side, as at D.

SOME SHADED SURFACES

Some shaded surfaces are shown in Fig. 12-68. An unshaded view is shown at A for comparison. Ruled-surface shading is shown at B, free-hand shading at C, stippled shading at D, and pressure-sensitive overlay shading at E and F.

Stippling, at D, consists of dots. Short, crooked lines can also be used to produce a shaded effect. It is a good method when it is well done, but it takes quite a bit of time. Pressure-sensitive overlays (used at E and F) are available in a great variety of patterns and can be applied quite easily.

AIRBRUSH RENDERING

Airbrush rendering produces illustrations that resemble photographs (Fig. 12-69). The airbrush (Fig. 12-70) is a miniature spray gun. It is used primarily to render illustrations and to retouch photographs. Compressed air is used to spray a solution (usually watercolor) that gives various shading effects.

Types of Airbrushes

Airbrushes may be classified according to the size of their spray pattern and by the type of spray-control mechanism. Care should be taken in the selection of an airbrush. The size and style should match the work that needs to be done.

The smallest spray pattern can be obtained from an *oscillating-* (vibrating-) needle airbrush. This type is

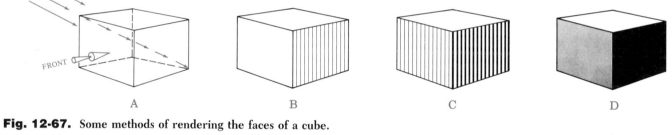

Fig. 12-67. Some methods of rendering the faces of a cube.

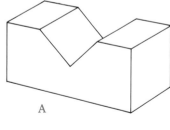

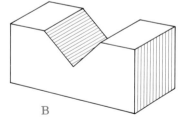
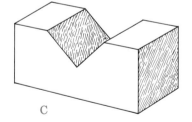
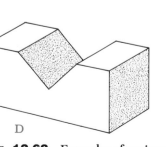
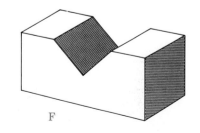

Fig. 12-68. Examples of various kinds of rendering.

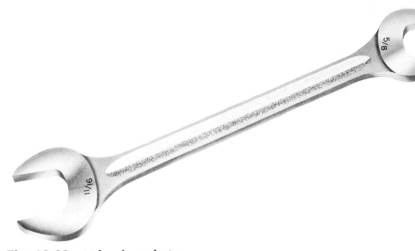

Fig. 12-69. Airbrush rendering.

Fig. 12-70. An airbrush. (R. J. Capece.)

capable of spraying very thin lines (hairlines) and small dots. It is the most expensive and is used only by professionals who do highly detailed rendering and retouch work. A slightly larger airbrush, often called the *pencil type*, is a general-purpose illustrator's airbrush. It can be adjusted to spray thin lines. In addition, it can be opened up to spray large surfaces and backgrounds. This airbrush is most popular for student use. The largest airbrush suitable for use in technical illustrating will spray a pattern large enough to do posters, displays, models, etc. This large brush is often called a *poster-type* airbrush.

Airbrushes are also classified by the type of spray-control mechanism. That is, an airbrush spray control is either single action or double action. Oscillating-needle and most pencil-type brushes have double-action mechanisms. Most poster-type brushes have single-action mechanisms. Single action simply means that when the finger lever is pressed, both color and air are expelled at the same time. Double action means that two motions are necessary. When the lever is pressed, air is released. When the lever is then pulled back, color is released. Much greater control is possible with a double-action airbrush. It is the kind usually used for rendering technical illustrations.

Air Supply

A constant supply of clean air is needed to produce a high-quality spray pattern. An air supply can be obtained from a carbonic gas unit (CO_2) or an air compressor.

An air transformer (regulator and filter) must be installed between the air supply and the airbrush. The regulated pressure for most airbrush rendering is 32 pounds per square

inch ($lb/in.^2$) [220 kilopascals (kPa)]. Less pressure may be used for special effects.

Supplies and Materials

A variety of supplies and materials is needed for airbrush work. Some are common art supplies. Others are special and can be obtained through an art or engineering supply house. The following list is in addition to standard drafting supplies and equipment:

- Airbrush and hose
- Rubber cement
- Rubber-cement pickup
- Razor knife (X-Acto or similar)
- White illustration board (not pressed)
- Frisket paper
- Watercolor brushes
- Medicine dropper
- Designer's watercolors (black and white)
- Palette
- Photo retouching set

Procedure for Airbrushing

The following procedure is generally used for airbrushing. Special effects are done well by experimenting with equipment and materials.

1. Prepare a line drawing of the desired object. Transfer it to the surface of the illustration board. Do not use standard typing carbon paper for transferring the image. Either buy a special transfer sheet or make one. This can be done by blackening one side of a sheet of tracing vellum with a soft lead pencil.

2. Cover the image area with frisket paper. This material is available in two forms: prepared and unprepared. Prepared frisket paper has one adhesive side protected

Fig. 12-71. Opening the area to be airbrushed.

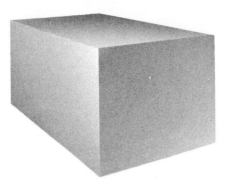

Fig. 12-72. Finished rendering of a cube.

with wax paper. Unprepared frisket paper must be coated with thinned rubber cement. Prepared frisket paper is more convenient to use. It is recommended for the beginner. Cut and remove frisket paper from the area to be airbrushed first (Fig. 12-71). Use the rubber-cement pickup to remove any particles of rubber cement left on the surface. Cover all other areas of the illustration board not covered by the frisket paper.

3. Mix the black watercolor in the palette. Water is placed in the palette cup with the medicine dropper. Squeeze a small amount of black watercolor onto the edge of the palette. Use a watercolor brush to mix the color into the water.

4. Transfer the mixed color from the palette to the color cup on the airbrush. You can use the watercolor brush to do this. Fill the cup about half full.

5. Render the exposed surface as needed.

6. Open a second portion of the frisket and cover the rendered surface. Continue in this way until all surfaces are rendered. Remove the frisket. Figure 12-72 shows the finished rendering.

Figure 12-73 shows examples of other objects rendered. Highlights may be added using white watercolor.

Photo Retouching

Photo retouching is a process used to change details on a photograph. Details may be added, removed, or simply repaired. This process is often needed in preparing photographs for use in publications. It can also be used for changing the appearance of some detail.

Photo retouching is usually done on a glossy photograph. The same basic procedure outlined above for standard airbrush work is used. Care must be taken not to damage the finish on the photo when cutting frisket paper. The gray tones are obtained

by using the tones of gray from a photo retouching kit. A watercolor brush may also be used for touching up fine details. Figure 12-74 shows a before and an after example of a retouched photograph.

■ WASH RENDERING

Wash rendering (also called *wash drawing*) is a form of watercolor rendering. It is done with watercolor and watercolor brushes. It is commonly used for rendering architectural drawings (Fig. 12-75). It is also used for advertising furniture and similar products in newspapers (Fig. 12-76). This technique is highly specialized and is usually done by a commercial artist. However, some technical illustrators and drafters are, at times, required to do this kind of illustrating.

■ SCRATCHBOARD

Scratchboard drawing is a form of line rendering. Scratchboard is coated with india ink, and a sharp instrument is used to make the lines. This is done by drawing the image on the inked surface and then scratching through the ink to expose the lines or surfaces (Fig. 12-77).

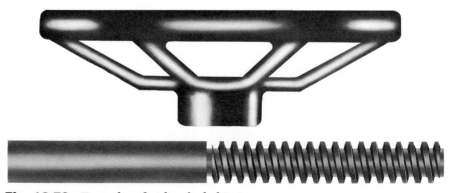

Fig. 12-73. Examples of airbrushed objects.

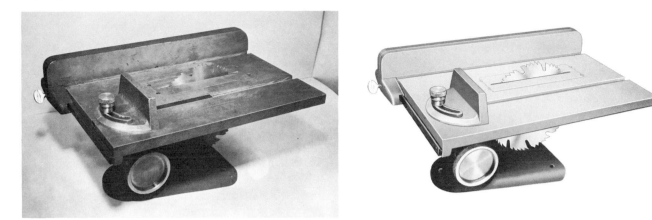

Fig. 12-74. Before and after retouching.

Fig. 12-75. Wash rendering of an architectural drawing.

Fig. 12-76. Wash rendering of a stereo cabinet.

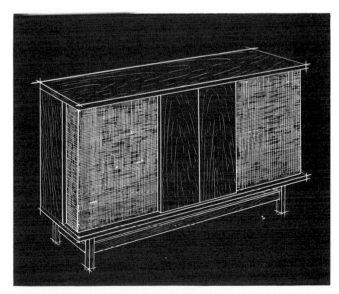

Fig. 12-77. Scratchboard rendering.

LEARNING ACTIVITIES

1. Collect technical illustrations from newspapers, magazines, owners' manuals, and service manuals. Select the best examples and make a bulletin board display.

2. Collect sample pictorial line drawings and try to identify the type (axonometric, oblique, perspective) represented in each case. These may also be used for display.

VOCABULARY

1. Axonometric projection
2. Isometric
3. Dimetric
4. Trimetric
5. Perspective
6. Oblique cavalier
7. Oblique cabinet
8. Isometric line
9. Nonisometric line
10. Ellipse
11. Revolution
12. Picture plane
13. Projector
14. Station point
15. Horizon line
16. Ground line
17. Vanishing point
18. Airbrush
19. Frisket paper
20. Photo retouch

REVIEW

1. Name three types of axonometric projection.

2. What are the three most common types of pictorial drawing?

3. What do you call a line that is not parallel to any of the normal isometric axes?

4. Is an isometric projection larger or smaller than an isometric drawing?

5. Name the two types of oblique drawings.

6. Surface shading is also called _____.

7. Changing details on a photograph is called _____.

8. Which type of pictorial drawing is most natural in appearance?

9. Which type is least natural in appearance?

10. A small spray gun used for rendering is called an _____.

Problems

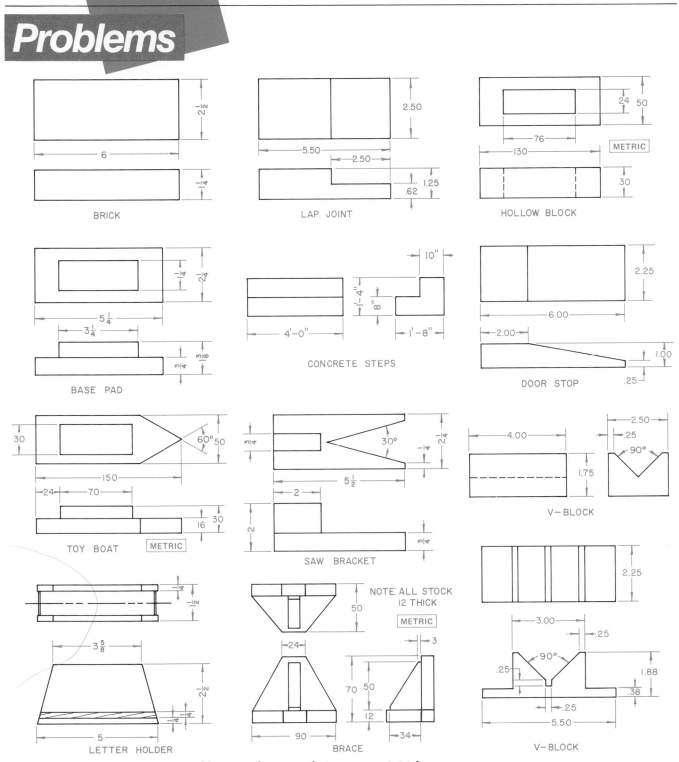

Fig. 12-78. Isometric drawing problems. Scale: optional. Assignment 1: Make an isometric drawing of the object assigned. **NOTE:** These problems may also be used for oblique and perspective drawing practice.

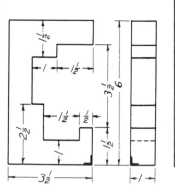

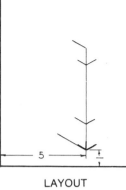

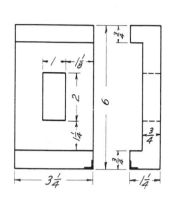

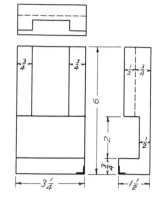

Fig. 12-79. Make an isometric drawing of the plate. Start at the corner indicated by thick lines.

Fig. 12-80. Make an isometric drawing of the notched block. Start at the corner indicated by thick lines.

Fig. 12-81. Make an isometric drawing of a babbitted stop. Start at the corner indicated by thick lines.

LAYOUT

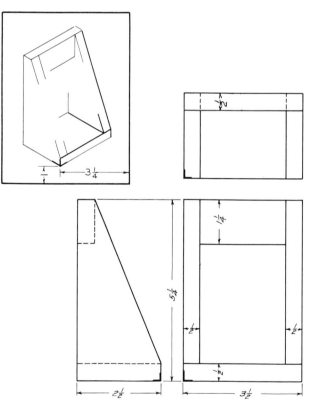

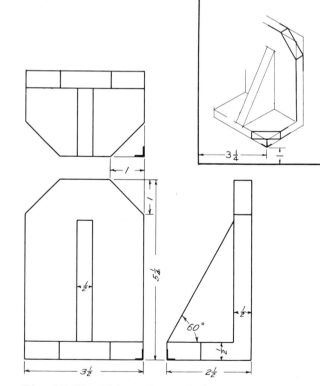

Fig. 12-82. Make an isometric drawing of the stirrup. The drawing is started on the layout at the upper left. Note the thick starting lines.

Fig. 12-83. Make an isometric drawing of the brace. The drawing is started on the layout at the upper right. Note the thick starting lines.

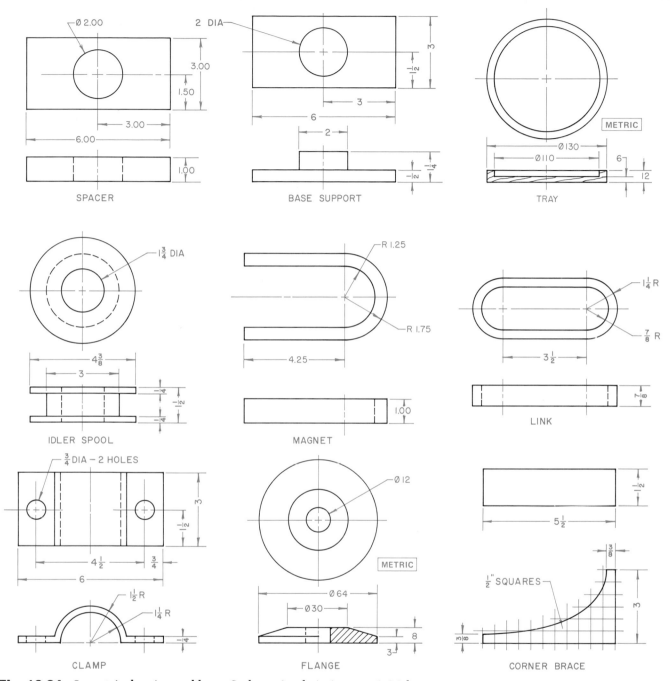

Fig. 12-84. Isometric drawing problems. Scale: optional. Assignment 1: Make an isometric drawing of the object assigned. Assignment 2: Make an isometric half- or full-sectional view as assigned. Do not dimension unless instructed to do so. NOTE: These problems may also be used for oblique and perspective drawing practice.

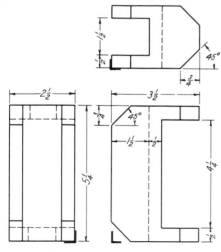

Fig. 12-85. Make an isometric drawing of the cross slide. Use the layout of Fig. 12-82.

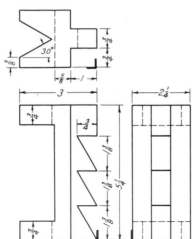

Fig. 12-86. Make an isometric drawing of the ratchet. Use the layout of Fig. 12-82.

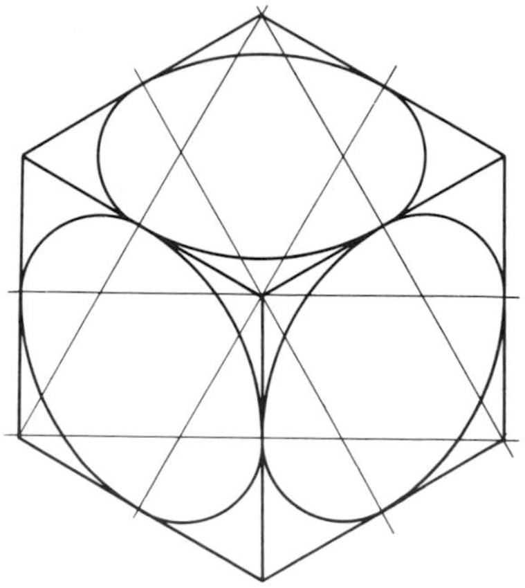

Fig. 12-87. Make an isometric drawing of a 3″ cube with an isometric circle on each visible side.

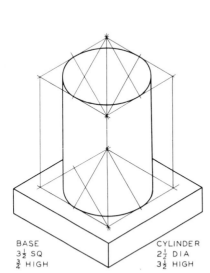

BASE
3½ SQ
¾ HIGH

CYLINDER
2½ DIA
3½ HIGH

Fig. 12-88. Make an isometric drawing of a cylinder resting on a square plinth (square base).

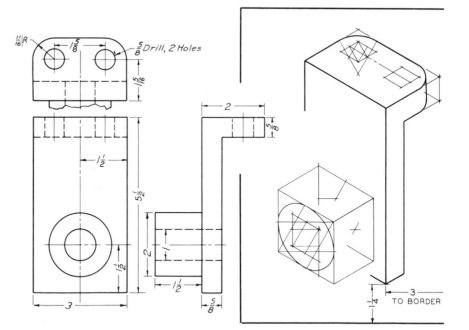

Fig. 12-89. Make an isometric drawing of the hung bearing. Most of the construction is shown on the layout. Make the drawing as though all corners were square, and then construct the curves.

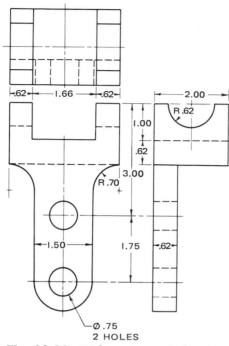

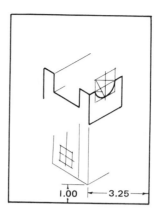

Fig. 12-90. Make an isometric drawing of the bracket. Some of the construction is shown on the layout. Make the drawing as though the corners were square, and then construct the curves.

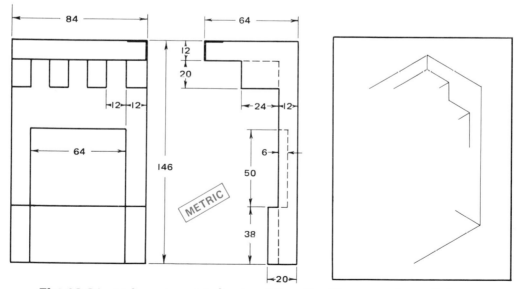

Fig. 12-91. Make an isometric drawing of the tablet. Use reversed axes. Refer to the layout to the right.

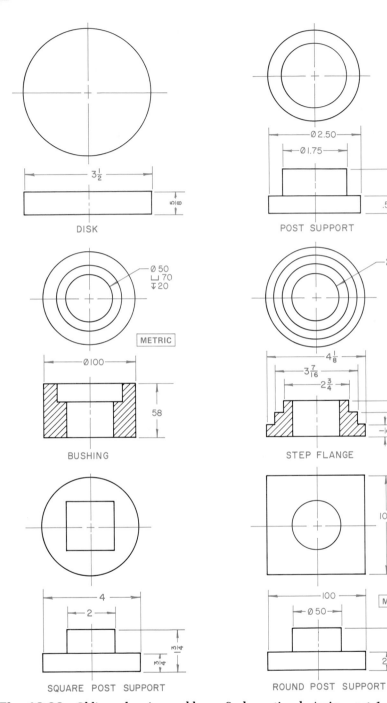
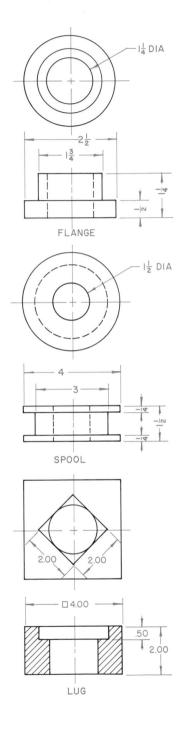

Fig. 12-92. Oblique drawing problems. Scale: optional. Assignment 1: Make an oblique drawing of the object assigned. Assignment 2: Make an oblique half- or full-sectional view as assigned. Do not dimension unless instructed to do so. NOTE: These problems may also be used for isometric and perspective drawing practice.

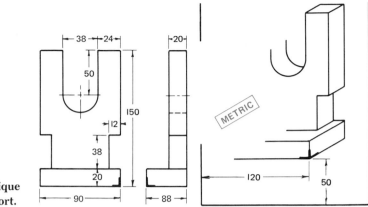

Fig. 12-93. Make an oblique drawing of the angle support.

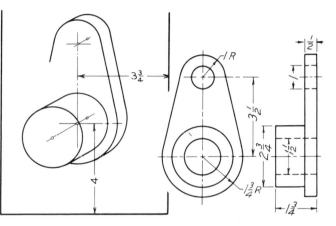

Fig. 12-94. Make an oblique drawing of the crank.

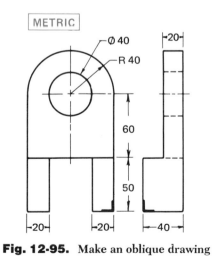

Fig. 12-95. Make an oblique drawing of the forked guide.

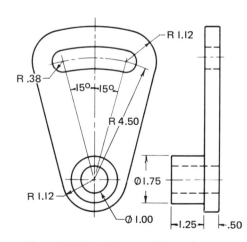

Fig. 12-96. Make an oblique drawing of the slotted sector.

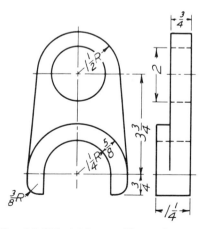

Fig. 12-97. Make an oblique drawing of the guide link.

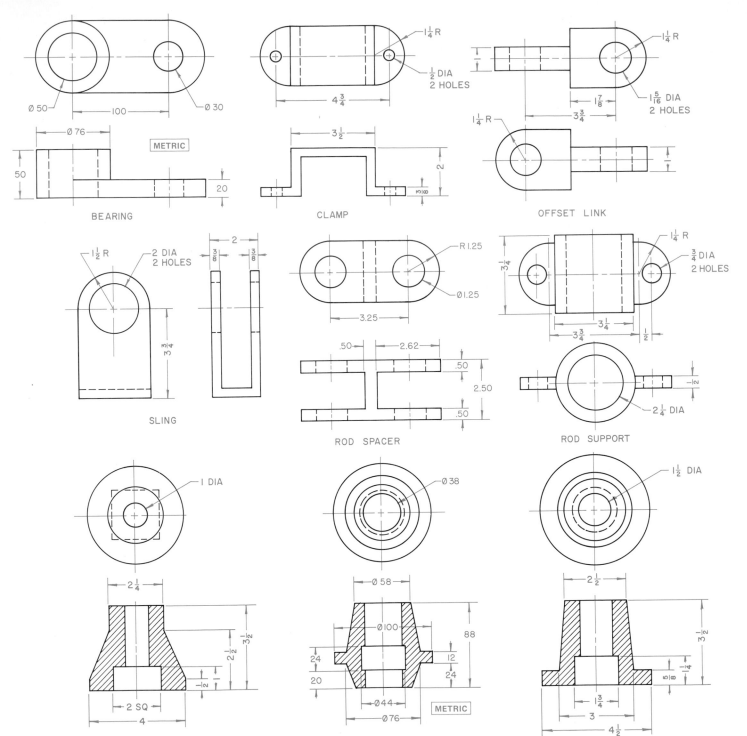

Fig. 12-98. Oblique drawing problems. Scale: optional. Assignment 1: Make an oblique drawing of the object assigned. Assignment 2: Make an oblique half- or full-sectional view as assigned. Do not dimension unless instructed to do so. NOTE: These problems may also be used for isometric and perspective drawing practice.

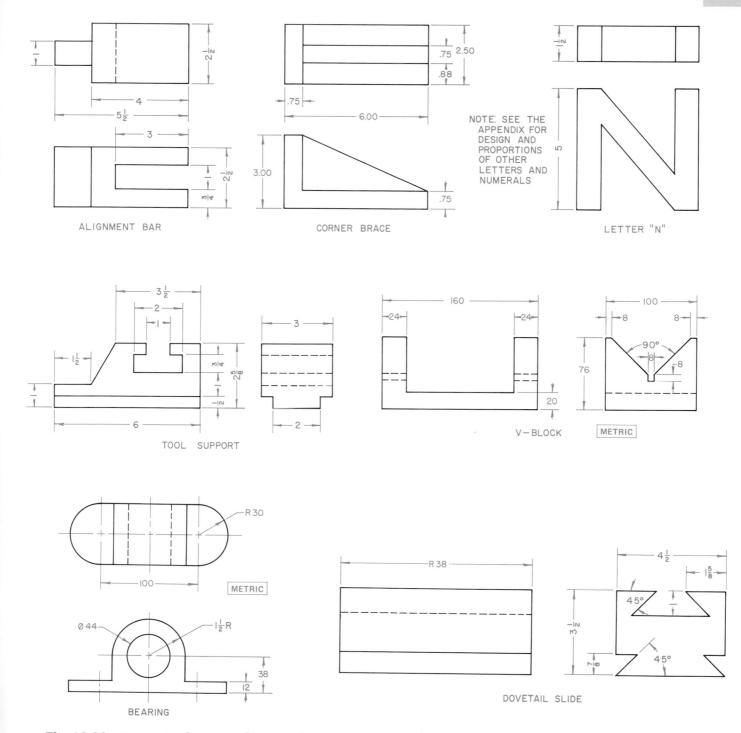

ALIGNMENT BAR

CORNER BRACE

NOTE: SEE THE APPENDIX FOR DESIGN AND PROPORTIONS OF OTHER LETTERS AND NUMERALS

LETTER "N"

TOOL SUPPORT

V—BLOCK

METRIC

METRIC

BEARING

DOVETAIL SLIDE

Fig. 12-99. Perspective drawing problems. Make a one-point-perspective or a two-point-perspective drawing of the object assigned. Use light, thin lines for the construction lines and do not erase them. Brighten the finished perspective drawing lines. Use any suitable scale. The locations of the PP, HL, GL, etc., are optional.

Fig. 12-100. Pictorial sketching and drawing. Assignment 1: Scale: optional. Most technical illustrations are basically pictorial line drawings. In order to develop a good understanding of the relationship of the various types, make an isometric, oblique-cavalier, oblique-cabinet, single-point-perspective, and two-point perspective sketch of the tool support in Fig. 12-99. Assignment 2: Scale: optional. Make instrument drawings of the same tool support (Fig. 12-99) in isometric, oblique-cavalier, oblique-cabinet, single-point perspective, and two-point perspective. Compare the sketches with the instrument drawings. Are they similar? Which type of pictorial drawing gives the most natural appearance?

Fig. 12-101. Exploded-view drawings. Make an isometric exploded-view drawing of the letter holder in Fig. 5-75. Scale: optional. Draw your own initials as an overlay $\frac{1}{32}''$ (1 mm) thick. Estimate the height and width.

Fig. 12-102. Surface shading. Scale: optional. Make a pictorial line drawing of the toy boat in Fig. 12-78. Redesign as desired. Refer to Fig. 12-67 and render the boat using the technique shown at C or D. Maintain sharp clean lines for contrast.

Fig. 12-103. Pictorial assembly drawing. Scale: optional. Make a pictorial assembly drawing of the trammel in Fig. 11-21. Number the parts and add a parts list for identification. Do not show sectional views. Redesign as desired. Add outline shading for accent if instructed to do so. Ink tracing is optional.

Fig. 12-104. Identification illustration. Scale: optional. Make an oblique-cavalier drawing of the mini-sawhorse in Fig. 5-78. Add part numbers and a parts list to make an identification illustration. Render if required. Trace in ink if required.

Fig. 12-105. Airbrush rendering. Make two-point-perspective line drawings of several basic geometric shapes (solids). Examples: cube, cylinder, sphere, cone. Scale: optional. Transfer each to a piece of hot-pressed illustration board and render them in watercolor, using an airbrush.

Fig. 12-106. Scratchboard rendering. Scale: optional. Make a two-point-perspective drawing of the knife rack in Fig. 5-76. Transfer the line drawing to a piece of scratchboard coated with india ink. Use any sharp instrument to scratch through the ink to expose the desired lines.

Fig. 12-107. Exploded-view drawing. Scale: optional. Make an isometric exploded-view drawing of the tic-tac-toe board in Fig. 5-81. Use outline shading to add contrast.

Fig. 12-108. Identification illustration. Make an isometric exploded-view drawing of the note box in Fig. 5-79. Scale: optional. Add part numbers and a parts list to make an identification illustration. Your initials should be designed as an inlay attached to a circular disk. Redesign the note box as desired. Render if required.

Fig. 12-109. Wash rendering. Make a one-point-perspective or a two-point-perspective drawing of any piece of wood furniture. Scale: optional. Transfer the line drawing to a piece of cold-pressed illustration board and render it, using watercolor and a watercolor brush. Use a touch of white or black to sharpen the edges. Keep wood grain and other fine detail lines sharp and clean.

Fig. 12-110. Assignment 1: Identification illustration. Scale: optional. Make an isometric assembly drawing of the trammel in Fig. 11-21. Show the full length of the beam and show two complete assemblies of the point, body, and knurled screw on the beam. Estimate any sizes not given. Redesign as desired. Add part numbers and a parts list to make an identification illustration. Assignment 2: Render the trammel in pencil or transfer the line drawing to hot-pressed illustration board and render in watercolor with an airbrush. Add the part numbers and parts list on an overlay sheet.

Fig. 12-111. Photo retouching. Obtain a glossy photograph of a simple machine or machine part. Study the individual features of the object and list all imperfections. Retouch the photograph using a hand watercolor brush or an airbrush, or both. If possible, have two copies of the original photograph so that a *before* and *after* comparison can be made.

Fig. 12-112. Airbrush rendering. Scale: optional. Make a two-point-perspective drawing of the hammer head in Fig. 5-82. Transfer the line drawing to a piece of hot-pressed illustration board and render it in watercolor, using an airbrush. Add a touch of white or black to edges for contrast.

Fig. 12-113. Surface shading. Scale: optional. Make a single-point-perspective drawing of one of your own initials. Render it using the technique shown at C or D in Fig. 12-67. This problem may also be used for practice in wash rendering or airbrush rendering. For wash rendering, the line drawing must be transferred to cold-pressed illustration board. For airbrush rendering, transfer the line drawing to hot-pressed illustration board.

Fig. 12-114. Exploded-view drawing. Scale: optional. Make an isometric exploded-view drawing of the garden bench in Fig. 5-80. Use ½″ DIA threaded rods, flat washers, and hex nuts to fasten the parts together. If assigned, add part numbers and a parts list to make an identification illustration. This problem may also be used for practice in wash rendering.

Fig. 12-115. Surface shading. Scale: optional. Make a two-point-perspective drawing of the V-block in Fig. 12-78. Refer to Fig. 12-68 and render the line drawing, using the technique assigned. Keep all edges crisp and sharp.

Fig. 12-116. Wash rendering. Make a two-point-perspective drawing of a house. Scale: 1:50 or 1:100 (¼″ = 1′-0″ or ⅛″ = 1′-0″). Transfer the line drawing to a piece of cold-pressed illustration board and render it, using watercolor and a watercolor brush.

Fig. 12-117. Scratchboard rendering. Scale: optional. Make a two-point-perspective drawing of any piece of wood furniture. Examples: coffee table, end table, desk, bench. Transfer the line drawing to a piece of scratchboard coated with india ink. Use any sharp instrument to scratch through the ink to expose the desired lines. See Fig. 12-77.

Fig. 12-118. Airbrush rendering. Scale: optional. Make a single-point-perspective drawing of the object of your choice in Figs. 5-75 through 5-82. Transfer it to a piece of hot-pressed illustration board and render it in watercolor, using an airbrush. This problem may also be used for practice in outline or surface shading.

Fig. 12-119. Wash rendering. Scale: 1:50 or 1:100 (¼″ = 1′-0″ or ⅛″ = 1′-0″). Make a line drawing of the front elevation of a house. Transfer the line drawing to a piece of cold-pressed illustration board and render it, using watercolor and a watercolor brush. This problem may also be used for practice in surface shading in pencil or charcoal.

Fig. 12-120. Surface shading. Scale: optional. Make a pictorial line sketch or instrument drawing of the edge protector in Fig. 5-65. Add surface shading as assigned by your instructor. Keep all edges sharp and crisp. This problem may also be used for practice in airbrush rendering.

Fig. 12-121. Design and surface shading. Scale: optional. Design and make a pictorial sketch of a "car of the future." Render in pencil.

Fig. 12-122. Airbrush rendering. Scale: optional. Make a pictorial assembly drawing of the coupler in Fig. 11-19. Transfer the drawing to hot-pressed illustration board. Render in watercolor, using an airbrush. Add a touch of white or black to keep the edges crisp and sharp.

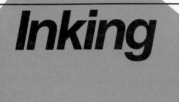

13 Inking

NATURE AND PURPOSE OF INKED DRAWINGS

Most technical drawings are made with pencil on a good-quality tracing paper (vellum) or on drafting film. However, ink is often used to make high-quality tracings that can be copied. As the use of ink drawing increases, better inking techniques and equipment are being developed.

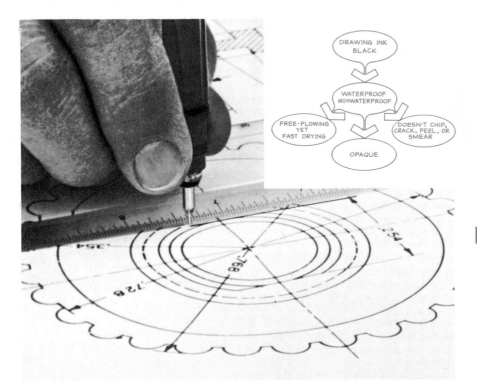

Fig. 13-1. The technical inking pen has made drafting easier and more efficient.

For example, the technical pen with cartridge ink supply, and all its accessories and attachments, has made inking more efficient (Fig. 13-1). It has also improved the quality of the finished drawing.

Microfilming and similar copying methods (also called micrographics) have made new inking methods and high-quality ink drawings necessary (see Chaps. 14 and 15). When technical drawings are reduced or enlarged photographically, some *definition* (detail and sharpness) is lost. A high-quality original always produces a high-quality reproduction (Fig. 13-2). If the original drawing is of poor quality, the reproduction may not be usable.

DRAWING INK

Ink used for technical drawings is also called *india ink*. It must have some special characteristics. For example, it must flow freely and yet dry fast. It must *adhere* (stick) well and must not chip, crack, peel, or smear after drying (Fig. 13-1). Also, it must be completely opaque in order to produce good uniform line

From original drawing

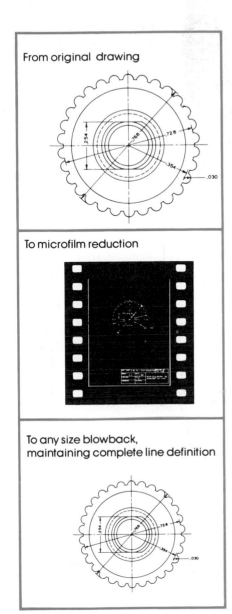

To microfilm reduction

To any size blowback,
maintaining complete line definition

Fig. 13-2. High-quality original ink drawings produce high-quality reproductions. (Koh-I-Noor Rapidograph, Inc.)

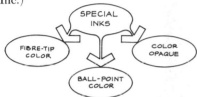

Fig. 13-3. Special inks for new methods of presentation on film, vellum, or paper.

tone and yet be erasable on all drafting media.

Both waterproof and nonwaterproof ink can be used on high-quality paper, cloth, polyester film, and illustration board. The waterproof ink is best suited for mechanical drawings. Nonwaterproof ink is best for fine-line pictorial drawings. Both types of ink produce opaque lines that reproduce well when they are photographed or copied by diazo machines (see Chap. 14).

Special inks come in colors. They include waterproof inks for ball-point pens that are used on computer or tape-controlled plotters. Different kinds of ink are made to be used on different kinds of surfaces (Fig. 13-3). Fiber-tip colored-ink cartridges are available for paper media on plotters.

■ BASIC INKING INSTRUMENTS

The ruling pen (Fig. 13-4) has blades that you can adjust to draw lines of different widths. It can be used to draw both straight and curved lines. Ruling pens are sometimes called *all-purpose pens*. They can be filled from a cartridge tube (Fig. 13-5), a squeeze bottle, or a dropper cap.

■ RULES FOR INKING WITH A RULING PEN

The simple rules listed below will help you produce high-quality ink drawings. Be sure to study them carefully before you try to use the ruling pen.

1. Do not hold the pen over the drawing when you are filling it.
2. Do not fill the pen too full; about ⅛ to ¼ in. (3 to 6 mm) is usually enough.

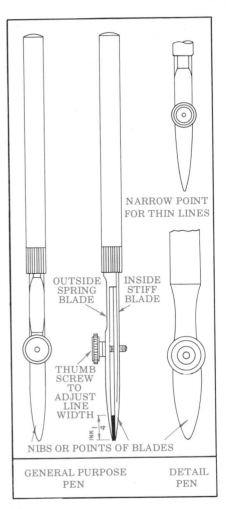

Fig. 13-4. The all-purpose ruling pen.

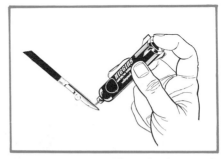

Fig. 13-5. Filling the ruling pen with an ink cartridge.

3. Never dip the pen into the ink bottle or allow ink to get on the outside of the blades.

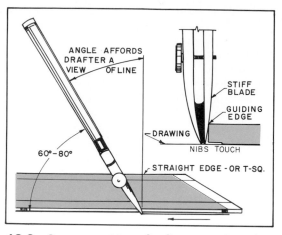

Fig. 13-6. Correct position of ruling pen when drawing lines.

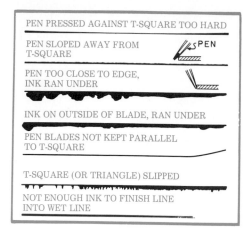

Fig. 13-7. Causes of faulty lines from misuse of a ruling pen.

4. Keep the blades clean by wiping them often with a soft towel, tissue, or pen cleaner.
5. While inking a line, keep both *nibs* (points) of the blades in contact with the drawing surface.
6. Keep the blades slanted in the direction of the pen stroke (Fig. 13-6).
7. Do not press the nibs hard against the straightedge. Pressing too hard will make line widths uneven.
8. Do not press the nibs too hard against the drawing surface. The blades may get dull or damage the drawing surface.

Faulty lines can result from many causes. Figure 13-7 shows some mistakes that can be avoided.

Technical pens have made the inking of technical drawings much easier (Fig. 13-8). Technical pens have points of different sizes to draw different line widths (Fig. 13-9). Customary and metric pen-point sizes are compared in Fig. 13-10. Technical pen points produce uniform line widths because their design provides a steady flow of ink that generally does not clog.

Some technical pens have a refillable cartridge for storing ink (Fig. 13-11). Others have a cartridge that is used once and then replaced.

Technical pen points are made of different materials for use on different surfaces. There are three main kinds of points:

1. Hardened stainless steel points, used on all surfaces
2. Self-polishing jewel points, used on very *abrasive* (rough) surfaces, such as drafting film
3. Wear-free tungsten-carbide points, used on drafting film and with high-speed, automated plotter drafting equipment

Technical pen points have a shoulder, as shown in Fig. 13-12, to prevent smudging. The shoulder is barrel-shaped for use with curved or straight guiding instruments.

The technical pen has some definite advantages over the ruling pen. They are the following:

1. It is easy to produce lines of uniform width (no line adjustments are required).
2. The barrel-shaped shoulder prevents smearing, because point and shoulder are offset (Fig. 13-12).
3. The cartridge makes it easy to

Fig. 13-8. A nine-pen technical inking set. (J. S. Staedtler, Inc.)

load ink. Also, it needs refilling less often than a ruling pen.

4. The ink laid down by a technical pen can dry more uniformly because it flows more evenly (Fig. 13-13).
5. The technical pen does not often need to be cleaned, because it has a weighted cleaning wire inside the capillary tube of the point (Fig. 13-12).
6. A syringe can be used to flush out a pen point in a cleaning solution (Fig. 13-14).

The technical pen can be stored in a humidified container that prevents the ink from drying out (Fig. 13-15). In some technical pens, a seal forms in the cap when the cap is tightened

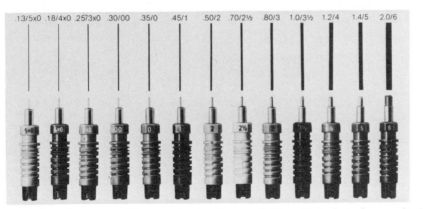

Fig. 13-9. The range of lines and point sizes available with technical pens. (J. S. Staedtler, Inc.)

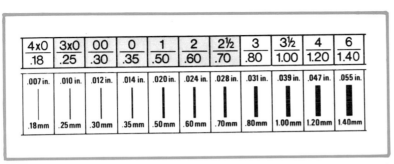

4x0	3x0	00	0	1	2	2½	3	3½	4	6
.18	.25	.30	.35	.50	.60	.70	.80	1.00	1.20	1.40
.007 in.	.010 in.	.012 in.	.014 in.	.020 in.	.024 in.	.028 in.	.031 in.	.039 in.	.047 in.	.055 in.
.18mm	.25mm	.30mm	.35mm	.50mm	.60mm	.70mm	.80mm	1.00mm	1.20mm	1.40mm

Fig. 13-10. Comparison of the widths of lines in inches and millimeters. (Koh-I-Noor Rapidograph, Inc.)

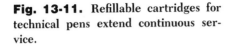

Fig. 13-11. Refillable cartridges for technical pens extend continuous service.

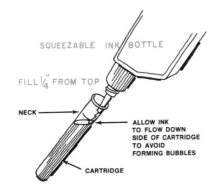

SQUEEZABLE INK BOTTLE

FILL ¼" FROM TOP

NECK

ALLOW INK TO FLOW DOWN SIDE OF CARTRIDGE TO AVOID FORMING BUBBLES

CARTRIDGE

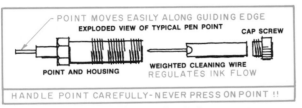

POINT MOVES EASILY ALONG GUIDING EDGE
EXPLODED VIEW OF TYPICAL PEN POINT
CAP SCREW
POINT AND HOUSING
WEIGHTED CLEANING WIRE REGULATES INK FLOW
HANDLE POINT CAREFULLY - NEVER PRESS ON POINT !!

Fig. 13-12. A clicking sound from the weighted cleaning wire ensures a smooth flow of ink.

RULING PEN   TECHNICAL PEN

Fig. 13-13. Section view of typical ink flow from a ruling pen and a technical pen.

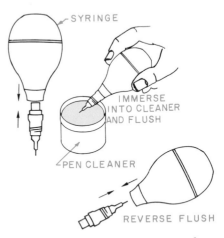

SYRINGE

IMMERSE INTO CLEANER AND FLUSH

PEN CLEANER

REVERSE FLUSH

Fig. 13-14. A syringe pressure cleaner may be used for cleaning and starting a pen. (Koh-I-Noor Rapidograph, Inc.)

Fig. 13-15. Humidified container for technical pens. (Keuffel & Esser Co.)

with a firm twist. If you are careful in filling and cleaning pens, your inking will be better.

INKING STRAIGHT LINES

Figure 13-16 shows the correct position for drawing lines with a techni-

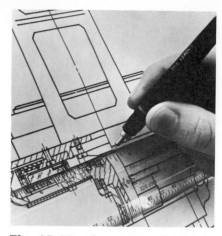

Fig. 13-16. The position of the technical pen is important when drawing lines. (Teledyne Post.)

Fig. 13-17. Technical pens are used easily with templates.

cal pen. Note the direction of the stroke and the angle of the ruling pen. Hold the technical pen in a nearly vertical position to get the most uniform line.

■ MAKING CIRCLES AND ARCS

Templates and compasses can be used with technical pens (Fig. 13-17). However, most templates cannot be used easily with ruling pens. Many compasses have ruling-pen attachments (Fig. 13-18) that are used for inking arcs and circles. Others have technical-pen adaptors, as shown in Fig. 13-19. To use a compass for inking, remove the pencil leg from the compass and replace it with the ruling-pen leg or the technical pen. Adjust the needle point carefully, as shown in Fig. 13-20, so that the legs are perpendicular to the drawing surface. Always draw the circle in one stroke. The compass can be slanted a little when a ruling pen

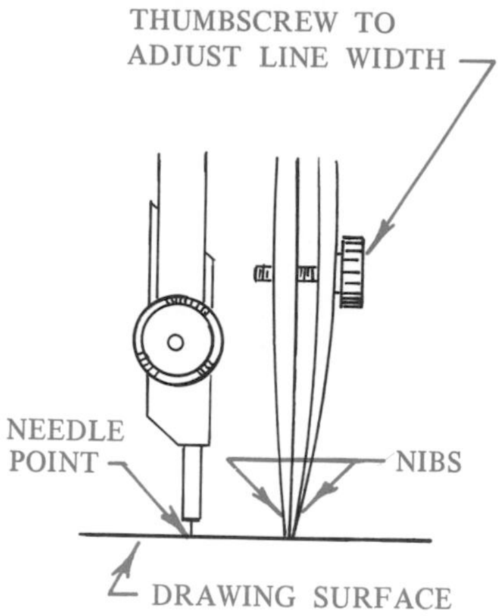

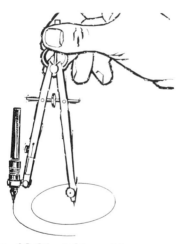

Fig. 13-20. Needle point of the compass is adjusted so that the ruling nibs and point are perpendicular to the drawing surface.

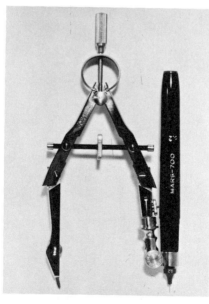

Fig. 13-21. Inking with a compass. Allow the compass to lean about 15° in the direction of the rotating motion. (Teledyne Post.)

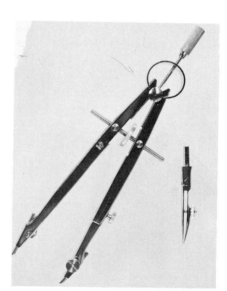

Fig. 13-18. Compasses are designed to hold a ruling pen. (Teledyne Post.)

Fig. 13-19. Technical-pen adaptors for compasses. (J. S. Staedtler, Inc.)

Fig. 13-22. Use of the lengthening bar for large circles.

is being used, but hold it vertically when using a technical pen (Fig. 13-21). The lengthening bar, or beam compass, can be used for large circles, as shown in Fig. 13-22.

■ INKING CURVED LINES

Irregular (French) curves can be used to guide the pen when curves that are not circular arcs are being inked. Use either fixed curved templates or irregular adjustable curves (Fig. 13-23). The following steps will help you in inking curves.

1. Locate the tangent points that change the direction of the curve or join it to a straight line (Fig. 13-24 at A).
2. Draw the curve lightly in pencil. Note where it changes from a smaller to a larger radius.
3. Ink over the middle of the pencil line (Fig. 13-24 at B).

■ LETTERING GUIDES AND EQUIPMENT

The lettering set shown in Fig. 13-25 has three basic tools for inking: a scriber, a set of lettering templates, and a set of technical pens. A scriber has a tailpin to follow the horizontal groove in the bottom of the tem-

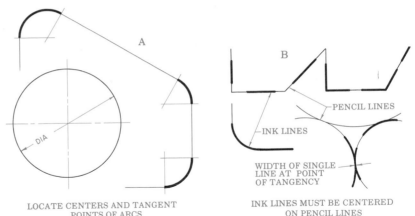

LOCATE CENTERS AND TANGENT POINTS OF ARCS

INK LINES MUST BE CENTERED ON PENCIL LINES

PENCIL LINES

INK LINES

WIDTH OF SINGLE LINE AT POINT OF TANGENCY

Fig. 13-24. (A) Ink curves to point of tangency first. (B) Ink lines must be centered over light pencil lines of layout.

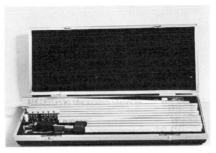

Fig. 13-25. A lettering set is one of the drafter's most useful tools. (Keuffel & Esser Co.)

plate, a tracer pin to follow the engraved letter, and a barrel slot. A technical pen is set in the slot. When the tracer pin is moved along the letter, the pen copies the letter above the template (Fig. 13-26). Lettering templates also come in many styles and sizes of lettering, from about 1.5 to 1/16 to 2 in. (1.5 to 50 mm) high. Some templates are used to draw symbols instead of letters (Fig. 13-27). The width of the pen point used

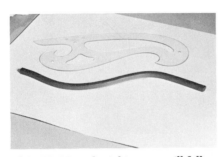

Fig. 13-23. The inking pen will follow curves.

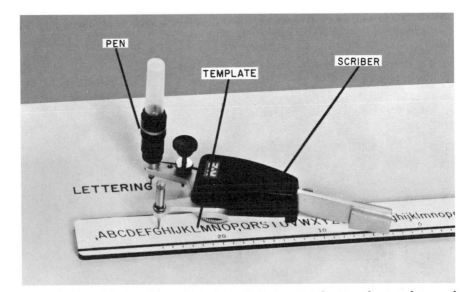

PEN

TEMPLATE

SCRIBER

LETTERING

Fig. 13-26. The three basic parts of a lettering set are the pen, the template, and the scriber. (Keuffel & Esser Co.)

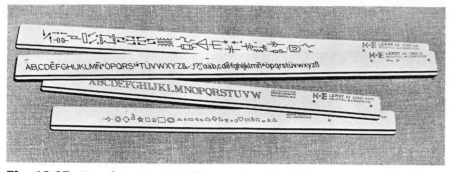

Fig. 13-27. Templates are available in many styles of lettering and kinds of symbols. (Keuffel & Esser Co.)

depends on the height of the letters and the size of the symbols.

ERASING TECHNIQUES

The ink used on polyester drafting film is waterproof. However, you can easily remove ink from the film by rubbing it with a moistened plastic eraser (Fig. 13-28). Do not use any pressure in rubbing. The polyester film does not absorb ink, and therefore all the ink dries on top of its highly finished surface (Fig. 13-29). Remove ink from other surfaces, such as tracing vellum or illustration board, with regular ink erasers. But be very careful. Press lightly with strokes in the direction of the line to remove ink caked on the surface. Too much pressure damages the surface and makes it hard to revise the drawing. The tooth surface shown in

Fig. 13-30 holds the ink more firmly and makes it harder to erase. Different erasing techniques work better on different surfaces. You should try different techniques to find the best one for each surface.

THE ALPHABET OF LINES

Line symbols, the different kinds of lines used to make drawings, are a kind of graphic alphabet. The *American National Drafting Standards Manual* recommends two widths of lines: thick and thin. Figure 13-31 shows the different line symbols and what they mean. Each kind of line has a definite meaning and must not be used for anything else. Detail

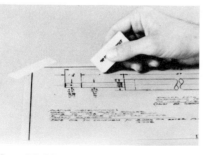

Fig. 13-28. Erasers will remove or absorb ink, depending on the media used. (Teledyne Post.)

INK   ADHESION INK

Fig. 13-29. The results of good (left) and poor (right) ink adhesion. The surface must be clean and the pen properly positioned.

INK MUST COME IN DIRECT CONTACT WITH SURFACE

Fig. 13-30. Section showing ink in contact with drawing surface.

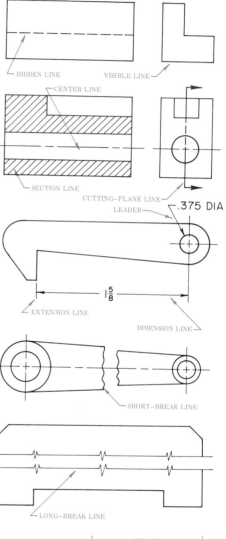

Fig. 13-31. Alphabet of lines for inking. Identify a technical-pen point for each type of line.

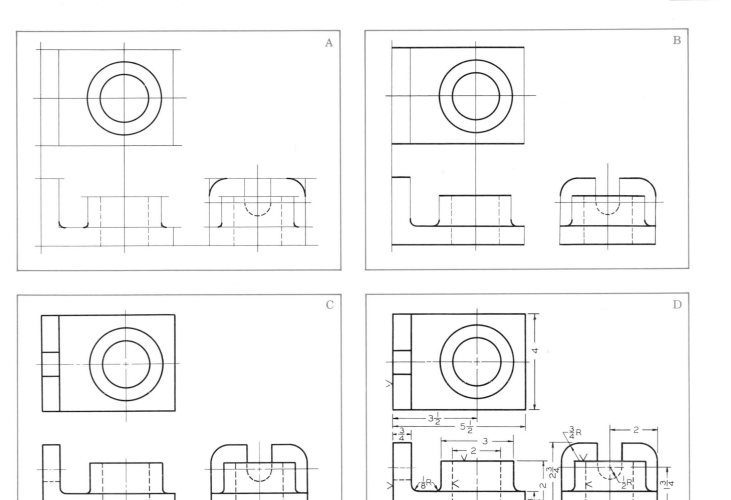

Fig. 13-32. Order of inking, or steps, in developing a finished drawing.

drawings should have fairly wide (thick) outlines. The center and dimension lines and hidden lines should be thin. In this way, the drawing will have contrast and be easy to read. If all the lines have the same width, the drawing looks flat and is hard to read.

■ ORDER OF INKING

Smooth joints and tangents, sharp corners, and neat fillets make a drawing look better and make it easier to read. Good inking, requires careful practice and a definite order of working.

The usual order of inking or tracing is shown in Fig. 13-32. First, ink the arcs, centered over the pencil lines, as in space A. Ink the horizontal lines next, as in space B. Complete the drawing with the vertical lines, as in space C. Then add the dimension lines, arrowheads, finish marks, and so on, and fill in the dimensions, as in space D. The order of inking is:

1. Ink main centerlines.
2. Ink small circles and arcs.
3. Ink large circles and arcs.
4. Ink hidden circles and arcs.
5. Ink irregular curves.
6. Ink horizontal full lines.
7. Ink vertical full lines.
8. Ink slanted full lines.
9. Ink hidden lines.
10. Ink centerlines.
11. Ink extension and dimension lines.
12. Ink arrowheads and figures.
13. Ink section lines.
14. Letter notes and titles.
15. Ink border lines.
16. Check drawing carefully.

■ ULTRASONIC CLEANER

The technical pen and the ruling pen sometimes need cleaning. The ultrasonic cleaner is very useful because it works so fast. The cleaning usually takes only 20 to 30 seconds, and the pen is ready to use again right away. To use an ultrasonic cleaner, just dip the pen point into a well containing cleaning fluid. From 60 000 to 80 000 cycles per second of sound energy act to loosen the dried ink on the point (Fig. 13-33). The normal high range of human hearing is 15 000 cycles per second.

Fig. 13-33. An ultrasonic cleaner for technical pens. (Keuffel & Esser Co.)

To clean a technical pen thoroughly, you may have to take it apart. Remove the cap and ink cartridge and soak the point in a special cleaning fluid. After removing the dried ink from the point, flush the parts in a stream of cold water. Dry all the parts before using the pen again. If the parts are not dry, the ink will be *diluted* (mixed with water) and lines will not be dark enough. If care is taken in handling and storing inking equipment, it will work well and need very little repair.

REVIEW

1. Name the two kinds of ink pens used in drafting.

2. What is the best position in which to hold a technical pen?

3. Name two ways to ink circles and arcs.

4. Name the instruments used to guide the pen when you are not drawing curved lines that are not arcs.

5. How do you erase india ink from drafting film?

6. How many widths of lines are recommended for use on ink drawings? Name them.

7. Why are circles and arcs inked before straight lines?

8. List three kinds of technical-pen points. Why are there different types?

9. Do not fill a ruling pen too full; about _____ to _____ inches (_____ to _____ mm) is usually enough.

10. Metric line widths are based on a _____ progression of the _____.

Problems

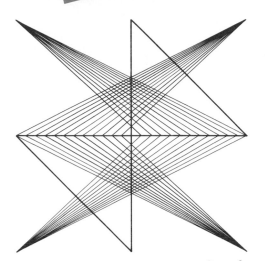

Fig. 13-34. Using light pencil construction lines, lay off a 6″ square and divide it equally into four 3″ squares. On the horizontal midline, mark off 24 ¼″ units. Proceed to ink in the lines to construct the given design.

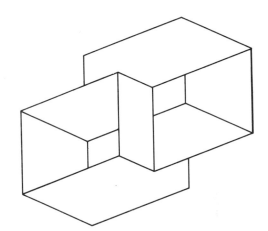

Fig. 13-35. Construct in ink the intersecting units that have an illusion of depth. The axis to the left is 20°, to the right 25°. The depth of a unit is 2″ to the left and 2¾″ to the right. The height is 2″.

Fig. 13-36. Develop the warped surface shown in pencil. (Light layout.) Place 15 ¼″ units on a 45° line to the left. Place 15 ⅜″ units on a 60° line to the right. Connect the units with straight lines, as shown, to form the curve.

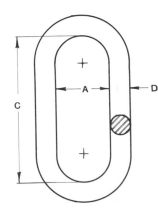

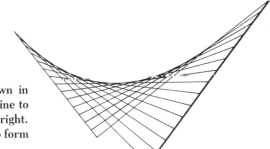

WIRE ROPE LINK		
A	**C**	**D**
1	3	⅜
1¼	4	½
1½	4	⅝
1¾	5	¾
2	5	⅞

Fig. 13-37. Prepare in ink one view of the wire rope short link with revolved section. Use dimensions selected by your instructor. Dimension your drawing. (United States Steel.)

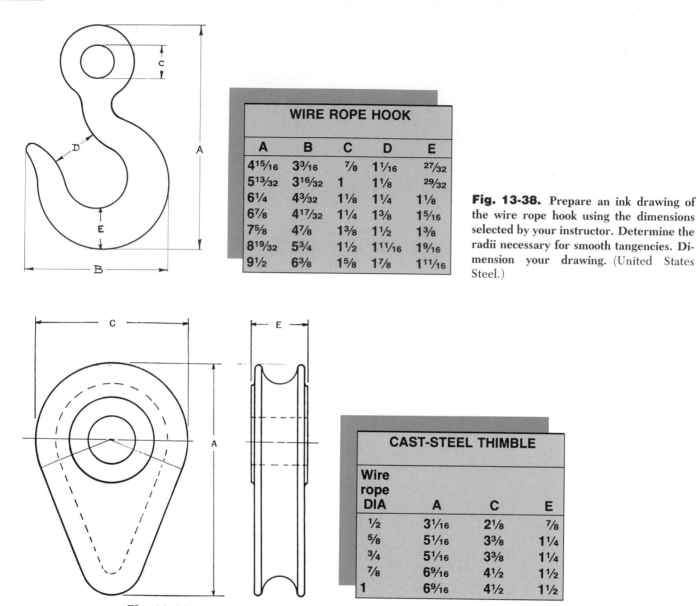

WIRE ROPE HOOK				
A	B	C	D	E
4¹⁵⁄₁₆	3³⁄₁₆	⁷⁄₈	1¹⁄₁₆	²⁷⁄₃₂
5¹³⁄₃₂	3¹⁶⁄₃₂	1	1¹⁄₈	²⁹⁄₃₂
6¹⁄₄	4³⁄₃₂	1¹⁄₈	1¹⁄₄	1¹⁄₈
6⁷⁄₈	4¹⁷⁄₃₂	1¹⁄₄	1³⁄₈	1⁵⁄₁₆
7⁵⁄₈	4⁷⁄₈	1³⁄₈	1¹⁄₂	1³⁄₈
8¹⁹⁄₃₂	5³⁄₄	1¹⁄₂	1¹¹⁄₁₆	1⁹⁄₁₆
9¹⁄₂	6³⁄₈	1⁵⁄₈	1⁷⁄₈	1¹¹⁄₁₆

Fig. 13-38. Prepare an ink drawing of the wire rope hook using the dimensions selected by your instructor. Determine the radii necessary for smooth tangencies. Dimension your drawing. (United States Steel.)

CAST-STEEL THIMBLE			
Wire rope DIA	A	C	E
¹⁄₂	3¹⁄₁₆	2¹⁄₈	⁷⁄₈
⁵⁄₈	5¹⁄₁₆	3³⁄₈	1¹⁄₄
³⁄₄	5¹⁄₁₆	3³⁄₈	1¹⁄₄
⁷⁄₈	6⁹⁄₁₆	4¹⁄₂	1¹⁄₂
1	6⁹⁄₁₆	4¹⁄₂	1¹⁄₂

Fig. 13-39. Prepare a two-view drawing of the solid cast-steel thimble with dimensions selected by your instructor. Select suitable radii for fillets and rounds. Note the points of tangency to determine the tapered shape. Dimension your drawing. (United States Steel.)

Fig. 13-40. Examine Fig. 13-32. Prepare an inked drawing of the figure. Create all three views and dimension. Scale: 2:1.

14 Drafting Media and Reproduction

■ DRAFTING MEDIA

The graphic communications usually prepared in the drafting room today are made on paper, illustration board, cloth, or film. The basic *media* (materials to draw on; singular, *medium*) have been greatly improved in recent years. This is because industry has demanded an ever-increasing standard of quality.

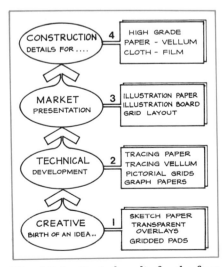

Fig. 14-1. Typical media for the four levels of graphic communication.

These materials are generally used in the four levels of communication (Fig. 14-1).

Level One: Creative Communication.
Sketch paper and transparent paper are used for refining the ideas captured on the original layout. Grid paper is used for controlling proportions and developing schematic layout diagrams.

Level Two: Technical Communication.
Sketches are refined on tracing paper. Charts and diagrams are made on grid paper, or special graph paper is used for pictorial drawings.

Level Three: Market Communications.
Illustration paper or board is used for good ink and color rendings of pictorial or cutaway views. These highly developed artistic sketches explain the details of the project to the client.

Level Four: Construction or Manufacturing Communication.
These drawings are made on high-grade tracing paper, vellum, cloth, or polyester film. Paper and film that

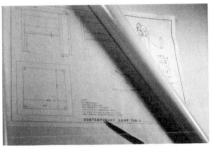

Fig. 14-2. Translucent drafting medium that is good for pencil and ink.

are *ageproof* (will not discolor or fall apart) are required for permanent records (Fig. 14-2).

The design-engineering team needs the basic drafting media with their differing capabilities for specific uses.

Paper
A wide range of opaque and translucent paper is available. Designers consider opaque drawing paper more limited than translucent paper for reproduction. High-grade opaque paper can be used for black and white and for colored pictorial renderings that later can be photographed for reproduction. The

opaque paper can be lined for drawing graphs or diagrams or for plotting mathematical solutions.

Translucent tracing papers are used for developing original detail drawings. This medium must have a good surface quality. It must take pencil well and withstand repeated erasing and also a lot of handling by the engineering team.

Translucent Tracing Papers

Natural tracing papers are made of rag. These are strong, but they are not really transparent enough to be used for all applications. The natural translucent paper without rag should be used only for sketching because it is not very durable. Advances have been made in papermaking to produce thin tracing paper that is reasonably transparent and retains pencil and ink well.

Transparentized Tracing Papers

These treated papers are sometimes called *vellum*. They are ideal for quick reproduction. This paper has rag in it and is very durable. It is treated with resins to make it transparent. It makes sharp prints quickly. Good erasability, long life, and high strength are other qualities of transparentized papers. These papers are also considered ageproof. Nevertheless, they still have one disadvantage. They show fold marks.

Cloth

Cloth, or linen as it used to be called, is of higher quality than paper. Today it is an expensive medium that is excellent for pencil and ink. It is made of cotton fiber sized with starch. It is durable and erasable, and it reproduces well.

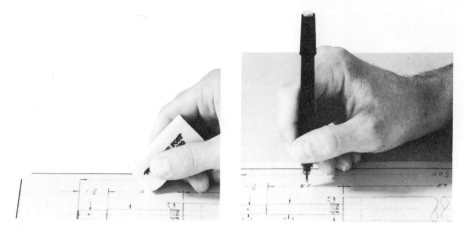

Fig. 14-3. Images can be removed with a moist vinyl eraser and corrections made on either side of double-matte film. (Teledyne Post.)

Film

The finest medium developed for the engineering team in recent years is polyester film. It will not shrink, and it is transparent, ageproof, and waterproof. Pencil or ink produce a clear image on the surface of film. That surface is stable even after repeated erasing. Film is classified by thickness and according to whether the surface is *matte* (not glossy) on one side or both.

In part, better results are being obtained from all the media because better lead, ink, and erasers have become available. Each new or improved medium has a surface that is *workable* (erasable and reproducible) (Fig. 14-3).

Fade-Out Grid and Lined Stock

High grades of tracing paper and film are available with a grid pattern. This pattern is very useful in drawing. It does not show up on copies (Fig. 14-4). The patterns generally have 10 squares, 8 squares, or 4 squares to the inch. There are also millimeter grids. This paper, called *cross-section paper*, is available in

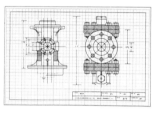

DRAWING PREPARED ON PREPRINTED GRID

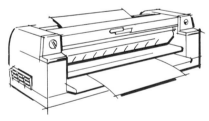

COPY MADE

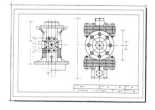

PRINT DOES NOT SHOW GRID LINES

Fig. 14-4. Fade-out grid and film are two types of media that have made drafting easier. (Bruning.)

Fig. 14-5. Illustration board is good for ink and presentations.

rolls, sheets, and pads. There is generally a choice of colored lines, though light blue is most commonly used. Special kinds of cross-section paper are made for orthographic and isometric drawings.

Illustration Board

Artists' drawing and illustration boards are used in preparing drawings for the client's approval. The surface quality of these boards varies. The hot-pressed surface is smooth and is used for ink renderings (Fig. 14-5). The medium and regular surfaces are preferred for pencil and color-wash drawings.

■ REPRODUCTION

Once a drawing is completed, many copies of it are often needed for various purposes. The design-engineering team needs copies they can refer to or use to develop or revise drawings. The factory needs copies to make and inspect parts. Subcontractors, which are other companies that supply parts, also need copies to

make and inspect those parts. The purchasing department needs copies to order standard parts not made in the shop. These might include bolts, nuts, washers, gaskets, and the like. If the item is *assembled* (put together) in a place other than the shop, additional copies are needed for assembly.

Fast, accurate, and economical methods of *reproduction* (making copies) have been available for many years (Fig. 14-6). It is important to choose the right reproduction process for a particular purpose. In making a choice, you need to consider such factors as these:

1. *Input of the originals*. What is the size, weight, color, and opacity of the medium used?
2. *Output quality*. How readable must the copies be? Are they to be transparent or opaque? Must they be workable?
3. *Size of reproduction*. Are the copies to be enlarged, reduced, or the same size? Are they to be folded by machine?
4. *Speed of reproduction*. How many copies of each original are needed? What is the speed of the copying machine?
5. *Color of Reproduction*. Is the process blue line, blueprint, or sepia brown?
6. *Cost of reproduction*. What are the costs of materials, operations, and overhead?

The equipment available today for copying engineering drawings includes blueprint, diazo, electrostatic, thermographic, and photographic.

Blueprint

This is an inexpensive process that makes a copy the same size as the

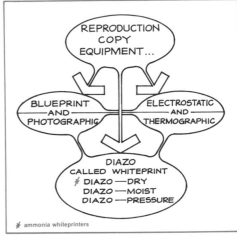

Fig. 14-6. The kinds of reproduction processes for copying drawings.

original on a light-sensitive contact photographic paper. The copy is like a negative, in that it has white lines on a blue background. Blueprint is the oldest copying process. Prints have to be washed in water after being exposed to light. They are then treated with a potash solution to make the blue darker, and rewashed. Blueprinting is a continuous machine process. The only hand work that it involves is trimming the finished copies.

Diazo

Sometimes the diazo process is called *whiteprint*. The diazo print has the opposite (positive) look to the blueprint. That is, it has dark lines on a white background. It is made from a drawing having dark lines (pencil or ink) on a translucent medium. Diazo printing is available in a few colors, including blue line, black line, and red line. The paper, vellum, or film stock has a light-sensitive diazo coating. This coating is developed by either a dry process (ammonia vapor), a semimoist process (liquid chemical), or a pressure process (amine chemical).

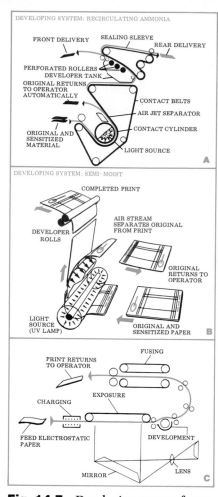

Fig. 14-7. Developing systems for reproduction. **(A)** Diazo dry ammonia. **(B)** Diazo moist chemical. **(C)** Electrostatic. (Bruning.)

Diazo Dry Process

The print paper in the diazo dry process has a dye coating. The paper is passed through a light source while in contact with the original translucent drawing (Fig. 14-7A). After this light exposure, the dye coating on the print paper remains only where the light did not pass through the original drawing. The print paper is then passed through ammonia *vapor* (gas). The remaining dye is thereby developed into a line image. The original drawing itself does not pass through the developer.

Diazo Moist Process

The print paper in the diazo moist process also has a light-sensitive coating. This coating is developed with a liquid chemical that only makes the paper damp. The drawing and print paper are exposed to light, while in contact, much as in the dry process. However, in the development phase, the print paper is fed through fine-grooved rollers where the developing liquid brings out the image. The print is passed over heating rods to dry. The original drawing does not pass through the developer (Fig. 14-7B).

Pressure Diazo

Diazo prints are made on paper, vellum, and film. Developing with an activator (chemical) can be fast, medium, or slow, depending on the coating on the print material. The speed must be adjusted to make prints with good contrast. The vellum and films are available in matte finish and are workable on either side. These copies are of such high quality that they can be used in place of originals or as new originals for design changes. They are called *intermediates*.

The pressure diazo (PD) process has been developed to make dry, odorless whiteprints (Fig. 14-8).

Electrostatic Reproduction

Xerography is a dry process. It uses an electrostatic force to deposit dry powder. The powder makes positive, black images on copy paper (Fig. 14-7C). The electrostatic process can

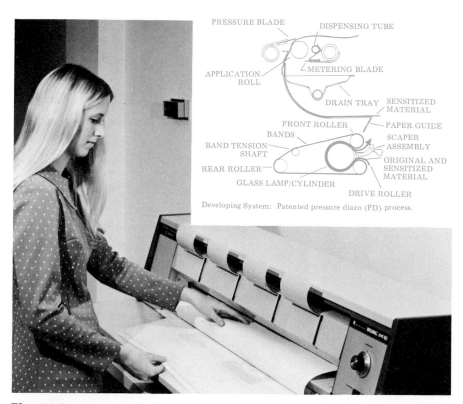

Fig. 14-8. Developing system for diazo dry-pressure activator. (Bruning.)

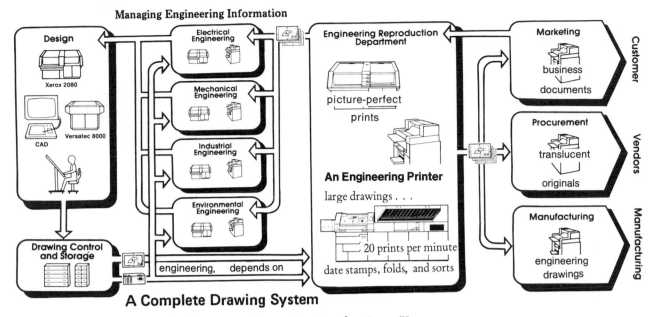

Fig. 14-9. Electrostatic reproduction system for engineering drawings. (Xerox Corp.)

produce copies that are enlarged, reduced, or the same size.

Xerography machines have an electrostatically charged plate coated with the metal selenium. Where light strikes the charged surface, the charge *dissipates* (goes away). But the surface of the plate stays charged where no light falls. The blank spaces on an original cause the charge to dissipate. The lines and lettering on an original let the charge remain.

An image is produced on the surface of the exposed plate by a *toner*, or developer, of magnetic powder. This powder sticks to the places on the plate where the image on the original has let the charge remain. The powder image is passed from the plate surface to copy paper by an electrical *discharge* (jump). The powder is *fused* (joined) to the paper by heat.

Drawings can be sent from city to city by electronic transmission. Originals, such as engineering drawings, can be changed by machine into electronic signals. These signals can be *transmitted* (sent) to other places. There, the signals can be changed by machine back into xerographic copies of the original. This process can be controlled by a computer. The signals can be transmitted by microwave radio or by telephone wires.

The newest reproduction processes are electrostatic (Fig. 14-9). These processes produce copies ranging from translucencies to photographs.

Thermographic Reproduction

Thermofax is a trade name of the company that first made thermographic copying machines. This kind of copying is based on the fact that a dark surface *absorbs* (soaks up) more heat than a light surface. The print paper is affected by heat instead of light. Thus the original engineering drawing does not have to be on translucent paper. However, the markings on the original must be carbon or metallic. Such markings are heated by *infrared light*, a kind of light that you cannot see. There is no developer in this process. Since the thermographic process is dry, copies can be colored or translucent.

Photographic Reproduction

The companies that make film have found ways to help reproduce engineering drawings. Copy cameras (Fig. 14-10) make images on light-sensitive material, such as film that is coated with the chemical silver halide. The images are made permanent with chemical solutions and other ways of fixing an image. Ways of reducing and enlarging images are made easy. High-contrast films are used to give lines uniform widths and blackness (Fig. 14-11).

Film makes better prints than the commonly used diazo process. The film has a matte finish. Changes can be made with pen and ink or pencil.

Fig. 14-10. Copy camera for photographic reproduction. (Keuffel & Esser Co.)

Fig. 14-11. This rendering has been photographically reproduced. Although the size has been reduced, clarity of line has not been lost. (E. Zavada.)

Film is most often used to make *intermediates* (copies used as originals), which you will learn about in Chap. 15.

Microfilm

This form of film is very good for keeping engineering records. Microfilm is excellent for handling engineering information. Records can be kept on different kinds of microfilm (Fig. 14-12). Roll film, jacketed film, micro-opaque microfilm, and aper-

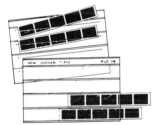

STORAGE

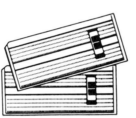

APERTURE CARD

RECORDS FOR FILE

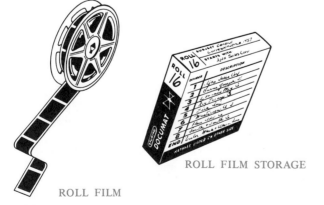

ROLL FILM

ROLL FILM STORAGE

Fig. 14-12. A microfilm system includes these elements. (Bruning.)

MICROFILM ROLL INDEX

Department: *National Sales Office*

Class of Records: *General Correspondence*

File Period	Subject	Roll No.	Divider
495.75	*Ajax Sales Corp.*	16	1
495.75	*Burns, Joseph L.*	16	2
495.74	*Donahue Mfg. Co.*	16	3
495.71	*Key Transportation Co.*	16	4
495.72	*Lark, Inc.*	16	5

Fig. 14-13. A microfilm reader. (Bruning.)

ture cards are all commonly used for engineering records. Roll film comes in widths of 16 mm, 35 mm, 70 mm, and 105 mm. There are scanning machines for studying and copying from the microfilm (Fig. 14-13).

Jacketed film is best for records that are used and changed often. The positive micro-opaque film is an opaque card stock. It is used with readers that can magnify it.

Microfiche is a sheet of transparent film that has many very small images in rows. Some of the sizes used for storage are 3 × 5 in. (75 × 125 mm), 4 × 6 in. (100 × 150 mm), and 5 × 8 in. (125 × 200 mm). Copies can be printed after the right image is chosen on a screen reader. Copies can be different sizes.

Aperture Card

An aperture card is a standard-size data file card. It has a rectangular slot that can hold a framed microfilm (Fig. 14-14). The most common card is 3¼ × 7⅜ in. (83 × 187 mm). Data is punched in the card except in the area near the film. This data card can give engineering information about the film and make it easy to store. Cards can be sorted, filed, and retrieved by machine. Images can be reproduced in the size needed.

■ DRAFTING MATERIALS AND SYSTEMS

On the engineering team, it is the drafter who chooses the right drawing medium for the level of communication needed to do the job. The technologies of the older media have been improved. New media have been invented. These advances have usually involved complicated new reproduction equipment that makes many copies. These machines have greatly lessened the amount of clerical work that takes place in the modern engineering office. All these changes give drafters wider and better choices than ever. Rapid change is likely to continue in this area. Drafters will always have to be learning how to use new tools of their trade.

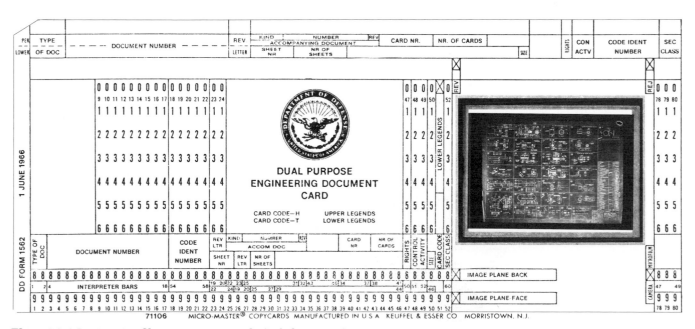

Fig. 14-14. A microfilm aperture card. (Teledyne Post.)

REVIEW

1. Name the best media for each of the four levels of communication.

2. What are the best uses for opaque paper?

3. Name five good qualities of vellum and one disadvantage of its use in drafting.

4. Define the purpose of an intermediate copy. Is the copy translucent?

5. Name some advantages of using drafting film.

6. List several things that affect the selection of a reproduction process.

7. What is the difference between the copy made on a blueprint machine and the one made on a diazo machine?

8. List several capabilities of electrostatic copiers that other reproduction processes do not have.

9. Microfilm is not a good storage system for original drawings because the image is too small to see properly. True or false?

10. List two reasons why a large company might put a microfilm storage and retrieval system in its engineering department.

15 Systems for Controlling Graphic Communication

■ ORGANIZING GRAPHIC COMMUNICATION

Drawings and reports done by the engineering team are usually processed in the drafting room. Two systems are used in processing design projects: written and *graphic* (drawn). They are the two keys for controlling projects from start to finish.

System One: Written Communication

Graphic forms of communication are generally controlled by a written communication system. This system is made up of technical reports. There are five kinds of technical reports:

1. *Design report*. Includes a statement of the needs of the client and an early *analysis* (study) of the project by the designers.
2. *Technical report*. A design analysis of all the things that go into researching, designing, and developing the project.

3. *Market report*. A formal report that includes recommendations (also called *client* report). This report comes out before it is decided whether or not to go ahead with manufacturing or construction on the project.
4. *Contract report*. Includes all the details of manufacturing or construction. This report is done after it has been agreed to start work on the project.
5. *Progress report*. A report made once in a while to check on the

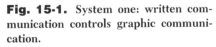

Fig. 15-1. System one: written communication controls graphic communication.

quality of the project and take another look at the specifications in the contract. This report usually appears at certain chosen dates or when certain stages of the project have been finished.

These reports are necessary to make the work of a design team on a project a success. They add to and support the graphic communication needed to design, start, finish, and sell a project (Fig. 15-1).

System Two: Graphic Communication

The four levels of graphic communication described in Chap. 1 are used over and over again as design teams work on design projects (Fig. 15-2).

Level One: Creative Communication. This is also called *the birth of an idea*. It usually takes the form of the designer's sketches showing new ideas to be studied.

Level Two: Technical Communication. Several technical people put together what they know about

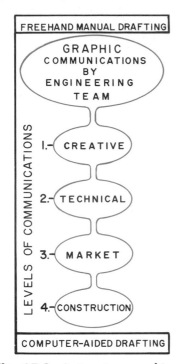

Fig. 15-2. System two: graphic communication captures and controls the design process.

materials and ways of doing things to suggest how to solve problems. These solutions are usually shown *graphically* (in drawings) because they are often difficult to put into words. Technical graphic communications include sketches, schematic diagrams, and graphs.

Level Three: Market Communication.
The technical study is refined and illustrated so that the client can judge the form, function, and style of the design. Pictorials, cutaways, and exploded views can help the client make a decision about the design.

Level Four: Construction Communication.
Detailed drawings are made for contracts, estimating, construction, or manufacturing. They must be full and accurate. Written specifications often go with detailed drawings.

Which levels of graphic communication are produced by the design team depends partly on how the team is put together. Large companies organize design teams to solve major problems. Their solutions help improve our technology and raise our standard of living. All the consumer products, as well as many services, result from the work of the design-engineering team. The scientist, engineer, technician, designer, and drafter are members of this team.

THE ENGINEERING TEAM

Two good examples of companies that need engineering teams would be a large architectural design firm and an automobile maker. The managers of large firms must use the design team to meet the needs of clients or future customers. The work of the team must be organized so that problems can be solved as quickly

and cheaply as possible. From the early study to the final working drawing, all work is done according to a set schedule. This is true whether the product is a skyscraper or an automobile (Fig. 15-3). The head of a design project may make a flowchart to help in managing the design process.

ORGANIZING THE LEVELS OF DRAFTING WORK

The organization chart in Fig. 15-4 shows a way of processing work through the drafting room. At all levels, the design process rests on the principles of mechanical drawing. The drafter can help the design team at all four levels of communication. The manager of a design team controls all levels of progress (Fig. 15-5).

■ *Level one.* The layout drafter helps the designer by working

Fig. 15-3. The manager controls the levels of work on the drawing board or at the computer station (**CAD**). (General Motors Corp.)

closely with the design sketches, changing them into mechanical drawings.

■ *Level two*. The experienced senior detailer helps study and judge sketches, layouts, and diagrams as design decisions and recommendations are made.

■ *Level three*. The drafter-illustrator generally assists in the presentation studies. This drafter specializes in making pictorial views and renderings for the client to study and approve.

■ *Level four*. The senior and junior detail drafters prepare the finished construction and manufacturing drawings based on the layout work. They are supervised by the design team manager. The final details are closely checked, often by a senior detailer. Designers and drafting technicians may use a computer-aided drafting system at all four levels.

SYSTEM FOR DRAFTING PROCEDURES

As the design team develops a project that uses drafting services, a record is kept of the progress under a contract number or job number. This number must be used on all reports and drawings. The number is a part of the information given in the title block of a drawing.

TITLE BLOCKS

The basic title block used in industrial drawings is shown in Fig. 15-6. The drawing-number block and the drawing title are needed to catalog and cross-index the drawing. The blocks giving the date, scale, signatures, initials, tolerance information,

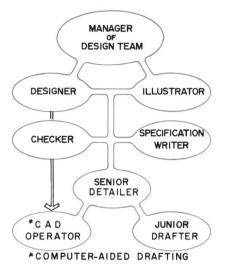

Fig. 15-4. The flow of work is managed for each project and the team members have various talents.

material specification, and revisions are needed to read the drawing and manufacture the product. All drawings, regardless of size, must have a title block.

REVISION BLOCK

Changes on a drawing are recorded in the revision block. It is usually found in the lower right-hand corner of the drawing sheet. Letters or numbers are used to identify the revised part of the drawing. These letters or numbers are needed to find out what change was made and who made it.

Fig. 15-5. The drafting technician and the supervisor are guided and supported by the team manager. (Bruning.)

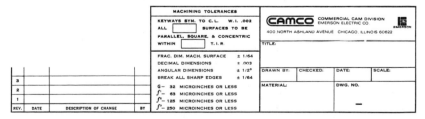

Fig. 15-6. The basic title block contains controlled information. (Camco.)

BASIC SHEET SIZES

Title blocks and revision blocks may differ a little in size and in where they are put. Their exact size and place depend on how big the drawing sheet is. Design teams in all industries have been using standard-size sheets. These come in multiples of 8½ × 11 in. and 9 × 12 in. The sizes are shown in Fig. 15-7. Both size ranges are identified by the letters A, B, C, D, and E.

The ISO "A" series is developed from a base sheet of paper with an area of 1 square meter (m²). The ratio of the sides of all the sheets of paper derived from the base is 1:√2. Therefore, the size of the base sheet is 841 × 1189 mm. Sizes are designated with the letter A followed by a number. This number tells how many times the base sheet has been divided. The number A1 means that the A0 sheet has been halved once. The number A2 means that it has been halved twice (Fig. 15-8). If sheets larger than A0 are required, they are designated by a number in front of the A0 designation. The next size larger than A0 would be designated 2A0, or 1189 × 1682 mm. The 4A0 sheet measures 1682 × 2375 mm.

CHECKING REPORT

Checking working drawings for accuracy is an important part of the drafting procedure system. The checking is usually done by an experienced detailer who reports to the chief drafter. The checker uses a print to look over the work, marking errors for correction or marking approval. This job is important because all the detail work must be accurate and use standard drafting practices.

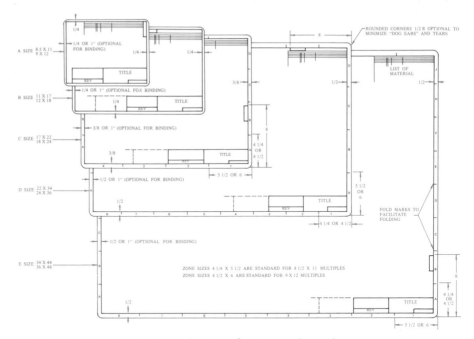

Fig. 15-7. Examine the basic drawing sheet sizes. (ANSI.)

PROGRESS REPORT

Drafting departments record the progress of the team detailing a project. Each drafter may have to report progress either daily or weekly. The report will record the job contract number and the number of hours spent working on each

ISO series	"A" paper sizes (mm)
A0	841 × 1189
A1	594 × 841
A2	420 × 594
A3	297 × 420
A4	210 × 297
A5	148 × 210
A6	105 × 148
A7	74 × 105
A8	52 × 74
A9	37 × 52
A10	26 × 37

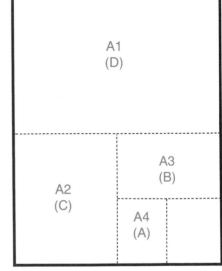

TOTAL SHEET A0
TOTAL SHEET E

A1 (D)

A2 (C)

A3 (B)

A4 (A)

Fig. 15-8. Metric paper sizes are established standards.

drawing. Engineering departments are generally given a certain length of time to finish a project. Therefore, the reports are important for keeping the job within the set time and for dividing the drawing work among the drafters.

DRAFTING STANDARDS

A major industry or a large company will generally set up a system for processing construction or manufacturing drawings. Some of the large manufacturing companies publish standard drafting practices. These help their own employees and their suppliers and customers as well. International Harvester, Caterpillar Tractor, and General Motors are just a few of the major companies that have published standards.

Early in the twentieth century, five engineering societies in the United States formed an organization to coordinate the development of national standards for engineering practices. The American Standards Association (ASA) has served U.S. industry for many years. In 1966, the ASA was reorganized as the United States of America Standards Institute and used the symbol USASI. This leading organization has been called the American National Standards Institute, Incorporated (ANSI) since an expansion in 1969. The institute has approved over 4000 standards. There are 16 standards and 2 supplements that make up the *American National Standard Drafting Manual*. Each of the standards was developed by national committees to give designers and drafters preferred drafting practices. The Y14 series of

standards is of particular interest to designers, architects, and drafters (Fig. 15-9).

The ANSI drafting standards are needed because of the many kinds of industrial work done across the country. Managements consider standards important in making communications the best possible. The Society for Automotive Engineers (SAE) has also set standards that are useful to the design and drafting profession. The great number of drawings done in the drafting room each day are not only standardized according to national practice, but they are also stored, filed, or controlled according to specific industrial standards.

CARE AND CONTROL OF DRAWINGS

In most companies, a system has been developed for cataloging and storing original drawings. Large companies keep the original drawings for a contract together under a contract number or job number. This makes the drawings easily available for copying or revising.

Some companies have a *closed catalog system* as part of the standards department. In this system, the number of people who can handle original drawings is tightly controlled. One person is responsible for keeping the drawings safe. That person may have a staff who file and copy the documents. Other companies favor an *open file system* within the engineering department. There, the original drawings are always available to anyone in the department. The open file system is generally not good because easy access to the original drawings can lead to mishandling.

AMERICAN NATIONAL STANDARD DRAFTING MANUAL	
Title of standard	Y14 number
Drawing Sheet Size and Format	Y14.1
Line Conventions and Lettering	Y14.2
Multiview and Sectional-View Drawings	Y14.3
Pictorial Drawing	Y14.4
Dimensioning and Tolerancing	Y14.5
Screw-Thread Representation	Y14.6
Gears, Splines, and Serrations	Y14.7
Gear-Drawing Standards—Part 1	Y14.7.1
Forgings	Y14.9
Metal Stampings	Y14.10
Plastics	Y14.11
Mechanical Assemblies	Y14.14
Electrical and Electronics Diagrams, including supplements Y14.15a and Y14.15b	Y14.15
Fluid Power Diagrams	Y14.17
Dictionary of Terms for Computer-Aided Preparation of Product Definition Data	Y14.26
Chassis Frames—Passenger Car and Light Truck— Ground Vehicle Practices	Y14.32.1

Fig. 15-9. The ANSI standards control all areas of drafting.

Some rules for storing originals efficiently and safely follow:

1. Develop a closed catalog system that requires expert handling.
2. If original drawings have to be revised a lot, handle them carefully. Work from a copy if possible. It is best to use an erasable intermediate copy. This is a print of the original drawing made on an erasable drawing paper. This paper is different from ordinary blueprint paper in that other prints can be made from it. (Prints, of course, cannot be made from other blueprints.)
3. The original drawing should be used only for making prints. It should never be used for desk reference.
4. Original drawings should be stored flat. They should not be folded because creases will show up as lines on prints.
5. Cleanliness in the drafting room is essential. Vellum and paper stock can absorb moisture and thereby pick up stains. Therefore, clean, dry hands and clean working surfaces are necessary.

A good drafting-room system provides easy, but controlled, access to the drafting files.

■ TYPICAL DRAWING STORAGE

Original drawings are stored in vertical files, as in Fig. 15-10. They can also be stored in horizontal file drawers, as in Fig. 15-11. The vertical files for drawings are as convenient as ordinary files for letters. All important documents should be flat and smooth. They should be protected from water and dust. Many drawings of sizes A ($8\frac{1}{2} \times 11$ in.), B (12×18 in.), and C (18×24 in.), as well as

Fig. 15-10. Vertical file storage with a capacity of 3000 to 5000 drawings. (Ulrich Plan File.)

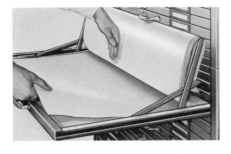

Fig. 15-11. Horizontal file storage for open or closed file systems. (Bruning.)

small ISO sheets, can be conveniently filed in vertical files.

Folding the Prints

Original drawings should not be folded. However, prints filed for reference are folded. The folded print is filed with the creased edge on top. This is done so that other prints will not be caught in an open fold. Figure 15-12 shows the ANSI system for

folding B-, C-, D-, and E-size prints. A number of systems are used to store prints of current work in the drafting room so that they can be referred to quickly. Figure 15-13 shows a compact quick-access filing system, sometimes called a *roll filing system*. This system can be moved around.

Storing folded prints can be avoided by microfilming them. Drawings must be of a high quality to be good for microfilming. The many reference details and supply catalogs for parts and equipment can also be microfilmed. In large companies, reader-printers like the one in Fig. 15-14 are used to retrieve design information.

Microfilming and Data Control

Storing drawings usually takes up a lot of space and is difficult and expensive. Microfilming was brought into the drafting room to make storing and copying drawings easier. The National Microfilm Association (NMA) sets standards for microfilming in the United States. This association has worked with the American National Standards Institute in setting its microfilming standards.

■ MICROFILMING PROCESS

A number of photographic systems are used in microfilming. However, 16mm and 35mm are the most popular film sizes. Film also comes in 70mm and 105mm. The film must be high contrast, high resolution, and fine-grained to be good for copying drawings. Only a small storage space is needed for microfilm. The data card that holds the microfilm can be taken out of storage by machine. The card may be used for printing a copy

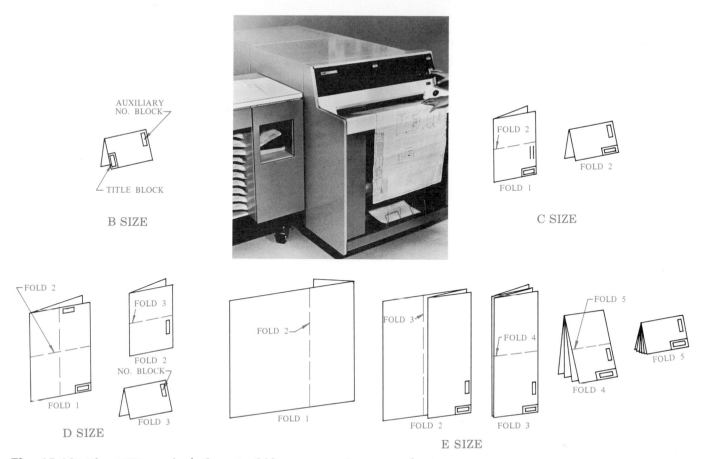

AUXILIARY
NO. BLOCK

TITLE BLOCK

B SIZE

FOLD 2

FOLD 2

FOLD 1

C SIZE

FOLD 2

FOLD 3

FOLD 2
NO. BLOCK

FOLD 1

FOLD 3

D SIZE

FOLD 2

FOLD 1

FOLD 3

FOLD 2

FOLD 4

FOLD 3

FOLD 5

FOLD 4

FOLD 5

E SIZE

Fig. 15-12. The ANSI standards for print-folding systems. Some reproduction machines can automatically fold prints. (Xerox.)

Fig. 15-13. A quick-access filing system for the office or job site. (Plan Hold Corp.)

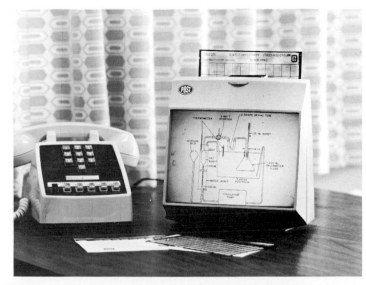

Fig. 15-14. A microfilm reader for the job site. (Teledyne Post.)

CREATE

DISTRIBUTE

STORAGE & RETRIEVAL

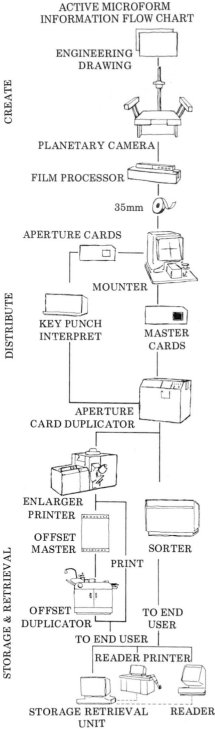

ACTIVE MICROFORM
INFORMATION FLOW CHART

ENGINEERING
DRAWING

PLANETARY CAMERA

FILM PROCESSOR

35mm

APERTURE CARDS

MOUNTER

KEY PUNCH
INTERPRET

MASTER
CARDS

APERTURE
CARD DUPLICATOR

ENLARGER
PRINTER

OFFSET
MASTER

SORTER

PRINT

OFFSET
DUPLICATOR

TO END
USER

TO END USER

READER PRINTER

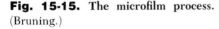

STORAGE RETRIEVAL
UNIT

READER

Fig. 15-15. The microfilm process.
(Bruning.)

or for studying in a microfilm reader-printer. Figure 15-15 shows the flow of the microfilm process. Copies from microfilm can be full size, half size, or original intermediates depending on the film or paper used. Half sizes are popular because they are easy to handle and cut paper costs in half (Fig. 15-16).

MICROFILMING TECHNIQUES FOR QUALITY REPRODUCTION

Microfilm images and copies from microfilm can be good only if the drafting in the original drawing is good. Microfilming standards have helped make drafting better. Good drafting has line work that is clear and clean. It has well-selected and well-positioned views. It has dimensions that can be read easily. The National Microfilm Association published Monograph No. 3, *Modern Drafting Techniques for Quality Microreproductions*, which has guidelines for preparing drawings for microfilming.

The major concerns in preparing high-quality original drawings for microfilming are listed as follows:

1. Selecting standard-size drawing media of high-quality vellum or film with good surface texture.
2. Selecting opaque pencil lead or ink that is right for the medium, that gives good line contrasts, and yet that can be erased easily.
3. Using lettering that can be easily read. An example is Microfont, a letter style set by NMA. The rules for height and spacing of letters are very important, as shown in Fig. 15-17.
4. Making the drawing easy to use

High-quality microfilm negative

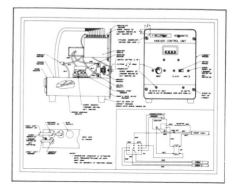

Blow back to a full-size second original.

Make a half-size negative from the microfilm and opaque if required.

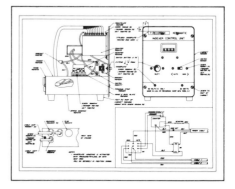

Use the negative to blow the drawing back to its original size. Additional drafting can be done at this time.

MICROFILM BLOWBACK

Fig. 15-16. The microfilm blowback process can provide intermediates for revision or modification. (Eastman Kodak Co.)

UNIT	COMMENTS	NOT THIS	POSSIBLE ERROR
B	Upper part small, but not too small.	B	8
H	Bar above center line.	H	————
M	Center portion extends below center line. Slight slant on uprights.	M or M	————
S	Lower part large, ends open. Slight angle on center bar.	S	8
T	Horizontal bar shall be full width of letter "E."	T	7
U	Full width.	V	V
V	Sharp point.	V	U
Z	All lines straight.	Z	2
I	Full height and heavy enough to be identified. No serifs.	1	7
2	Upper section curved with open hook. Bottom line straight.	2	8 or Z
3	Upper portion same as lower. Never flat on top.	3 or 3	8 or 5
4	Body open, ends extended.	4	7 or 9
5	Body large, curve dropped to keep large opening. Top fairly wide.	5 or 5 or 5	6 or 3 or S
6	Large body, stem curved and open.	6	8
8	Lower part larger than upper, full and round to avoid blur.	8 or 8	B
9	Large body, stem curved but open.	9	8

MICROFONT
ABCDEFGHIJKLMNOPQR
STUVWXYZ 1234567890

Fig. 15-17. Microfont, a lettering style developed by NMA. (National Microfilm Association.)

by spacing the views well and showing only as much detail as is needed.

These microfilming standards are important to industrial progress.

Good camera work is wasted on poor drafting. Microfilm copies cannot be any better than the original drawings. Several systems have been developed that make secondary originals cheaply.

■ SYSTEMS OF PREPARING INTERMEDIATES

An *intermediate* is a copy of a drawing that is used in place of the original. It is often called a *secondary original*. The designer and drafter can change the intermediate while leaving the original alone. The basis for working with intermediates is that if an original has been drawn before, it does not have to be drawn again (Fig. 15-18). Using intermediates avoids tracing or redrawing. The seven methods that follow are only some of the ways that time can be saved on professional drafting jobs.

1. *Scissors drafting.* The unwanted sections of an intermediate are removed with scissors, a knife, or a razor blade. This is called *editing*. A new print or new intermediate is then made. More changes can then be drawn on this new original.
2. *Correction fluids.* Unwanted data can be removed from an intermediate by putting on correction fluid. New data can then be drawn in the cleaned area. Prints are made from the corrected intermediates.
3. *Erasable intermediates.* This kind of copy can be erased with ease. Thus, corrections can be made quickly without fluid. Redrawing can be done equally well with pencil or ink. The original remains the same.
4. *Block-out, or masking.* This system is best when there is little time for reworking a drawing. The area to be blocked out is covered with opaque tape. A print is then made in which the masked areas produce blank spots.

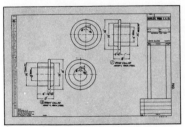

Say you have an existing drawing:

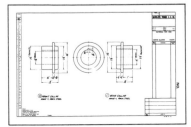

And you want to revise it like this.

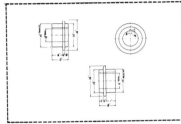

First you make a clear film reproduction of the original and <u>cut out</u> the elements.

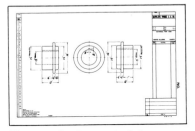

Then <u>tape</u> the elements in their new positions on a new form.

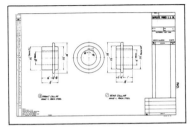

Photograph it on film with a matte finish (the tapes and film edges will disappear). Draw in whatever extra detail you want—and you have a new original drawing.

PASTEUP DRAFTING

Fig. 15-18. Intermediates, called *second-generation originals*, can save drawing time. Paste-up is only one of the ways. (Eastman Kodak Co.)

Changes are then entered in these areas.

5. *Transparent tape.* Added data can be put on paper, cloth, or film and then taped in place on an intermediate. Dimensions, notes, or other data symbols can be added in this way. The notes can be typewritten on transparent press-on material, which is then put in place.

6. *Composite grouped intermediates.* When several drawings are to be combined, a composite grouping should be cut from intermediates and taped in place. A final intermediate is then made.

7. *Composite overlays.* Translucent original drawings can be combined into a composite intermediate by placing the originals in the desired places. The composite is then photographed.

The drafter often must decide which intermediate system gives the best-quality secondary original for the least amount of time and materials. The quality of the copies made by the system chosen must meet the needs of the job. Along with intermediate systems, computer-layered drafting has become a working tool of the design team.

■ PIN-BAR DRAFTING

This is an important overlay system that can shorten the time it takes to produce a set of drawings. Drafters in different disciplines (plumbing,

Fig. 15-19. Overlay drafting can coordinate the work of many drafting technicians with pin-bar registration. (Eastman Kodak Co.)

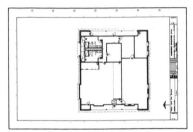

Take the basic floor plan.

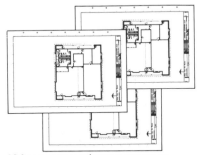

Make as many copies as necessary on clear punched film; distribute to the involved drafters.

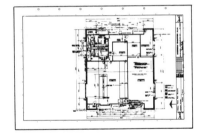

Each drafter puts this base drawing on a pinbar, superimposes a punched blank matte overlay, and drafts a particular assignment.

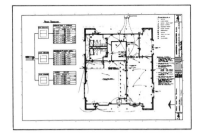

Reduced-size negatives are made of the base drawing and overlays. These negatives are then combined to produce enlarged positives of the required disciplines.

PIN BAR LAYERED OVERLAY DRAFTING

electrical, and mechanical) can work on the same job at the same time.

The basic principle of overlay drafting is a pin-bar registration system in which every piece of drafting

Make a halftone print of the photograph.

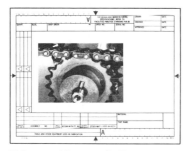

Tape the halftone print to a drawing form, and photograph it to produce a negative.

Make a positive reproduction on matte film.

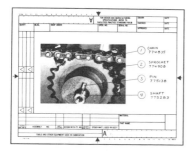

Now draw in your callouts— and the job is done.

PHOTO DRAFTING

Fig. 15-20. Photodrafting can complement and clarify technical data. (Eastman Kodak Co.)

media (film, paper, and negatives) is identically punched to match a row of pins (Fig. 15-19) on an alignment bar. Each member of the drafting team is provided with a pin bar and a reproduction of the base drawing, such as a floor plan. The drafter places a sheet of punched matte film over the base drawing (registration is automatic) and proceeds to draw the particular assignment.

Various overlays from related drawings can be combined to create composites. The computer-aided drafting system has a similar overlay, or layered, ability to build a drawing in sequential patterns. That is, the object is drawn on one layer, dimensions on another layer, and so on.

■ PHOTODRAFTING

A picture not only saves a lot of words but also limits the time needed for drafting. Architectural and engineering teams use photodrawings because they are generally easier to understand (Fig. 15-20).

This intermediate drafting process combines line work with photos and is similar to the solid modeling used in computer graphics.

■ COMPUTER-AIDED DRAFTING SYSTEMS

The interactive graphic systems, which you will learn about in Chap. 16, require procedural systems for effective management. The basic drafting standards apply all over again, and similar procedures are effectively performed.

Similarities are easy to identify between manual and computer-aided drafting. Some similarities include levels of communication, title blocks, basic sheet sizes, folding standards, closed and open filing systems, storage systems, microfilming abilities, overlay (or layered) drafting, editing, modifying, photodrawing, and three-dimensional modeling (Fig. 15-21).

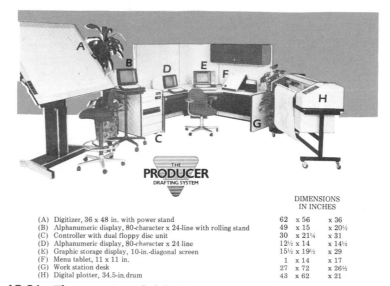

THE
PRODUCER
DRAFTING SYSTEM

		DIMENSIONS IN INCHES		
(A)	Digitizer, 36 x 48 in. with power stand	62	x 56	x 36
(B)	Alphanumeric display, 80-character x 24-line with rolling stand	49	x 15	x 20½
(C)	Controller with dual floppy disc unit	30	x 21¼	x 31
(D)	Alphanumeric display, 80-character x 24-line	12½ x 14		x 14½
(E)	Graphic storage display, 10-in.-diagonal screen	15½ x 19½		x 29
(F)	Menu tablet, 11 x 11 in.	1	x 14	x 17
(G)	Work station desk	27	x 72	x 26½
(H)	Digital plotter, 34.5-in. drum	43	x 62	x 21

Fig. 15-21. The computer-aided drafting process is similar to manual drafting procedures, and standards are the same. (Bausch & Lomb.)

REVIEW

1. Why are drafting standards important?

2. If a drafter's job is to make drawings, why should he or she be required to provide information for reports?

3. List several reasons why a drafter should not check her or his own drawing.

4. Why is the proper care and control of drawings important?

5. List advantages and disadvantages of using intermediates.

6. What is the main purpose of the American National Standards Institute?

7. How is the ISO sheet size expanded beyond an A0 sheet?

8. Which of the four levels of graphic communication is related to client approval?

16 Computer-Aided Design and Drafting

■ INTRODUCTION

The engineering drawing has been an integral part of industry for many years. It is the link between engineering design and manufacturing. Information is quickly communicated to manufacturing in the form of drawings prepared according to prescribed drafting standards. An engineering drawing may be prepared by means other than using the conventional tools. Traditionally, drafting instruments have been used to apply lead or ink to vellum or Mylar. The preferred alternative now is to prepare the drawing with the aid of a computer. This method is known as *computer-aided drafting* or *computer-aided design and drafting*. It is efficient and is rapidly replacing the handmade drawing. Computer-aided design and drafting is abbreviated CAD or CADD. (These abbreviations are formed by using the first letters of each word of the phrases.) A CAD system may be suitable for a wide variety of applications, including drafting for mechanical, architectural, landscape, interior design, and electrical, electronic, structural, and civil engineering (Fig. 16-1).

The desired drawing or design that is displayed on the cathode-ray tube (CRT) screen, or monitor, is converted into a drawing by a CAD system. However, not all companies are interested in drawing output. In chemical plant design, for example, liquid- and gas-piping system design requires nongraphic analysis. Thus,

Fig. 16-1. The major hardware components of a small CAD system. (Cascade.)

some CAD systems are not graphics-based. Other product manufacturing companies may not always require drawings. Instead, the information is transmitted from the CAD system directly to product manufacturing equipment. This method is known as CAD/CAM (computer-aided design/computer-aided manufacturing). CAD/CAM increases accuracy since the actual step of drafting preparation where errors may occur is eliminated. The manufacturing equipment may include a wide range of machinery types, such as lathes, milling machines, and so on. The instructions may also be sent to robotic equipment for automatic product handling. CAD/CAM is particularly adaptable to the mechanized segment of industry and is now improving production capabilities.

Machines used in the computer-aided manufacturing process may be referred to as numerically controlled machines. They can be given instructions with a series of numbers that provide (1) the proper tool selec-

Fig. 16-2. Industries are served by CAD/CAM (computer-aided drafting/computer-aided manufacturing). (Computervision.)

tion, (2) the proper machine movements, and (3) the appropriate cutting dimensions for shaping a product (Fig. 16-2).

A TURNKEY SYSTEM

A turnkey CAD system consists of all the hardware and software equipment necessary to create engineering drawings. It responds to each user's specialized commands. The options available for drafting details on a CAD system are designed to be easy to learn and easy to use. It allows the drafter to *interact* with a menu to create and edit (change or redefine) drawings of any size and at any scale. A menu is a device you can use to input commands to the system quickly. Basic tasks such as lines of any style, circles, arcs, lettering, and dimensioning are used to define the product. Advanced tasks such as copy repeat and layering are also available on most turnkey systems.

INTERACTIVE COMPUTER-AIDED DRAFTING

The development of powerful tabletop microcomputers allows drafting technicians to work at an interactive single or series of interconnected workstations. The term "interactive" is used since a human must "interact" with the machine for it to be operable (Fig. 16-3). CAD systems are referred to as user-friendly. A prompt, or instruction, is given for every drawing step. If one can read and follow instructions, learning how to operate a CAD system becomes very easy. The systems are very "friendly"; drafters and designers do not have to learn computer programming or understand a computer language. CAD users interact with a series of terms using on-screen (monitor) prompts. Various input devices are used in the interactive process. A specialized drafting software program permits *interaction* (talking to) with the computer. The computer in turn reacts (talks back) on the monitor with written (annotated) or drawn (graphic) results (Fig. 16-4).

TRADITIONAL DRAFTING VS. COMPUTER-AIDED DRAFTING

The principles of drafting are common to both traditional and computer-aided drafting. *Drafting standards necessary for shaping mechanical drawings are unchanged.* The knowledge of drafting principles from the alphabet of lines to sectioning and dimensioning procedures continues to be necessary. In traditional drafting, pencil skills for lettering and drawing line weights are essential. The proper use of hand-held tools and supplies such as compasses and leads are necessary and vital to good results.

In automated drafting, the basic skills for lettering and drawing lines are not required. CAD is an automated process or a tool that replaces drafting tables and hand-held tools. It "automatically" produces consistent lettering and regulates line work. CAD is able to improve the production of drawings better than any other tool, without replacing the role of the engineer, architect, designer, or drafting technician.

The system consists of a series of electronic equipment that has the ability to store and recall drafting data. It can only store and recall the

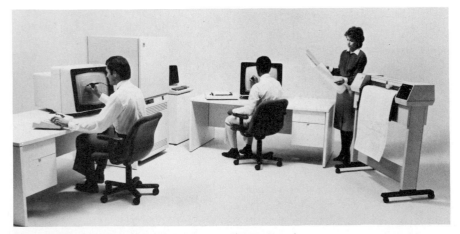

Fig. 16-3. An interactive CAD system. (IBM Corp.)

Fig. 16-4. Interactive workstation with display monitor and input devices. (Computervision.)

the operator. Projects will be selected to establish a benchmark. A benchmark establishes a work load of current working drawings and allows comparisons to be evaluated. Productivity will start to pick up on the system. The first projects reveal that time is saved, and new record keeping procedures tend to show continued progress. During the third month, the CAD operator is very familiar with the software program and has added work to the weekly schedule. Productivity is becoming more measurable. The staff is adjusting to the change and there are indications that the CAD transition team is for-

commands of an operator in control of the system. It is advantageous, however, because it stores data accurately by the use of a high-speed digital computer. The information contained in the first 12 chapters of this textbook is the basis for mechanical drawing. That kind of knowledge is stored as data in the computer. The advanced chapters, such as architectural drafting and electrical drafting, would be classified as special software applications because additional data is required for CAD operation (Fig. 16-5).

CAD PRODUCTIVITY AND LEARNING CURVE

The CAD system will perform repetitive geometric tasks that are tedious work for the skilled drafting technician. The system can draw, redraw, correct, enlarge, dimension, file, and revise drawings more economically than the manual drafting pro-

cess. The high cost of producing drawings manually has resulted in creative software programs that decrease turnaround time. Detail and assembly drawings are produced faster and more accurately. Thus projects flow more quickly through the design and detailing process with CAD. Drawing production should increase by 30 to 40 percent.

When a CAD system is implemented in an engineering design office a learning curve is established. That is, how much time will it take to train CAD drafters to be productive? During the first month, pre-installation planning will establish normal operational procedures while self-paced and formal training will acquaint the selected staff with the functions of the system. The productivity for this period is restricted to developing commands and sample work that has the geometric characteristics of future assignments. In the second month, formal training has been completed and examples of future assignments are evaluated for

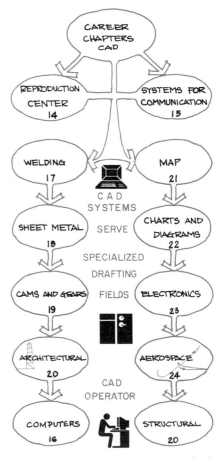

Fig. 16-5. Specialized fields of drafting are served by software data stored in CAD systems.

mally selecting the best projects for development on the CAD program. By the fourth month, the CAD manager and drafters are aware of faster response time in the interactive process, and productivity is increasing weekly.

The actual time frame will vary. It depends on the type of system and how user-friendly it is, the CAD background of the employees, and the type of drawings. The learning curve will also have to be developed around the type of drafting work and the complexity of the system. Solid growth patterns will be developed over a year or two for most systems as operators are trained and systems are expanded.

Fig. 16-6. This microcomputer CAD system is designed for engineering and architectural offices. (Cascade.)

■ *TABLE-TOP SYSTEM CAPABILITY*

A microcomputer CAD system for a mechanical or machine drafting technician can offer the following characteristics (Fig. 16-6):

1. Draw construction lines as needed, at any convenient spacing, through any points, at any angles, and tangent to one or more arcs.
2. Draw lines as needed, any type or width, for example, hidden or phantom, etc.
3. Draw circles and arcs of any size. Several options exist to create arcs, including center and radius, diameter, tangencies, and three points.
4. Set the scale of final drawings and remember the scale when zooming-in to work on details or redrawing on the plotter.
5. Calculate actual distances, whether the drawing is scaled or actual size.

6. Group geometric editing within a drawing: move, erase (delete), and copy any group of parts.
7. Move any part of a drawing to any location. Correct or change as additions are made.
8. Erase all or any part of a line, arc, or any geometric form.
9. Make a mirrored image of a drawing. Symmetrical designs are created.
10. Draw and display images separately and then combine the work on "layers."
11. Add straight line or arc pointers to a drawing (leaders).
12. Label drawings and add notes with various lettering styles.
13. Perform automatic dimensioning with various styles of lettering.
14. Perform associative dimensioning, chained or accumulative.
15. Check dimensions and qualify accuracy in decimal inch, fractional inch, or metric sizes.
16. Save the entire drawing or any part of it for later use.
17. Retrieve and use stored drawings to revise and build new drawings.

18. Perform automatic sectioning or cross-hatching.

There are additional options that have been added to the state-of-the-art drafting. As you think of one more there is a programmer who can write the program.

■ *INTERACTIVE AND PASSIVE CAD*

In an interactive system, the CAD operator uses direct commands to interact with the system. A command for a specific line and line weight can be called for and altered easily with quick computer response. The line may be stored, recalled to the screen, modified or edited (changed), and then reproduced on a printer or plotter. As an object is drawn it is graphically generated on the CRT, or monitor. When it is in its final form the CAD operator may call for a copy on the plotter. This is called *interactive graphics*. If the object is drawn after all the data is assembled and stored and then called at a later date to be plotted, it is called *passive graphics*. The major components for an interactive system are shown in Fig. 16-7. The central hardware unit is called a graphics processor (GP), which is the system's computer. This computer is connected to a tape unit, disk unit, and interactive workstation shown in the foreground of Fig. 16-7.

■ THE INDIVIDUAL WORK STATION (FIG. 16-8)

Graphics creation occurs by using input devices, which include: (1) alphanumeric keyboard, (2) function board, (3) light pen, (4) graphics tablet (digitizer), (5) joystick, and (6) other specialized devices.

The keyboard is used to communicate with the system. The graphics tablet contains a specified area for "digitizing" locations on the monitor. It also contains a menu area which is used to input commands. This menu can be activated with an electronic digitizing pen rather than by typing on a keyboard. Digitizing is a process of pen activation on the graphics tablet to locate X-Y coordinates on the screen. The pen can identify a location or command a menu to become available. This system has an image control unit (ICU) which provides a variety of functions to control the image on the monitor. The monitor, or graphics terminal, displays the visual image or drawing. At any stage of development the printer will provide you with a paper ("hard") copy of the drawing created on the monitor.

Alphanumeric Keyboard

The keyboard has keys which are arranged like the keys of a typewriter (Fig. 16-9). It is an extended version with a few extra keys that perform

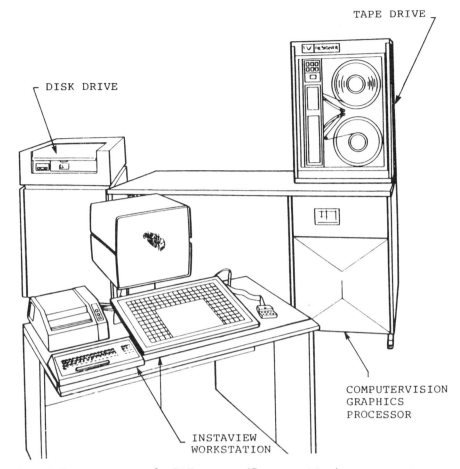

Fig. 16-7. Components of a CAD system. (Computervision.)

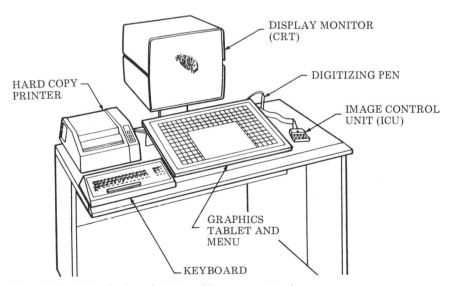

Fig. 16-8. Individual workstation. (Computervision.)

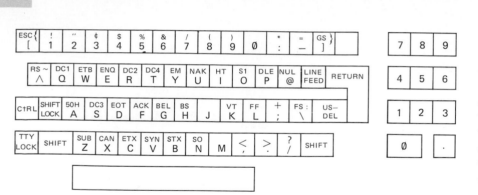

ESC { [	! 1	" 2	¢ 3	$ 4	% 5	& 6	/ 7	(8	) 9	Ø	* :	= –	GS }]	

Fig. 16-9. Alphanumeric keyboard. (Computervision.)

special functions. On the right-hand side, for example, is a numeric pad. The numeric pad enters numbers quickly in response to command functions. The keys can function in three different ways: (1) press the key alone for initial response, (2) press the key and the shift key together to use the character shown on the top half of the key, or (3) press the key and the control key together to specify the internal function of the key. The return key can function to execute a command.

The alphanumeric keyboard is indispensable when adding notes or text to a drawing. Every CAD system has some form of alphanumeric key input.

Graphics Tablet

The graphics tablet is a drawing table on which the CAD operator can input information to the monitor using an electronic pen, or stylus. It is the most popular method used to locate a position on the monitor. This position is indicated by a bright mark called a "cursor." Note that the graphics area is surrounded by the menu blocks in Fig. 16-10.

The squares on the menu pad define a function on the digitizing surface of the tablet. The menu items within the squares are selected by activating the stylus. Each square

can execute one command or a series of commands. This saves time by eliminating the need to type each command in its own code. For each area of drafting specialization there is a menu that can be placed on all or part of the tablet. Menus are interchangeable with each system. Some of the menu items, however, are common to all areas of specialization and these are permanently adhered to the tablet surface (Fig. 16-11).

The electronic pen, or stylus, is used to locate specific positions on the monitor (Fig. 16-12). It will move a cursor on the monitor to select a desired point on the graphic area after the desired menu item has been selected (Fig. 16-13). When the point is determined, the pen is activated. During menu selection

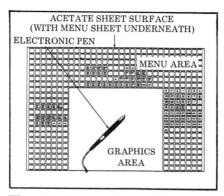

Fig. 16-10. Graphics tablet, menu, and stylus. (Computervision.)

the pen should be pointed to the middle of the menu box before it is activated.

The monitor allows the CAD operator to see the visual responses activated with the stylus and keyboard when appropriate. The monitor will display the texts of the commands that have been activated or entered.

Other Input Equipment

Drawing creation by graphics tablet input is the most popular means. Other input devices, however, are available for use. A function keyboard, which is also called a function board or menu pad, is a popular unit on some larger CAD systems. It is used to "call up" a particular part of a program. To draw lines, for example, a button labeled LINES would be pressed. The effect is the same as "digitizing" the LINES box on a tablet; that is, part of the program is used to create lines.

Other popular devices used to locate the cursor are joysticks, light pens, or thumb wheels. Each locates the cursor a little differently. A joystick is pointed in a direction to quickly "steer" the cursor to a new location. A light pen is touched directly to the screen of the monitor at the desired location. Thumb wheels position horizontal and vertical cross hairs, and locate the cursor at their intersection. Any of these devices may be found on a particular CAD system.

■ OUTPUT

A printer or hard copy unit provides a quick copy of the images displayed on the monitor. It is not a final drawing but it is a hard copy on paper. It is used as a "check print" to be examined and reviewed.

weights will be of high quality. The plot can be made on Mylar or vellum as desired.

PROCESSING

Processing equipment includes the media that the programs are stored on and the units that drive each. The four types of storage devices used to process computer data are hard disk, floppy disk, magnetic tape, and bubble memory.

The disk storage device in the unit shown in Fig. 16-14 is a hard disk. It has the ability to store data and make it readily accessible. The data, or information, provides a drafter or designer modeling geometry and drafting details for the machining processes. The high-speed disks are similar to a phonograph record, but they have one continuous concentric band called a *track*. There are 100 to 1000 tracks on a disk. The hard disks

Fig. 16-11. A closeup of a partial graphics tablet menu. (Computervision.)

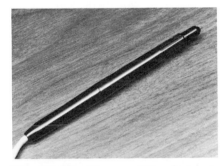

Fig. 16-12. Closeup of the electronic pen, or stylus. (Cascade.)

A final drawing is normally produced accurately on a plotter (Fig. 16-14). Various plotter pens are used to produce working production drawings. The lettering and line

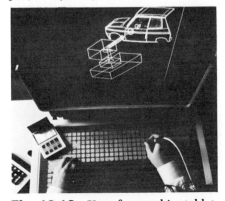

Fig. 16-13. Use of a graphics tablet. A robot path is generated on the screen with an electronic pen, and the image control device guides the motion. (Computervision.)

Fig. 16-14. The plotter on this CAD system is located to the left rear. It is preparing a drawing of the display on the monitor. (Bruning CAD.)

will hold several million bytes of information (Fig. 16-15). Floppy disks come in varying sizes. An 8-in. disk can hold over 1M byte of information. A 5¼-in. disk can hold 180K bytes on one side.

Magnetic tape devices allow the operator to transfer part designs and detailed standards from a disk to a magnetic tape. Magnetic tapes are used for transferring software programs into a CAD system. The magnetic tape, which consists of thin reels of metallic oxide-coated tape, is similar to audio recording tape. The tapes are controlled through high-speed reels that read and write with mechanical mechanisms. The bits are translated to magnetized oxide tape in tracks to form characters. A bit is a BInary digiT that has a memory location in an electronic device. A byte is a group of adjacent bits. High-density tapes control 800 to 1600 bits/in. (BPI).

The single greatest effort in designing storage units is to develop an increased storage capacity and decrease the time of accessibility.

You may also think of random access memory (RAM) storage as storage. It can be used as a library of information about drafting and design that can be retrieved. The microprocessor has to remember the location of every drafting detail in the library.

■ MECHANICAL CAD-GENERATED DRAWINGS

Chapter 4 in this textbook, on geometry for mechanical drawing, is the basis for all computer-generated drawings. The computer projects two-dimensional geometry in the X and Y directions. The origin of a drawing would be (X, Y) generally located at the lower left corner of the

monitor or media (paper). Figure 16-16 at A illustrates two-dimensional coordinates and three-dimensional model geometry with X, Y, and Z coordinates. The computer locates information on a coordinated informational base. The point at which the three axes intersect, XO, YO, ZO, is the origin of a spatial model called *three-dimensional drafting*. One way of seeing the model or concept more clearly is to imagine the model being created within a transparent three-dimensional box (Fig. 16-16 at B). The model's height (Y), width (X), and depth (Z) are created on the faces of the transparent box. When the images of the model move from the model to the face of the box, the geometry is said to be two-dimensional.

Lines and circles are located on a drawing by the coordinate system or by an input device. Figure 16-17, for

Fig. 16-15. This powerful computer CAD system offers 64 million characters of main memory and 48 channels. (IBM Corp.)

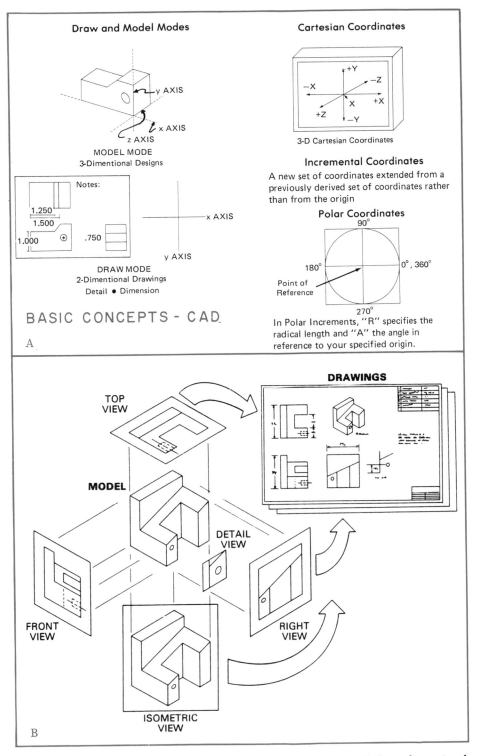

Draw and Model Modes

y AXIS
x AXIS
z AXIS

MODEL MODE
3-Dimentional Designs

Notes:
1.250
1.500
1.000   .750

x AXIS

y AXIS

DRAW MODE
2-Dimentional Drawings
Detail ● Dimension

BASIC CONCEPTS - CAD.

A

Cartesian Coordinates

+Y
−X   −Z
X   +X
+Z
−Y

3-D Cartesian Coordinates

Incremental Coordinates

A new set of coordinates extended from a
previously derived set of coordinates rather
than from the origin

Polar Coordinates

90°

180°   0°, 360°

Point of
Reference

270°

In Polar Increments, "R" specifies the
radical length and "A" the angle in
reference to your specified origin.

TOP
VIEW

DRAWINGS

MODEL

DETAIL
VIEW

FRONT
VIEW

RIGHT
VIEW

ISOMETRIC
VIEW

B

Fig. 16-16. Basic drafting concepts are the basis for two- and three-dimensional CAD development. (Computervision.)

example, illustrates an object with three circles of a specified diameter. The operator's command to create the graphics might be:

1. Define grid pattern.
   a. Select the DEFINE GRIDS task.
   b. Key in the desired interval (for example, 1.00) and press the RETURN key.
   c. Select the DISPLAY GRIDS task. This procedure will display a grid pattern as shown. It helps you to draw more quickly and accurately. The cursor (bright mark on the monitor) will only have to be positioned *near* a desired grid dot. When the cursor is activated it will "snap to" the closest dot.

2. Draw the outline.
   a. Select the LINES task. Most systems will "default" to solid lines. This means that only solid lines will be created unless changed by you.
   b. Locate and set the cursor successively at points 1, 2, 3, 4, 5, 6, 7, and 8. The outline of the object will be created as shown.

3. Create the holes.
   a. Select the CIRCLE or ARC task.
   b. Key in the desired interval (for example, 1.00) and press the RETURN key.
   c. Locate and set the cursor at points $D1$, $D2$, and $D3$. The circles will be located as shown.

4. Create the centerlines.
   a. Select the CENTERLINE task.
   b. Locate and set the cursor at each endpoint of the centerline to be created.

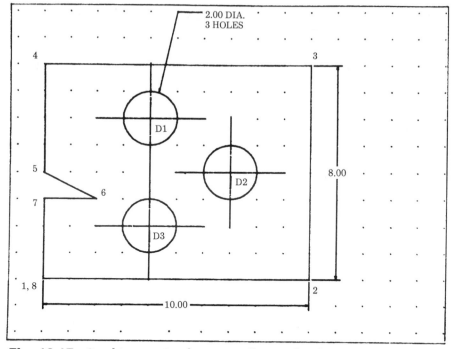

Fig. 16-17. Graphics creation for a partially completed CAD-generated drawing.

c. Select MOVE NEXT POINT. This is the same step as picking up your pencil and moving it to a new start location.

d. Repeat steps b and c for the remaining centerlines.

5. Create the leaderline.
   a. Select the LEADER or ARROW task.
   b. Locate and set the cursor at the three endpoints of the leader. The leader will be drawn as shown.

6. Create text.
   a. Select the TEXT or NOTE task.
   b. Locate and set the cursor at the desired starting point for the note.
   c. Key in the note 2.00 DIA and press RETURN.
   d. Locate and set the cursor at the desired starting point for the second note.

   e. Key in 3 HOLES and press RETURN.

7. Automatic dimensioning.
   a. Select the DIMENSION task.
   b. Position and set the cursor at points 1 and 2. This defines the segment to be dimensioned.
   c. Position and set the cursor at the location to place the dimension. The 10.00 dimension will be created as shown.
   d. Position and set the cursor at points 2 and 3.
   e. Position and set the cursor at the location to place the dimension. The 8.00 dimension will be created as shown.

The exact command structure depends on the type of CAD system used. This illustration, however, shows that user-friendly CAD systems allow the creation of geometric forms automatically and instantly.

In general, basic functions, commands, and operational procedures of CAD systems are similar. The programmer prepares the program to (1) command the geometry generators (which produce geometric forms), (2) control the size generators (which determine paper size and problem size), and (3) modify the drawing with various instructions such as rotating or zooming-in.

COMPUTER TYPES

The variations in CAD systems are generally based on the size of the processing unit. It is referred to as the central processing unit (CPU) and is the "brains" of the computer.

There are two types of computers used in the engineering office. *Analogue computers* are used extensively for solving differential equations and scientific research. *Digital computers* are used for engineering and business computations and are the type commonly used for CAD systems. Digital computers are classified into three categories: (1) mainframe computers, (2) minicomputers, and (3) microcomputers.

Mainframe CAD Systems

All digital computers can perform only one task at a time. So the large computers are very fast and need high-speed peripherals such as hard disks and magnetic tapes that can respond to an array of processors or subcomputers. They also need a high-speed language that is efficient. The most common mainframe engineering language is FORTRAN (FORmula TRANslation). When a computer graphics system is not interactive, a program or problem may be developed on tape. This data is used for input on a batch mode and prioritized to be accepted when the computer has time to accept the program. Therefore the person awaiting computer results experiences a *turnaround time*. This is the difference between passive graphics and interactive graphics. So to accommodate interactive graphics a minicomputer was developed.

Minicomputers

Minicomputers are smaller than mainframe computers, yet they are very powerful. Their development was made possible because of the microprocessor. They emerged because of the expense and limited interaction of large mainframes. In fact, minicomputers are now the dominant type used in industry. The minicomputer provides a *real-time* operation. This is a data process that displays graphics during the actual input operation time. A network of systems allows interaction of several work stations that interact with a large minicomputer.

Microcomputers

The microcomputer industry was standardized on 8 bits = 1 byte. A byte forms a sequence in the computer known as a "character" of information. Each symbol in a computer has its own byte arrangement. The size of a computer's memory is expressed in terms of bytes rather than bits.

Originally, the microcomputer started by Intel was a 4-bit processor. The first microcomputers were slow, single-purpose systems. The 16-bit computer has advanced status of the microcomputer's power.

A 16-bit microcomputer is a machine that is controlled by a 16-bit microprocessor. A microprocessor is a microelectronic circuit called a chip. The chip is a silicon wafer smaller than the size of a fingernail that is densely packed with circuitry. Each one can hold many electronic components.

The chip is the basis of a microcomputer. By adding memory (RAM and ROM) to the logic system (ALU—arithmetic logic unit) and connection ports for input devices (like keyboards) and output devices (like a graphics terminal), the microprocessor becomes a microcomputer system. The ability of the microcomputer to interact with graphic terminals, line printers, plotters, and digitizers is called peripheral compatibility. Each of the devices are called peripheral devices.

The microcomputer can be compared to the whole human body. The body has eyes, hands, feet, and a brain to control them. The microprocessor is similar to the brain. It is the center for interpretation and provides experiences (programs) to assist in the decision-making process.

The program is controlled and stored through electrical impulses. To the microcomputer a pulse beat of electric charge is a 1; the absence of the electric charge is a 0. The fundamental number system in digital computers is the binary (base 2) ON-OFF system.

The 16-bit computer can store more information than an 8-bit computer in user memory. Computers have two types of memory. User memory is information built into the working area of the computer and is called RAM (random access memory). Peripheral memory is called ROM (read-only memory) and is stored on tapes or disks for high-speed interaction.

The CAD operator can call up every detail in the RAM library. It is like making a telephone call to a specific drafting detail called an entity. In the 8-bit microprocessor 65,535 calls can be made for information stored in RAM. A 16-bit microprocessor can call upon 1,048,560 numbers of details. This is 16 times more than the 8-bit processor.

■ TECHNOLOGY OVERVIEW

Architects, engineers, designers, and drafting technicians have shaped the heritage of humans with reed pens, wood-cased pencils, technical pens, and lead holders. They have pushed and pulled the T square, parallel rule, straightedge, and other drafting instruments across highly technical designs and details. Skyscrapers and space shuttles have emerged from this. The photostat camera, blueprint, blueline ozalid print, and sepias, along with scissors drafting pasteups have evolved as technical advances in the drafting room. Thumb tacks, and drafting tape along with pin bar registration overlays have aligned the media for tomorrow's designs. The tracing cloth, transparent rag vellums, fade-out grid, polyester film (Mylar), and aperature cards for microfilm have been the standards for databases and storage. The powerful microcomputer has arrived to complement and update our drafting standards. It replaces many of the items above but it only enhances the design drafting process.

■ COMPUTER ARCHITECTURE

The arrangement of the hardware components in a CAD system is called *computer architecture*. Systems may be configured to serve a single drafting station or a network of workstations (Fig. 16-18) for a design drafting team. As mentioned previously, a system that is ready for operation is referred to as turnkey. The open-ended operation is a hardware and software system that has upward compatibility. It has the ability to interface with additional hardware and software and is expandable. The new standards, Initial Graphic Exchange Specifications (IGES), as defined by the National Bureau of Standards, will enhance compatibility. Computers and software will be capable of communicating with each other as industry interrelates with telewriters and phone interfaces.

Fig. 16-18. A network of CAD systems. (Cascade.)

■ INVESTIGATING SYSTEM CAPABILITY

There are two considerations involved in investigating the differences in CAD systems. The first is the type of hardware used. The second is the specific tasks performed by the drafting software working in the system (Fig. 16-19).

Some questions to ask about hardware are

1. Is there a mainframe, minicomputer, or microcomputer driving the system?
2. How much memory is available: 32K, 64K, 128K, or 2M bytes?
3. What type of peripheral equipment will work in the system?
4. How many compatible components are available to build the system?

5. Can data entry be made with more than one component?
6. Does the supplier provide a user's manual that lists software availability?
7. What type of power supply is required?
8. Are special environmental conditions required?
9. How easy are the start-up, log-on, and log-off procedures?
10. How is the system driven? Hard disks, floppy disks, bubble memory, or magnetic tapes?
11. What is the procedure for log-on and security of data storage?
12. How many controlling features are available?
13. Is there a service contract available? A warrantee?
14. Is the system hardware turnkey? Can the software be updated?
15. Can the system be networked? Is it upwardly compatible?

Some questions to ask about software for drafting and design are

1. What specialized drafting fields are available on software?
2. What size computer is required for the software?
3. Are three-dimensional wire frame and solid model software programs available? Does the program contain hidden line removal and rotation options?
4. How fast are the monitor responses to commands for drawing entities from the menu?
5. Will the software work on more than one system?
6. Is there software design for overlays? How many layers?
7. Can you move quickly from one menu to another?
8. Will the system group entities? Erase a group? Will it copy and store a subassembly group?

9. How many line styles and widths are available? Are pen widths programmed?

10. Is there a help key to provide routine guidance procedures or are you prompted with each instruction?

11. Does the system have automatic dimensioning?

12. Does the system have automatic sectioning?

13. How sophisticated is the line editing task?

There are certainly additional questions that could be formulated for both hardware and software. Could you specify the requirements of a CAD system in order of preference for a technical education center? The brochures from the major hardware and software suppliers will help determine if a particular system meets your goals. The effective use of a CAD system starts with the primitive tasks used for shape description.

■ GETTING STARTED

1. Define an acceptance process that will evaluate a system's performance.

2. Establish a benchmark for work to be achieved in the specialized drafting program. Example of a benchmark is a previous project that typifies the work you expect to perform.

3. Determine upward compatibility. Can additional hardware and software be interfaced?

When the CAD system is analyzed to meet the needs of a specialized drafting field the benchmarks will set the examples of progress and should be updated to provide new learning curves.

A typical job description for a drafting technician in the year 1990 will include:

1. Operate various computer and peripheral equipment.

2. Prepare input for and execute utility programs.

3. Maintain files of technical information and verify the correctness of the file input.

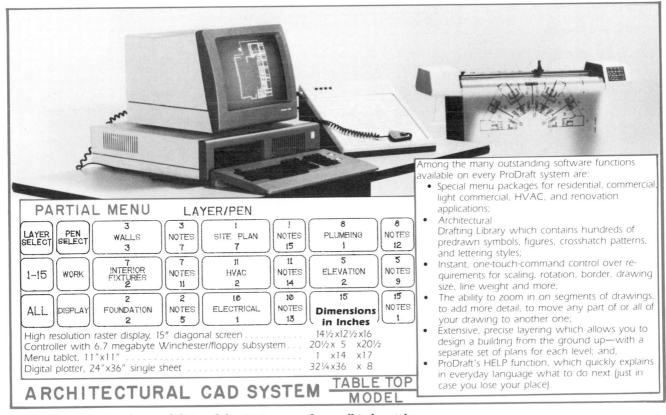

PARTIAL MENU   LAYER/PEN

LAYER SELECT	PEN SELECT	3 WALLS 3	3 NOTES 7	1 SITE PLAN 7	! NOTES 15	8 PLUMBING 1	8 NOTES 12
1-15	WORK	7 INTERIOR FIXTURES 2	7 NOTES 11	11 HVAC 2	11 NOTES 14	5 ELEVATION 2	5 NOTES 9
ALL	DISPLAY	2 FOUNDATION 2	2 NOTES 5	10 ELECTRICAL 1	10 NOTES 13	15 Dimensions in Inches	15 NOTES 1

High resolution raster display, 15" diagonal screen 14½ x12½ x16
Controller with 6.7 megabyte Winchester/floppy subsystem . . . . 20½ x 5 x20½
Menu tablet, 11"x11" 1 x14 x17
Digital plotter, 24"x36" single sheet . . . . 32¼ x36 x 8

ARCHITECTURAL CAD SYSTEM   TABLE TOP MODEL

Among the many outstanding software functions available on every ProDraft system are:
- Special menu packages for residential, commercial, light commercial, HVAC, and renovation applications;
- Architectural Drafting Library which contains hundreds of predrawn symbols, figures, crosshatch patterns, and lettering styles;
- Instant, one-touch-command control over requirements for scaling, rotation, border, drawing size, line weight and more;
- The ability to zoom in on segments of drawings, to add more detail, to move any part of or all of your drawing to another one;
- Extensive, precise layering which allows you to design a building from the ground up—with a separate set of plans for each level; and,
- ProDraft's HELP function, which quickly explains in everyday language what to do next (just in case you lose your place).

Fig. 16-19. Examine the capabilities of the CAD system for small industrial operations. This system is designed for architectural offices. (Bausch & Lomb.)

4. Monitor work flow of production systems.
5. Participate in the enhancement of system and development efforts.
6. Interface with other departments about possible enhancements and determine computer program error.
7. Provide technical liaison and assistance to users.
8. Evaluate and maintain new software and hardware.
9. Provide training in the above tasks for less experienced personnel.

This knowledge will be in addition to traditional design and drafting theory and application.

■ INDUSTRIAL COMPUTER-GENERATED DRAWINGS

Typical computer-generated drawings illustrate various areas of drafting:

1. *Chapter 18.* Flat-pattern generation-development from a three-dimensional design has been a major problem for sheet-metal fabricators. Determination of bend allowances and calculation of the bending and stretching are automatic with any specified material. The computer program determines appropriate tools and punches that can be entered into the data base for graphic solution. Examine the drawings in Fig. 16-20.
2. *Chapter 23.* Electronic drafting with CAD has provided a new state of the art. Software programs produce block and schematic diagrams for electrical networks (Fig. 16-21). The powerful command language provides

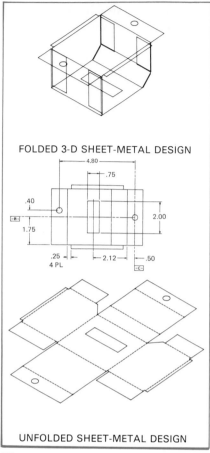

FOLDED 3-D SHEET-METAL DESIGN

UNFOLDED SHEET-METAL DESIGN

Fig. 16-20. CAD systems can support design and drafting of tooling for sheet-metal development. (Computervision.)

predefined components in a library of graphic primitives that can build complex design units. Circuit design has automatic routing and editing features that respond extremely fast. Circuit design elements that can be created and manipulated include blocks, components, wires, bundles, names, parameters, and notes.
3. *Chapter 20.* Architectural drawings are conveniently developed on the computer system and quality designs emerge as illustrated in Fig. 16-22. The use of layering (layers) in architectural detailing

helps put all the pieces together fast. Architectural styled detailing is available with architectural styled lettering on some systems.
4. *Chapters 1 to 15.* Various mechanical CAD drawings are illustrated in Fig. 16-23. The illustrations include:
   a. Detail drawing
   b. Three-dimensional model of an assembly
   c. Two-dimensional assembly and balloons

These serve to show that CAD has virtually unlimited application in the mechanical design drafting discipline.

■ THE CAD LANGUAGE—INTRODUCTION

CAD systems are menu-driven. That is, a particular menu item must be selected before creating a specific type of graphics. Command menus vary from system to system. In Fig. 16-24 you are provided with modes of drafting that are familiar to the CAD drafter. Examine the unique character of the menu system and assimilate the need to know drafting. Can you imagine the task or command that could be generated on the screen by using the electronic pen on any given square on the menu pad?

There are basic assumptions that are common to the positioning of images on the screen. You must know what it is that is to be positioned. This requires a *thorough knowledge of design and drafting principles.* Only then can you begin to build images on the screen for orthographic (multiview) or pictorial (three-dimensional modeling) projection.

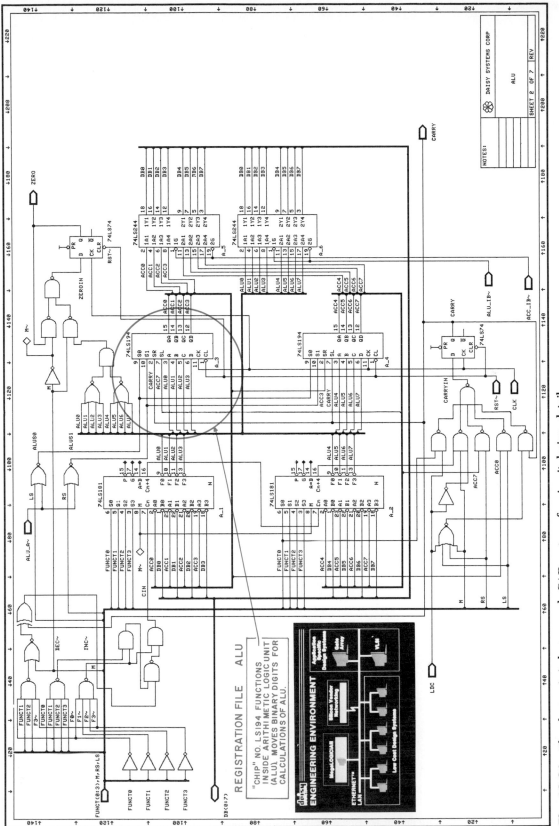

Fig. 16-21. The electronic industry needs CAD systems for circuit design, detailing, and testing. (Daisy Systems Corp.)

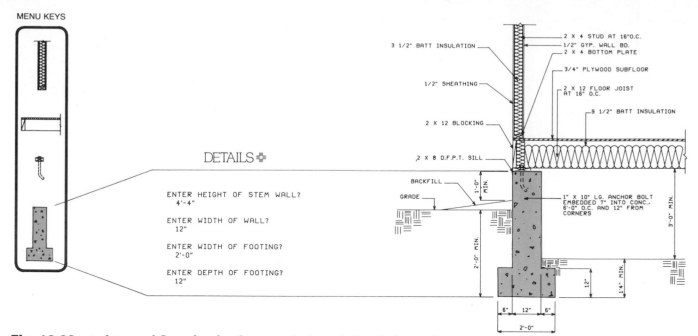

Fig. 16-22. Architectural floor plan details are easily formed. Detailed wire diagram perspective, interiors, and elevations are efficiently developed with CAD. (Sigma Design.)

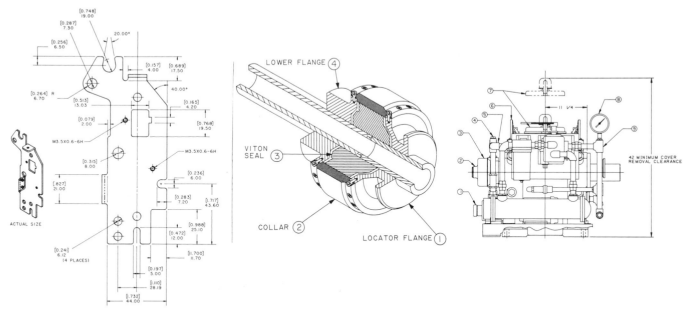

The dimentional values in detail drawings are automatically derived from the part geometry and drawn to specified standards which include English/metric dimensioning.

From the 3D model, sectional views are easily generated for precise visualization of part relationships and insertion of assembly information.

Assembly drawings are readily derived from layout and detail drawings. Automatic bill of material generation provides additional time savings.

Fig. 16-23. Three CAD-generated mechanical drawings. (Computervision.)

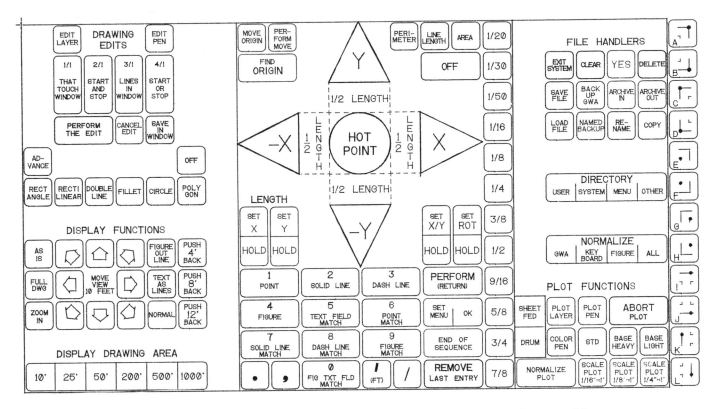

Fig. 16-24. Examine the menus from two different systems. Try to understand how these are used. (Computervision, Bausch & Lomb.)

Instead of manually inking the finished art, CAD systems interface with a variety of electrostatic and inked plotters to produce constantly high-quality production copies of details or pictorials at any scale. This is the primary difference between a traditional and CAD drafter. The high technology of new CAD systems available today supports investment because of the following benefits:

1. Increased productivity
2. Shorter product development time
3. Lower costs
4. Better designs
5. Greater accuracy
6. Improved control of changes
7. Higher standardization of designs

Now, skilled designers and drafters can put aside their triangles and pens and increase their productivity. CAD can offer up-to-date hardware, software, training, service, and support to change the training of the drafting technician to a CAD operator for tomorrow's jobs.

VOCABULARY

Boot up Term describing restarting of system including a log-in or log-on.

CAD (computer-aided design) Process which uses a computer to assist in the creation or modification of a design.

CAM (computer-aided manufacturing) Process employing computer technology to manage and control the operations of a manufacturing facility.

Cartesian coordinates The entering of explicit coordinates along the X, Y, or Z axis.

Catalog Index which may be used to locate a particular file or group of files stored on disk or tape. More generally, a group of files belonging to one family of files.

Clipping Establishing view boundaries.

Configuration Directory of computer and peripheral devices at a single installation.

Construction space Space which uses the coordinate system defined by the construction plane.

Cross-hatching Process of filling in an outline with a series of symbols to highlight a part.

Cursor A cross hair on the monitor that corresponds to the location of the electronic pen.

Database Comprehensive collection of information having structure and organization suitable for communication, interpretation, or processing by the system.

Default A value assumed by the system unless specifically replaced by user.

Delete Command used to negate a particular construction. Commonly used with entities, group, part, or view.

Diameter dimension (radial) Automatic measurement of circular entities.

Digitize The act of specifying a location or selecting an entity.

Drawing A graphic representation of data. A part may consist of a model and many drawings. Each drawing may be of a different size or unit, and the appearance of the model in each view contained within a drawing may be independently tailored. The association between model and drawing is maintained, however, so that any changes to the model will be reflected automatically in all drawings.

Dynamic menu A graphic representation of programmable commands which are defined and formulated by the user and displayed on an instaview screen. Command keywords are accessed by digitizing.

Dynamics The ability to move or insert a set of graphics on the screen under control of the cursor.

Echo The command that allows user to see certain system parameters (that is, grid, layer, clipping boundaries).

Edit Function that allows for changes, additions or deletions in text.

End point A coordinate (X, Y, Z) value representing the end of an entity.

Entity Fundamental building blocks which a designer uses to represent a product, for example, arc, circle, line, text, point, spline, figure, or nodal line.

Escape Special key allowing user to interrupt computer operation.

Execute file A file program containing a set of system commands which will automatically be executed by the system.

Exit To terminate work on a part.

Extents The amount of space or surface that something (for example, part, entity, view) occupies or the distance it extends on the CRT.

File A logical collection of data treated as a unit which occupies space on a storage device such as a disk.

Fonts Kind of type; all of one size and style. Text font: a complete character set of one style and size. Line font: a repetitive pattern used to give meaning to a line.

Grid Network of uniformly space reference points.

Group A number of entities identified by the system as one.

Hard copy A paper copy of computer output.

Incremental coordinates A new set of coordinates extended from a previously derived set of coordinates rather than from the origin.

Input The data to be processed, the process of putting data into the system.

Insert The action of placing entities, figures, or information.

Justification Specification of placement of text and line fonts. Text and line fonts may be left-, center-, or right-justified relative to a specified point.

Key menu Preset list of written commands on digitizing tablet enabling user to digitize commands rather than typing them.

Label Allows for inserting of text with a leader line from entity to text.

Layer Logical concept used to discriminate (separate) group(s) of data within a given drawing. Layering enables the operator to specify derived display elements to be visible. May be thought of as a series of 255 transparencies arranged in any order yet having no depth.

Linear dimension Automatically measures distance and/or length of lines.

Log-in Process whereby user enters system.

Log-out Process and command whereby user exits from system.

Measure Command to determine distance between two points, a point and an entity, or two entities. May also measure area or angle.

Menu Input device consisting of command squares on a digitizing surface. A graphic representation of programmable keys that are stored on the disk.

Mirror Allows the creation of a mirrored image of an entity along a vector.

Model A collection of data and/or geometry representing an object or product being designed.

Model space Coordinate system in which a three-dimensional database is defined.

On-line Equipment or devices in a system which are directly connected to and under the control of the computer.

Operating system (system level) Software program which controls the execution and implementation of subprograms.

Output The end result of a computer operation.

Parameter A constant whose values determine the operation or characteristics of a program.

Part A graphic construction which is referenced by a symbolic name.

Parts library Database collection of often used parts.

Placement (move) Assignment of an entity to a place or location.

Precision When used with dimensioning allows user to indicate number of places to right of decimal point.

Program Set of machine instructions or symbolic statements combined to perform a task.

Prompt Any message or symbol from the computer system informing or asking the user about possible actions or operations.

Properties User-established attributes of an entity or subfigure within the model.

Radius dimension Allows for the measurement of an arc or circle.

Repaint Command used to refresh the CRT display.

Restore image Returns to a portion of a drawing saved before zooming, scrolling, etc.

Rotate The turning of entities or groups of entities about a single axis.

Run Continuous execution of a program.

Save Saves the current display status. Allows operator to zoom or scroll without fear of losing current status.

Scale Ratio of the current display with respect to the database.

Scroll Temporary movement of objects relative to a fixed border.

Select Specifies the entity insertion parameters. Commonly used with grid, layer, text, view, etc.

Software Set of programs, procedures, rules, and associated documentation concerned with the operation of a data processing system.

Syntax Structure of expressions in a language and the rules governing the structure of a language.

Tag Allows for automatic assignment of alphanumeric names to selected entities for easier identification.

Text Written information in a particular font.

Text file File containing text. Text file may be created by the user with the Edit command.

Translate Relocation of an entity or group of entities to a specified point.

Trap size An invisible area within whose boundaries the electronic pen can identify entities.

Trim Allows user to alter (stretch or shorten) the existing end points of an entity.

Verify Describes the entity selected as to its length, radius, layer, and all relevant information.

View Pictorial representation of the model positioned within the confines of a drawing.

Window Bounded area which user determines for entity selection or verification.

Zoom To enlarge or decrease proportionately the size of display.

REVIEW

1. What is the meaning of the phrase CAD/CAM?

2. What are the major pieces of hardware that make up a CAD system?

3. Explain how CAD/CAM systems control numerically controlled machines.

4. What is a turnkey system?

5. Explain the difference between an interactive CAD system and a passive CAD system.

6. What is a learning curve for a CAD system?

7. What do you consider to be the five most important characteristics of a microcomputer CAD system?

8. Explain what is meant by input devices and describe two functions of the peripheral hardware.

9. How would you explain the difference between a microprocessor and a microcomputer?

10. Which five questions are the most important when considering the hardware for a CAD system?

11. List what you would consider the five important characteristics of software.

12. Define the word *benchmark* for a CAD system.

13. What is the basic difference between a mainframe and a microcomputer?

14. What is a menu? Is it considered an interactive device?

15. What are memory devices and how do they function?

17 Welding Drafting

JOINING METALS

Welding is a way of joining metal parts together. The art of welding is very old. In prehistoric times, it was used to make rings, bracelets, and other jewelry. Today it is very important in industry. Standard-shaped steel pieces, such as plates and bars, are welded together to make machine bases, frames, and other parts. Buildings are also assembled by welding. Welded steel parts are lighter and stronger than parts made by forging or casting. Figure 17-1 shows a pulley housing made by casting. Compare it with the similar part made by welding shown in Fig. 17-2.

Welding has become a major assembly method in industries that use steel, aluminum, and magnesium to build cars, trucks, airplanes, ships, or buildings (Fig. 17-3). In a single dumptruck there are hundreds of welds. The basic framework of the truck is fastened together by welding. Welding makes products stronger and longer-lasting. It also makes them look better. A drafter who is drawing parts to be welded works with a design engineer who knows what kinds of welding to use with different metals. Standard drafting symbols for welding have been established by the American Welding Society (Appendix Table A-34).

Welding joins materials by either heat or pressure or a combination of the two. There are two basic kinds of welding. Fusion welding uses only heat. Resistance welding combines heat and pressure.

Resistance welding was developed in 1857 by James Prescott Joule. Industry began using it after the 1880s when electric power became available in large quantities. One kind of fusion welding, arc welding was first performed by De Meritens in France in 1881. Another kind, gas welding, was effectively developed in 1885, when two gases—oxygen, from liquid air, and acetylene, from calcium carbide—were brought into use.

WELDING PROCESSES

Welding processes include fusion, gas, arc, thermit, gas and shielded arc, and resistance welding. Soldering and brazing, although called by their separate names, are also forms of welding.

Fusion Welding

Fusion welding is done with welding materials in the form of a wire or rod. This material is heated with a gas flame or a carbon arc. When it melts, it fills in a joint and combines with the metal being welded.

Gas Welding

Gas welding is fusion welding done with a flame of oxygen and acetylene, air and acetylene, or any other combination of gases that produce enough heat (Fig. 17-4). Burning acetylene supported by oxygen gives temperatures of between 5000 and 6500°F (2760° and 3595°C).

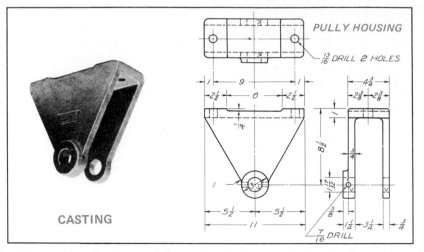

Fig. 17-1. Pulley housing made by casting. (Wellman Engineering Co. and Lincoln Electric.)

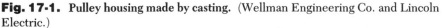

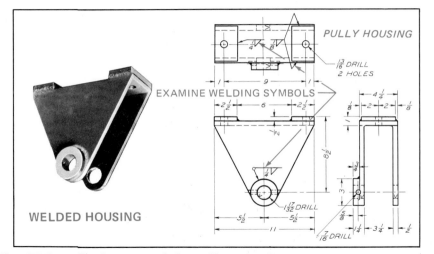

Fig. 17-2. Pulley housing made by welding. (Wellman Engineering Co. and Lincoln Electric.)

Arc Welding

Arc welding is done by forming an electric arc between the *work* (part to be welded) and an electrode (Fig. 17-5). This arc is made with a direct current (dc) generator or other power source. It causes intense heat to develop at the tip of the electrode. This heat is then used to melt a spot on the work and on a rod of welding filler material, so that both can be fused together.

Thermit Welding

Thermit welding is based on the natural chemical reaction of aluminum with oxygen. A mixture, or charge, made of finely divided aluminum and iron oxide is ignited by a small amount of special ignition powder. The charge burns rapidly, producing a very high temperature. This melts the metal, which then flows into molds and fuses mating parts.

Gas and Shielded Arc Welding

Aluminum, magnesium, low-alloy steels, carbon steels, stainless steel, copper, nickel, Monel, and titanium are some of the metals that can be welded with this kind of process. There are two forms of gas and shielded arc welding. They are *tungsten-inert gas* (TIG) and *metallic in-*

Fig. 17-3. Welding building components for Disney World. (United States Steel Co.)

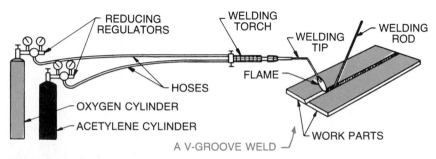

Fig. 17-4. The gas-welding process. (General Motors.)

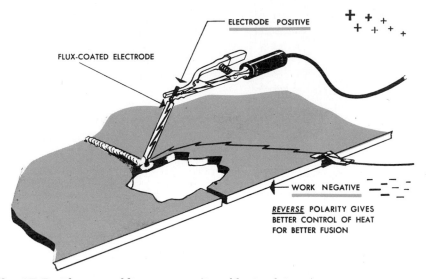

Fig. 17-5. The arc-welding process. (Republic Steel Corp.)

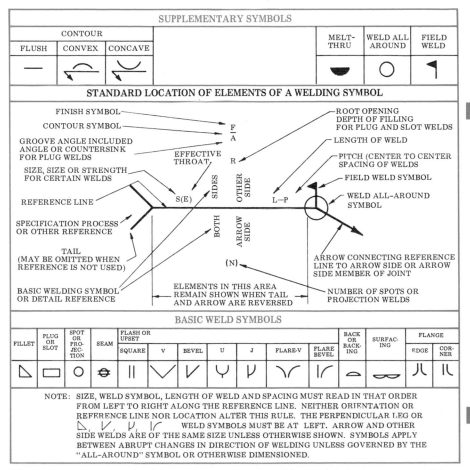

Fig. 17-6. Location of welding information on a welding symbol. (American Welding Society.)

ert gas (MIG). In TIG welding, the electrode that provides the arc for welding is made of tungsten. Since it provides only the heat for fusion, some other material must be used with it for filler. In MIG welding, the electrode contains a consumable metallic rod. It provides both the filler material and the arc for fusion.

Resistance Welding

This kind of welding is done with heat and pressure. It is a good way to fuse thin metals. To join two metal pieces, an electric current is passed through the points to be welded. At those points, the resistance to the charge heats the metal to a plastic state. Pressure is then applied to complete the weld. When current and pressure are confined to a small area between electrodes, the resulting weld is called a *spot weld*.

■ WELDING DRAWINGS

Standard drafting symbols for welding have been established by the American Welding Society. They let the drafter give the type, size, location, and all other specifications for a weld. There are many combinations available. Weld symbols are shown in Fig. 17-6. Also shown is the way in which other welding information is given along with these symbols. Every drafting room should have a copy of the latest edition of the American Welding Society Standard Welding Symbols. (See Appendix Table A-34.)

■ BASIC WELDED JOINTS

There are five basic kinds of joints used in welding (Fig. 17-7). Each can have many variations. The many

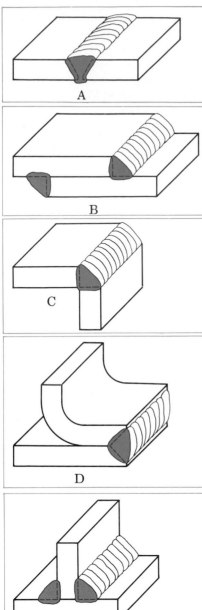

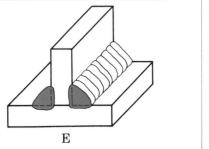

Fig. 17-7. Five basic types of welded joints: (**A**) butt joint, (**B**) lap joint, (**C**) corner joint, (**D**) edge joint, (**E**) T-joint.

combinations and varieties of joints are needed because welding is used in so many different situations. To choose the right joint, one must be familiar with the materials and their

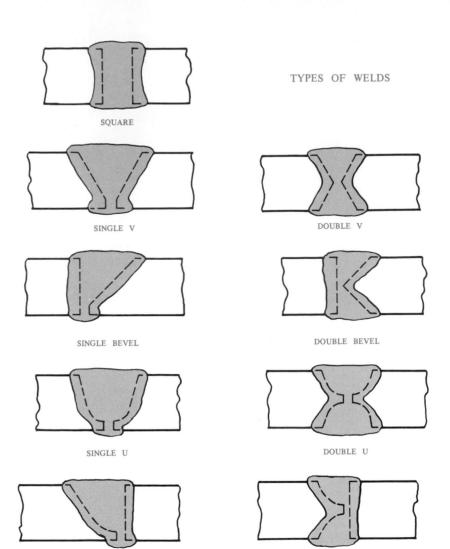

Fig. 17-8. Nine basic types of grooved welds as applied to a butt joint.

conditions and have practical welding experience. Choosing the right weld depends on the type of material, the tools to be used, and the cost of preparation.

■ BASIC TYPES OF WELDS

Figure 17-8 shows the basic types of grooved welds. Note that they may be single or double in form. These

basic types of welds are shown as applied to a butt joint. They can also be applied to all the other basic types of joints. Typical dimensions for a butt joint with a V-grooved weld are shown in Fig. 17-9. The dimensions for a T-joint with a bevel-grooved weld are shown in Fig. 17-10. Typical dimensions for a U-grooved weld on a butt joint are given in Fig. 17-11. Figure 17-12 shows the typical dimensions for a J-grooved weld on a T-joint.

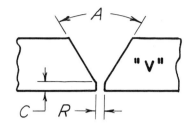

Fig. 17-9. Dimensions for a V-joint weld. *A* = 60° min., *C* = 0 to ⅛ in., *R* = ⅛ to ¼ in. Stock: ½ to ¾ in.

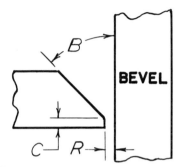

Fig. 17-10. Dimensions for a bevel-grooved weld. *B* = 45° min., *C* = 0 to ⅛ in., *R* = ⅛ to ¼ in.

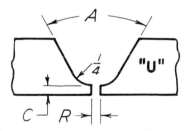

Fig. 17-11. Dimensions for a U-grooved weld. *A* = 45° min., *C* = 1/16 to 3/16 in., *R* = 0 to 9/16 in.

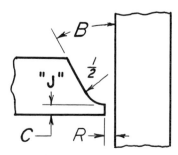

Fig. 17-12. Dimensions for a J-grooved weld. *B* = 25° min., *C* = 1/16 to 3/16 in., *R* = 0 to 9/16 in.

■ SYMBOLS FOR BASIC ARC AND GAS WELDS

The symbols in Fig. 17-6 can be used to describe any desired weld. By combining them, you can describe the most complex or simple welded joint. Figure 17-6 showed the standard way in which welding information is given, along with welding symbols. The notes in the illustration told how to place symbols and data in relation to the reference line. When drawing a fillet, bevel, or J-grooved weld symbol, always place the perpendicular leg of the symbol to the left (Fig. 17-13).

The welding symbol tells what kind of weld to use at a joint and where to place it. The symbol can be drawn on either side of the joint as space permits. Always draw an arrow leading from the symbol's reference line to the joint. The side of the joint to which the arrow points is called the *arrow side*. The opposite side is called the *other side*. If the weld is to be on the arrow side of the joint, draw the type-of-weld part of the symbol below the reference line (Fig. 17-14 at A and D). If the weld is to be on the other side, draw the type-of-weld part of the symbol above the reference line (Fig. 17-14 at B and E). If the weld is to be on both sides of the joint, draw the type-of-weld part of the symbol both above and below the reference line (Fig. 17-14 at C).

When the weld is to be a J-grooved weld, the arrow from the welding symbol must be placed correctly or it can be confusing. For example, in Fig. 17-15 at A, it is not clear which piece is to be grooved. At B, the arrow has been redrawn to show clearly that it is the vertical piece that is to be grooved (as at C). At D, two welds are called for. The symbol below the reference line indicates a J-grooved weld on the ar-

Fig. 17-13. The perpendicular leg on a weld symbol is always drawn to the left.

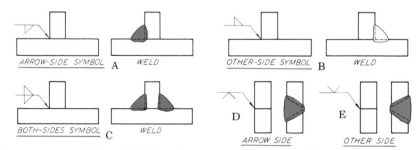

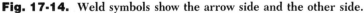

Fig. 17-14. Weld symbols show the arrow side and the other side.

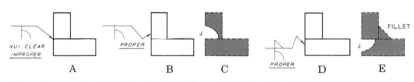

Fig. 17-15. Welding symbols show the J-grooved weld.

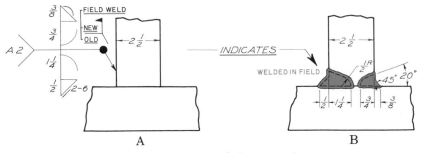

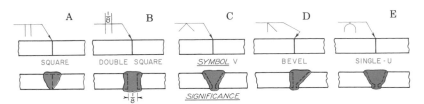

Fig. 17-16. Interpretation of welding symbols or specifications.

Fig. 17-17. Symbols for five typical groove welds.

row side. The arrow shows that it is the horizontal piece that is to be grooved. The symbol above the reference line indicates a fillet weld on the other side. The drawing at E shows how the completed welds would look.

In Fig. 17-16 at A, the reference dimensions have been included with the welding symbols. Also, the typical specification A2 has been placed in the tail of the reference line. The drawing at B shows how this data is used.

Figure 17-16 at B shows the joint made according to the reference specifications at A. This joint can be described as follows: a double filleted-welded, partially grooved, double-J T-joint with incomplete penetration. The J-groove is of standard proportion. The radius R is ½ in. (13 mm) and the included angle is 20°. The penetration is ¾ in. (19 mm) for the other side and 1¼ in. (32 mm) deep for the arrow side. There is a continuous ⅜ in. (10 mm) fillet weld on the other side. There is a ½ in. (13 mm) fillet weld on the arrow

side. The fillet on the arrow side is 2 in. (50 mm) long. The pitch of 6 in. (150 mm) indicates that it is spaced 6 in. (150 mm) center to center. All fillet welds are standard at 45°.

SUPPLEMENTARY SYMBOLS

In Fig. 17-16 at A, there is a black, solid dot in the elbow of the reference line. This dot or a black flag is a supplementary symbol for a field weld. This means that the weld is to be made in the field or on the construction site rather than in the shop.

In the tail of the reference line in Fig. 17-16 at A is the typical specification A2. Its meaning is as follows: The work is to be metal-arc process, using a high-grade, covered, mild-steel electrode; the root is to be unchipped and the welds unpeened, but the joint is to be preheated.

In Fig. 17-16 at A, there is a flush contour symbol over the ½ in. (13 mm) fillet weld symbol. This indicates that the contour of this weld is

to be flat-faced and unfinished. Over the ⅜ in. (10 mm) fillet weld on the same reference line there is a convex contour symbol. This indicates that this weld is to be finished to a convex contour. Figure 17-6 shows the supplementary welding symbols to be used for finished welding techniques.

TYPICAL GROOVE WELDS

Figure 17-17 shows the five typical groove welds. Each is shown by its symbol and by a picture as it would apply to a butt joint. At A is a square-grooved joint, at B is a squared both-sides joint, at C is a V-grooved joint, at D is a bevel-grooved joint, and at E is a U-grooved joint.

TYPICAL PLUG AND SLOT WELDS

Examples of plug and slot welds are shown in Fig. 17-18. The welding symbol is the same for both types.

BASIC RESISTANCE WELDING

Resistance welding, as described above, is done with heat and pressure. Electric current is passed through the points to be welded. Re-

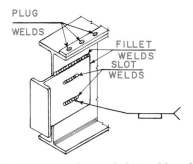

Fig. 17-18. Plug and slot welds. The symbol shown applies to both kinds.

BASIC RESISTANCE WELD SYMBOLS			
TYPE OF WELD			
Spot	Projection	Seam	Flash or upset
◯	◯	⊖	||

SUPPLEMENTARY SYMBOLS			
Weld all around	Field weld	Contour	
		Flush	Convex
◯	⏋	—	⌒

Fig. 17-19. Basic resistance-weld symbols.

sistance to the charge generates the welding heat. Then pressure is used to complete the weld. *Flash welding* is a special kind of resistance welding. It is done by placing the parts to be welded in very light contact or by leaving a very small air gap. The electric current then flashes, or arcs. This melts the ends of the parts, and the weld is made.

BASIC RESISTANCE-WELDING SYMBOLS

There are four basic resistance-weld symbols (Fig. 17-19). They signify spot, projection, seam, and flash or upset welds. The basic reference line and arrow are used with resistance-weld symbols as with arc- and gas-weld symbols. However, in general, there is no arrow side or other side. The same supplementary symbols also apply, as Fig. 17-19 shows.

TYPICAL SPOT WELDS

Resistance spot welding is done on lapped parts. The welds are relatively small in area. Figure 17-20 at A shows a reference symbol in the top view. The arrow points to the working centerline of the weld. At A, the minimum diameter of each weld is specified at .30 in. (7.6 mm). At B, the minimum shearing strength of each weld is specified at 800 lb [3.558 kilonewtons (kN)]. At C and

D, the reference data indicates that the first weld is centered 1 in. (25 mm) from the left end and the welds are spaced 2 in. (50 mm) from centerline to centerline.

TYPICAL PROJECTION WELDS

A projection weld is identified by strength or size. Figure 17-21 at A and D shows parts set up for such a weld. In each case, one part has a boss projection. At B, the reference

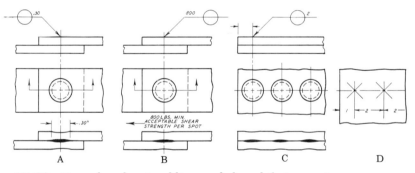

Fig. 17-20. Examples of spot-welding symbols and their meanings.

700 lb (3.10 kN) means that the acceptable shear strength per weld must be at least 700 lb (3.10 kN). At C, the reference data 500 lb (2.22 kN) specifies the strength of the weld, the 2 means that the first weld is located 2 in. (50 mm) from the left side, and the 5 specifies a weld every 5 in. (125 mm) center to center. At E, the number .25 in. (6.0 mm) specifies the diameter of the weld. At F, the diameter of the weld is .25 in. (6.0 mm), there is a weld every 2 in. (50 mm) beginning 1 in. (25 mm) from the left side, and the (5) specifies that there is to be a total of five welds. Notice the arrow-side and other-side indications.

TYPICAL SEAM WELDS

Figure 17-22 at A shows butt-seam and lap-seam welds. The symbol for

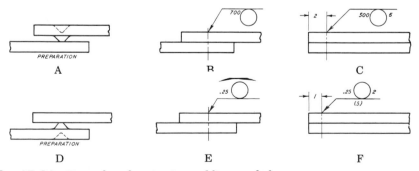

Fig. 17-21. Examples of projection-welding symbols.

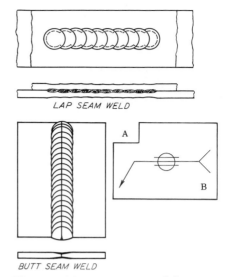

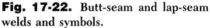

LAP SEAM WELD

BUTT SEAM WELD

Fig. 17-22. Butt-seam and lap-seam welds and symbols.

seam welding is shown at B. The side view shows the two pieces positioned edge to edge for butt-seam welding and overlapping for lap-seam welding, which is done with a series of tangent spot welds.

■ WELDING SYMBOLS

Figure 17-23 shows the use of welding symbols in technical drawings. At A, welding symbols are included on a machine drawing. At B, they are included on a structural drawing.

A

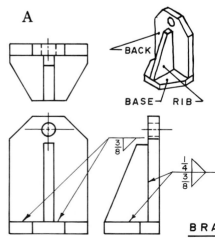

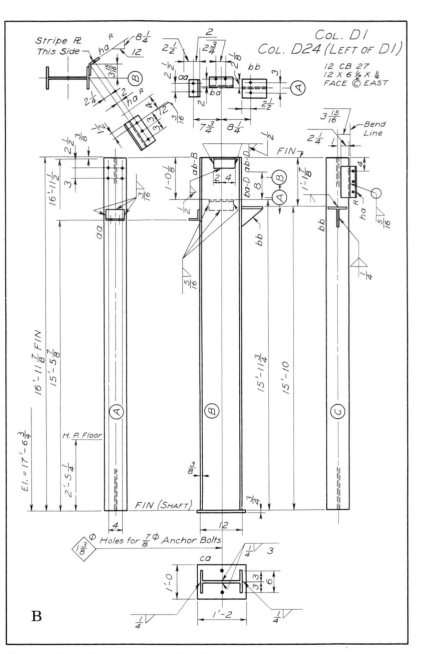

Fig. 17-23. The application of welding symbols: **(A)** on a machine part, **(B)** on a structural drawing.

REVIEW

1. Which kind of welding was developed first, resistance or fusion, and who was the developer?

2. Describe briefly the difference between gas welding and arc welding.

3. Name the gases that are normally used in gas welding.

4. What is the basic principle of arc welding?

5. What do MIG and TIG mean?

6. Name the five basic welding joints.

7. Name nine basic types of groove welds.

8. Draw a welding symbol and identify five features.

Problems

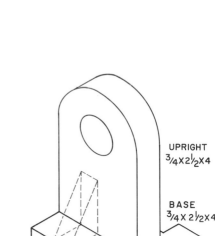
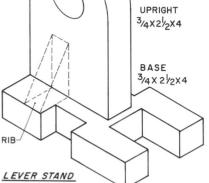

UPRIGHT
$3/4 \times 2\frac{1}{2} \times 4$

BASE
$3/4 \times 2\frac{1}{2} \times 4$

RIB

LEVER STAND

Fig. 17·24. Make a three-view drawing of the lever stand. Estimate sizes not given. Provide a support rib for the upright member. Provide dimensions and welding symbols.

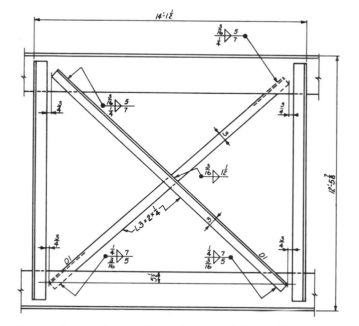

Fig. 17·25. Make a welding drawing of the double-angle cross bracing. Select a suitable scale. The horizontal members are 18 in.

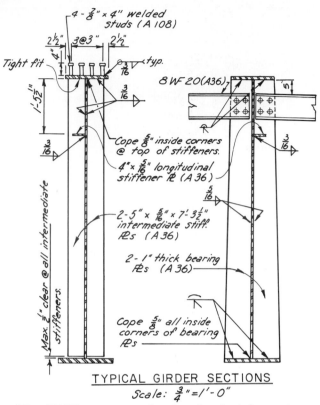

TYPICAL GIRDER SECTIONS

Scale: $\frac{3}{4}$"=1'-0"

Fig. 17-26. Prepare a drawing of each girder section and place the symbols for welding in appropriate locations with dimensioning.

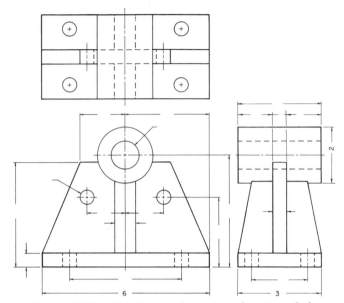

Fig. 17-27. Develop a three-view drawing of the bearing support. Determine dimensions and draw them in decimals (two places). Apply appropriate welding symbols to assemble the five parts that form the bearing support.

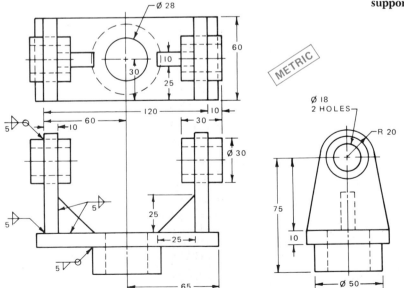

Fig. 17-28. Prepare three views of the double-bearing swivel support. Prepare a parts list. Dimension in decimals or in millimeters.

Fig. 17-29. Prepare a detailed welding drawing of the pulley housing in Fig. 17-2.

Fig. 17-30. Prepare a working drawing of column *D*1 in Fig. 17-23B. Apply appropriate welding symbols and dimensions.

18

Surface Developments and Intersections

SURFACE DEVELOPMENT

The book cover shown in Fig. 18-1. is an example of a surface development. At A, the cover is laid out flat. At B, it is wrapped around a book to make a protective covering. Notice that it fits neatly around all surfaces. It does so because each part has been carefully measured and laid out in relation to other parts. The layout is full size and made on a single flat plane. A surface development is also called a *stretchout*, a *pattern*, or simply a *development*.

Making surface developments is an important part of industrial drafting. Many different industries use surface developments. Familiar items including pipes, ducts for hot- or cold-air systems, parts of buildings, aircraft, automobiles, storage tanks, cabinets, office furniture, boxes and cartons, frozen food packages, and countless other items are designed using surface development.

To make any such item, a surface development is first drawn as a pat-

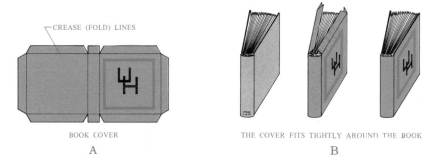

Fig. 18-1. A book cover is an example of a surface development.

tern. This pattern is then cut from flat sheets of material that can be folded, rolled, or otherwise formed into the required shape. Materials used include paper, various cardboards, plastics and films, metals (such as steel, tin, copper, brass, aluminum), wood, fiberboard, fabrics, and so on.

THE PACKAGING INDUSTRY

Packaging is a very large industry that uses surface developments. Creating packages takes both engineering and artistic skill. The packages

must be designed so that they protect their contents during shipment. They must also look attractive for sales appeal. Some packages are meant to be used just briefly and then thrown away. Others are made to last a long time.

Packages and containers for industrial goods are designed to be mass produced at a reasonable cost. A familiar example is the soft-drink carrier (Fig. 18-2). A flat pattern is first made and the first carton is then cut from strong kraftboard and is decorated, folded, and assembled. Its surface is white to permit attractive printing and design. In industry,

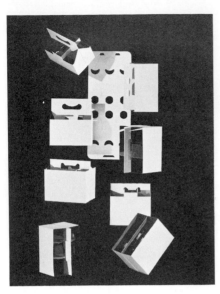

Fig. 18-2. Soft-drink cartons are cut in a flat pattern and then folded. (Olin Packaging Division, Olin Mathieson Chemical Corp.)

these jobs are all done by specially designed machines (Fig. 18-3 and 18-4).

Packages and cartons are made of many materials and in many thicknesses. Some are made of thin or medium-thickness paper stock (Figs. 18-5 and 18-6). This material can be folded easily into the desired form. Some are designed so that no glue is needed. Others may need glue on their tabs.

Packages made of cardboard, corrugated board, and many other materials require allowances for thickness. Examples are boxes made in two parts, a container and a cover (Fig. 18-7), and a slide-in box (Fig. 18-8).

Designing package pattern layouts, lettering, color, and artwork

Fig. 18-4. Phase just prior to glue-lap contact. (Olin Packaging Division, Olin Mathieson Chemical Corp.)

Fig. 18-3. Twenty-up die on cylinder bed, showing makeready on cylinder. Dies are used to cut the sheet material for making packages. (Olin Packaging Division, Olin Mathieson Chemical Corp.)

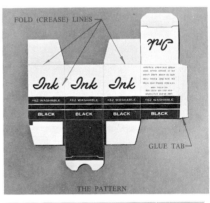

Fig. 18-5. A familiar container made by cutting and folding a flat sheet.

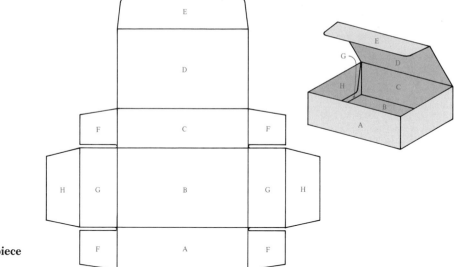

Fig. 18-6. Pattern for a one-piece package with fold-down tabs.

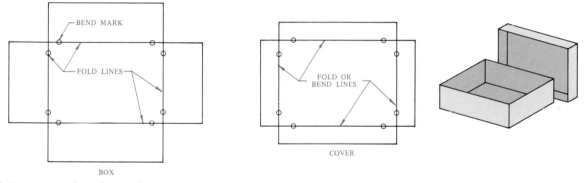

Fig. 18-7. Pattern for a box and cover.

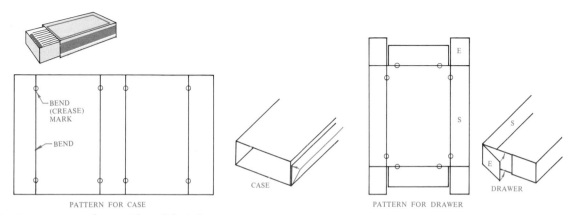

Fig. 18-8. A two-part package with a slide-in box.

provides jobs for many people. Look at various cartons and packages. You will see many interesting and challenging problems in the development of surfaces.

SHEET-METAL PATTERN DRAFTING

Many different metal objects are made from sheets of metal that are laid out, cut, formed into the required shape, and fastened together. The shaping is done by bending, folding, or rolling. The fastening is done by riveting, seaming, soldering, or welding.

For each sheet-metal object, two drawings are usually made. One is a pictorial drawing of the finished product. The other shows the shape of the flat sheet that, when rolled or folded and fastened, will form the finished object (Fig. 18-9). This second drawing is called the *development,* or *pattern,* of the piece. Drawing it is called *sheet-metal pattern drafting* or *surface development.*

A great many thin-metal objects without seams are formed by *die stamping* (pressing a flat sheet into shape under heavy presses). Examples range from brass cartridge cases and household utensils to steel wheelbarrows and parts of automobiles and aircraft. Other kinds of thin-metal objects are made by spinning. Such objects include some brass- and aluminumware. Stamping and spinning both stretch the metal out of its original shape.

Making sheet-metal objects can involve many operations. These include cutting, folding, wiring, forming, turning, beading, and so forth. All are done with machines. Some machines and operations are shown in Fig. 18-10. The machines shown

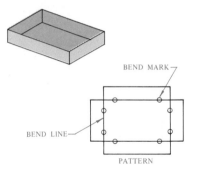

Fig. 18-9. Pictorial drawing and stretchout of a sheet-metal object.

are for hand operations. For industrial mass production, large, complex automatic equipment is generally used to perform these operations.

DEVELOPMENT

There are two general classes of surfaces: plane (flat) and curved. The six faces of a cube are plane surfaces. The top and bottom of a cylinder are also plane surfaces. However, the side surface of the cylinder is curved (Fig. 18-11). There are also different kinds of curved surfaces. Those that can be rolled in contact with a plane surface, such as cylinders and cones, are called *single-curved surfaces.* Exact surface developments can be made for them.

Another kind of curved surface is the *double-curved surface.* It is found on *spheres* and *spheroids.* Exact surface developments cannot be made for objects with double-curved surfaces. However, drafters can make approximations.

Figure 18-12 shows how to cut a piece of paper so that it can be folded into a cube. The shape cut out is the pattern of the cube. Figure 18-13 shows the patterns for all five of the regular solids. If you wish to understand surface development better, lay these patterns out on rather stiff

drawing paper. Then cut them out and fold them to make the solids. Secure the joints with tape.

Any solid that has plane surfaces can be made in the same way. However, each plane must be drawn in its proper relationship to the others in the development of the pattern.

SEAMS AND LAPS

Drawing developments is only part of sheet-metal pattern drafting. Drafters must also know about the processes of wiring, hemming, and seaming. In addition, they must know how much material should be added for each. *Wiring* involves reinforcing open ends of articles by enclosing a wire in the edge. How this is done is shown in Fig. 18-14 at A. To allow for wiring, a drafter must add a band of material to the pattern equal to 2.5 times the diameter of the wire. *Hemming* is another way of stiffening edges. Single- and double-hemmed edges are shown at B and C. Edges are fastened by soldering on lap seams (D), flat lock seams (E), or grooved seams (F). Other types of seams and laps are shown at G and H. Each has its own general or specific use. How much material is allowed in each case depends on the thickness of the material, method of fastening, and application. In most cases, the corners of the lap are notched to make a neater joint.

PARALLEL-LINE DEVELOPMENT

Parallel-line development is a simple way of making a pattern. It is done by drawing the edges of an object as parallel lines. The patterns in Figs. 18-12 and 18-15 are made in this way. Figure 18-16 is a pictorial view of a rectangular prism. In Fig. 18-17,

Squaring Shears

Used for trimming and squaring sheet metal.

Folding Machine

The Folding Machine is used extensively for edging sheet metal or the forming of locks or angles.

Wiring Machine

Works the metal completely and compactly around wire. Depending on shape of work, seats to receive wire are prepared on Folder, Brake or Turning Machine.

Box and Pan Brake

A small bench-mounted brake for straight bends up to full length of machine or for box and pan work up to 3 inch depth.

AA — Showing edges turned by Pexto Burring Machine. Note right-angle burr on body of can and a still more pronounced burr on the bottom piece. The edge on bottom is turned smaller than on the body.

Burring Machine

A difficult operation to master, but practice will produce uniform flanges on sheet metal bodies. Prepares the burr for bottoms preparatory to setting down and double seaming.

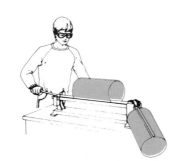

Forming Machines

Used for forming flat sheets into cylinders of various diameters, such as stove pipe, the bodies of vessels, cans, etc. Made in a variety of sizes and capacities.

A — This shows how bottom edge of body and bottom of can are prepared by Burring Machine for Setting-Down.

B — The Pexto Setting-Down Machine closes the seam as shown here. It works both speedily and accurately.

Setting-Down Machine

The Setting-Down Machine prepares the seams in body of vessels for double seaming.

AA — Seats for Wire — made on the Turning Machine.

Turning Machine

Used to prepare a seat in bodies to receive a wire. The operation is completed with use of Wiring Machine.

Doubling-Seaming Machine

Offering in various styles and follows the setting-down operation.

Beading Machine

For ornamenting and stiffening sheet metal bodies.

Fig. 18-10. Some of the machines used in sheet-metal working. (The Peck, Stow, and Wilcox Co.)

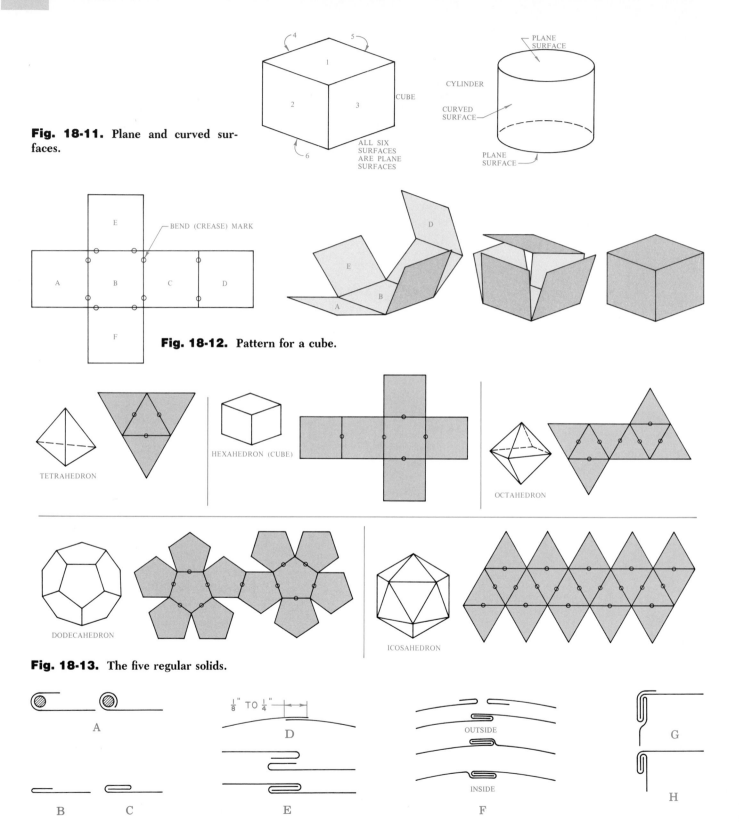

Fig. 18-11. Plane and curved surfaces.

CUBE

ALL SIX SURFACES ARE PLANE SURFACES

PLANE SURFACE

CYLINDER

CURVED SURFACE

PLANE SURFACE

BEND (CREASE) MARK

Fig. 18-12. Pattern for a cube.

TETRAHEDRON

HEXAHEDRON (CUBE)

OCTAHEDRON

DODECAHEDRON

ICOSAHEDRON

Fig. 18-13. The five regular solids.

$\frac{1}{8}$" TO $\frac{1}{4}$"

OUTSIDE

INSIDE

Fig. 18-14. Wiring, seaming, and hemming.

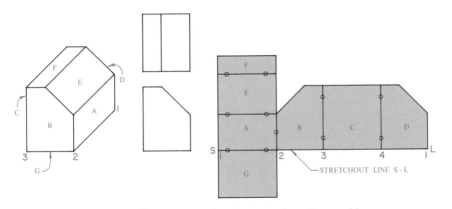
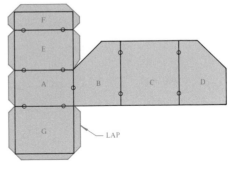

Fig. 18-15. A pattern for a prism, showing stretchout line and lap.

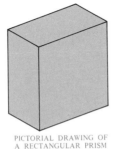

PICTORIAL DRAWING OF
A RECTANGULAR PRISM

Fig. 18-16. Pictorial drawing of a rectangular prism.

a pattern for this prism is made by parallel-line development. To draw this pattern, proceed as follows:

A. Draw the front and top views full size. Label the points as shown (18-17A).

B. Draw the *stretchout line* (SL). This line may also be called the *measuring line* or the *development line*. Find the lengths of

sides 1-2, 2-3, 3-4, and 4-1 in the top view. Measure off these lengths on SL (Fig. 18-17B)

C. At points 1, 2, 3, 4, and 1 on the stretchout line, draw vertical *crease* (fold, bend) *lines*. Make them equal in length to the height of the prism (Fig. 18-17C).

D. Project the top line of the pattern from the top of the front

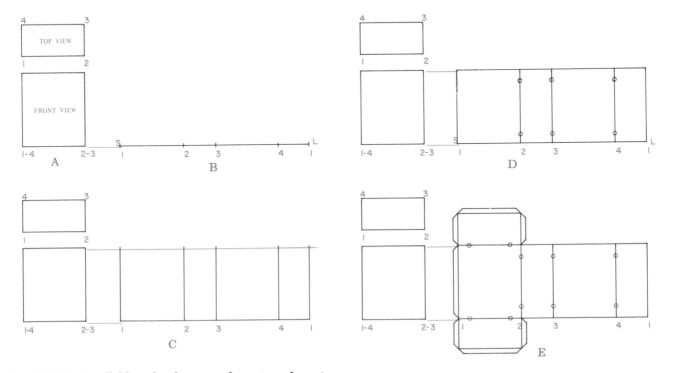

Fig. 18-17. Parallel-line development of a rectangular prism.

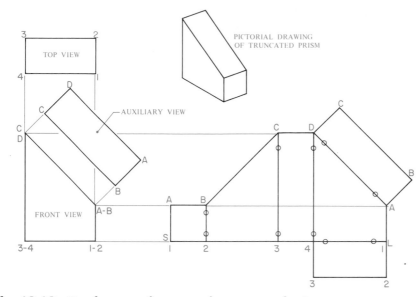

Fig. 18-18. Development of a pattern for a truncated prism.

view. Make it parallel to SL. Darken all outlines until they are thick and black. You can use a small circle or X to identify a fold line (Fig. 18-17D).

E. Add the top and bottom to the pattern by transferring distances 1-4 and 2-3 from the top view, as shown in Fig. 18-17E. Laps or

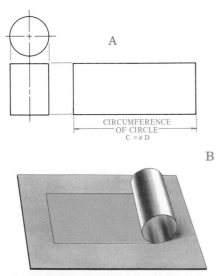

Fig. 18-19. Developed surface of a right circular cylinder.

tabs may be added for assembly of the prism. The size of the laps will vary depending on how they are to be fastened and the kind of material used.

A slight variation is the pattern for a truncated prism shown in Fig. 18-18. To draw it, first make front, top, and auxiliary views full size. Label points as shown. The next two steps are the same as steps B and C in Fig. 18-17. Then project horizontal lines from points *A-B* and *C-D* on the front view to locate points on the pattern. Connect the points to complete the top line of the pattern. Add the top and bottom as shown.

■ CYLINDERS

Figure 18-19 shows a surface development for a cylinder. It is made by rolling the cylinder out on a plane surface.

In the pattern for cylinders, the stretchout line is straight and equal in length to the circumference of the cylinder (Fig. 18-19B). If the base of

the cylinder is perpendicular to the axis, its rim will roll out to form the straight line. If the base is not perpendicular to the axis, you will have to make a right section to get the stretchout line.

■ DEVELOPMENT OF CYLINDERS

Imagine that the cylinder is actually a many-sided prism. Each side forms an edge called an *element*. On the surface of the cylinder, however, these elements seem to form a smooth curve. This is because there are so many of them and they are so close together. Imagining the cylinder this way will help you find the length of the stretchout line. This length will equal the total of the distances between all the elements. Technically, of course, the elements are infinite in number. But, for your purposes, you need only mark off elements at convenient equal spaces around the circumference of the cylinder. Then add up these spaces to make the stretchout line.

Figure 18-20 is a pictorial view of a truncated right cylinder. Figure 18-21 shows how to develop a pattern for this cylinder. To draw this pattern, proceed as follows:

1. Draw the front and top views full size. Divide the top view into a convenient number of equal parts

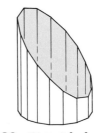

Fig. 18-20. Pictorial drawing of a truncated right cylinder.

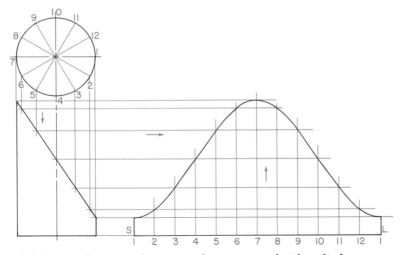

Fig. 18-21. Development of a pattern for a truncated right cylinder.

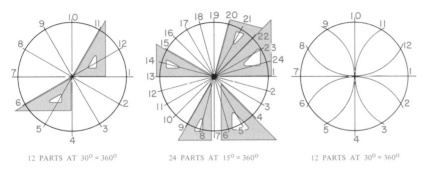

12 PARTS AT 30° = 360° 24 PARTS AT 15° = 360° 12 PARTS AT 30° = 360°

Fig. 18-22. Dividing a circle.

(12 in this case). (See Fig. 18-22 for various methods of dividing a circle.) This will locate a set of equally spaced points (the tips of the elements around the edge in the top view).

2. Draw the stretchout line. Its actual length will be determined later, when you mark off the elements.

3. Using a divider, find the distance between any two elements in the top view. Then mark off this distance along SL as many times as there are parts in the top view. Label the points thus found as shown. Then, from each point, draw a vertical construction line upward.

4. Project other lines downward from the elements on the top view to the front view. Label the points where they intersect the front view.

5. From these intersection points, project horizontal construction lines toward the development.

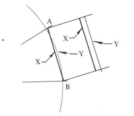

Fig. 18-23. A straight line is the shortest distance between two points.

6. Locate corresponding points where the horizontal construction lines meet the vertical lines from SL. Connect these points in a smooth curve.

7. Darken outlines and add laps as necessary. The colored arrows on the figure show the direction in which the various lines are projected.

Since the surface of a cylinder is a smooth curve, your pattern will not be wholly accurate. This is because it was made by measuring distances on a straight line (chord) rather than on a curve. Figure 18-23, which looks like part of the top view of the cylinder discussed above, shows that the distance from point to point is slightly less along a chord than along the arc (radial distance). In most cases, however, the difference is so slight that the inaccuracy is not critical. The difference can be found by figuring the actual length of the arc using the formula circumference = πD. Then measure this distance out along the stretchout line as shown in Fig. 18-19A.

A slightly different method for developing a cylinder is shown in Fig. 18-24. In this case, a front and a half-bottom view are used. Attaching the two views saves time and increases accuracy. Notice that both methods produce the same results.

■ TO DRAW THE PATTERN FOR A TWO-PIECE, OR SQUARE, ELBOW

This elbow consists of two cylinders cut off at 45°. Therefore, only one pattern is needed, as shown in Fig. 18-25. Allow a lap for the type of seam to be made. If a lap is not needed on the curved edges, both

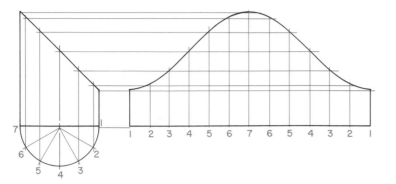

Fig. 18-24. Development of a pattern for a truncated right cylinder, using a front and half-bottom view.

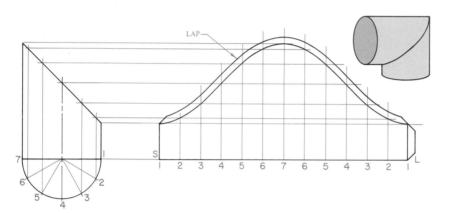

Fig. 18-25. Pattern for a square elbow.

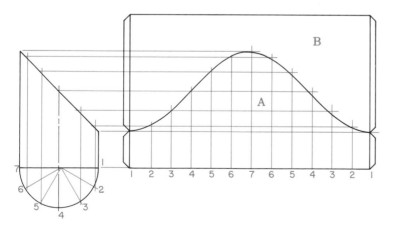

Fig. 18-26. Both parts of the pattern may be made on one stretchout.

parts can be developed on one stretchout, as shown in Fig. 18-26. Notice that the seam on part A is on the short side, while on part B it is on the long side. In Fig. 18-25, the seam on both pieces would be on the short side. In most cases, this is not critical.

THE DEVELOPMENT OF A FOUR-PIECE ELBOW

The pattern for a four-piece elbow is shown in Fig. 18-27. To draw it, proceed as follows:

A. Draw arcs with the desired inner and outer radii to produce an elbow to fit the desired pipe size, as shown at A. Divide the outer circle into six equal parts. Draw radial lines from points 1, 3, and 5 to locate the joints (seams). Draw tangents to the arcs through points 2 and 4 on both arcs to complete the front view.

B. Draw a half-bottom view and divide it as shown at B. Project the elements from this view to the front view.

C. Develop the pattern for the part *A* just as the pattern for the square elbow was developed in Fig. 18-25 and 18-26. If you do not have to allow for laps on the curved edges, you can draw the patterns for all four parts as at C. To find what lengths to mark off on the vertical lines, work from the front view. In the front view of part *B*, for example, continue the contruction lines from the half-bottom view, but make them parallel to the surface lines. Then measure all the lines within part *B*. These lengths, starting with the longest, are the ones to use in the pattern of part *B*. The curves for all the parts are the same. Therefore, you need plot only one of them. It can then be used as a *template* (pattern) for the others.

RADIAL-LINE DEVELOPMENT

In the patterns for prisms and cylinders, the stretchout line is straight,

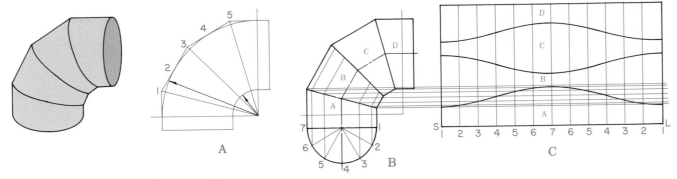

Fig. 18-27. Pattern for a four-piece elbow.

and the measuring lines are perpendicular to it and parallel to each other. Hence the name parallel-line development. On cones and pyramids, however, the edges are not parallel. Therefore the stretchout line will not be a continuous straight line. Also, the measuring lines, in-

Fig. 18-28. Developed surface of a cone.

stead of being parallel to each other, will *radiate* (project) from a single point. This type of development is called *radial-line development*.

■ CONES

Imagine the curved surface of a cone as being made up of an infinite number of triangles, each running the height of the cone. To understand the development of the surface, imagine rolling out each of these triangles, one after another, on a plane. The resulting pattern would look like a sector of a circle. Its radius would be equal to an element of the cone,

that is, a line from the cone's tip to the rim of its base. Its arc would be the length of the rim of the cone's base. The developed surface of a cone is shown in Fig. 18-28.

■ TO DRAW THE PATTERN FOR A RIGHT CIRCULAR CONE

A *right circular cone* is one in which the base is a true circle and the tip is directly over the center of the base (Fig. 18-29A). A *frustum* of a cone is made by cutting through the cone on a plane parallel to the base (Fig. 18-

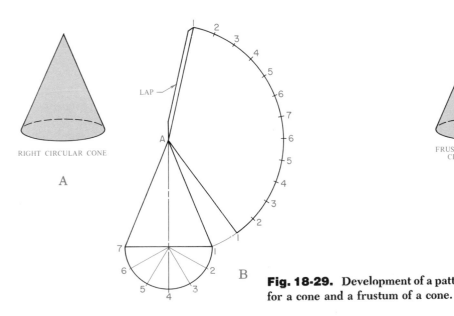

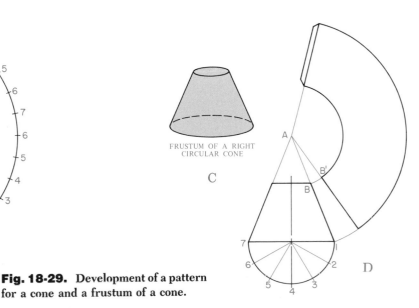

Fig. 18-29. Development of a pattern for a cone and a frustum of a cone.

29C). The pattern for a cone is shown in Fig. 18-29B. To draw it, proceed as follows:

1. Draw front and half-bottom views to the desired size.
2. Divide the half-bottom view into several equal parts. Label the division points as shown.
3. On the front view, measure the *slant height* of the cone, that is, the true distance from the *apex* (tip) to the rim of the base (line A1). Using this length as a radius, draw an arc of indefinite length as a measuring arc. Draw a line from the apex *A* to the arc at any point a short distance from the front view.
4. With a divider, find the straight-line distance between any two division points on the half-bottom view. Then use this length to mark off spaces 1-2, 2-3, 3-4, and so forth, along the arc. Label the points to be sure none have been missed. Complete the development by drawing line *A*1 at the far end.

5. Add laps for the seam as desired. How much to allow for the seam depends on the size of the development and the type of joint to be made.

Figure 18-29D shows the development for a frustum of a cone. To draw it, use the same method as in Fig. 18-29B, but draw a second arc *AB* from point *B* on the front view.

THE PATTERN FOR A TRUNCATED CIRCULAR CONE

A *truncated circular cone* is a circular cone that has been cut along a plane that is not parallel to the base (Fig. 18-30A). The pattern for such a cone is shown in Fig. 18-30B. To draw it, proceed as follows:

1. Draw the front and top, bottom, or half-bottom views.
2. Proceed as in Fig. 18-29 to develop the overall layout for the pattern.
3. Project points 1 through 6 from

the bottom view to the front view and then to the apex. Label the points where they intersect the *miter* (cut) line to avoid mistakes. These lines, representing elements of the cone, do not show in true length on the front view. Their true length shows only when they are projected to the points on the arc.

4. Project the elements of the cone from the apex to the points on the arc.
5. In the front view, find the points on the miter line that were located in step 3. Project horizontal lines from them to the edge of the front view. Continue these lines as arcs through the development. Mark the points where they intersect the element lines. Join these points in a smooth curve. Complete the pattern by adding a lap.

TO FIND THE TRUE LENGTH OF AN EDGE OF A PYRAMID

In Fig. 18-31, the pyramid at A is shown at B in top and front views. In neither view does the edge *OA* show in true length. However, if the pyramid were in the position shown at C, the front view would show *OA* in true length. At C, the pyramid has been revolved about a vertical axis until *OA* is parallel to the vertical plane. At D, the line *OA* is shown before and after revolving.

The construction at D is a simple way to find the true length of the edge line *OA*. First, draw the top view of *OA*. Revolve this view to make the horizontal line *OA'*. Project *A'* down to meet a base line projected from the original front view. Draw a line from this intersec-

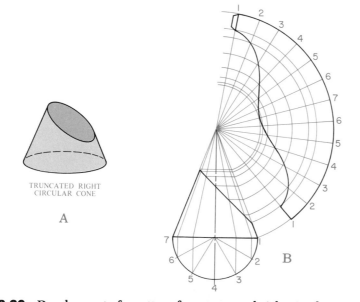

TRUNCATED RIGHT
CIRCULAR CONE

A

B

Fig. 18-30. Development of a pattern for a truncated right circular cone.

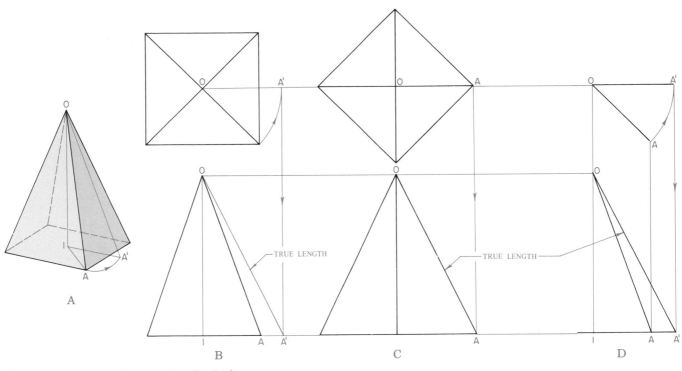

Fig. 18-31. To find the true length of a line.

tion point to a new front view of O. This line will show the true length of OA.

■ TO DRAW THE PATTERN FOR A RIGHT RECTANGULAR PYRAMID

Figure 18-32 shows the pattern for a right rectangular pyramid. To draw it, proceed as follows:

1. Find the true length of one of the edges ($O1$ in this case) by revolving it until it is parallel to the vertical plane ($O1'$).
2. With the true length as a radius, draw an arc of indefinite length to use as a measuring arc.
3. On the top view, measure the lengths of the four base lines. Mark these lengths off as straight-line distances along the arc.

4. Connect the points and draw crease lines. Mark the crease lines.

■ TO DEVELOP THE PATTERN FOR AN OBLIQUE PYRAMID OR A TRUNCATED OBLIQUE PYRAMID

Figure 18-33 shows a development of an oblique pyramid. To draw it, proceed as follows:

1. First, find the true lengths of the lateral edges. Do this by revolving them parallel to the vertical plane, as is shown for edges $O2$ and $O1$. These edges are both revolved in the top view, then projected to locate $2'$ and $1'$. Lines $O2'$ and $O1'$ in the front view are the true lengths of edges $O2$ and $O1$. Edge $O2$ = edge $O3$. Edge $O1$ = edge $O4$.

2. Start the development by laying off 2-3. Since edge $O2$ = edge $O3$, you can locate point O by plotting arcs centered on 2 and 3 and with radii the true length of $O2$ ($O2'$). Point O is where the arcs intersect.
3. Construct triangles O-3-4, O-4-1, and O-1-2 with the true lengths of the sides to complete the development of the pyramid as shown.

Figure 18-34 shows a pattern for a truncated oblique pyramid. The pyramid has an inclined surface $ABCD$. If it were not truncated, it would extend to the apex point O. To draw the pattern, first find the true lengths of OA, OB, OC, and OD. For this pyramid, $OA = OD$ and $OB = OC$. To locate these lengths, locate B' and A' in the front view. Lines OB' and OA' will give the true lengths. Then make a development

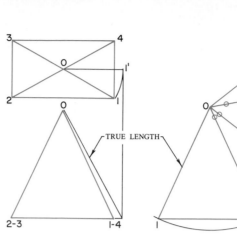

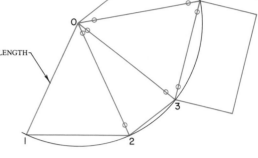

Fig. 18-32. Development of a pattern for a right rectangular pyramid.

TRUE LENGTH

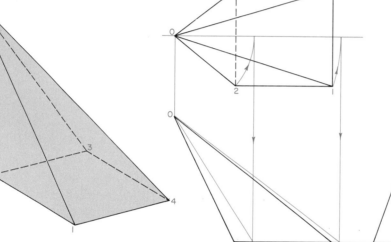

Fig. 18-33. To develop a pattern for an oblique pyramid.

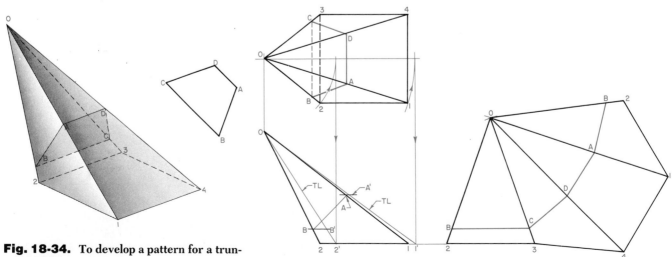

Fig. 18-34. To develop a pattern for a truncated oblique pyramid.

of the pyramid as if it extended to point O. Now lay off the true lengths found on the corresponding edges of this development. Join them to make the pattern for the edge of the frustum. The inclined surface is shown in the auxiliary view and could be attached to the rest of the pattern as indicated.

TRIANGULATION

Triangulation is a method used for making approximate developments of surfaces that cannot be exactly developed. It involves dividing the surface into triangles, finding the true lengths of the sides, and then constructing the triangles in regular order on a plane. Because the triangles will have one short side, on the plane they will approximate the curved surface. Triangulation may also be used for single-curved surfaces.

Figure 18-35 shows how triangulation is used in developing an oblique cone. To draw this development, proceed as follows:

1. Draw elements on the top- and front-view surfaces to create a series of triangles. Number the elements 1, 2, and so forth. For a better approximation of the curve, use more triangles than are shown here.
2. Find the true lengths of the elements by revolving them in the top view until each is horizontal. From the tip of each, project down to the front-view base line to get a new set of points 1, 2, and so forth. Connect these with the front view of point O to make a true-length diagram, as at C.
3. To plot the development at D, construct the triangles in the order in which they occur. Take the distances 1-2, 2-3, and so forth,

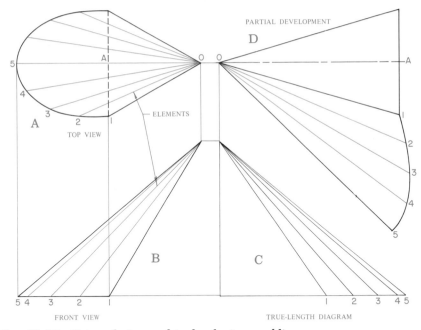

Fig. 18-35. Triangulation used in developing an oblique cone.

from the top view. Take the distances $O1$, $O2$, and so forth, from the true-length diagram.

TRANSITION PIECES

Transition pieces are used to connect pipes or openings of different shapes, sizes, or positions. Transition pieces have a surface that is a combination of different forms, including planes, curves, or both. A few transition pieces are shown in Fig. 18-36.

Figure 18-37 shows the making of a square-to-round transition piece. According to the manufacturer, the piece is easily made in two parts. The illustration shows how one of the parts is positioned for making the last of eight partial bends (creases) on the second conical corner. For this conical bending, one end of the plate is moved out the proper distance for each bend. At the other end, meanwhile, the point for the square corner remains fixed.

DEVELOPMENT OF A TRANSITION PIECE

As you have seen, transition pieces have a surface that is a combination of different forms. These forms are developed by triangulation. This consists of dividing a surface into triangles (exact or approximate), and then laying them out on the development in regular order.

Figure 18-38 shows a transition piece connecting two square ducts, one of which is at 45° with the other. This piece is made up of eight triangles, four of one size and four of another.

To draw the development, proceed as follows:

1. On the top and front views, number the various points as shown. Lines 1-2, 2-3, 3-4, and 4-1 show in their true size in the top view. Lines AB, BC, CD, and DA show in their true size in both views.
2. Find the true length of one of the

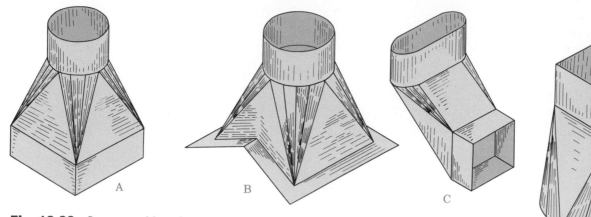

Fig. 18-36. Some transition pieces.

Fig. 18-37. Forming a square-to-round transition piece. (Dreis & Krump Manufacturing Co.)

slanted lines, in this case A4. Do this by revolving it in the top view until it is parallel to the vertical plane. Then project down to the front view, where the true length will show in line 4A′. This will also be the true length of D4, D3, and so forth.

3. Start the development by drawing line DA.

4. Using A and D as centers and a radius equal to the true length of A4 or any of the other slanted lines, draw intersecting arcs to locate point 4 on the development of the piece.

5. Draw another arc using point 4 as a center and a radius equal to line 4-3. Where this arc intersects the arc drawn in step 4 with D as a center, point 3 is located.

6. Proceed to lay off the remaining triangles until the transition piece is completed.

Figure 18-39 shows a transition piece made to connect a round pipe with a rectangular one. This piece contains four large triangles. Between them are four conical parts with apexes at the corners of the rectangular opening and bases, each one quarter of the round opening. To

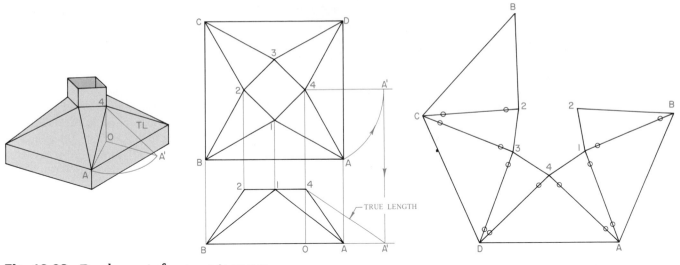

Fig. 18-38. Development of a square-to-square transition piece.

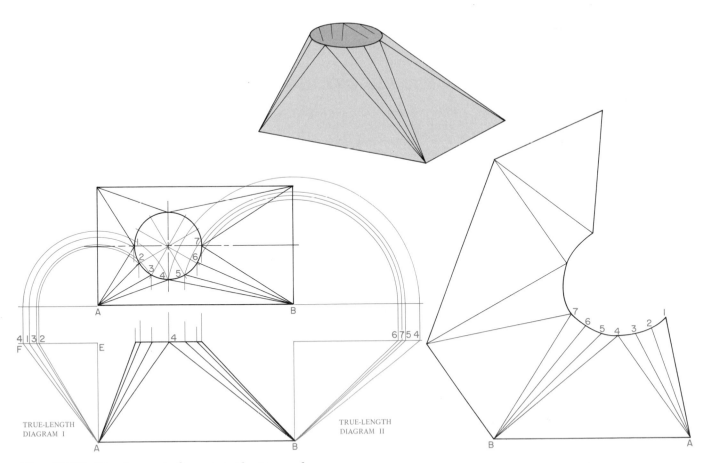

Fig. 18-39. Development of a rectangular-to-round transition piece.

draw the development, proceed as follows:

1. Start with the partial cone whose apex is at *A*. Divide its base 1-4 into a number of equal parts, as 1-2, 2-3, 3-4. Then draw the lines *A*2 and *A*3 to give triangles approximating the cone.
2. Find the true length of each of these lines. In practice, this is done by constructing a separate diagram, diagram I. The construction is based on the fact that the length of each line is the hypotenuse of a triangle whose altitude is the altitude of the cone and whose base is the length of the top view of the line.
3. Begin diagram I by drawing the vertical line *AE*, the altitude of the cone, on the front view. On the base *EF*, lay off the distances *A*1, *A*2, and so forth, taken from the top view. In the figure, this is done by swinging each distance about the point *A* in the top view, then dropping perpendiculars to the base *EF*.
4. Connect the points thus found with point *A* in diagram I to find the desired true lengths. Diagram II, constructed in the same way, gives the true lengths of the triangle lines *B*4, *B*5, and so forth, on the cone whose apex is at *B*.
5. Next, start the development with the seam line *A*1. Draw line *A*1 equal to the true length of *A*1 (taken from true-length diagram I).
6. With 1 as a center and a radius equal to the distance 1-2 in the top view, draw an arc. Then draw another arc using *A* as a center and a radius equal to the true length of *A*2. Where the

two arcs intersect will be point 2 on the development.

7. With 2 as center and radius 2-3, draw an arc. Then draw another with center *A* and radius of the true length of *A*3. Where these two arcs intersect will be point 3.
8. Proceed similarly to find point 4. Then draw a smooth curve through the points 1, 2, 3, and 4 thus found.
9. Next, attach triangle *A*4*B* in its true size. Find point *B* by drawing one arc from *A* with a radius *AB* taken from the top view and another arc from 4 with a radius the true length of *B*4. Where these arcs intersect will be point *B*.
10. Continue until the development is completed.

■ INTERSECTIONS

When a line pierces a plane, the point where it does so is called the

point of intersection (Fig. 18-40). When two plane surfaces meet, the line where they come together, or where one passes through the other, is called the *line of intersection* (Fig. 18-41). When a plane surface meets a curved surface, or where two curved surfaces meet, the line of intersection may be either a straight line or a curved line, depending on the surfaces and/or their relative positions. Package designers, sheet-metal workers, and machine designers all must be able to find the point at which a line pierces a surface and the line where two surfaces intersect.

■ INTERSECTING PRISMS

Figures 18-42 and 18-43 show some of the ways that different surfaces intersect.

To draw the line of intersection of two prisms, first start the regular top and front views. In Fig. 18-44, a square prism passes through a hexag-

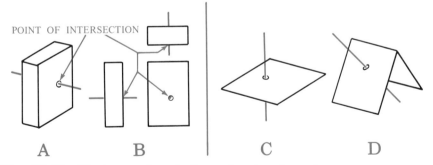

POINT OF INTERSECTION

A   B   C   D

Fig. 18-40. The intersection of a line and a plane is a point.

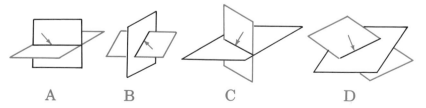

A   B   C   D

Fig. 18-41. The intersection of two planes is a line. The arrow points to the line of intersection.

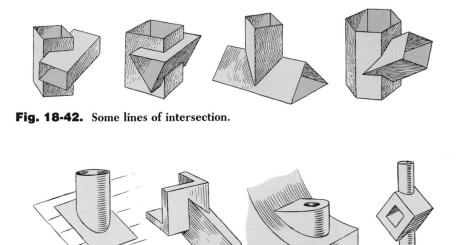

Fig. 18-42. Some lines of intersection.

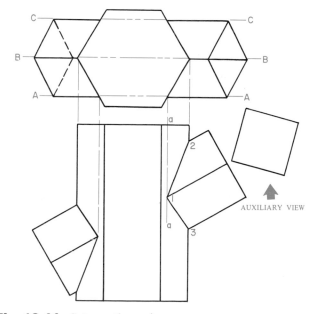

Fig. 18-43. Some lines of intersection.

onal prism. Through the front edge of the square prism, pass a plane parallel to the vertical plane. The top view of this plane appears as a line *AA*. The cut this plane makes in one of the faces of the hexagonal prism shows in the front view as the *cutting line aa*. This line meets the front

edge of the square prism at point 1. Point 1 is a point on both prisms and, therefore, a point in the desired line of intersection. Next, through the top edge of the square prism, pass plane *BB*, also parallel to the vertical plane. This plane will make one cutting line through the lateral edges of

the hexagonal prism. It will make another through the top edge of the square prism. Where the two lines meet will be point 2 on the front view. Points 1 and 2 are on both prisms. Therefore, a line joining them will be on both prisms and thus a part of the line of intersection. Plane *BB* also determines point 3.

The planes in the above example are called *cutting planes*. You can use them to solve most problems in intersections. For intersecting prisms, pass a plane through each edge on both prisms that touches the line of intersection. Each such plane will make cutting lines on both prisms. Where the cutting line on one prism meets the cutting line on the other, there is a point on the required line of intersection. In Fig. 18-45, four cutting planes are required. No planes are needed in front of *AA* or behind *DD*, because there they would cut only one of the prisms. Planes *AA* and *DD* are thus called *limiting planes*.

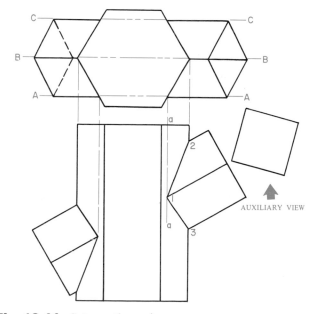

Fig. 18-44. Intersecting prisms.

AUXILIARY VIEW

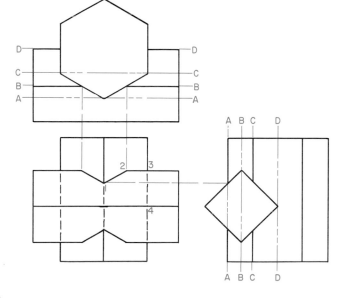

Fig. 18-45. Intersection of prisms.

INTERSECTING CYLINDERS

Figure 18-46 shows how to draw the line of intersection of two cylinders. Since there are no edges on the cylinders, you have to assume positions for the cutting planes. Draw plane *AA* to contain the front line (element) of the vertical cylinder. This plane will also cut a line (element) on the horizontal cylinder. Where these two lines meet in the front view, there is a point on the required curve. Similarly, the other planes will cut other lines on both cylinders that intersect at points common to both cylinders. The figure also shows the development of the vertical cylinder. Figure 18-47 shows how to draw the line of intersection of two cylinders joined at an angle. Here the cutting planes are located by an auxiliary view. To make the development of the inclined cylinder, take the length of the stretchout line from the circumference of the auxiliary view. If the cutting plane chosen divides this circumference into equal parts, the measuring lines will be equally spaced along the stretchout line. Project the lengths of the measuring lines from the front view. Join their ends with a smooth curve. This may be done with a French curve.

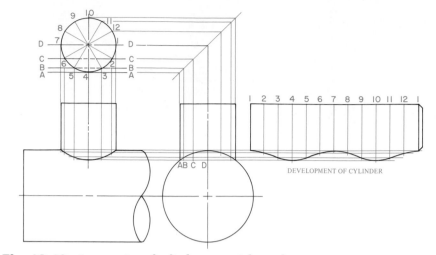

Fig. 18-46. Intersection of cylinders at a right angle.

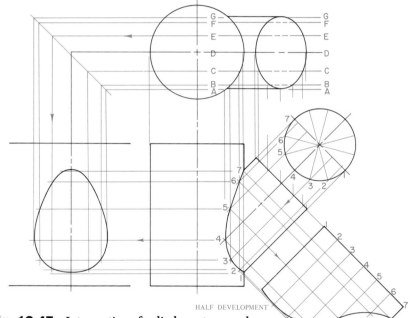

Fig. 18-47. Intersection of cylinders at an angle.

INTERSECTION OF CYLINDERS AND PRISMS

Finding the intersection line of a cylinder and a prism is also done with cutting planes. In Fig. 18-48, a triangular prism intersects a cylinder. The planes *A*, *B*, *C*, and *D* cut lines on both the prism and the cylinder. These lines cross in the front view at points that determine the curve of intersection. To make the development of the triangular prism, take

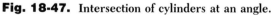
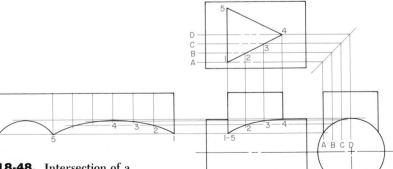

Fig. 18-48. Intersection of a prism and a cylinder.

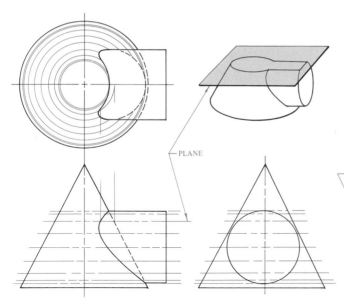

Fig. 18-49. Intersection of a cylinder and a cone.

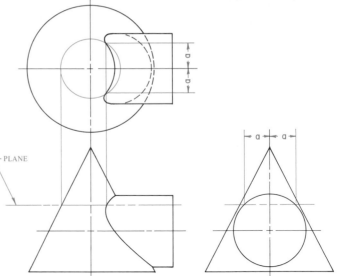

Fig. 18-50. A cutting plane.

the length of the stretchout line from the top view. Take the lengths of the measuring lines from the front view. Note that one plane of the triangular prism (line 1-5 in the top view) is perpendicular to the axis of the cylinder. The curve of the intersection line on that face is made with the radius of the cylinder. The other curve is made with a French curve.

■ INTERSECTION OF CYLINDERS AND CONES

To find the intersection line of a cylinder and a cone, use horizontal cutting planes, as shown in Fig. 18-49. Each plane will cut a circle on the cone and two straight lines on the cylinder. Where the straight lines of the cylinder cross the circles of the cone in the top view, there are the points of intersection. Project these points onto the front view to get the intersection line. Figure 18-50 shows this construction for a single plane. Use as many planes as are needed to make a smooth curve.

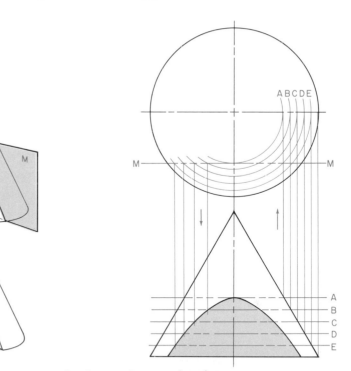

Fig. 18-51. Intersection of a plane and a curved surface.

■ INTERSECTION OF PLANES AND CURVED SURFACES

Fig. 18-51 shows the intersection of a plane *MM* and the curved surface of a cone. To find the line of intersection, use the horizontal cutting planes *A*, *B*, *C*, and *D*. Each plane cuts a circle from the plane *MM*. Thus, you can locate points common to both *MM* and the cone, as in the

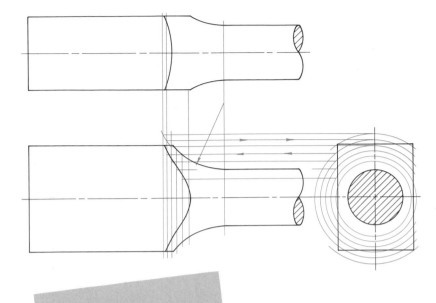

top view. Project these points onto the front view to get the curve of intersection.

Figure 18-52 shows the intersection at the end of a connecting rod. To find the curve of intersection, use cutting planes perpendicular to the rod's axis. These planes cut circles as shown in the end view. The circles, in turn, cut the "flat" at points that can be projected back as points on the curve.

Fig. 18-52. Intersection of a plane and a turned surface.

LEARNING ACTIVITIES

1. Collect a variety of packages (cartons). Take them apart and lay them out flat to see how the original pattern was developed.

2. Select several odd-shaped objects and design packages to suit. Trace the patterns onto kraftboard, cut them out, and assemble them. Be sure to provide tabs for fastening corners.

VOCABULARY

1. Stretchout
2. Pattern
3. Development
4. Die stamping
5. Parallel-line development
6. Stretchout line
7. Development line
8. Measuring line
9. Template
10. Radial-line development
11. Right circular cone
12. Frustum of a cone
13. Truncated circular cone
14. Miter line
15. Triangulation
16. Point of intersection
17. Line of intersection

REVIEW

1. Name the three basic types of surface developments.

2. Name the two general classes of surfaces.

3. A sphere has a _____-curved surface.

4. A development that goes from round to square is called a _____ piece.

5. Lines on a stretchout that show where to make a fold are called _____ lines.

6. Using a series of triangles to develop a pattern is called _____.

7. Cone shapes are produced by using _____-line development.

Problems

Fig. 18-53. Developments. Scale: full size. Problems A through L are planned to fit on an 11″ × 17″ or 12″ × 18″ drawing sheet. Draw the front and top views of the problem assigned. Develop the stretchout (pattern) as shown in the example at the right. For problems A through F, add the top in the position it would be drawn for fabrication. Include dimensions and numbers if instructed to do so. Patterns may be cut out and assembled.

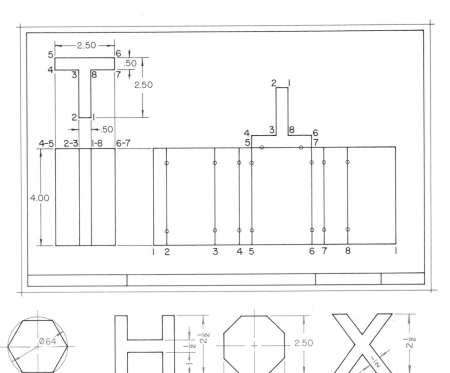

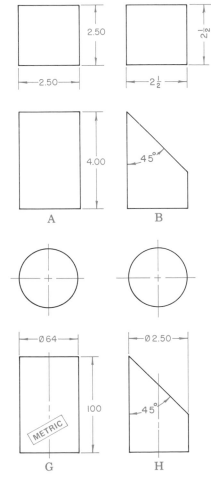

A   B   C   D   E   F

G   H   I   J   K   L

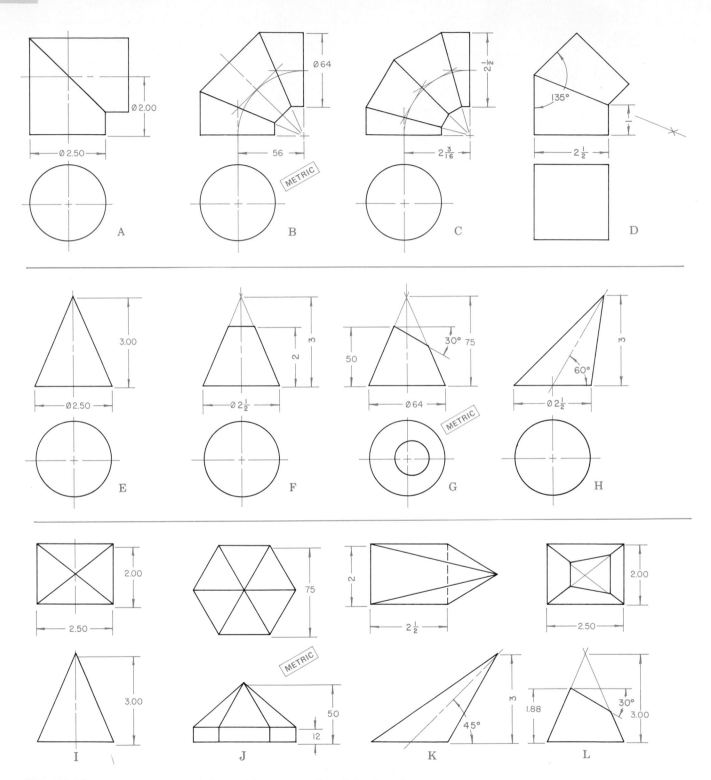

Fig. 18-54. Make two views of the problem assigned and develop the pattern.
Scale: full size.

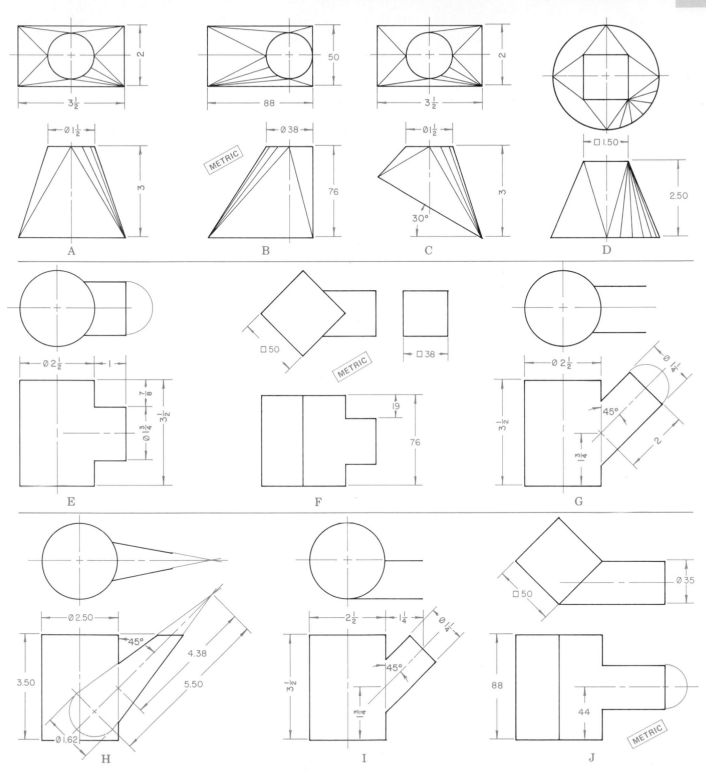

Fig. 18-55. Make two views of the problem assigned. In problems A through D, complete the top views and develop the pattern. In problems E through J, complete views where necessary by developing the line of intersection and completing the top views in G, H, and I. Develop patterns for both parts in problems E through J.

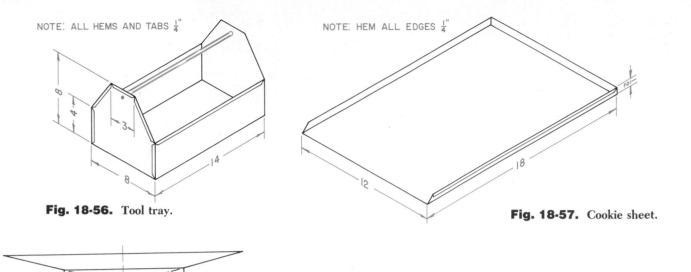

NOTE: ALL HEMS AND TABS ¼"

Fig. 18-56. Tool tray.

NOTE: HEM ALL EDGES ¼"

Fig. 18-57. Cookie sheet.

Fig. 18-58. Brass candy tray.

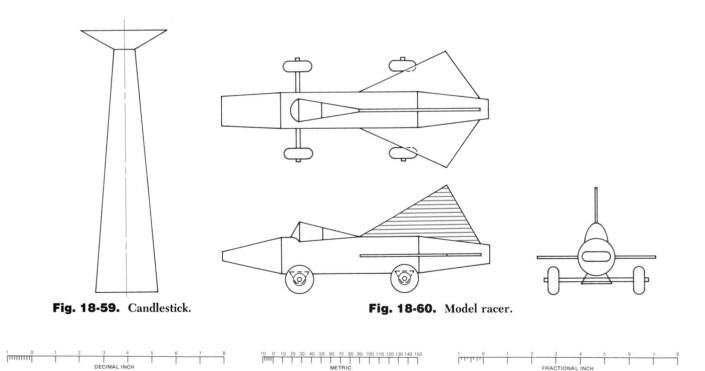

Fig. 18-59. Candlestick.

Fig. 18-60. Model racer.

DECIMAL INCH

METRIC

FRACTIONAL INCH

Make pattern drawings for Figs. 18-56 and 18-57. No other views are necessary. Make complete working drawings of Figs. 18-58 through 18-60, including all necessary views and patterns. Take dimensions from the printed scales at the bottom of the page for Figs. 18-58 through 18-60.

19 Cams and Gears

CAMS AND GEARS

Cams and gears are machine parts that do specific jobs. Cams and gears can transmit motion, change the direction of motion, and change the speed of motion in a machine (Fig. 19-1). The ability to draw cams and gears and to understand their function is a first step in becoming a machine designer. This chapter will introduce the skills that are basic to drawing and understanding these important machine parts.

Fig. 19-1. Cams and gears are necessary parts of machines. (USI-Clearing, a U.S. Industries Co.)

CAMS

A *cam* is a machine part that usually has an irregular curved outline or a curved groove. When the cam *rotates* (turns), it gives a specific motion to another machine part that is called the *follower*. The cam and the follower together make up the cam mechanism. The cam drives the follower. The plate cam in Fig. 19-2 would have a follower that moves up and down as the cam turns on the shaft. The cylindrical (can-shaped) cam in Fig. 19-3 would have a follower that moves back and forth parallel to the axis of the shaft. The grooved cam in Fig. 19-4 would have a follower that moves in an irregular pattern as the cam turns on a shaft.

It is the design of a cam that provides the predetermined and continuous motion of the follower. This kind of motion is needed in all automatic machinery. Therefore, the cam is important to the automatic control and accurate timing found in many kinds of machinery. The unlimited variety of shapes makes the cam useful to the designer.

Fig. 19-2. A plate cam. (Camco.)

Fig. 19-3. A cylindrical cam. (Camco.)

Fig. 19-4. A grooved cam. (Camco.)

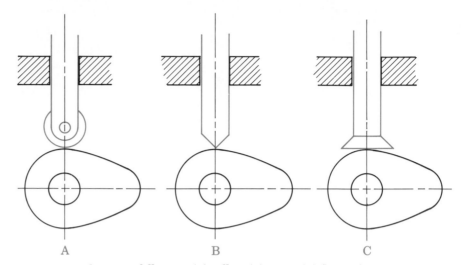

Fig. 19-5. Plate cam followers: **(A)** roller, **(B)** point, **(C)** flat surface. (Camco.)

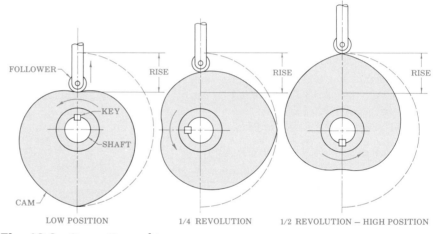

Fig. 19-6. Cam action and terms.

CAM FOLLOWERS

Three types of followers for plate cams are shown in Fig. 19-5. The roller follower is used to reduce friction to a minimum. Therefore, the roller follower transmits force at high speeds. The flat-surface follower and the point follower are made with a hardened surface to reduce wear from friction. Flat-surface and point followers are generally used with cams that rotate slowly.

Plate cams require a spring-loaded follower so that contact can be made throughout a full revolution. The *rise* (lifting) of the follower by the cam is made through direct contact of cam and follower. However, contact of cam and follower during a *drop* (fall) or *dwell* (rest) cannot be made unless contact is brought about by an outside pressure. Outside pressure given by a spring pushes the follower against the cam and ensures direct contact.

CAM TERMS

The illustrations in Fig. 19-6 are pictorial descriptions of how the cam works. The stroke, or rise, takes place within one half a revolution, or 180°. The movement is repeated every 360°, or one full revolution.

KINDS OF CAMS

A cam for operating the valve of an automobile engine is shown in Fig. 19-7. This cam has a flat follower that rests against the face of the plate cam. In Fig. 19-8, the slider cam at A

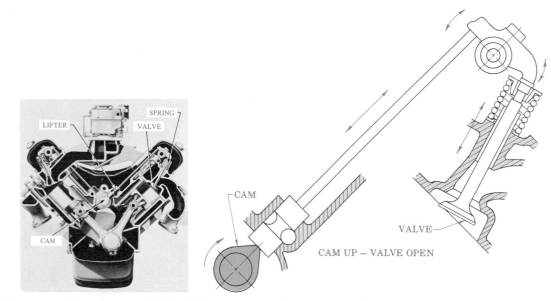

Fig. 19-7. Automobile valve cam. (Oldsmobile Div., General Motors Corp.)

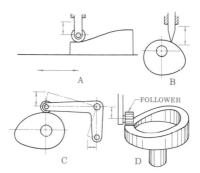

Fig. 19-8. Kinds of cams: (A) slider, (B) offset plate, (C) pivoted roller, (D) cylindrical roller.

moves the follower up and down as the cam moves back and forth from right to left. An offset plate cam with a point follower off center is shown at B. A pivoted roller-follower cam is shown at C. the cylindrical-edge cam with a swinging follower is shown at D. All cams can be thought of as simple inclines that produce predetermined motion.

◼ *CAM LAYOUT*

The shape of the cam determines the direction of motion in the follower. The displacement diagram in Fig.

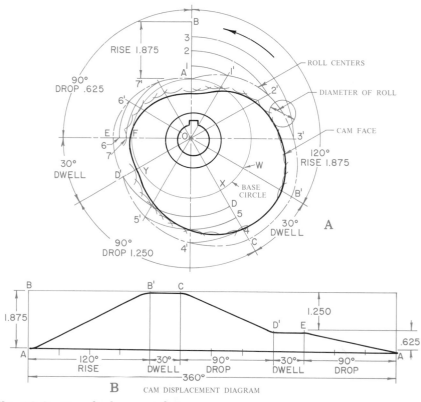

Fig. 19-9. Cam displacement diagram.

19-9 shows the shape of the cam. it also shows the motion the cam will produce through one revolution. The length of the displacement diagram represents one revolution of 360°. The height of the diagram represents the total displacement stroke, 1.875 in., of the follower from its lowest position. The cam is assumed to rotate with a constant speed. The length or time of a revolution is divided into convenient parts. The parts are proportional to the number of degrees for each action. The divisions of the revolution into convenient parts are called *time periods*. These proportional parts are identified A, 1^1, 2^1, B^1, C, 4^1, 5^1 D^1, E, 6^1, 7^1, A, as shown. Each proportional part or time period is a 30° angular division of the base circle.

THE PROBLEM

Point O is the center of the cam shaft, and point A is the lowest position of the center of the roller follower. The center of the roller follower must be raised 1.875 in. with uniform motion during the first 120° of a revolution of the shaft. It must then dwell for 30°, drop 1.250 in. for 90°, dwell for another 30°, and drop .625 in. during the remaining 90°. The shaft is assumed to revolve *uniformly* (with constant speed).

DISPLACEMENT DIAGRAM

The displacement diagram at B in Fig. 19-9 illustrates the solution to the problem. Points A, B^1, C, D^1, E, and back to A relate to the travel pattern that will take place in one complete revolution.

TO DRAW THE PROFILE OF THE CAM

The profile of the cam in Fig. 19-9A has five important features.

1. *Rise*. Divide the rise AB, or 1.875 in., into a number of equal parts. Four parts are used, but eight parts could make the layout

more accurate. The rise occurs from A to W (120°). Divide it into the same number of equal parts as the rise (four at 30°) with radial lines from O. Using center O, draw arcs with radii $O1$, $O2$, $O3$ and OB until they locate 1^1, 2^1, 3^1, and B^1 on the four radial lines. Use an irregular curve to draw a smooth line through these four points.

2. *Dwell*. First draw an arc B^1C (30°) using radius O^1B^1. This will allow the follower to be at rest, because it will stay the same distance from center O.

3. *Drop*. At C, lay off CD, 1.250 in., on a radial line from O. Divide it into any number of equal parts (three are shown). Next, divide the arc XY (90°) on the base circle into the same number of equal parts (three). Draw three radial lines from center O every 30°. Then draw arcs with center O and radii $O4$, $O5$, *and* OD to locate points 4^1, 5^1, and D^1 on the three radial lines. Using an irregular curve, draw a smooth line through points 4^1, 5^1, and D^1.

4. *Dwell*. Draw an arc D^1E^1 (30°) with radius OD. This will provide the constant distance from center O to let the roller follower be at rest.

5. *Drop*. In the last 90° of a full revolution, the roller will return to point A. It will move through a distance EF, or .625 in. First divide EF into any number of equal parts (three are shown). Then divide arc FA into the same number of equal parts (three). Next, draw radial lines every 30°. Then draw arcs with radii $O6$ and $O7$ to locate points 6^1 and 7^1. Using the irregular curve as a guide, draw a smooth curve through points E, 6^1, 7^1, and A. This finishes the roll centers.

Using the line-of-roll centers as a centerline, draw *successive* arcs (one after the other) with the radius of the roller, as shown. Then use an irregular curve to draw the cam profile. The profile will be a smooth curve tangent to the arcs you drew representing the roller.

CAM MOTION

A cam can be designed so that the followers can have three types of motion. The displacement diagram is used to plot the different kinds of motion.

Uniform Motion

Uniform-motion (steady-motion) cams are suitable for high-speed operations. In Fig. 19-10 at A, the thin line represents uniform motion. Equal distances on the rise (8 units) are made for equal distances on the travel (equal intervals of time, 8 units). To avoid a sudden jar at the beginning and end of the motion, use arcs to change it slightly. The small change is shown by the heavy line formed by arcs E and F.

Harmonic Motion

Harmonic-motion (smooth-acting-motion) cams are also good to use with high speeds. They are used when a smooth start-and-stop motion is needed. The method for plotting a harmonic curve is shown in Fig. 19-10 at B. To draw harmonic motion, first draw a semicircle with the rise as the diameter. Divide the semicircle into eight equal parts, using radial lines. Then divide the travel into the same number of equal parts. Project the eight points horizontally from the semicircle until they intersect the corresponding vertical projections as shown. Finally, draw a smooth curve through all eight points with an irregular curve.

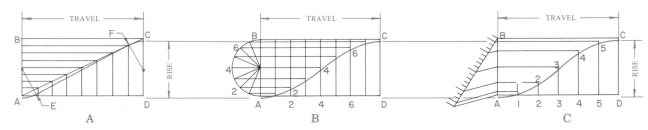

Fig. 19-10. Kinds of cam motion: (**A**) uniform, (**B**) harmonic, (**C**) uniformly accelerated or decelerated.

Uniformly Accelerated and Decelerated Motion

Cams with *uniformly accelerated and decelerated motion* (steadily increasing and decreasing motion) do not operate at a constant speed. The motion produced is very smooth.

In Fig. 19-10 at C, we are plotting a curve called a parabola, or parabolic curve. A parabola is formed when a cone is sliced vertically at any place other than through the center. See Fig. 18-51. A cam designed in the following manner will have a uniform acceleration and deceleration curve. First, divide the rise into parts proportional to 1, 3, 5, . . ., 5, 3, 1. Note that the size parts are not equal. Then divide the travel into the same number of parts, but divide it equally. Project the six points horizontally from the rise until they intersect the corresponding vertical projections. Finish by drawing a smooth curve through all the points of intersection, as shown.

■ CAM DRAWINGS

A drawing for a face, or plate, cam is shown in Fig. 19-11. Note that the amount of movement, or rise, is given by showing the radii for the dwells. These are a 4.5-in. radius and a 7.0-in. radius. Harmonic motion is used, and there seem to be two rolls working on this cam. In Fig. 19-12, a drawing for a *barrel* (cy-

lindrical) cam is shown with a displacement diagram. The diagram shows two dwells and two kinds of motion. Note that the distance traveled from center to center is 1.5 in.

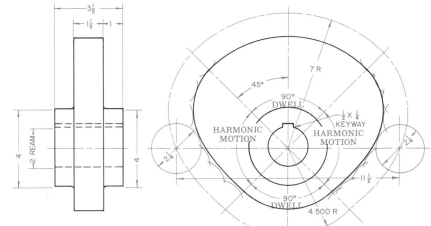

Fig. 19-11. A drawing of a plate cam.

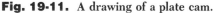

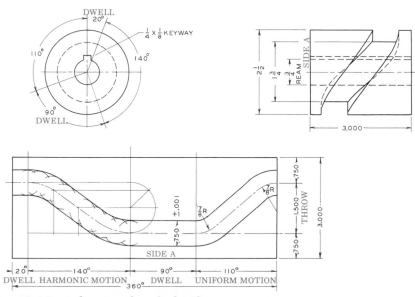

Fig. 19-12. A drawing of a cylindrical cam.

Fig. 19-13. Several gears that are used as typical machine elements: (A, B, and C) spur gears, (D) bevel gear, (E) pinion (spur gear), (F) rack (E and F together are called a *rack and pinion*), (G) internal gear. (The Fellows Gear Shaper Co.)

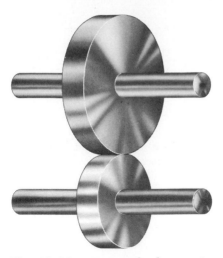

Fig. 19-14. Friction wheels are a simple means of transmitting rotary motion from one shaft to another.

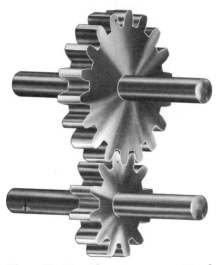

Fig. 19-15. The spur gear. Teeth added to friction wheels provide a more efficient means of transmitting rotary motion.

GEARS

There are many kinds of gears, a few of which are illustrated in Fig. 19-13. However, one of the most practical and dependable machine parts for transmitting rotary motion from one shaft to another is the *spur gear*. The operation of simple spur gears can be explained in this way. If the rims of two wheels are in contact, as in Fig. 19-14, both will revolve if only one is turned. If the small friction wheel is two thirds the diameter of the larger wheel, it will make one and one-half revolutions for every one revolution of the larger wheel. This assumes no slipping occurs. When the load on the driven wheel gets larger and the wheel is hard to turn, slipping begins to occur. Friction wheels cannot be counted on for a smooth transfer of rotary motion. When teeth are added to the wheels in Fig. 19-14, they become gears (Fig. 19-15). Teeth added to the wheels provide the same kind of motion as rolling friction wheels. Now, however, there is no slipping.

GEAR TEETH

The basic forms used for gear teeth are *involute* and *cycloidal* curves. These curves are explained in the following paragraphs.

The spur gear and pinion gear with parallel shafts shown in Fig. 19-15 are good examples of involute gears. Note that the small spur gear is called the *pinion*. Gear teeth have a special shape that lets them mesh smoothly. This shape is an involute curve. Figure 19-19 explains an involute curve that is used in drawing gear teeth. An involute of a circle can be thought of as a curve made by a taut string as it unwinds from around the circumference of a cylinder (circle).

A cycloidal curve can be thought of as the path of a curve formed by a point on a rolling circle (Fig. 19-16). The information given in this chapter is for the 14½° involute system. It can be used for a 20° system as well,

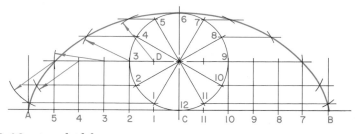

Fig. 19-16. A cycloidal curve.

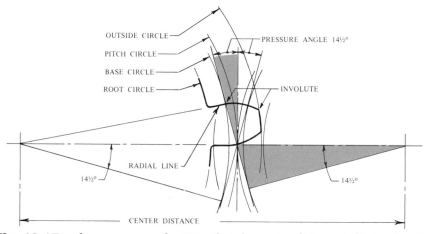

Fig. 19-17. The pressure angle. Note that the center distance indicates the distance between shafts of mating gears.

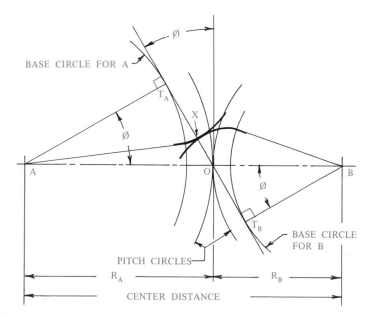

Fig. 19-18. Gear tooth interaction; the rolling nature of surface contact.

since the only practical difference is the number of degrees of the pressure angle. The 14½° or 20° refers to the pressure angle (Fig. 19-17). The pressure angle and the distance between the centers of mating spur gears determine the diameters of the base circles (Fig. 19-17). Note that the point of tangency of the gears is their pitch diameter. These are equal to the diameters of the rolling friction wheels that are replaced by the gears. The involute is drawn from the base circle, which is smaller than the pitch circle.

In Fig. 19-18, R_A is the radius of the pitch circle of the gear with center at A. R_B is the radius of the pitch circle of the pinion with the center at B. The distance between gear centers is $R_A + R_B$. The line of pressure $T_A T_B$ is drawn through O (which is the point of tangency of the pitch circles). It makes the pressure angle ϕ (Greek letter phi) with the perpendicular to the line of centers. This angle is 14½°. Note that lines AT_A and BT_B are drawn from centers A and B perpendicular to the pressure-angle line $T_A T_B$. A point X on a cord (line of pressure $T_A T_B$) will describe the points that form the involute curve as the cord winds and unwinds. This represents the outlines of gear teeth outside the base circles. The profile of the gear tooth inside the base circle is a radial line. Figure 19-19 illustrates the cord unwinding off the surface of the base circle. To simplify the drawing of the involute curve, note that the radius used in Fig. 19-20 is one eighth the pitch diameter.

■ SPUR-GEAR TERMS AND FORMULAS

Some parts of a spur gear are named and illustrated in Fig. 19-21. Use the

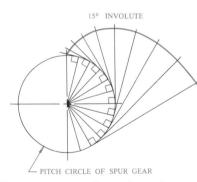

Fig. 19-19. Involute of a circle.

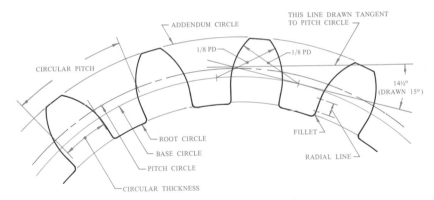

Fig. 19-20. Simplified method of drawing a gear tooth.

following terms and formulas to find the dimensions you need for standard 14½° involute spur-gear problems.

Spur-Gear Terms

1. N = number of teeth
   N_G = number of teeth of gear
   N_P = number of teeth of pinion
2. D = pitch diameter—diameter of pitch circle
   D_G = pitch diameter of gear
   D_P = pitch diameter of pinion
3. P = diametral pitch—number of teeth per inch of pitch diameter
4. a = addendum—radial distance the tooth extends above the pitch circle
5. b = dedendum—radial distance the tooth extends below the pitch circle
6. D_O = outside diameter—pitch diameter plus twice the addendum gives the overall gear size
7. D_R = root diameter—pitch diameter minus twice the dedendum gives the root diameter
8. h_t = whole depth—radial distance from the root diameter to the outside diameter, equal to addendum plus dedendum
9. p = circular pitch—distance from a point on one tooth to the

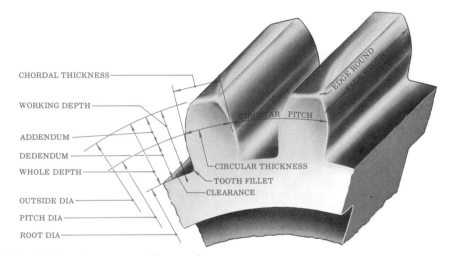

Fig. 19-21. Gear terms illustrated.

same point on the next tooth measured along the pitch circle; the distance of one tooth and one space
10. t_c = circular thickness—thickness of a tooth measured along the pitch circle
11. c = clearance—difference between the addendum and the dedendum; the distance between the top of a tooth and the bottom of the mating space when gear teeth are meshing
12. h_K = working depth—the mating distance a tooth projects into the mating space; twice the radial distance of the addendum

13. o = pressure angle—the direction of pressure between teeth at the point of contact
14. D_b base-circle diameter—the circle from which the involute profile is developed

Spur-Gear Formulas

1. N = number of teeth = DP
   $= \dfrac{\pi D}{P} = D_O \times P - 2$
2. D = pitch diameter = $\dfrac{N}{P}$
   $= D_O - a$
3. P = diametral pitch = $\dfrac{N}{D} = \dfrac{\pi}{p}$

4. a = addendum = $\dfrac{1}{P} = \dfrac{p}{\pi}$

5. b = dedendum
$= \dfrac{1.157}{P} = \dfrac{1.157p}{\pi}$

6. D_O = outside diameter
$= \dfrac{N + 2}{P} = D + 2a$
$= \dfrac{(N + 2)p}{\pi}$

7. D_R = root diameter = $D - 2b$
$= D_O - 2(a + b)$

8. h_t = whole depth = $a + b + c$
$= \dfrac{2.157}{P} = \dfrac{2.147p}{\pi}$

9. p = circular pitch = $\dfrac{\pi D}{N} = \dfrac{\pi}{P}$

10. t_c = circular thickness = $\dfrac{p}{2}$

11. c = clearance = $\dfrac{0.157}{P}$
$= \dfrac{0.157p}{\pi}$

12. h_k = working depth = $2 \times a$
$= 2 \times \dfrac{1}{P}$

■ APPLICATION OF GEAR FORMULAS

Before the formulas can be applied, written descriptions must be given. Typical information given to the drafter is as follows: A pair of involute gears is to be drawn according to the specifications that follow.

■ The pressure angle will be 14½°.
■ The distance between parallel shaft centers will be 12 in.
■ The driving shaft will turn at 800 revolutions per minute (r/min) clockwise.
■ The driven shaft will turn at 400 r/min.
■ The diametral pitch equals 4 (number of teeth per inch of pitch diameter).

Typical Computations

The following computations must be done before the formulas can be applied.

1. Pitch radius of the *pinion* (smaller gear)
$R_p = \dfrac{400}{400 + 800} \times 12$ in.
$= 4$-in. radius

2. Pitch radius of the spur gear
$R_s = \dfrac{800}{400 + 800} \times 12$ in.
$= 8$-in. radius

Velocity ratio ½ (4 in.: 8 in.)

The first two computations are based on the ratio of the velocity of the two cylinders. One cylinder drives the other. Thus, the ratio is obtained by dividing the velocity (r/min) of the driver by the velocity (r/min) of the driven member.

3. Number of teeth on the pinion
$= N_p = DP = 4 \times 8 = 32$

4. Number of teeth on the spur gear
$= N_s = DP = 4 \times 16 = 64$

5. Addendum = $a = \dfrac{1}{P} = \dfrac{1}{4}$ in.

6. Dedendum = b
$= \dfrac{1.157}{p} = \dfrac{1.157}{4}$
$= 0.289$ in.

■ STANDARD INVOLUTE GEARS

Involute gears are interchangeable when they have set conditions that allow them to mesh properly. The four conditions for interchanging involute gears are the following: the same diametral pitch, the same pressure angle, the same addendum, and the same dedendum.

■ GEAR DRAWINGS

It is not necessary to show the teeth on typical gear drawings. A drawing for a cut spur gear is shown in Fig. 19-22. The gear blank should be drawn with dimensions for making the pattern and for the machining

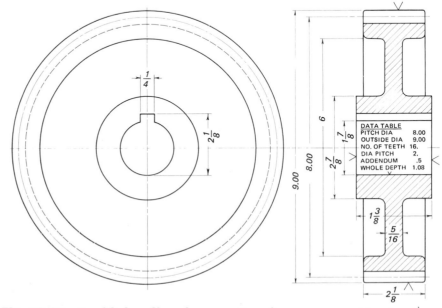

Fig. 19-22. Simplified profile and cross section of a spur-gear working drawing.

DATA TABLE	
PITCH DIA	8.00
OUTSIDE DIA	9.00
NO. OF TEETH	16.
DIA PITCH	2.
ADDENDUM	.5
WHOLE DEPTH	1.08

operations. The spur-gear drawing should include information for cutting the teeth. It should also include information for the tolerances required and a notation of the material to be used. On assembly drawings, a simplified gear may be used, as shown in Fig. 19-23. Even though the drawing is simplified, however, it should include all necessary notes for making the gear.

Involute Rack and Pinion

A rack and pinion is shown in Fig. 19-24. A *rack* is simply a gear with a straight pitch line instead of a circular pitch line. The tooth profiles become straight lines. These lines are perpendicular to the line of action.

Worm and Wheel

Figure 19-25 shows how the worm and wheel mesh at right angles. The *worm gear* is similar to a screw. It may have single or multiple threads. This system would be used to transmit motion between two perpendicular nonintersecting shafts. The *wheel* is similar to a spur gear in design except that the teeth must be curved to engage the worm gear.

Bevel Gears

When two gear shafts intersect, bevel gears are used to transfer motion. Sometimes bevel gears are referred to as *miter gears*. This is the case when the gears are the same size and the shafts are at right angles. Figure 19-26 shows four rolling cones. Think of bevel gears as replacing the friction cones, just as the spur gear replaced the circular friction wheels. Figure 19-27 illustrates mating bevel gears. The smaller gear is called the pinion. More information on bevel gears follows.

Fig. 19-24. A rack and pinion. (Brad Foote Gear Works, Inc.)

Fig. 19-25. Application of a worm and wheel. (Industrial Drives Division, Eaton Corp.)

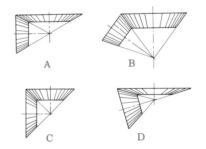

Fig. 19-26. Rolling cones that represent bevel gearing.

Fig. 19-27. Mating bevel gears. (Arrow Gear Co.)

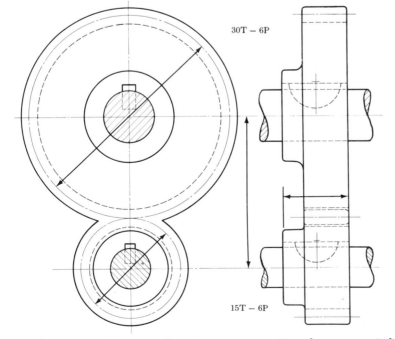

30T — 6P

15T — 6P

Fig. 19-23. Simplified drawing of mating spur gears. Note the tangent pitch circles, the number of teeth, and the pitch.

Bevel-Gear Terms

Some basic information about bevel gears is given in Fig. 19-28. There are three Greek letters used in the following list: α = alpha, δ = delta, and Γ = gamma. Examine Fig. 19-28 for the similarity of design found in spur and bevel gears. Note the differences. The pitch diameter is the diameter of the pitch cone in the bevel-gear design. The circular pitch and the diametral pitch are based on the pitch diameter just as in spur gears. The important features, such as the angles, are listed here.

α = addendum angle
δ = dedendum angle
Γ = pitch angle
Γ_R = root angle
Γ_o = face angle
a = addendum
b = dedendum
a_n = angular dedendum
A = cone distance
F = face
D = pitch diameter
D_O = outside diameter
N = number of teeth
P = diametral pitch
R = pitch radius

Bevel-Gear Drawing

On working drawings of bevel gears, the needed dimensions for machining the blank must be given, as well as all the gear information. An example of a working drawing for a cut bevel gear is shown in Fig. 19-29.

Gear Information

The American National Standards Institute has established standards for satisfactorily detailing gear drawings. ANSI Y14.7, Section Seven, and B6.1 and B6.5 are standard references that can be used for further study of the subject of gear detailing.

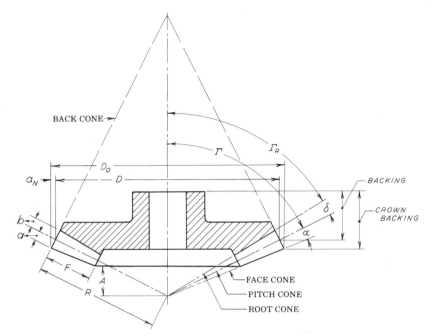

Fig. 19-28. Bevel-gear terms. Note the cone shape.

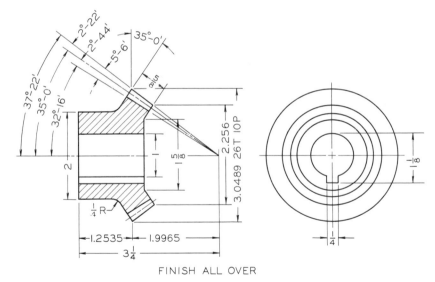

Fig. 19-29. Simplified profile and detailed section of a bevel gear.

REVIEW

1. Other than time, what is important in the design of a cam?

2. What is the purpose of a displacement diagram?

3. List three types of cams and three types of followers.

4. List three uses of a cam and the kinds of cam used.

5. Why is harmonic motion used in a cam?

6. How is the curve on a spur-gear tooth developed?

7. What technical phrase describes the ratio of the number of teeth to the pitch diameter?

8. Rolling cones are used to describe the meshing of what kind of gears?

9. What two bits of information are needed to find circular pitch?

10. List two applications of bevel gears.

11. Name the four circles used by the drafter in drawing gears.

Problems

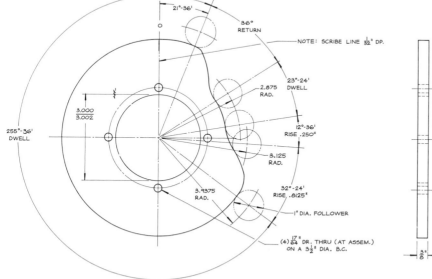

Fig. 19-30. Make a profile of the radial plate cam. Prepare a displacement diagram similar to the one in Fig. 19-9 to illustrate the travel patterns from the base circle.

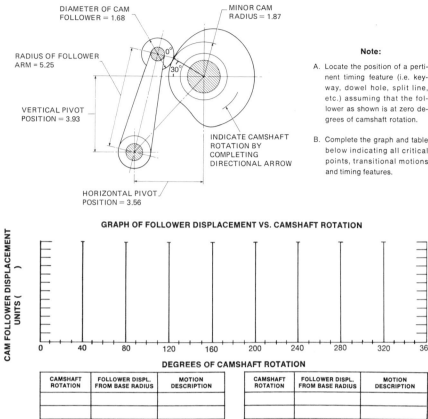

DIAMETER OF CAM
FOLLOWER = 1.68

MINOR CAM
RADIUS = 1.87

RADIUS OF FOLLOWER
ARM = 5.25

VERTICAL PIVOT
POSITION = 3.93

INDICATE CAMSHAFT
ROTATION BY
COMPLETING
DIRECTIONAL ARROW

HORIZONTAL PIVOT
POSITION = 3.56

Note:

A. Locate the position of a pertinent timing feature (i.e. keyway, dowel hole, split line, etc.) assuming that the follower as shown is at zero degrees of camshaft rotation.

B. Complete the graph and table below indicating all critical points, transitional motions and timing features.

GRAPH OF FOLLOWER DISPLACEMENT VS. CAMSHAFT ROTATION

CAM FOLLOWER DISPLACEMENT
UNITS (   )

0   40   80   120   160   200   240   280   320   360

DEGREES OF CAMSHAFT ROTATION

CAMSHAFT ROTATION	FOLLOWER DISPL. FROM BASE RADIUS	MOTION DESCRIPTION	CAMSHAFT ROTATION	FOLLOWER DISPL. FROM BASE RADIUS	MOTION DESCRIPTION

Fig. 19-31. Complete the travel pattern for the displacement diagram. Note that the minor radius is given. Move 75° clockwise in a dwell position from 30° above the horizontal centerline. Determine the rise and travel in the next 285° in displacement units of .125 in. Then develop the new plate cam profile. (Camco.)

Fig. 19-32. Prepare a drawing of the displacement diagram and the cam profile in Fig. 19-9. Shaft: .62″ DIA. Roller: .56″ DIA. Base circle: 2.50″ DIA. Hub: 1.12″ DIA. Diagram line = 2.5 × 3.14.

Fig. 19-33. Prepare the three diagrams in Fig. 19-10 to illustrate the three types of cam motion. Travel: 2½″. Rise: 1¼″.

Fig. 19-34. Draw the two views of the plate cam in Fig. 19-11. Prepare a displacement diagram to illustrate the rises.

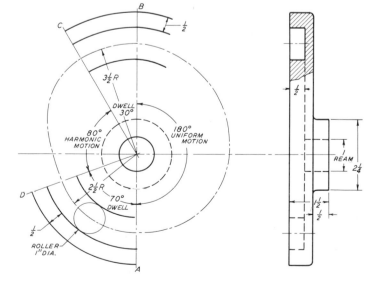

Fig. 19-35. Draw a box (grooved) cam for the conditions given in the illustration. Draw the path of the roll centers and lay off the angles. From *A* to *B*, rise from 2½″ to 3½″, with modified uniform motion. From *B* to *C*, dwell, radius 3½″. From *C* to *D*, drop from 3½″ to 2½″, with harmonic motion. From *D* to *A*, dwell, radius 2½″. Draw the groove, using a roller with a 1″ diameter in enough positions to fix outlines for the groove. Complete the cam drawing. Keyway = ½″ × ⅛″. Leave construction lines.

Fig. 19-36. Prepare a two-view drawing of the spur gear in Fig. 19-22. Make a simplified profile and section as shown. Using the given data, draw in two gear teeth after calculating circular thickness.

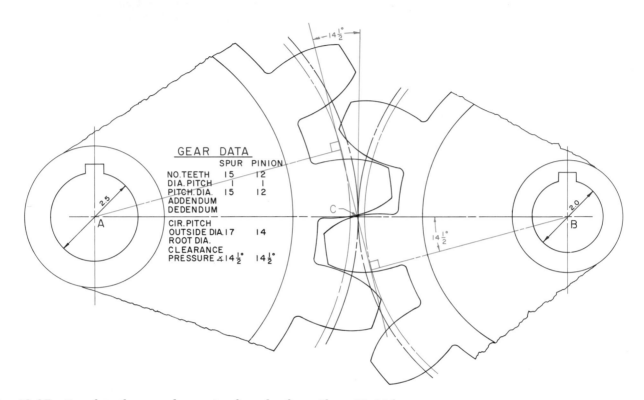

GEAR DATA

	SPUR	PINION
NO. TEETH	15	12
DIA. PITCH	1	1
PITCH. DIA.	15	12
ADDENDUM		
DEDENDUM		
CIR. PITCH		
OUTSIDE DIA.	17	14
ROOT DIA.		
CLEARANCE		
PRESSURE ∡	14½°	14½°

Fig. 19-37. Complete the gear data, using formulas from Chap. 19. Make an enlarged drawing of a mating spur and pinion as shown. Select a suitable scale. Use an involute to draw the gear tooth profile or 1/8 PD, as directed by the instructor.

Fig. 19-38. Prepare a working drawing of the mating gears in Fig. 19-23. Calculate data necessary to draw the gear teeth and prepare a data table.

Fig. 19-39. Prepare the two views of the bevel gear in Fig. 19-29. Review the simplified profile and insert the proper pitch circles as centerlines in the front view. Dimension the drawing.

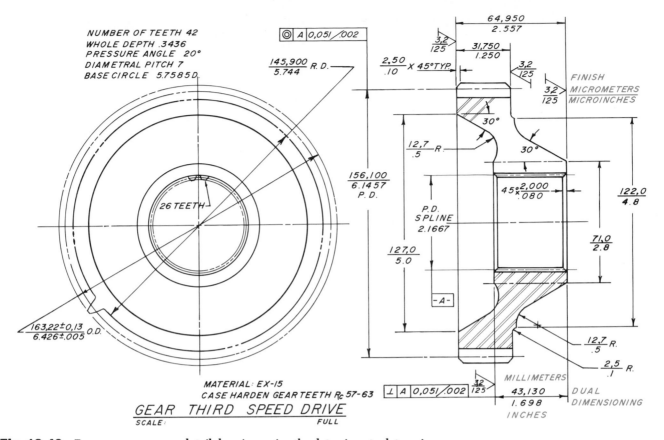

NUMBER OF TEETH 42
WHOLE DEPTH .3436
PRESSURE ANGLE 20°
DIAMETRAL PITCH 7
BASE CIRCLE 5.7585 D.

$\frac{145,900}{5.744}$ R. D.

$\frac{156,100}{6.1457}$ P. D.

26 TEETH

$\frac{163,22 \pm 0,13}{6.426 \pm .005}$ O.D.

MATERIAL: EX-15
CASE HARDEN GEAR TEETH R$_C$ 57-63

GEAR THIRD SPEED DRIVE
SCALE: FULL

⊚ | A | 0,051 | .002

$\frac{64,950}{2.557}$

$\frac{3,2}{125}$

$\frac{31,750}{1.250}$

$\frac{3,2}{125}$

$\frac{2,50}{.10}$ X 45° TYP.

$\frac{3,2}{125}$ FINISH
MICROMETERS
MICROINCHES

30°

$\frac{12,7}{.5}$ R

30°

45° $\frac{2,000}{.080}$

$\frac{122,0}{4.8}$

P.D.
SPLINE
2.1667

$\frac{71,0}{2.8}$

$\frac{127,0}{5.0}$

-A-

$\frac{12,7}{.5}$ R.

$\frac{2,5}{.1}$ R.

$\frac{32}{125}$

⊥ | A | 0,051 | .002

MILLIMETERS

$\frac{43,130}{1.698}$

INCHES

DUAL
DIMENSIONING

Fig. 19-40. Prepare a spur-gear detail drawing, using the data given to determine necessary dimensions. Note that the gear is designed in metric dimensions (using a comma in place of a decimal point) and is dual-dimensioned with decimal inches. (International Harvester Co.)

20 Architectural and Structural Drafting

■ TODAY'S ENVIRONMENT

The work that architects do greatly affects the way we live. Our everyday lives take place within an *environment* (setting) designed by architects and city planners. Architectural drawings are the plans for three-dimensional forms called *architecture*. Architects design buildings and the spaces formed around them. These buildings (sometimes called structures) in a planned environment may be arranged to form rural communities, suburbs, or cities. Architects have been planning, designing, and supervising the construction of buildings around the world for centuries. Today they continue to provide functional (useful) buildings that encourage and heighten all human activity. The architect's job is to give people a better and richer life—to design a better tomorrow.

You can easily examine architects' work. Look all around your community. Evaluate some of the new and old styles of housing. Look closely at the apartments, condominiums, churches, banks, stores, and offices in your neighborhood. Are they useful and good looking? Do you think the architect did a good job of planning? Did he or she select durable materials that are attractive?

■ CAREER OPPORTUNITIES

There are many jobs related to the building industry. The rapid growth of small towns and big cities demands the creation of new environments (Fig. 20-1). The design team is faced with an exciting and difficult job. The team includes architects, city planners, landscape architects, interior designers, specification writers, drafting technicians, illustrators, construction supervisors, structural engineers, mechanical engineers, and office personnel, all working together on large architectural projects. Some employees must work with the clients. Others must plan design services. Someone has to figure out costs. A large architectural firm has many different jobs for many different kinds of workers.

The team must consider new materials and ways of building. Major concerns today are energy conservation and environmental safety. Passive solar design implies that the design team can orient the building to the site and use materials that save energy.

Professional organizations can provide information on career opportunities. One example is the American Institute of Architects, located at 1735 New York Avenue N.W., Washington, DC 20006.

■ COMMUNITY ARCHITECTURE

You can see the effects of the architect's planning by looking carefully at any town or city. Many residential areas have a traditional or contemporary (modern) architectural style. Others combine a number of styles. One style may be built in different materials.

There may be wooden, or frame, cottages and brick bungalows. The larger residential buildings usually have rental apartments. Some may

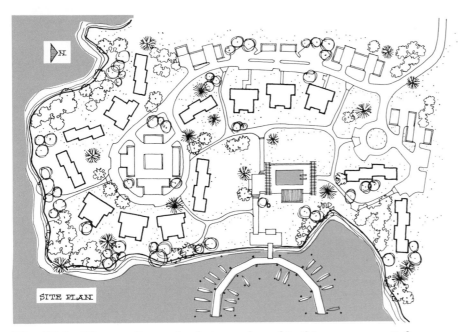

Fig. 20-1. What are four major features planned in this new community?

be *condominiums*. These are buildings in which the units are individually owned, not rented (Fig. 20-2). Buildings with many stories are usually made of materials such as brick, stone, heavy timber, steel, and concrete. These materials make large buildings strong and capable of lasting for a long time.

Different architectural styles can be identified by geometric forms, size, roof shape, and materials. Several kinds of houses are shown in this chapter. California contemporary, ranch, French contemporary, Colonial, French Chateau, Georgian, and English Tudor are some of the common styles of residential buildings.

NEIGHBORHOOD ARRANGEMENTS

Residential buildings are arranged in geometric patterns. The pattern is generally rectangular. The basic unit is called a *block*. Religious buildings may be located on an important street in or near the center of the neighborhood. Many churches and synagogues are on the corner of a

block or in the center of a community. Parks or recreation centers are conveniently placed throughout each neighborhood. Figure 20-3 shows a contemporary recreation center for a neighborhood that has a lot of space. The more important, busy streets often form the boundaries of a neighborhood. These streets are frequently lined with small stores and with small office and apartment buildings.

URBAN DEVELOPMENT

Neighborhoods combine to make uptowns and cities. The main force affecting the human environment today is the movement of people from *rural* (farm) areas to *urban* (city) areas.

In the United States, many people have also moved from "inner" cities and farms to *suburbs*. These are communities on the edges of cities.

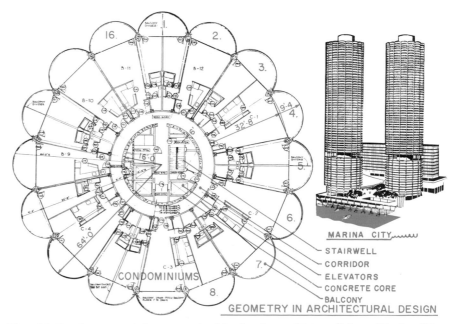

Fig. 20-2. Geometry is important in this plan for multistory living. (Marina City, Chicago, Bertterand Goldberg Associates.)

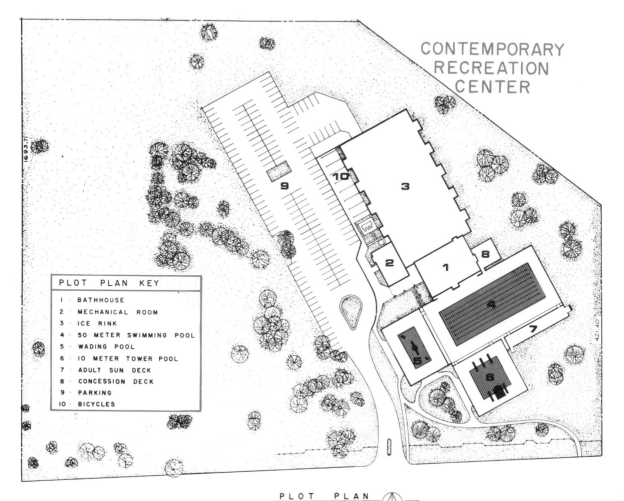

CONTEMPORARY
RECREATION
CENTER

PLOT PLAN KEY

1 · BATHHOUSE
2 · MECHANICAL ROOM
3 · ICE RINK
4 · 50 METER SWIMMING POOL
5 · WADING POOL
6 · 10 METER TOWER POOL
7 · ADULT SUN DECK
8 · CONCESSION DECK
9 · PARKING
10 · BICYCLES

PLOT PLAN
SCALE   0   40

Fig. 20-3. The plot plan and major features of a recreation center in a proposal to a park board. (J. E. Barclay, Jr., & Associates.)

The growing cities and suburbs need *municipal* (public) buildings to serve their new residents. Municipal buildings often include the city hall and facilities for services such as police, fire protection, water, and sanitation. Health centers such as medical offices and hospitals are also important. Cultural facilities such as libraries, museums, and theaters are significant needs. Architectural services are needed to plan and develop these communities.

TRAFFIC PATTERNS

The mobility of a community's residents is reflected in transportation systems. The streets around rectangular blocks or the wandering lanes with "dead ends" that form a community are connected to major thoroughfares. These are lined with large stores, shopping centers, and office buildings. Apartment buildings, banks, and hospitals are also generally located on or near key roads that make it easy to get to them. Government bodies known as building and zoning commissions decide what kinds of buildings will be in different areas. Traffic patterns are designed so as to let industrial employees get to and from work easily. The success of a community may rest on good planning for pedestrians and vehicles.

ARCHITECTURE DEFINED

Architecture is not just buildings. It is people—how they live, work, play, and worship. Good architectural environments can make any human activity better. Thus, architecture is thought of as any physical

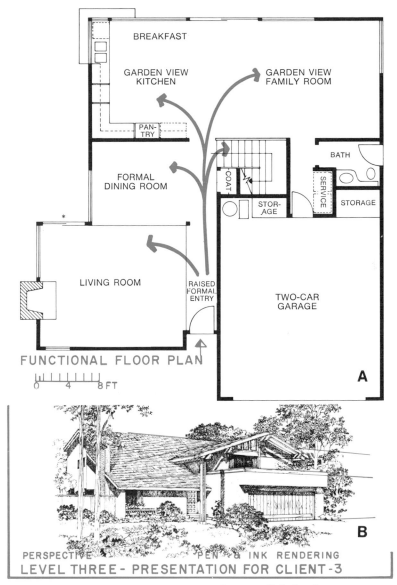

Fig. 20-5. A functional traffic pattern matches a well-defined contemporary California residence. (Larwin Co.)

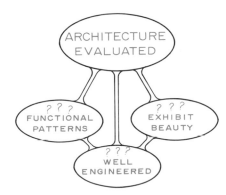

Fig. 20-4. The basis for evaluating architecture focuses on three questions.

ARCHITECTURE EVALUATED

Architecture affects everyone's life. Those who study an architectural environment. It is the sprawling suburb, the crowded slum, the mighty industrial complex, the bright lights of Broadway.

structure look for three things in it (Fig. 20-4):

1. The structure has to satisfy a social purpose. That is, it must show *functional* (useful) patterns for human activity. Begin by studying how people move through the structure to see how its spaces meet their needs (Fig. 20-5A).

2. The structure must be well engineered. Structural members have to be well constructed. Materials must be well selected (Fig. 20-6).

3. The structure must have *aesthetic value* (beauty). Successful architecture has to have a pleasing form based on appealing design qualities (Fig. 20-5B).

Architectural form is easily evaluated by trained architects. Designing a human environment means merging function, structure, and beauty.

■ THE ARCHITECT'S OFFICE

The typical architectural firm is a *partnership*. It has two to four main partners. The partners employ six to twelve persons. However, some firms may have a hundred or more employees. The large firm may offer all the basic architectural services. These include architectural design, structural design, mechanical engineering, civil engineering, landscape design, interior design, and urban planning. These services may be used in the design and building of one structure or of an entire city. Some firms provide only one service. They may make subcontracts with other specialized firms in working on a project.

Not all architectural firms are partnerships. One kind has just one owner. This is called a *proprietorship*. Another kind is owned by people who buy stock in the business. This kind is a *corporation*. Whatever the form of business organization, however, an architectural group works as a team. Each member of the architectural team has a well-defined position. An office usually has an overall project director. Individual

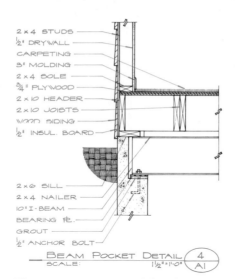

2 x 4 STUDS
½" DRYWALL
CARPETING
3" MOLDING
2 x 4 SOLE
¾" PLYWOOD
2 x 10 HEADER
2 x 10 JOISTS
WOOD SIDING
½" INSUL. BOARD

2 x 6 SILL
2 x 4 NAILER
10" I-BEAM
BEARING PL.
GROUT
½" ANCHOR BOLT

BEAM POCKET DETAIL   4
SCALE: 1½" = 1'-0" A1

Fig. 20-6. A structural detail that defines good engineering. (Douglas Strom.)

team leaders might help find clients or work on a particular project. The National Council of Architecture Licensing Boards in Washington, D.C., promotes professional registration of architects within each state.

■ THE BASIC DRAWINGS

For centuries, architects have been preparing four basic kinds of drawings. These drawings are considered necessary for communicating the details of a project that is to be built. The four kinds of drawings are plan, elevation, perspective, and section. The projection methods of representation used in mechanical drawing are also used in architectural drawing.

The Plan. This is a drawing that shows the horizontal plane on which spaces are arranged for human use (Fig. 20-7). The three main areas of a residential plan are living, sleeping, and service. The plan is a section that

cuts through walls and shows room arrangements.

The Elevation. The elevation, or facade, defines the structural form and architectural style. The roof shape, sides, and side openings are shown in the architectural style chosen and in relationship to the plan. With such drawings, the need for changes to provide balance or symmetry can be easily studied. The elevations, however, are only two-dimensional. This limits in-depth study (Fig. 20-8).

The Perspective Drawing. It is easiest to see architectural form in a pictorial presentation drawing. One-, two-, or three-point perspective drawings show a building in a realistic way. Architectural perspective studies do not have to be mechanically accurate. Therefore, free-hand techniques, along with good proportions, can be used in making these drawings (Fig. 20-9).

The Cross-Detail Section. In this kind of drawing, a vertical plane might be chosen to cut a section to show construction details. Typically, material details are labeled. A section might be cut across the entire structure to aid in interpreting the relationships of the important spaces. In full sections, it is easier to see the proportions between spaces and how these relate to construction and use (Fig. 20-10).

Levels of Communication

The four levels of communication discussed in Chap. 1 apply to architectural drafting (Fig. 20-11).

Level One: Creative Communication. The preliminary architectural

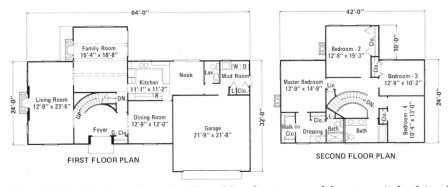

FIRST FLOOR PLAN   SECOND FLOOR PLAN

Fig. 20-7. The floor plan is the first of four basic types of drawings. (Inland Steel Co.)

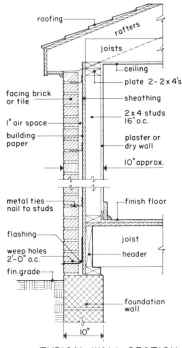

TYPICAL WALL SECTION

Fig. 20-10. A cross-section detail, the fourth of the four basic drawings.

SOUTHERN COLONIAL

Fig. 20-8. An elevation, the second of the four basic drawings. (Inland Steel Co.)

Fig. 20-9. A perspective, the third of the four basic drawings. (Inland Steel Co.)

studies are sketched or laid out lightly so that they can be studied and changed.

Level Two: Technical Communication. The sketches of refined ideas are approved by professionals.

Level Three: Market Communication. The pictorial drawings and colored perspective presentation drawings are evaluated and approved by the client.

Level Four: Construction Communication. Detailed plans, elevations, and sections, together with specifications, are studied and approved by builders. Working drawings from the suppliers of materials supplement the architect's detailed drawings.

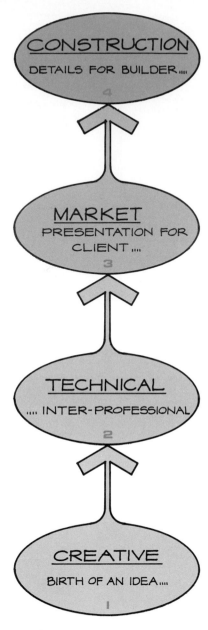

COMMUNICATIONS

Fig. 20-11. The four levels of graphic communication used in the architect's office.

Fig. 20-12. Architectural line technique.

ARCHITECTURAL DRAFTING TECHNIQUES

A good line technique is the most important skill for architectural designing and detailing. In Fig. 20-12, a series of lines are shown that intersect and stop after forming a flared corner. The visual weights of lines express spatial relationships. Good lines are needed for reproduction. They are needed to make the design appealing. They are needed so that the third dimension of depth can be imagined on a two-dimensional medium, as shown in Fig. 20-13. In well-prepared architectural details, line tone and texture complement each other.

Lettering

Both traditional and contemporary styles of lettering are commonly and acceptably used by architects. The traditional lettering is based on the Old Roman alphabet, as shown in Fig. 20-14. These letters have *serifs*. Serifs are small lines extending from the main strokes. Elaborate and important titles are often designed with Old Roman letters. However, plain, single-stroke Gothic lettering is better for architectural working drawings (Fig. 20-15). The condensed and extended styles shown are often used by experienced professionals. They use the right style for inserting notes or balancing the drawing.

Preparing Title Blocks

An architectural firm can create a quickly identified image with a *logo* (symbol), as shown in Fig. 20-16. Some firms use forms with pre-printed borders, title, and revision blocks. The blocks are ready to be filled in with appropriate pencil or ink lettering. The typical title blocks shown in Fig. 20-17 were created by students preparing for professional roles. The title block generally gives the following information: the owner's name and address, the type of structure, the architect's name and address, the title of the sheet and the sheet number, the date, the scale,

Fig. 20-13. Line techniques that feature the major proportions and add three dimensions. (Justus Company, Inc.)

OLDTOWN PUBLIC LIBRARY				
NINTH AVENUE AND STANTON STREET				
DRAWN BY	DATE	YURIKO OHASHI	REVISED BY	DATE
TRACED BY	DATE	ARCHITECT	JOB NO	
CHECKED BY	DATE	LEA BLDG. – OLDTOWN, TEX	SHEET **2** OF 8	

Fig. 20-14. Old Roman lettering applied in an architectural style.

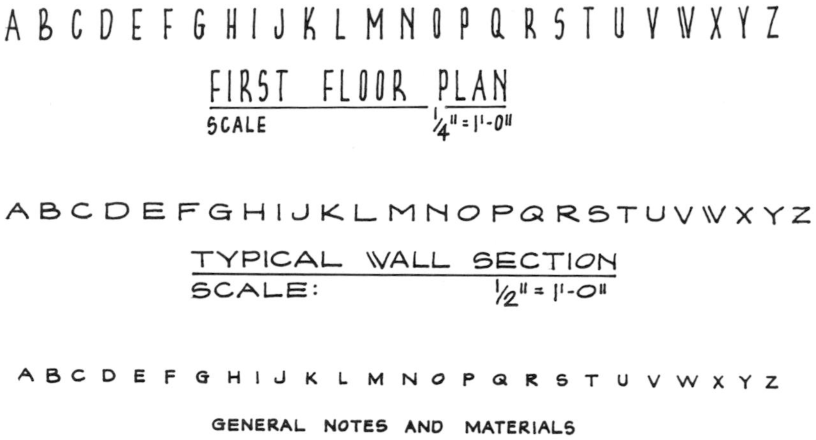

A B C D E F G H I J K L M N O P Q R S T U V W X Y Z

FIRST FLOOR PLAN
SCALE   ¼" = 1'-0"

A B C D E F G H I J K L M N O P Q R S T U V W X Y Z

TYPICAL WALL SECTION
SCALE: ½" = 1'-0"

A B C D E F G H I J K L M N O P Q R S T U V W X Y Z
GENERAL NOTES AND MATERIALS

Fig. 20-15. Single-stroke architectural lettering. Condensed (narrow), extended (bold expanded), and general Gothic stroke.

PROPOSED ADDITION AND ALTERATIONS
TO THE RESIDENCE OF
DALE AND LESLIE ALLAN
145 PARK AVE.   GLEN ELLYN, ILLINOIS

DOUGLASS STROM AND ASSOCIATES
ARCHITECTURAL DESIGNERS
ONE SALT CREEK ROAD   HINSDALE, ILLINOIS

SHEET NO. **A1** OF TWELVE

DRAWN: *DAS CHECKED: APPROVED: SCALE: ¼"=1'-0" DATE: 4 MAY 1986

Fig. 20-16. Title block and bold logo for Douglass Strom and Associates, Architectural Designers.

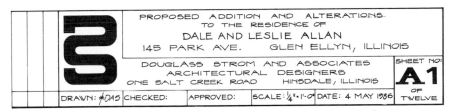

CARLOS MARTINEZ AND ASSOCIATES
SEARS TOWER, CHICAGO
DRAWN: CM   DATE: 10-16-85   JOB NO.
CHECKED: SCALE: ¼"=1'-0"   A-10-79

GAIL SUNG ASSOCIATES : ARCHITECTS
STANDARD OIL BUILDING, CHICAGO, ILLINOIS
DR BY GS   DECEMBER 14, 1984   JOB NO.
OK BY   GRADE: 127814

JEAN N. SCOVIC AND ASSOCIATES
3476 NORTH BEACON ST. - BOSTON, MASSACHUSETTS
DRAWN: J.N.S.   DATE- OCT. 15, 1985   SHEET
CHECKED: SCALE- ⅛" = 1'-0"   A-33

MICHAEL J. LEVIN AND ASSOCIATES
ARCHITECTURAL DESIGNERS
JOHN HANCOCK CHICAGO, ILLINOIS
DRAWN: DATE: SHEET NO.
CHECKED: SCALE: ½"=1'-0"   A-3

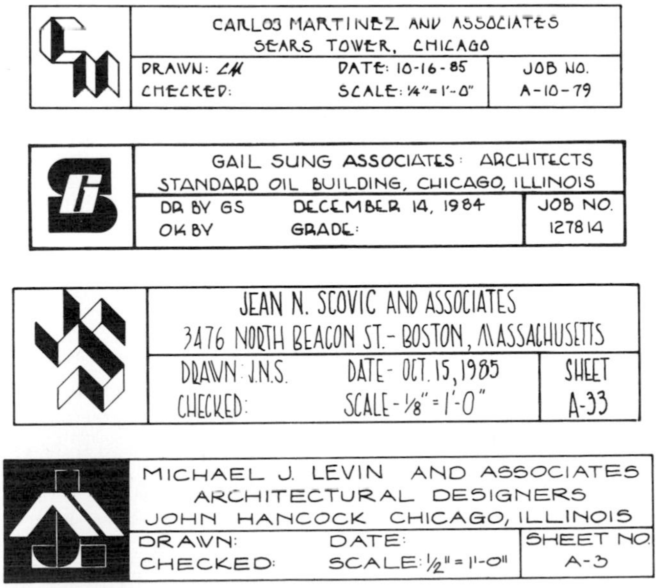

Fig. 20-17. Title blocks created by students to convey a designer's image.

and the initials of the detailing drafter and the supervisor.

Material Symbols

Architectural symbols indicate the materials used in building. They are shown in Fig. 20-18. These symbols are used in plan, section, and elevation drawings.

Pressure-Sensitive Symbols

There are many graphic aids that help the designer meet professional standards on both plan and elevation drawings. Examples of these heat-resistant and pressure-sensitive symbols are shown in Fig. 20-19. Many landscape symbols, such as shrubs, hedges, and trees, are available in both plan and elevation forms. There are also doors, furniture, plumbing and electrical symbols, and many other items for presentation and detail drawings. You transfer these symbols to the vellum, film, or paper by burnishing them with a smooth, blunt instrument.

Architectural Wall Symbols

The details with dimensions in the plan have technical meaning in the building industry. Typical wall symbols for a residential structure are shown in Fig. 20-20. The cross-hatched masonry wall at A is dimensioned to the outside face of the wall. This is done with solid brick, concrete, and concrete-block walls. The frame wall at B is dimensioned to the face of the stud. The brick-veneer wall at C with crosshatched brick and a sole plate is dimensioned to the face of the stud. The remaining material is assumed to be exterior wall facing.

MATERIAL INDICATION

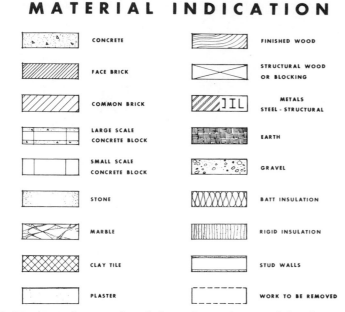

CONCRETE	FINISHED WOOD
FACE BRICK	STRUCTURAL WOOD OR BLOCKING
COMMON BRICK	METALS STEEL - STRUCTURAL
LARGE SCALE CONCRETE BLOCK	EARTH
SMALL SCALE CONCRETE BLOCK	GRAVEL
STONE	BATT INSULATION
MARBLE	RIGID INSULATION
CLAY TILE	STUD WALLS
PLASTER	WORK TO BE REMOVED

Fig. 20-18. Typical material symbols used in architectural detail.

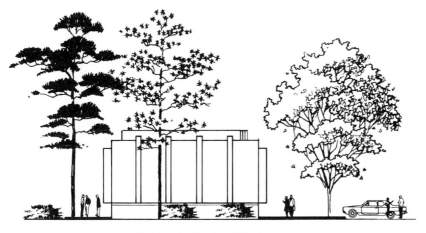

NORTH ELEVATION

Fig. 20-19. Examples of pressure-sensitive symbols used for architectural presentation. (Para-tone, Inc.)

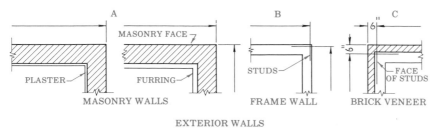

EXTERIOR WALLS

Fig. 20-20. Wall symbols and four dimensioning techniques.

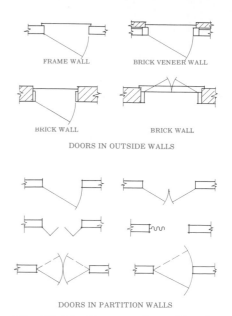

FRAME WALL   BRICK VENEER WALL

BRICK WALL   BRICK WALL

DOORS IN OUTSIDE WALLS

DOORS IN PARTITION WALLS

Fig. 20-21. Door openings with swing symbols put in place using architect's template (see Fig. 20-23).

Symbols for Door and Window Openings

Some door symbols are shown in Fig. 20-21. The plan shows the outside door ajar in a single line. An arc indicates the direction in which the door swings. Interior doors are shown in a variety of ways. Typical windows are shown in Fig. 20-22. Symbols help the reader interpret window movement. They also show what type of wall material frames the window.

Templates for Design and Detail

Templates have improved the quality of the drawings developed today in the architect's office. Templates are usually made of lightweight plastic with openings that form structural shapes or fixtures. You follow the outline of the opening with your pencil or pen to make the graphic image, as in Fig. 20-23. Templates are available in ⅛″ and ¼″ scales.

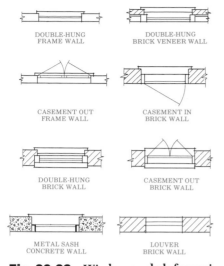

Fig. 20-22. Window symbols for various types of window openings (see Fig. 20-43).

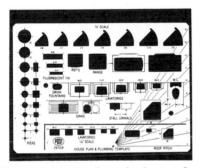

Fig. 20-23. Architect's house-planning template with fixtures and appliances. (Teledyne Post.)

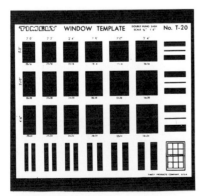

Fig. 20-24. Window template to simplify drawing openings in plans and elevations. (Teledyne Post.)

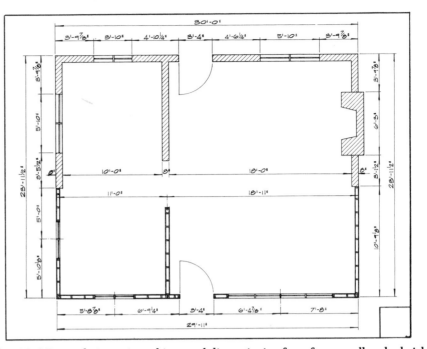

Fig. 20-25. Techniques in architectural dimensioning for a frame wall and a brick wall.

Figure 20-24 shows one of the many special-purpose templates. These have door, window, landscape, furniture, structural steel–shape, plumbing, and electrical symbols.

◼ DIMENSIONING TECHNIQUES

Structural wall openings are dimensioned in similar ways on the plan drawings for frame and brick-veneer structures. The general dimensioning techniques discussed in Chap. 6 are used. However, the dimension line is unbroken. In addition, the dimensions appear above the dimension line, as shown in Fig. 20-25. Dimensions end with the usual arrowhead or with several other symbols, as shown in Fig. 20-26. Note that dimension lines may cross when the sizes of interior rooms are given. The plan has only width and depth

dimensions, as shown. Any needed heights are shown as floor-to-ceiling dimensions on one of the elevations or on a suitable wall section.

To help simplify and standardize designing buildings, the building industry and the American National Standards Institute established a standard, Modular Coordination,

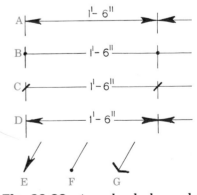

Fig. 20-26. Arrowheads, heavy dots, and heavy 45° slashes are used in architectural dimensioning.

A62.1. This standard was sponsored by the American Institute of Architects. The basic customary-inch module is 4 in. for all U.S. building materials and products. The International Standards Organization has established 100 mm (almost 4 in.) as the module for countries using the metric system. Modular components are typically designed on a 4-in. or 100-mm centerline. All buildings currently designed for customary modular specifications are planned with multiples of 4 in. Modular planning reduces building costs and time. This is because materials do not have to be cut to many different sizes.

The dimensions used on architectural drawings are either finish or construction details. The elevations generally have the expected finished dimensions. The floor plan and sectional details have either construction or finish dimensions, as shown in Fig. 20-27.

Dimension lines are placed about 3/8 in. (10 mm) apart. They are needed for overall finished building and room dimensions. The structural openings and wall offsets are defined with detailed construction dimensions. Dimension lines showing the distances between centerlines (for modules) or extension lines (for general design) usually end in arrowheads. Dimensions are placed above the line and given in feet and inches, as shown. Note the different ways of showing dimensions that are acceptable for architectural drawings (Fig. 20-28).

SCALES AND DIMENSIONING RELATIONSHIP

Residential designs and details are usually developed with the aid of reduction scales. The architect's scale

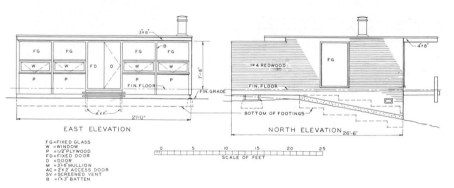

Fig. 20-27. Elevations have very few essential dimensions. (Adapted from drawings of Henry Hill, architect, AIA, and John W. Kruse, associate, AIA, San Francisco, CA.)

is discussed in Chap. 3. The 1/4″=1′-0 (1:50 in metric) scale is best suited for plans for houses and small buildings. The usual scale for larger buildings is 1/8″=1′-0 (1:100 in metric). Plot plans may be drawn at 1/10″=1′-0 or 1/32″=1′-0, but it is better to use an engineer's scale at 1″=20′, 1″=30′, or 1″=40′ (1:200, 1:300, or 1:400 in metric). Enlarged details are developed at 1″=1′ or 1½″=1′. Sectional details are defined at 1/2″=1′ or 3/4″=1′. Some details may require half-size or full-size drawings. Metric

scale for enlarged details may be 1:5, 1:10, or 1:20.

MATERIALS DESIGN AND DETAIL

The new environments designed by architects are created from materials. Detailed drawings and specifications must show how these materials are to be fabricated and constructed. The *Sweets Architectural Catalog* for materials manufac-

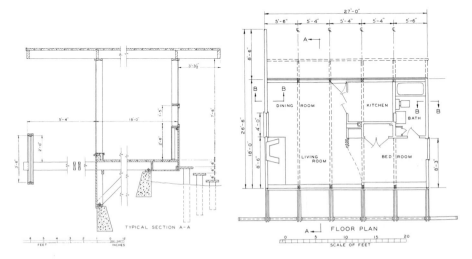

Fig. 20-28. Dimensions that apply to section and plan drawings. (Adapted from drawings of Henry Hill, architect, AIA, and John W. Kruse, associate, AIA, San Francisco, CA.)

Lumber				
Nominal size	2 × 4	2 × 6	2 × 8	2 × 10	2 × 12
Dressed size	1½″ × 3½″	1½″ × 5½″	1½″ × 7½″	1½″ × 9½″	1½″ × 11½″
Nominal size	4 × 6	4 × 8	4 × 10	6 × 6	6 × 8
Dressed size	3⁹⁄₁₆″ × 5½″	3⁹⁄₁₆″ × 7½″	3⁹⁄₁₆″ × 9½″	5½″ × 5½″	5½″ × 7½″
Nominal size	6 × 10	8 × 8	8 × 10		
Dressed size	5½″ × 9½″	7½″ × 7½″	7½″ × 9½″		

Boards				
Nominal size	1 × 4	1 × 6	1 × 8	1 × 10	1 × 12
Actual size, common boards	¾″ × 3⁹⁄₁₆″	¾″ × 5⁹⁄₁₆″	¾″ × 7½″	¾″ × 9½″	¾″ × 11½″
Actual size, shiplap	¾″ × 3″	¾″ × 4¹⁵⁄₁₆″	¾″ × 6⅞″	¾″ × 8⅞″	¾″ × 10⅞″
Actual size, tongue-and-groove	¾″ × 3¼″	¾″ × 5³⁄₁₆″	¾″ × 7⅛″	¾″ × 9⅛″	¾″ × 11⅛″

Fig. 20-29. Standard sizes of lumber and boards (customary U.S. measure).

turers is one of the most important tools of the design team. Many materials are available in standard units for the building trades. Others are custom-designed. A few standards are discussed here.

Lumber

Lumber may be specified by *nominal* dimensions. These differ from the *actual* dimensions of the surfaced wood. For example, most of the lumber and boards for residential construction are surfaced on four sides. The *dressed* (finished) sizes are noted in Fig. 20-29.

Masonry

Figure 20-30 gives the sizes of brick building materials. The common brick has modular dimensions. Brick, block, stone, and stucco may be used for exterior and interior walls and floors. Most of these materials can also serve as structural load-bearing walls for supporting

floors and roofs. They generally can be *bonded* (overlapped) so as to increase their structural strength, as shown in Fig. 20-31. Bonding also

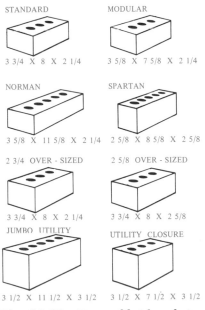

STANDARD
3 3/4 X 8 X 2 1/4

MODULAR
3 5/8 X 7 5/8 X 2 1/4

NORMAN
3 5/8 X 11 5/8 X 2 1/4

SPARTAN
2 5/8 X 8 5/8 X 2 5/8

2 3/4 OVER - SIZED
3 3/4 X 8 X 2 1/4

2 5/8 OVER - SIZED
3 3/4 X 8 X 2 5/8

JUMBO UTILITY
3 1/2 X 11 1/2 X 3 1/2

UTILITY CLOSURE
3 1/2 X 7 1/2 X 3 1/2

Fig. 20-30. Types of brick and sizes available.

forms decorative patterns. Well-designed masonry needs little upkeep. Its colors and patterns are important parts of the overall architectural design.

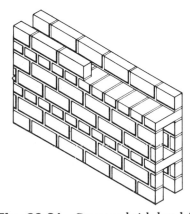

8" ALL ROLOK WALL
COMMON BOND

Fig. 20-31. Common brick bond for building a wall. (Structural Clay Products Institute.)

1. Gable end	22. Shutters
2. Louver	23. Exterior trim
3. Interior trim	24. Waterproofing
4. Shingles	25. Foundation wall
5. Chimney cap	26. Column
6. Flue linings	27. Joists
7. Flashing	28. Basement floor
8. Roofing felt	29. Gravel fill
9. Roof sheathing	30. Heating plant
10. Ridge board	31. Footing
11. Rafters	32. Drain tile
12. Roof valley	33. Girder
13. Dormer window	34. Stairway
14. Interior wall finish	35. Subfloor
15. Studs	36. Hearth
16. Insulation	37. Building paper
17. Diagonal sheathing	38. Finish floor
18. Sheathing paper	39. Fireplace
19. Window frame and sash	40. Downspout
20. Corner board	41. Gutter
21. Siding	42. Bridging

Fig. 20-32. The typical parts of a frame house. (From National Bureau of Standards.)

Concrete

Concrete may be thought of as a structural unit. As footing, foundation walls, and poured-in forms, it supports floor loads. Both interior and exterior walls can be formed from concrete. Concrete can be precast to give certain finishes. Color, texture, and pattern all work together to create a desired "look" for a wall.

■ PARTS OF A HOUSE

The main parts of a house are shown in Fig. 20-32. Every house does not have all of these parts. Some parts may be made of different materials. The typical wood framing of an exterior wall is shown in Fig. 20-33A. The framing begins on the foundation wall with the sill, header, and floor plate. Then the stud wall is erected. Next, sheathing, plywood, or insulation board is put on. Many kinds of facing can be used. Horizontal or vertical siding is typical. In addition, shakes or shingles are often used. Also common today is brick veneer, shown in Fig. 20-33B.

Housing Frame

The framework of a building must be strong and rigid to ensure low main-

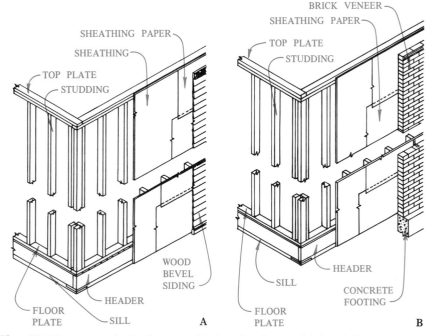

Fig. 20-33. Pictorial of a frame wall (A) with siding and (B) with brick veneer.

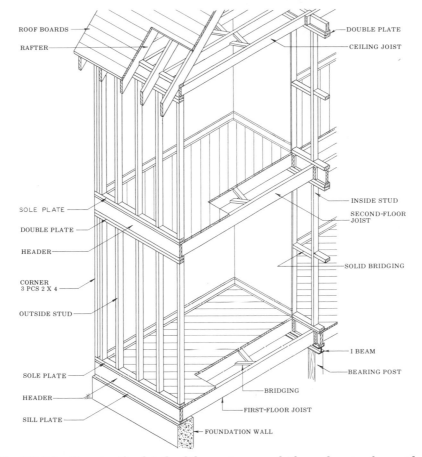

ROOF BOARDS
RAFTER
DOUBLE PLATE
CEILING JOIST
SOLE PLATE
DOUBLE PLATE
HEADER
CORNER
3 PCS 2 X 4
OUTSIDE STUD
INSIDE STUD
SECOND-FLOOR
JOIST
SOLID BRIDGING
SOLE PLATE
HEADER
SILL PLATE
I BEAM
BEARING POST
BRIDGING
FIRST-FLOOR JOIST
FOUNDATION WALL

Fig. 20-34. Examine the details of the western, or platform, framing for residential design.

tenance costs over many years. Even a prefabricated home and a custom-designed home have some features in common.

Western Framing

In western, or platform, framing, each floor is framed separately (Fig. 20-34). The first floor is built on top of the foundation wall as a platform. Studs are one story high. They are used to develop and support the framework for the second story and the load-bearing interior walls.

Balloon Framing

In balloon framing, the studs are two stories high, as in Fig. 20-35. A false

girt inserted in the stud wall carries the second-floor joists. A box sill is used. Diagonal bracing brings rigidity to the corners. This system is not commonly used today. However, architects must know about it for remodeling older homes.

Plank and Beam Framing

Plank and beam framing (Fig. 20-36) uses heavier posts and beams than the other systems. These members carry a deck of continuous planking. This kind of framing allows ceilings to be higher and more open, with fewer supporting members. It also generally costs less to build.

Sill Construction

Figure 20-37 shows different types of sill and wall construction. At A, the frame wall is set up on a box-sill construction. Note the metal termite shield atop the foundation wall. At B, a brick veneer starts below the floor line on a stepped foundation wall. At C, the slab construction is reinforced with a wire mesh.

Corner Studs and Sheathing

Some typical corner bracing is shown in Fig. 20-38. Diagonal sheathing was formerly the most common kind of bracing. Today, however, to save labor, builders use horizontal

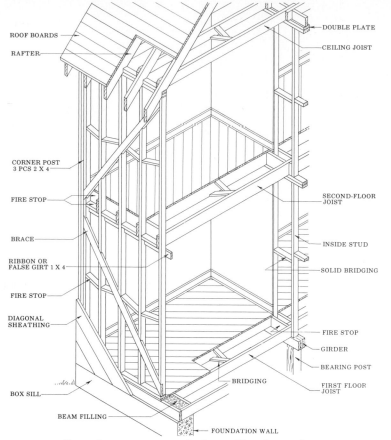

ROOF BOARDS

RAFTER

DOUBLE PLATE

CEILING JOIST

CORNER POST
3 PCS 2 X 4

FIRE STOP

BRACE

RIBBON OR
FALSE GIRT 1 X 4

FIRE STOP

DIAGONAL
SHEATHING

BOX SILL

BEAM FILLING

SECOND-FLOOR
JOIST

INSIDE STUD

SOLID BRIDGING

FIRE STOP

GIRDER

BEARING POST

BRIDGING

FIRST FLOOR
JOIST

FOUNDATION WALL

Fig. 20-35. Balloon framing is typical of an older form of two-story construction.

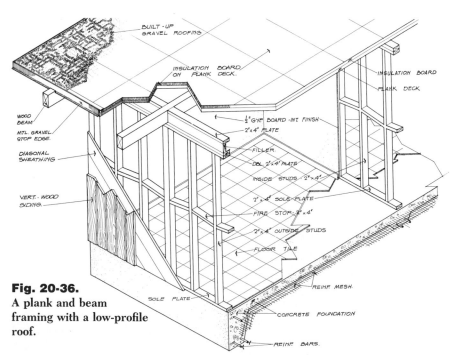

BUILT-UP
GRAVEL ROOFING

INSULATION BOARD
ON PLANK DECK.

INSULATION BOARD

PLANK DECK

WOOD
BEAM

MTL. GRAVEL
STOP EDGE.

DIAGONAL
SHEATHING

VERT. WOOD
SIDING.

½" GYP BOARD-INT FINISH.

2"x 4" PLATE

FILLER

DBL. 2"x 4" PLATE

INSIDE STUDS 2"x 4"

2"x 4" SOLE PLATE

FIRE STOP-2"x 4"

2"x 4" OUTSIDE STUDS

FLOOR TILE

REINF. MESH.

SOLE PLATE

CONCRETE FOUNDATION

REINF BARS.

Fig. 20-36.
A plank and beam framing with a low-profile roof.

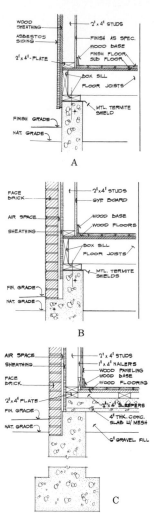

A

WOOD SHEATHING
ASBESTOS SIDING
2"x 4"-PLATE
2"x 4" STUDS
FINISH AS SPEC.
WOOD BASE
FINISH FLOOR
SUB FLOOR
BOX SILL
FLOOR JOISTS
MTL. TERMITE SHIELD
FINISH GRADE
NAT. GRADE

B

FACE BRICK
AIR SPACE
SHEATHING
2"x 4" STUDS
GYP BOARD
WOOD BASE
WOOD FLOORS
BOX SILL
FLOOR JOISTS
MTL. TERMITE SHIELDS
FIN. GRADE
NAT. GRADE

C

AIR SPACE
SHEATHING
FACE BRICK
2"x 4" PLATE
FIN. GRADE
NAT. GRADE
2"x 4" STUDS
1"x 4" NAILERS
WOOD PANELING
WOOD BASE
WOOD FLOORING
2"x 4" SLEEPERS
4" THK. CONC. SLAB W/ MESH
6" GRAVEL FILL

Fig. 20-37. Various types of sill construction. (**A**) Box-sill with a frame wall, (**B**) stepped footing with a brick-veneer wall, (**C**) stepped footing with a brick-veneer wall and poured-concrete slab floor.

sheathing and plywood along with modular insulation board. Plywood is used not only on exterior walls but also for *interior decking* (subflooring) and roof sheathing. Anchor bolts secure the superstructure to the substructure. These are normally ½- to ¾-in. (12- to 19-mm) bolts. They are placed about 8 ft (2440 mm) apart and extend 18 in. (457 mm) into the concrete.

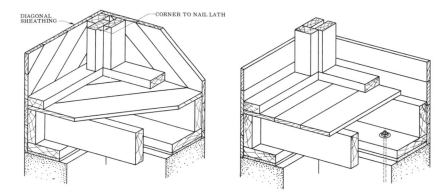

Fig. 20-38. Arrangement of corner studs in forming a frame wall.

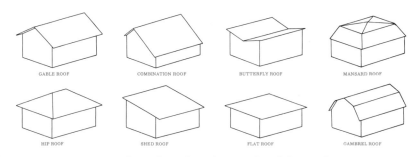

Fig. 20-39. Some typical residential roof types that define style.

Roof Designs

Some basic roof types are shown and named in Fig. 20-39. The shape of the roof is often the key to a building's architectural style. The common shapes are the gable, hip, flat, and shed. The mansard, gambrel, butterfly combination, clerestory, and A-frame shapes are used more with specific design styles. The verti-cal measurement of a roof is called the *rise*. The horizontal dimension is called the *run*. Together, they determine the *roof pitch* (slope).

Roof-Framing Sections

Figure 20-40 shows some typical cornice details. Note the terms used for the members that finish off the joints between the wall and roof. Some rafters are open at the ends. Others are boxed. There is also built-up flat roofing for plank and beam construction. Aluminum and galvanized gutters remain common. One detail shows a built-in, metal-lined gutter. This is a costly detail often included on formal designs. The note COND. @ BRICK (condition at brick) means that a brick-veneer construction may be added.

Stairway Framing and Detail

Three types of stairs are (1) the straight run, (2) the platform, and (3) the circular. Stairs are made up of *risers* (the vertical part of each step) and *treads* (the horizontal part of each step). These parts are illustrated in Fig. 20-41. The *rise* of a

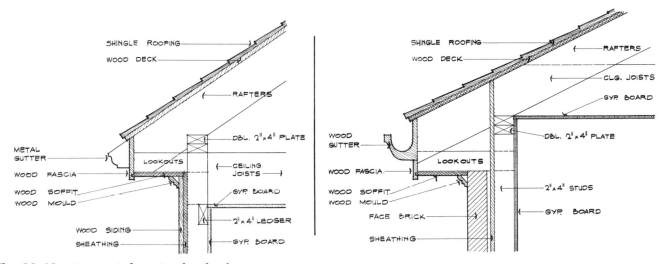

Fig. 20-40. Four typical cornice details, showing construction.

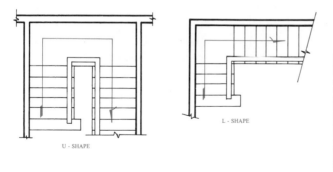

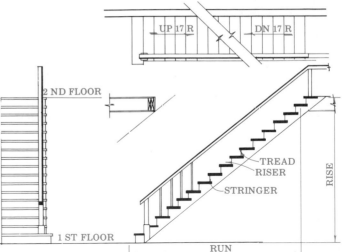

Fig. 20-41. Stair details with terms and layouts.

flight of stairs is the height measured from the top of one floor to the top of the next floor. The *run* is the horizontal distance from the face of the first riser to the face of the last riser in one stairwell. It also equals the sum of the width of the treads.

Risers are generally 6½ to 7½ in. (165 to 190 mm) high. The width of treads is such that the sum of one riser and one tread is about 17 to 18 in. (430 to 480 mm). A 7-in. (180-mm) riser and an 11-in. (280-mm) tread is considered a general standard. You can easily use a scale to divide the floor-to-floor height into the number of risers. A good stairway feels comfortable and safe to use. A simple rule for building a safe stairway is to keep the angle of incline between 28° and 35°.

On working drawings, stairways are usually not drawn in their entirety. Instead, break lines are used, and the drawing shows what is on the level beneath the stairs.

Doors

Doors are usually 6 ft 8 in. or 7 ft 0 in. (2000 or 2100 mm) high. The width may vary from 2 ft 0 in. to 3 ft 0 in. (600 to 900 mm), but it is usually 2 ft 8 in. or 3 ft 0 in. (800 to 900 mm). The thickness varies from 1⅜ to 1¾ in. (35 to 45 mm) for interior doors, and from 1¾ to 2½ in. (45 to 65 mm) for exterior doors. The head, jamb, and sill details may vary depending on whether a swinging, sliding, or folding door is used. The door must fit its frame closely. Yet it must also open and close easily. Figure 20-42 shows various patterns for doors.

Doors to the outside are usually larger than others to allow for heavier use and for bringing in furniture. These doors are usually 3 ft 0 in. (900 mm) wide. Bedroom doors are usually 2 ft 6 in. (762 mm). Bathroom doors run 2 ft 4 in. (710 mm). Bifold and folding accordian doors have special features.

Windows

The style of a house determines what style of window is used and how windows are placed. Double-hung and casement windows are practical in most kinds of houses. However, horizontal sliding windows are very popular. Figure 20-43 shows the various types available to the architect.

Casement windows are popular for French and English designs. They are hinged at the sides to swing open (in or out). Hopper windows are hinged at the bottom. Awning windows are hinged at the top. Both hopper and awning windows are also called *projected windows*. Sliding windows move sidewise. Double-hung windows are commonly used for Colonial and American-style structures. Each double-hung window contains two independent sashes that can move in a vertical track. A counterbalance holds them at any position desired. Newer windows can have a press-in, spring-loaded track that is very convenient. Fixed windows and *jalousies* (windows made of adjustable glass slats) are special kinds of windows. Figure 20-44 shows sectional details and some technical terms for a window. Normally, windows are placed in walls so that their tops line up with the top of the door.

■ WORKING DRAWINGS

These are the most important class of drawings. They include plans, elevations, sections, schedules, schematics, and details. Along with the

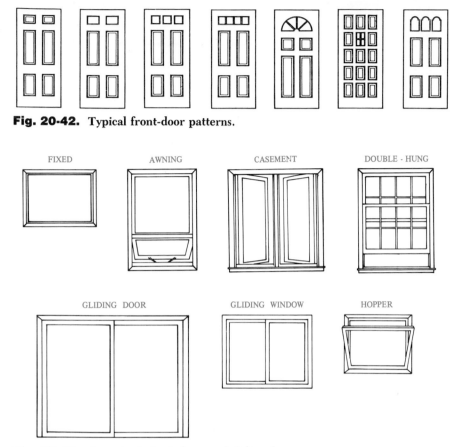

Fig. 20-42. Typical front-door patterns.

FIXED AWNING CASEMENT DOUBLE - HUNG

GLIDING DOOR GLIDING WINDOW HOPPER

Fig. 20-43. Some typical windows and sliding doors.

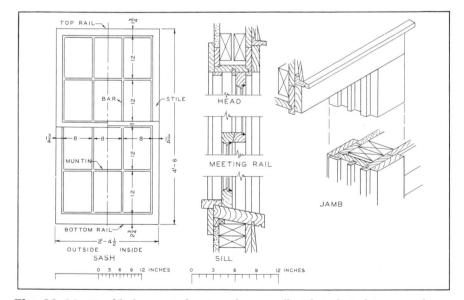

TOP RAIL

BAR

STILE

MUNTIN

BOTTOM RAIL

OUTSIDE INSIDE
SASH

HEAD

MEETING RAIL

SILL

JAMB

0 3 6 9 12 INCHES   0 3 6 9 12 INCHES

Fig. 20-44. Double-hung window in a frame wall with technical terms and sectional details.

specifications for materials and finish, they are the guides used in the construction and erection of a building. Figure 20-45 is a preliminary design for a ranch house. Figures 20-46 through 20-51 are the working drawings developed from this design. They form a complete set of plans for the house.

Plans

The basement plan in Fig. 20-46 serves as a guide for constructing the foundation. Therefore, it must be completely dimensioned. It should be checked with the first-floor plan and can be developed from it. Note the foundations for the porch and garage. Window placement depends on structural needs.

The first-floor plan in Fig. 20-47 is a horizontal section taken above the floor. It shows all walls, doors, windows, and other structural features. It also shows fixed features such as the cabinets, stairways, heating and plumbing fixtures, lighting outlets in the walls, and ceilings. Frame walls are drawn to what is shown on the scale as 6 in. thick. Windows and doors are located properly, then indicated by conventional symbols. Their sizes are listed on schedules that accompany the plans.

Elevations

Figure 20-48 shows front and side elevations. A complete set of plans would include four elevations. Elevations show the exterior look of a house, floor and ceiling heights, openings for windows, doors, roof pitch, and selected materials. To draw an elevation, start with the grade line. Then locate the center lines that indicate the finished working dimensions. Plot the doors, windows, and other openings from the floor plan.

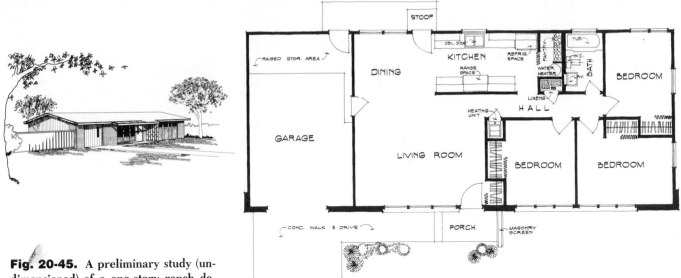

Fig. 20-45. A preliminary study (un-dimensioned) of a one-story ranch developed for client approval.

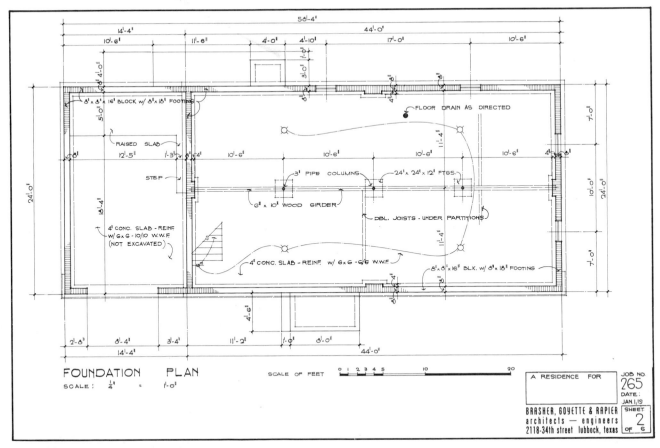

Fig. 20-46. A foundation plan with structural notes and dimensions ready for contractor approval.

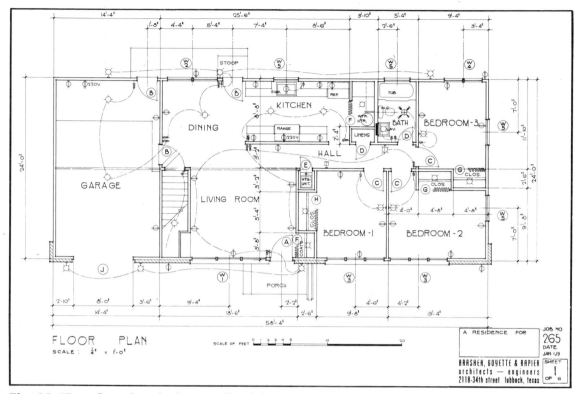

Fig. 20-47. A floor plan of a frame wall with brick veneer on the front wall and around the corners.

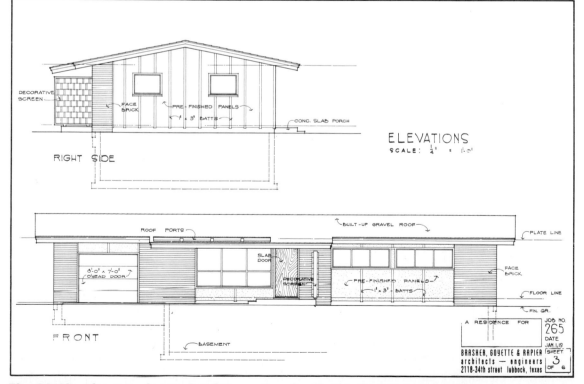

Fig. 20-48. Elevations showing foundation walls and roof pitch of 2.5:12 (gravel roof).

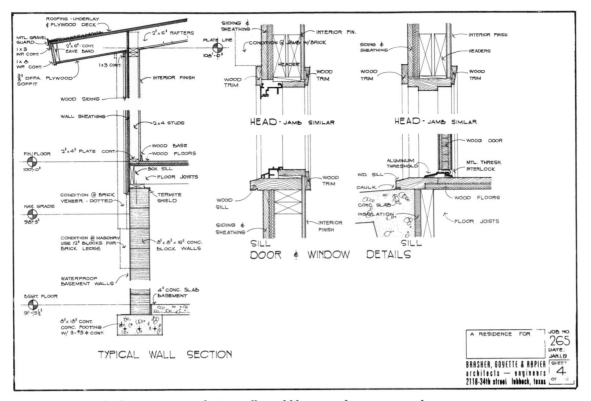

Fig. 20-49. Typical sections. Foundation walls could be poured concrete, as shown in Figure 20-37 at B.

Sections

Figure 20-49 details a typical wall section from substructure to super-structure. This drawing, to a larger scale than that of the elevation, shows the wall in a much clearer and more detailed view. Door and window details are also included. They are developed as the building progresses. That way the millwork and custom framework will assemble readily.

Electrical Planning

Figure 20-50 is a plan of the electrical wiring. It shows the circuits for 110-V and 220-V service. It also shows the electric outlets and switches located for the major appliances. A key is included for the various symbols. Additional data would

then be listed in the formal specifications.

Site Plan and Schedules

Figure 20-51 is the site plan. A site plan shows the lot and locates the house on it. It should give complete and accurate dimensions. It should also show all driveways, sidewalks, and other pertinent information required by the building inspector. The site plan in the figure is for an ordinary urban lot. Note the roof plan. Its center ridge represents a gable roof. What scale would be best for drawing this site plan?

Figure 20-51 also includes schedules describing the five types of windows and nine types of doors used in the house. Another, perhaps more

important, schedule lists the finishes required for the floors, walls, and ceilings.

Plumbing Symbols

Some of the standard symbols are shown on the preceding working drawings. There are also standard schematics and diagrams for plumbing. These are listed in the latest edition of *Architectural Graphic Standards*. These standards are periodically revised by the American Institute of Architects.

Electrical Symbols

Additional electrical symbols are shown in Fig. 23-37. These are typically used on house plans to show the basic circuit layout needed.

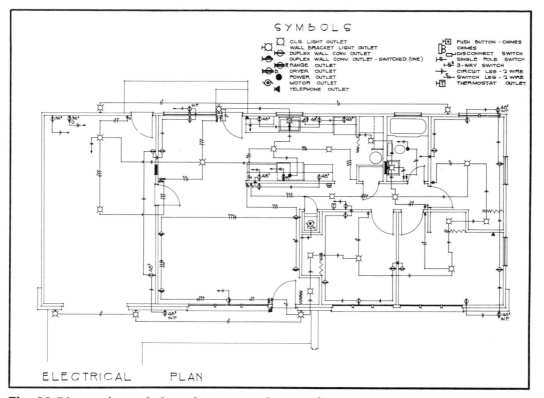

Fig. 20-50. An electrical plan indicates circuit layout and services.

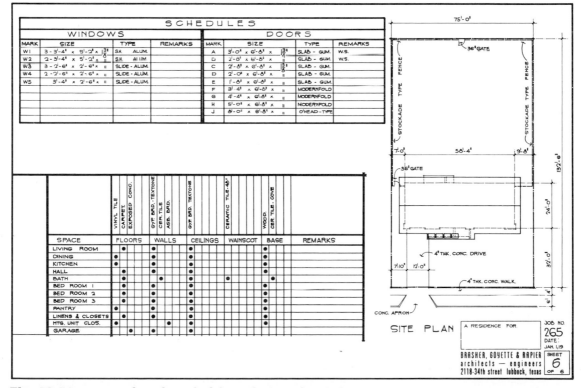

Fig. 20-51. Door and window schedules with site and roof plan.

Fig. 20-52. Contemporary French mansard–styled residence. Can you imagine this house with a hip roof? (Orrin Dressler, designer-contractor.)

■ WORKING DRAWINGS FOR A CONTEMPORARY MULTILEVEL RESIDENCE

Figure 20-52 shows a home with a French mansard roof. This kind of roof is shaped to fit and hug a building. It is textured in natural wood shakes to give a feeling of warmth. The site plan (Fig. 20-53) is contoured to show the gentle roll of the land. The family-room (lower-level), main-level (Fig. 20-54), and upper-level (Fig. 20-55) plans are arranged

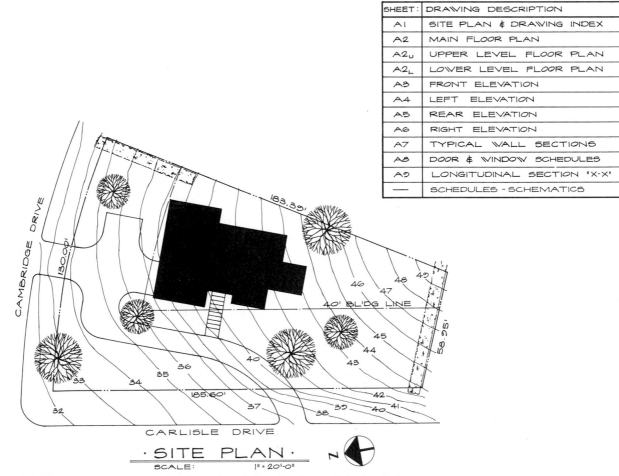

SHEET:	DRAWING DESCRIPTION
A1	SITE PLAN & DRAWING INDEX
A2	MAIN FLOOR PLAN
A2$_U$	UPPER LEVEL FLOOR PLAN
A2$_L$	LOWER LEVEL FLOOR PLAN
A3	FRONT ELEVATION
A4	LEFT ELEVATION
A5	REAR ELEVATION
A6	RIGHT ELEVATION
A7	TYPICAL WALL SECTIONS
A8	DOOR & WINDOW SCHEDULES
A9	LONGITUDINAL SECTION "X-X"
—	SCHEDULES - SCHEMATICS

· SITE PLAN ·
SCALE: 1" = 20'-0"

Fig. 20-53. Site plan with contour lines that define the slope of the land. (Orrin Dressler, designer-contractor.)

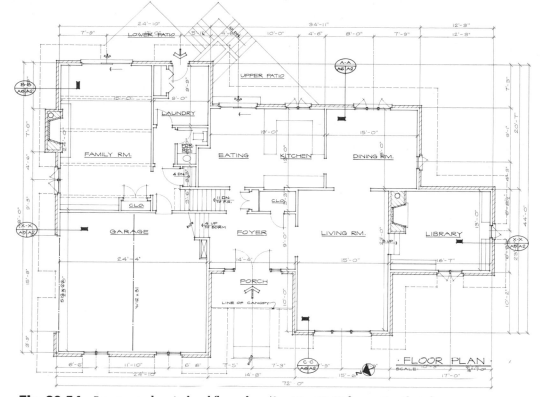

Fig. 20-54. Lower- and main-level floor plan. (See Fig. 20-60 for section that shows levels. (Orrin Dressler, designer-contractor.)

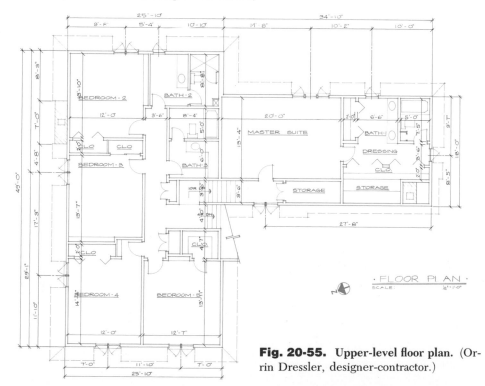

Fig. 20-55. Upper-level floor plan. (Orrin Dressler, designer-contractor.)

Fig. 20-56. Front elevation designed into sloping site. (Orrin Dressler, designer-contractor.)

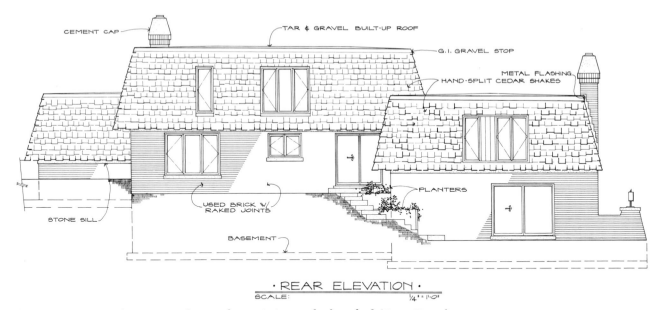

Fig. 20-57. Rear elevation. Is there a door missing to the laundry? (Orrin Dressler, designer-contractor.)

gracefully around a central core of stairs and hallways. The front elevation (Fig. 20-56) shows the rolling contour of the site. Note the line weights on the cedar shakes that make up the roof. These add depth to the view. The rear elevation (Fig. 20-57) shows contoured stairs between the main and lower levels. The right and left elevations are shown in Figs. 20-58 and 20-59. The longitudinal section (Fig. 20-60) explains the arrangement of levels and rooms. The sections in Fig. 20-61 are identified by symbols on the main floor plan (Fig. 20-54).

Architectural specifications are more detailed written instructions

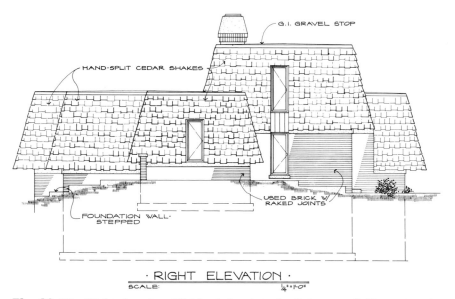

G.I. GRAVEL STOP

HAND-SPLIT CEDAR SHAKES

USED BRICK W/ RAKED JOINTS

FOUNDATION WALL- STEPPED

· RIGHT ELEVATION ·

SCALE: ¼" = 1'-0"

Fig. 20-58. Right elevation. Which window is in the dining room? (Orrin Dressler, designer-contractor.)

G.I. GRAVEL STOP

TAR & GRAVEL BUILT-UP ROOF

USED BRICK- RAKED JOINTS

TOP OF FOUND.

ROUGH-SAWN VERT. CEDAR SID.

16'-0" x 7'-0" FLUSH OVERHEAD DOOR

· LEFT ELEVATION ·

SCALE: ¼" = 1'-0"

Fig. 20-59. Left elevation. (Orrin Dressler, designer-contractor.)

that should accompany any set of plans. They are essential for turning these plans into a building. They note the general conditions of the site. They also specify the materials to the client's and contractor's mu-

tual agreement. Guidelines for specifications have been drawn up by the American Institute of Architects. These guidelines are available through local state chapters of the AIA.

■ THE STRUCTURAL DRAFTER

The structural drafter is usually a member of an engineering team. This team often works together with

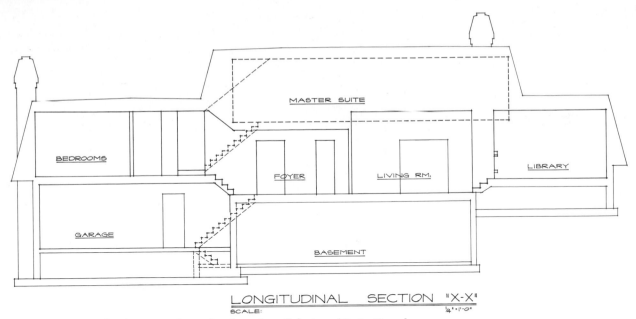

LONGITUDINAL SECTION "X-X"
SCALE: ¼" = 1'-0"

Fig. 20-60. Longitudinal section shows levels and roof design. (Orrin Dressler, designer-contractor.)

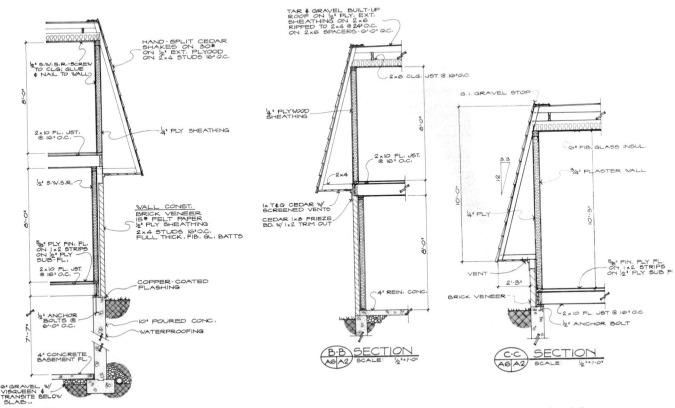

Fig. 20-61. Sectional views, A-A, B-B, and C-C, as shown on the main-level floor plan. (Orrin Dressler, designer-contractor.)

other teams under the direction of a project manager or a job superintendent. As an example of this teamwork, structural and architectural designers often combine their efforts. The architectural designer designs the form of a building based on the function it will have (Fig. 20-62). Then the structural designer designs the frame of the building to fit this form (Fig. 20-63). The work of the structural drafter is very important in engineering. The construction of buildings, bridges, and other structures depends on the detailed instructions in structural drawings.

A structural drafter usually works at one of the following five jobs:

1. Drafting details in an architect's or engineer's office
2. Preparing construction details for a contractor (making the shop drawings for a construction company)
3. Drafting structural details for a manufacturer of structural materials
4. Working for the engineering department of a manufacturing plant that maintains its own engineering operations
5. Preparing drawings for government or other agencies that regulate the construction and design of public buildings, bridges, dams, and other structures

Career opportunities for structural drafters generally depend on how much practical experience they gain as junior members of engineering teams. Structural drafters can be promoted to such jobs as structural detail checker, estimator, chief drafter, construction supervisor, or building inspector (Fig. 20-64).

Further career opportunities can be gained through continuing education. Formal training at a junior col-

Fig. 20-62. The John Deere & Company administrative center, designed by Eero Saarinen and Associates, required many large sheets of structural details. (John Deere & Co.)

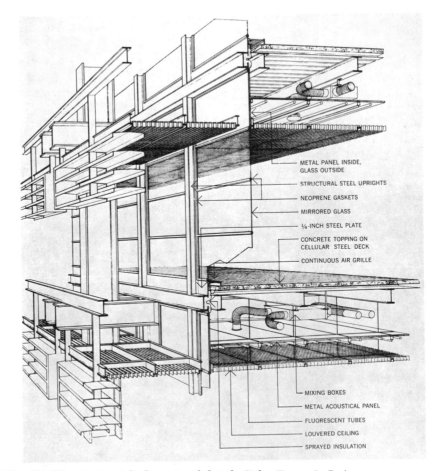

METAL PANEL INSIDE, GLASS OUTSIDE
STRUCTURAL STEEL UPRIGHTS
NEOPRENE GASKETS
MIRRORED GLASS
¼-INCH STEEL PLATE
CONCRETE TOPPING ON CELLULAR STEEL DECK
CONTINUOUS AIR GRILLE

MIXING BOXES
METAL ACOUSTICAL PANEL
FLUORESCENT TUBES
LOUVERED CEILING
SPRAYED INSULATION

Fig. 20-63. A pictorial of structural detail. (John Deere & Co.)

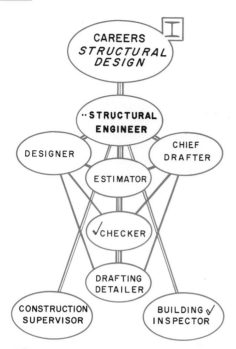

Fig. 20-64. Career opportunities in structural design.

lege or university is very helpful. To succeed as a structural drafter, you must have the ability to design and detail structural components and have an aptitude for mechanics.

STRUCTURAL MATERIALS

Designers and detail drafters must be familiar with a great many structural materials. In addition, all members of the engineering team must learn about new construction materials and systems as they are developed. All structural materials have special ways of being assembled and fastened together. These ways must be considered whenever accurate drafting details are prepared (Fig. 20-65).

The basic structural materials used today are steel, wood, concrete, structural clay products, and stone

masonry. Different materials have different characteristics. It is the designer's job to choose the right combination of materials to bear the stresses imposed by a building.

STEEL AS A STRUCTURAL MATERIAL

Steel makes a good construction material because of the shapes into which it can be formed at the mill. Steel shapes are produced in rolling mills. They are then shipped to fabrication shops where they are cut to specific lengths and where connections are prepared. Some of the basic steel shapes are the wide-flanged W-shape used as a beam or column; the S-shape, formerly known as an I-beam; the C-shape, formerly known as the channel; and the L-shape, formerly known as the angle.

These steel shapes have framed the skyscrapers of our cities for nearly three quarters of a century. The American Institute of Steel Construction maintains regional offices from coast to coast to help provide guidelines for designing and building steel structures.

A Steel-Framed Village

A completed A-frame building is shown in Fig. 20-66. Each of the 13 steel A-frames is approximately 220 ft (67 000 mm) across at the base, 135 ft (41 150 mm) across at the top, and 15 ft (4570 mm) across the vertical bents. It is built of tubes, wide-flanged sections, and chords made of 18 × 26 in. (455 × 660 mm) tubes. Each A-frame was assembled on the site and erected in five pieces. All the steel members are connected with high-strength bolts (HSB).

The A-frames were designed with the aid of a computer to hold up under many different combinations of *loads* (weights or pressures borne by a structure). These include wind loads, temperature changes, the *dead load* (weight of the building), and the *live load* (weights added temporarily, such as furniture). If the building is designed correctly, these loads are carried through the structure to the ground. For the building in Fig. 20-66, which is located in Florida, the designer had to allow for wind loads from hurricane winds. Allowance must always be made for loads created by geographical location, the prevailing weather, the function of the building, and the

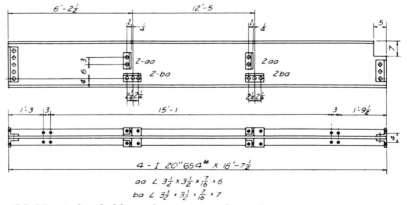

Fig. 20-65. A detailed beam has accurate dimensions.

Fig. 20-66. The structural shapes are dependent on structural steel detail. (United States Steel Corp.)

size and shape of the form determined by the architectural designer.

A Steel-Framed Bridge

Steel forms the framework of a bridge over a deep gorge of the Rio Grande near Taos, New Mexico (Fig. 20-67). The bridge had a rigid structural framing made of high-strength steel fastened with high-strength bolts and welds. More than 1900 customary tons, or 1725 metric tons (t), of structural steel were used in its construction. The center span is 600 ft (183 m) long. The two side spans are each 300 ft (91.5 m) long. The distance from the canyon floor to the bridge floor is 600 ft (183 m).

A typical welded-steel member of this bridge appears in Fig. 20-68.

■ STEEL SYSTEMS

Using steel, engineers have developed a number of new structural systems. The Unistrut Space Frame shown in Fig. 20-69 consists of four or five modular units in a geometric pattern. The basic unit is made of four or five parts that are bolted together. The system is used mainly in canopies and roofs. Note how the geometric pattern of the exposed steel becomes a part of the overall design. This roof is in a mall in Columbia, Maryland.

■ STRUCTURAL DOME SYSTEMS

Figure 20-70 shows a dome over a theater in Reno, Nevada. The geometry used in this structure is called

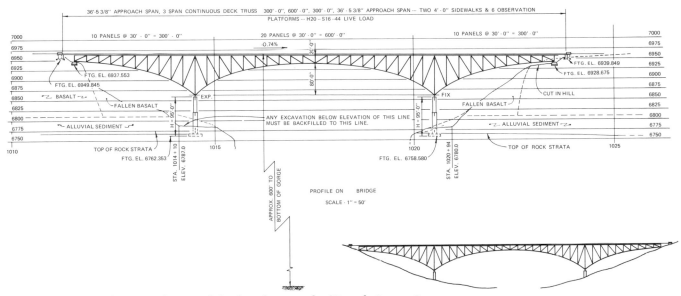

Fig. 20-67. A schematic design of bridge framework. (United States Steel Corp.)

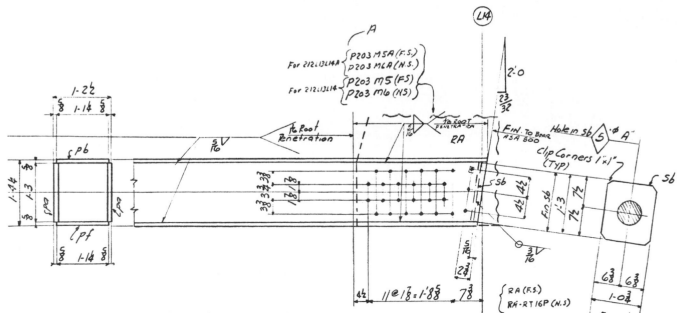

Fig. 20-68. A welded beam for the bridge. (United States Steel Corp.)

geodesic. It is different from other dome geometry in that its strength is in all directions (omnidirectional). The three-dimensional triangulated framing of the dome (Fig. 20-71) makes it exceptionally strong. It also uses less material than other dome frames. Geodesic geometry was invented by Dr. R. Buckminster Fuller.

Fig. 20-69. Space forms are geometrically developed to shape buildings. (Unistrut Corp.)

Fig. 20-70. A flexible geometric steel dome for a theater. (Temcor.)

■ STRUCTURAL DRAFTING OF STEEL SHAPES

Earlier in this chapter, we learned some of the basic structural steel shapes. The American Institute of Steel Construction (AISC), 101 Park Avenue, New York, NY 10017, pub-

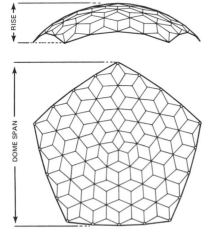

Fig. 20-71. The dome has geometric characteristics. (Temcor.)

lishes the *Manual of Steel Construction (AISC Manual)*. The seventh edition of the *AISC Manual*, or any handbook published by a major steel company, lists all the major shapes available and the great variety of their sizes and weights. The *AISC Manual* contains tables for designing and detailing the various shapes in any combination. Figure 20-72 shows cross sections of various plain steel shapes. These shapes are grouped by the AISC as follows:

1. American Standard beams (S).
2. American Standard channels (C).
3. Miscellaneous channels (MC). These include special-purpose channels that are not standard.
4. Wide-flange shapes (W). These are used as both beams and columns.
5. Miscellaneous shapes (M). These lightweight shapes look in cross section like W shapes.
6. Structural tees (ST, WT, MT). These are made by splitting S, W, and M shapes, usually along the middepth of their webs.
7. Angles (L). These consist of two legs of equal or unequal widths. The legs are at right angles to each other.
8. Plates (PL) and flat bars (Bar). These are rectangular in cross section.

These plain shapes are basic to structural detailing. You must be familiar with them in order to make adequate drawings.

■ SCALES

Structural details are prepared at a scale of 1″ = 1′-0″ 1:10 for beams up to 21 in. (533 mm) in depth. For beams of greater depth, a ¾″ = 1′-0″

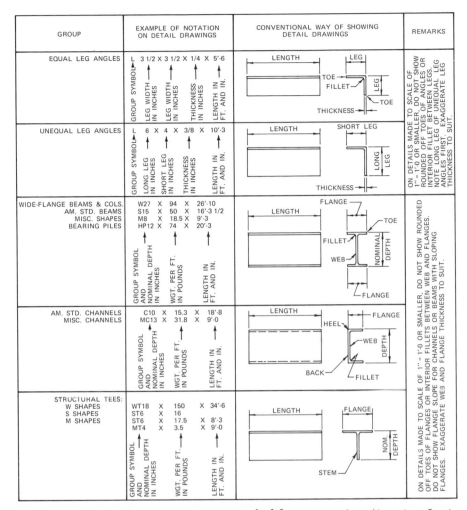

Fig. 20-72. Steel shapes in section are stocked for construction. (American Institute of Steel Construction.)

1:20 scale is preferred. The overall length of structural members can be shortened as long as details are shown adequately. Also, very small dimensions, such as a clearance, can be exaggerated to clarify views.

■ TYPICAL STEEL DETAILS ON THE DESIGN DRAWINGS

Designers always place on their design drawings all the information needed to prepare shop drawings.

Figure 20-73 shows a small part of a designed floor plan. The view is from above. There are enough notes and dimensions on the plan to prepare a shop drawing of the wide-flanged beam (W).

The 20-ft dimension is presumed to be the structural bay, or distance from *A* to *B*. The structural members are at right angles to one another unless noted. The height given on the line diagram of a beam is significant. Height elevations are assumed to be level at the figure given. The

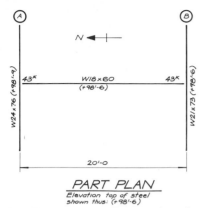

PART PLAN
Elevation top of steel
shown thus: (+98'-6)

Fig. 20-73. A small part of a plan, arranging steel members between beams A and B. (Adapted from *AISC Handbook*, with permission of the publisher.)

figure shows two elevations, 98'-6" and 98'-9".

■ SHOP DRAWINGS

Detail drawings like the one in Fig. 20-74 are essential for making structural pieces. The figure is a detail of a beam. This kind of drawing seldom describes the connections of mating parts (Fig. 20-75). Instead, it just shows those features (for example, connection angles) that will be involved when the piece is used in building.

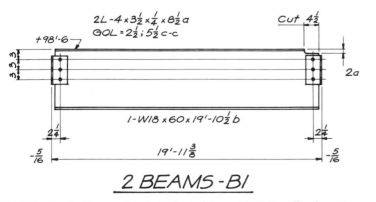

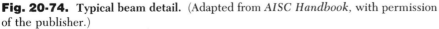

2 BEAMS - B1

Fig. 20-74. Typical beam detail. (Adapted from *AISC Handbook*, with permission of the publisher.)

In preparing structural details, the drafter refers to the handbook and the dimensions for detailing. Figure 20-76 shows both frame and seated connections.

■ RIVETING

One way to join beams is with rivets. The standard symbols for rivets are seen in Fig. 20-77 at A. A typical buttonhead rivet is shown at B. If the riveting is to be done in the shop, the drafter will use shop rivet symbols. These are open circles the diameter of the rivet head. If the riveting is to be done in the field (on the construction site), the drafter will use field rivet symbols. These are blacked-in circles the diameter of the rivet hole. Lines on which rivets are spaced are called *gage lines*. The distance between rivet centers on the gage lines is called the *pitch*.

■ STRUCTURAL BOLTING

High-strength steel bolts are rated by the American Society for Testing and Materials (ASTM). The bolt can be applied in the field or in the shop. The hole into which it fits is normally

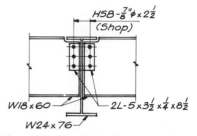

Fig. 20-75. Typical connection or framing of mating beams. (Adapted from *AISC Handbook*, with permission of the publisher.)

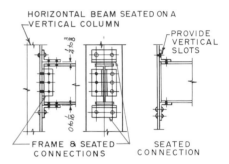

Fig. 20-76. Frame and seated connections. (Adapted from *AISC Handbook*, with permission of the publisher.)

¹⁄₁₆ in. (2 mm) larger than the bolt. Figure 20-78 shows two kinds of bolted connections: frame and seated. The bolt transmits the force of the beam load to the column. The stress this creates in the bolt is *shear* (cutting). The stress created in the column is *compressive* (pushing together).

■ WELDING STRUCTURAL MEMBERS

Structural steel members are usually welded with the metal-arc process. The fillet weld is the most common on structural connectors. (See Chap. 17 for a review of the standard welding symbols.)

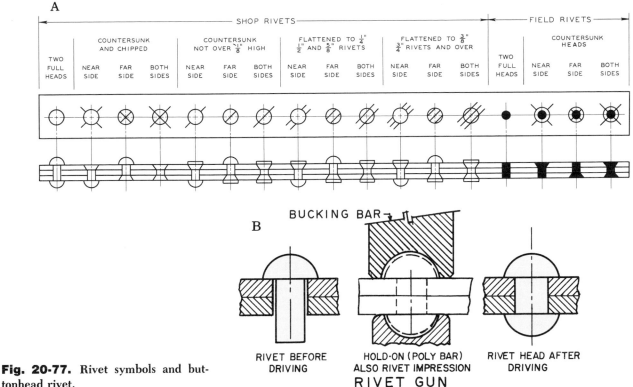

Fig. 20-77. Rivet symbols and buttonhead rivet.

DIMENSIONING

In structural drawings, dimensions are given primarily to working points. For beams, give dimensions to the centerline. For angles and, normally, channels, give dimensions to the backs. Give vertical dimensions on beams and channels to the tops or bottoms. Generally, do not dimension the edges of flanges and the toes of angles. Make the dimension lines continuous and unbroken. Place the dimensions above the dimension lines.

When dimensions are in feet and inches, use the foot symbol but not the inch symbol.

Structural Drawings

Figure 20-79 is a structural drawing of a small steel roof truss. This symmetrical piece is detailed about the left of a centerline. Study the drawing closely. On the drawing, each member is completely dimensioned or described. In addition, the dimensions shown adequately relate the fixed location of each structural member.

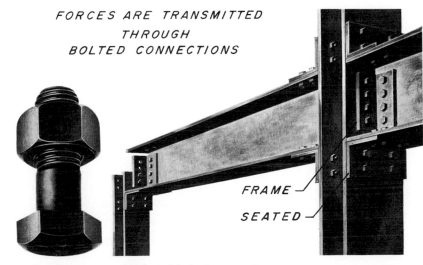

FORCES ARE TRANSMITTED THROUGH BOLTED CONNECTIONS

FRAME

SEATED

Fig. 20-78. Structural bolt and bolted connections.

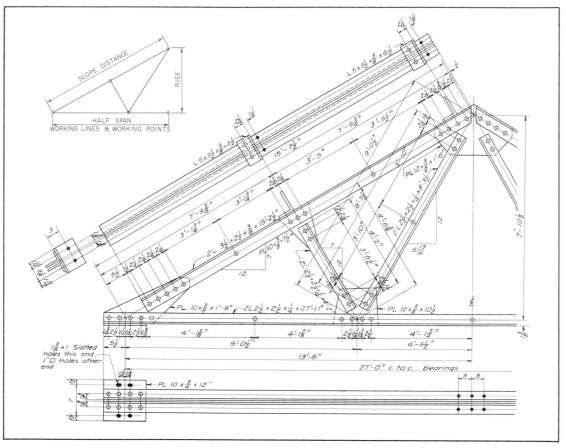

Fig. 20-79. Roof-truss detail symmetrical about centerline.

Roof Trusses

Figure 20-80 shows diagrams and names for some roof trusses and bridge trusses. The ones shown are only a few of those available. Each type can also be modified to carry specific loads.

■ WOOD FOR CONSTRUCTION

Wood is commonly used for the frames of homes and other small structures. Details for wood construction have been drawn up by the National Forest Products Association. They are now used as a standard method of construction.

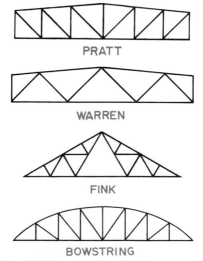

Fig. 20-80. Roof- and bridge-truss diagrams.

Structural timber can be manufactured in many forms by *laminating* (cutting wood into thin slabs and gluing them together). Builders can buy "factory grown" timbers in any size or shape (Fig. 20-81). Some of the forms available include tudor arches, radial arches, parabolic arches, A-frames, and tapered beams. The American Institute of Timber Construction (AITC) has set up guidelines for makers of structural glued laminated timber.

Figure 20-82 shows some of the construction details that must be used with structural timber. These detail drawings show how timbers are joined together and how struc-

Fig. 20-81. Laminated wood forms take many shapes. (American Institute of Timber Construction.)

tural members are anchored to foundations.

CONCRETE SYSTEMS

Many of our buildings, bridges, and dams have only been made possible by concrete. The American Concrete Institute has prepared a manual of standard practice for concrete structures.

Concrete has only limited strength unless it is specially prepared. It is made from a mixture of gravel, sand, water, and portland cement. Various grades are produced, depending on the proportions of these ingredients. The concrete can also be reinforced or prestressed.

Reinforced concrete has steel bars embedded in it. These bars are arranged to bear the structural loads that the concrete could not support by itself. When concrete and steel are combined, they can be used in the monolithic form shown in Fig. 20-83. Figure 20-84 shows a typical reinforced concrete detail.

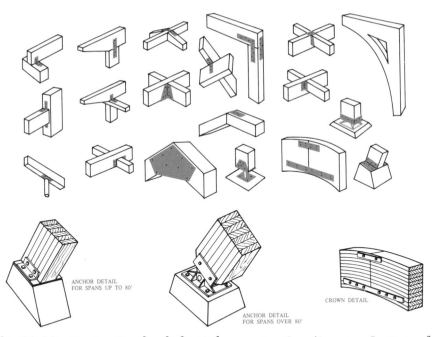

ANCHOR DETAIL FOR SPANS UP TO 80'

ANCHOR DETAIL FOR SPANS OVER 80'

CROWN DETAIL

Fig. 20-82. Construction details for timber construction. (American Institute of Timber Construction.)

Fig. 20-83. Monolithic concrete forms shape new structures. (Ceco Steel Products Corp.)

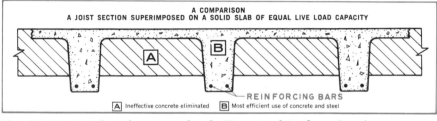

A COMPARISON
A JOIST SECTION SUPERIMPOSED ON A SOLID SLAB OF EQUAL LIVE LOAD CAPACITY

A

B

REINFORCING BARS

A Ineffective concrete eliminated   B Most efficient use of concrete and steel

Fig. 20-84. Reinforced concrete detail. (Ceco Steel Products Corp.)

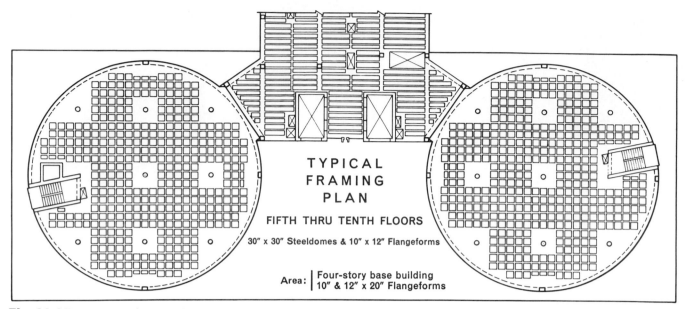

Fig. 20-85. Concrete forms and concrete forms for cavities in structure are shown on placement drawing. (Ceco Steel Products Corp.)

In *prestressed* concrete, the reinforcing bars are stretched before the concrete is poured over them. The prestressed form will then accept a predesigned load. This combination of concrete and stretched steel is stronger than either plain concrete or reinforced concrete.

Detailed Drawings

Concrete forms designed by a structural engineer are drawn for the manufacturer's use only. Construction drawings of these forms are prepared by the manufacturer. These drawings are made to show the contractor the location, placement, and connections. See Fig. 20-85 for typical details.

■ STRUCTURAL CLAY SYSTEMS

The solid brick wall is the oldest type of masonry construction known. Bricks are made from different types of clay in many shapes, forms, and colors. The common brick size is 2¼ × 3¾ × 8 in. (57 × 95 × 203 mm). Brick walls are made to support floors and roofs. For the vertical walls to be able to carry the horizontal floors and roofs, the bricks must have high compressive strength.

Structural Bonding

Bricks in construction are arranged in overlapping and interlocking patterns and fastened together with connecting mortar joints. This produces a structural assembly that acts as a single structural unit. Figure 20-

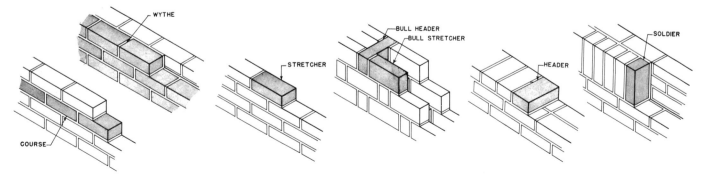

Fig. 20-86. Brick bonding forms structural walls. (Structural Clay Products Institute.)

86 shows some of the common bonds and structural patterns used with bricks. Note the terms applied to the brick.

How strong a structural clay system is generally depends on the strength of the mortar joints. When the design limits this strength, the designer can call for *reinforced masonry*. This is masonry with steel rods or wire embedded in the mortar.

Brick or concrete masonry can also be used to enclose a structural steel framework. This provides enough fireproofing to meet standard building codes.

REVIEW

1. Name five jobs for structural drafters.
2. Name five basic structural steel shapes.
3. What is the major advantage of a geodesic dome?
4. How do you dimension beams, angles, and channels?
5. How do you draw dimension lines on structural drawings?
6. What is the difference between reinforced concrete and pre-stressed concrete?
7. What symbols are used in structural dimensions, and when do you use them?
8. What does the strength of structural clay systems depend on?
9. What is the difference between a frame and a seated steel-beam connection?
10. What are the metric dimensions of a common brick?
11. Name the two styles of lettering used in architecture.
12. List four house styles.
13. What are the three main ways to judge a building's architecture?
14. What are the four basic drawings that an architect makes?
15. What are working drawings?
16. What major services does an architect render?
17. Define architecture in your own words.
18. Draw four material symbols used by the architect.
19. Illustrate a line technique important to architectural style.
20. What are pressure-sensitive symbols?
21. List six types of windows used in houses.
22. What are the preferred scales for plan and elevation drawings of small buildings?
23. What scale is preferred for enlarging details?
24. List three types of wall constructions common in residential design.
25. What is modular coordination?

Problems

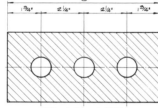

INDEX TO DRAWINGS

ARCHITECTURAL

A1 — SITE PLAN, INDEX TO DRAWINGS, ARCHITECTS SYMBOLS
A2 — GROUND FLOOR PLAN
A3 — FIRST FLOOR PLAN
A4 — TYPICAL FLOOR PLAN (2ND THRU 10TH FLRS.)
A5 — PENTHOUSE & ROOF PLANS
A6 — LARGE SCALE CORE PLANS, DETAILS, SECTIONS
A7 — LARGE SCALE APARTMENT PLANS, ELEVATIONS, DETAILS
A8 — EXTERIOR ELEVATIONS
A9 — EXTERIOR ELEVATIONS, STAIR ENTRANCE, DETAILS
A10 — WALL SECTIONS, CROSS SECTION, MISCELLANEOUS DETAILS
A11 — ENTRANCE SECTIONS, MISCELLANEOUS DETAILS

STRUCTURAL

S1 — FOUNDATION & GROUND FLOOR PLAN
S2 — GENERAL NOTES, SECTIONS, DETAILS
S3 — FIRST FLOOR FRAMING PLAN
S4 — TYPICAL FLOOR FRAMING PLAN (2ND THRU 10TH FLRS.)
S5 — ROOF & PENTHOUSE FRAMING PLAN
S6 — COLUMN SCHEDULE & DETAILS
S7 — BEAM SCHEDULE & DETAILS
S8 — STAIR SECTIONS & DETAILS
S9 — PARKING DECK GRADES, BEAM & SLAB DETAILS

Fig. 20-87. Practice lettering the architectural style shown in the "Index to Drawings." The lettering sizes should be ¼″, ³⁄₁₆″, and ⅛″. Use guide lines and show the box around lettering.

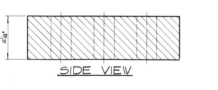

PLAN VIEW

SIDE VIEW

END VIEW

DR'WG SCALE: 6″=1′-0″

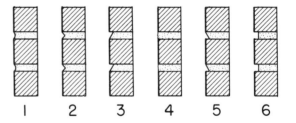

1 2 3 4 5 6

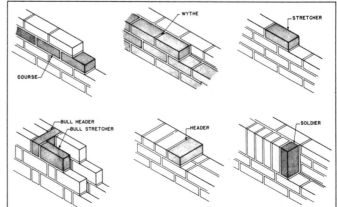

Fig. 20-88. Make a three-view drawing of the brick at the scale shown. Dimension and label. Label the stretcher, header, and face. Make a pictorial in isometric of the brick bonds. The nominal dimensions are 2¼″ × 4″ × 8″ for easy drafting (½″ mortar joints). Scale: 1½″ = 1′-0″. Draw the six views of the bricks in section. Label the types of joints: (1) concave, (2) V-joint, (3) weathered, (4) flush, (5) struck, (6) raked.

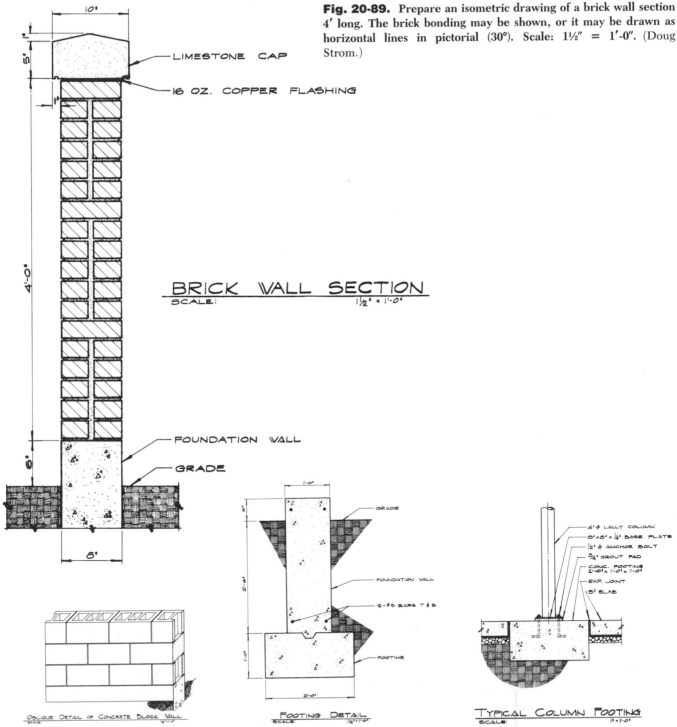

Fig. 20-89. Prepare an isometric drawing of a brick wall section 4' long. The brick bonding may be shown, or it may be drawn as horizontal lines in pictorial (30°). Scale: 1½″ = 1′-0″. (Doug Strom.)

LIMESTONE CAP

16 OZ. COPPER FLASHING

BRICK WALL SECTION
SCALE: 1½" = 1'-0"

FOUNDATION WALL

GRADE

Oblique Detail of Concrete Block Wall
SCALE:

Fig. 20-90. Prepare an oblique or isometric detail of a concrete block wall as assigned by the instructor. Investigate the sizes of block. Scale: 1½″ = 1′-0″.

GRADE

FOUNDATION WALL

2-#5 BARS T & B

FOOTING

FOOTING DETAIL
SCALE:

Fig. 20-91. Draw a footing detail at the scale shown and label parts.

4" ∅ LALLY COLUMN

8"x8"x ¼" BASE PLATE

½" ∅ ANCHOR BOLT

¾" GROUT PAD

CONC. FOOTING
2'-0"x 1'-0"x 1'-0"

EXP. JOINT

3" SLAB

TYPICAL COLUMN FOOTING
SCALE: 1"=1'-0"

Fig. 20-92. Prepare a section detail of the column footing, using the given sizes of the structural items. (Doug Strom.)

Figs. 20-93 through 20-98. These are all part of the same structure. When preparing the details (Figs. 20-95 through 20-98), refer to the framing plan (Fig. 20-94). (Doug Strom.)

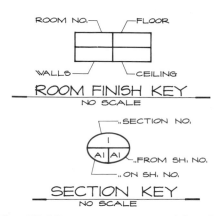

ROOM FINISH KEY
NO SCALE

SECTION KEY
NO SCALE

Fig. 20-93. Prepare a sectional key diagram with an ellipse template and a room finish symbol. Note the section notes on framing diagram.

Fig. 20-94. Prepare a framing diagram as shown. Outside building dimensions are 24′-0″ × 30′-0″. Joists are 16″ on center. Include your sectional diagrams as ellipses or rectangles.

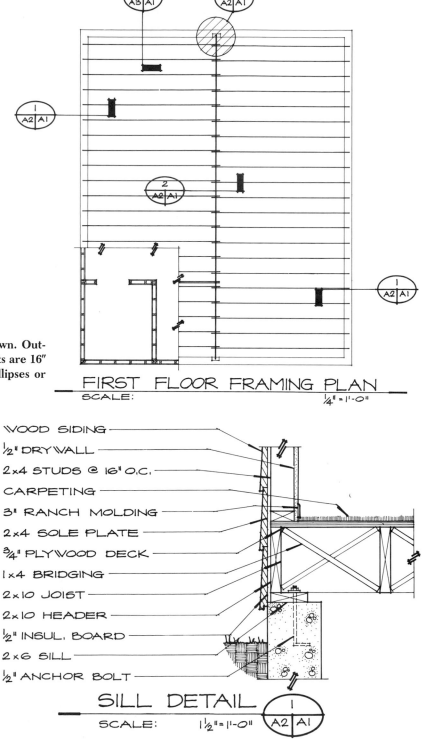

FIRST FLOOR FRAMING PLAN
SCALE: ¼″ = 1′-0″

WOOD SIDING
½″ DRYWALL
2×4 STUDS @ 16″ O.C.
CARPETING
3″ RANCH MOLDING
2×4 SOLE PLATE
¾″ PLYWOOD DECK
1×4 BRIDGING
2×10 JOIST
2×10 HEADER
½″ INSUL. BOARD
2×6 SILL
½″ ANCHOR BOLT

SILL DETAIL
SCALE: 1½″ = 1′-0″

Fig. 20-95. Using the nominal dimensions of the materials listed, construct the sill detail in section. The lapped, 1″ siding has an 8″ exposure. Scale as shown.

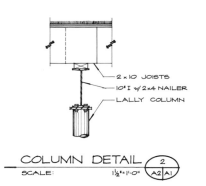

COLUMN DETAIL (2) A2|A1
SCALE: 1½"=1'-0"

Fig. 20-96. Prepare a partial girder-column detail. The column is 4″ with a 6″ plate welded to the top. The 10″ I-beam is bolted to the plate. The 2 × 4 rests on the 4½″ flange of the beam.

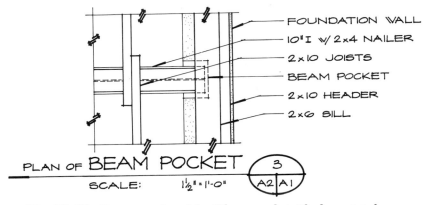

- FOUNDATION WALL
- 10" I w/ 2x4 NAILER
- 2x10 JOISTS
- BEAM POCKET
- 2x10 HEADER
- 2x6 SILL

PLAN OF BEAM POCKET (3) A2|A1
SCALE: 1½"=1'-0"

Fig. 20-97. Prepare a plan of the 6″ beam pocket. The beam is to have a 4″ bearing on the concrete wall.

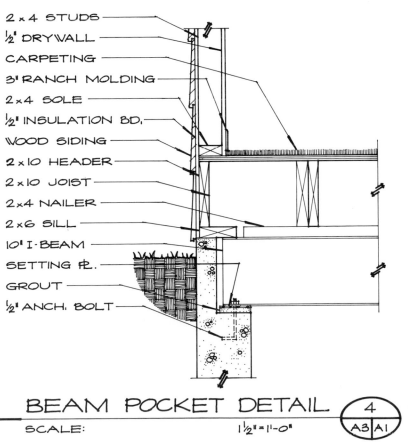

- 2x4 STUDS
- ½" DRYWALL
- CARPETING
- 3' RANCH MOLDING
- 2x4 SOLE
- ½" INSULATION BD.
- WOOD SIDING
- 2x10 HEADER
- 2x10 JOIST
- 2x4 NAILER
- 2x6 SILL
- 10' I-BEAM
- SETTING PL.
- GROUT
- ½" ANCH. BOLT

BEAM POCKET DETAIL (4) A3|A1
SCALE: 1½"=1'-0"

Fig. 20-98. Prepare a sectional beam-pocket detail showing the 10″ I-beam on the stepped foundation wall. Allow the beam flange to appear as a nominal ½″ thickness.

Figs. 20-99 and 20-100.   Prepare a pictorial of a typical roof detail. Studs, ceiling joists, and rafters are 16″ on center. Joist and rafters: 2″ × 6″. Studs and top plates: 2″ × 4″. Ridge beam: 2″ × 8″. Scale: 1½″ = 1′-0″. Draw a pictorial of the built-up girder (3) 2″ × 10″.

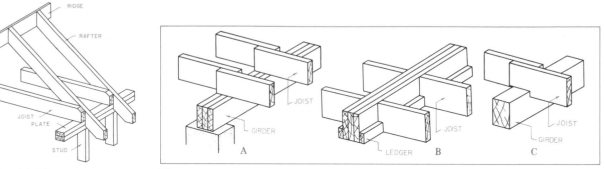

Fig. 20-99.   **Fig. 20-100.**

Fig. 20-101 through 20-105.   These are a partial set of plans for a two-story residence. Prepare drawings as assigned by the instructor. (A. W. Wendell & Sons, architect-contractor.)

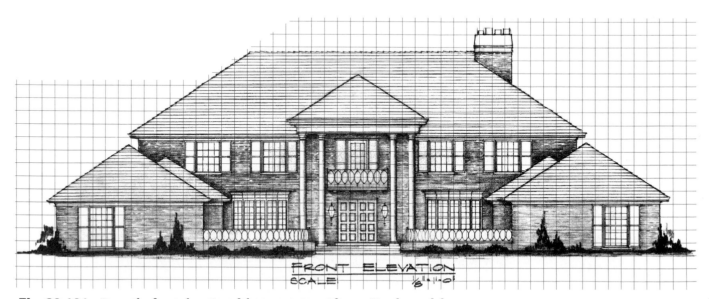

FRONT ELEVATION
SCALE   ⅛″=1′-0″

Fig. 20-101.   Draw the front elevation of the two-story residence. Use the modular grid to establish the rectangular sizes (closest 2″). Note that the double front doors equal 6′-0″ wide. Draw at a scale of ⅛″ = 1′-0″. Illustrate the brick, masonry walls, and asphalt roofing with horizontal lines. Add windows, shutters, and decorative appointments to suit your own design.

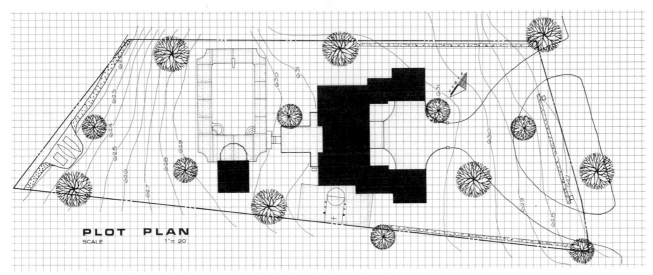

PLOT PLAN
SCALE   1" = 20'

Fig. 20-102. Examine the gridded plot plan, and develop the boundary lines at a scale of 1″ = 20′, on a C-size sheet. Dimension the length of each boundary line. Complete the plan of the house on the site, and landscape with trees and apply contour lines. All radii on the driveway must be 18′. The pool, cabana, and basketball court are optional.

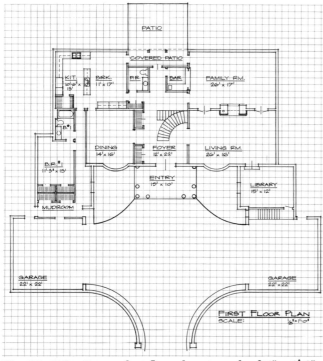

FIRST FLOOR PLAN
SCALE: ⅛"=1'-0"

Fig. 20-103. Draw a first-floor plan at a scale of ⅛″ = 1′-0″. By examining the grid, locate windows on the plan from elevations. Label the rooms and the appropriate sizes for this preliminary architectural study for your client.

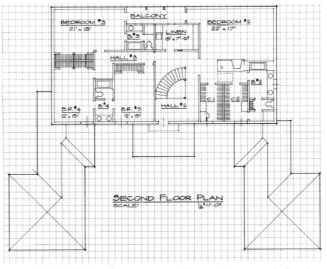

SECOND FLOOR PLAN
SCALE: ⅛"=1'-0"

Fig. 20-104. Prepare a second-floor plan to include the roof plan over the single-story area. Scale: ⅛″ = 1′-0″. Label the rooms with names and dimensions. Note the hip roof design and intersecting inclined planes forming the roof.

Fig. 20-105. Draw a right-side elevation by examining the proportions on the modular grid. Note the centerlines that locate the finished floor and ceiling lines. Find the common roof pitch and label it on the elevation.

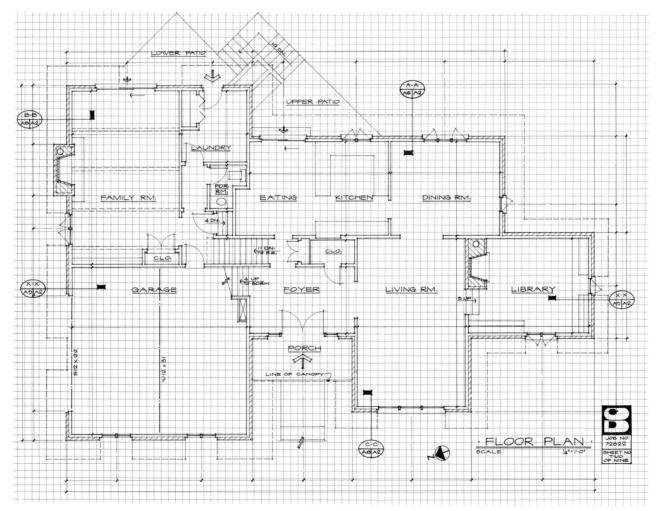

Fig. 20-106. Note that this floor plan is Fig. 20-54 in the text. Prepare the first-floor plan at a scale of ¼″ = 1′-0″. Add the dimensions to the plan. Examine the modular grid for room sizes and compare them to Fig. 20-54.

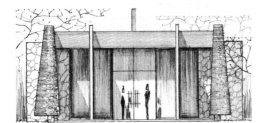

Fig. 20-107.

Figs. 20-107 through 20-110. Examine the front, west, and south elevations of the chapel in relation to the floor plan. Draw the plan and lay out the narthex, nave, and chancel at a scale of ⅛″ = 1′-0″. Use dividers and the scale on the floor plan to determine sizes. Render the plan with appointments and add site landscape. Draw elevations with an 18′-high roof line using shingles. This is a preliminary study and dimensions are not accurate at this stage of presentation. Render elevations.

Fig. 20-108.

Fig. 20-109.

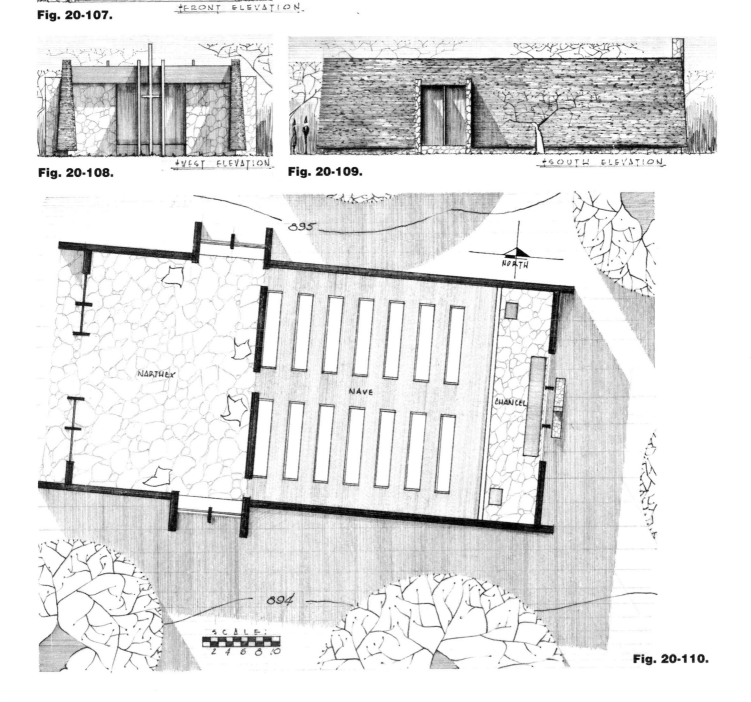

Fig. 20-110.

Fig. 20-111. Develop an isometric drawing of the western, or platform, framing in Fig. 20-34. Rafters: 2″ × 6″. Joists: 2″ × 8″. Studs: 2″ × 4″.

Fig. 20-112. Develop a detailed floor plan of the ranch-styled house in Fig. 20-47, with complete dimensions. (Scale: ¼″ = 1′-0″). Optional: Complete a set of plans, elevations, and sections for the house. NOTE: The cross-corner technique adds style to the drawings.

Fig. 20-113. Develop a complete set of plans for the contemporary multilevel house in Fig. 20-54. Include (1) site plan, (2) floor plans, (3) elevations, and (4) sections.

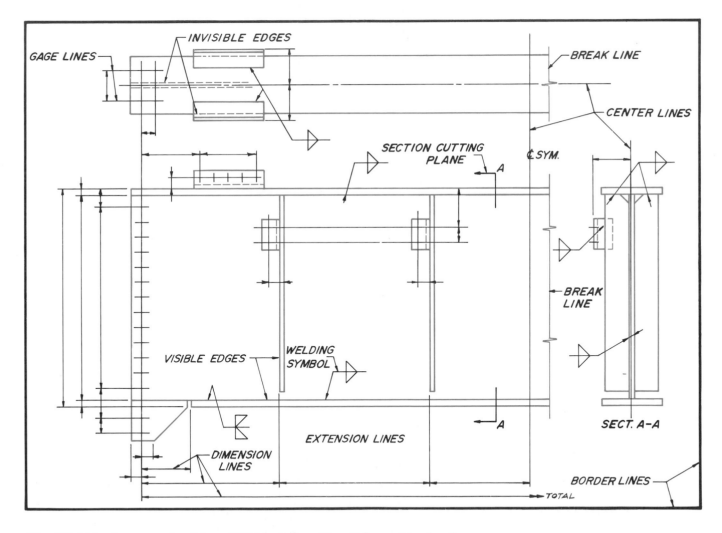

Fig. 20-114. Prepare a detail for a 22″-high girder with a 6″ flange. Develop the girder showing the conventional lines (shown in turquoise) with heavy line weights. Identify angles, welds, stiffeners, and field bolts. Fill in missing dimensions with guidance of the instructor. Scale: 1½″ = 1′-0″.

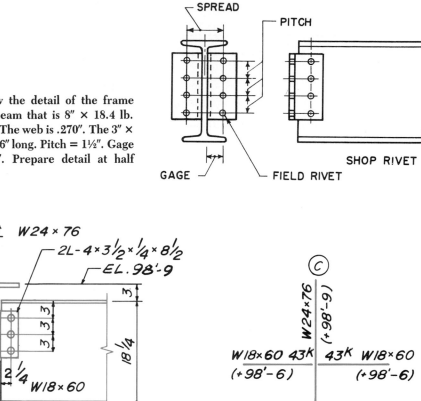

Fig. 20-115. Draw the detail of the frame connection on an S-beam that is 8″ × 18.4 lb. The flange is 4″ wide. The web is .270″. The 3″ × 3″ connection angle is 6″ long. Pitch = 1½″. Gage = 1¾″. Rivet = ½″. Prepare detail at half scale.

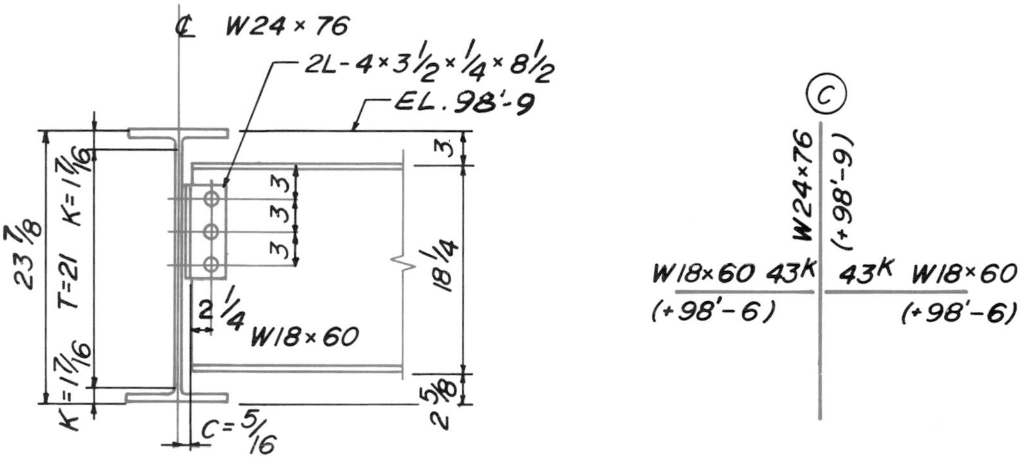

Fig. 20-116. Prepare a partial detail of the framed connection of the 18″-wide flanged beam with a 24″-wide flanged beam. Scale: 3″ = 1′-0″. Prepare a partial part-plan diagram about column centerline *C*, with noted members, and elevations.

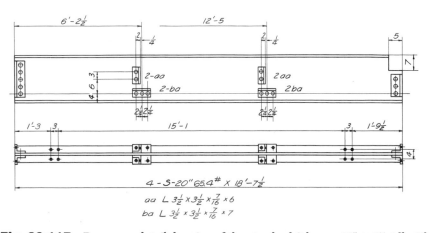

Fig. 20-117. Prepare a detail drawing of the standard S-beam, 20″ × 65.4 lb. The flange is 6¼″ wide with web thickness of ½″. Scale: ¾″ = 1′-0″.

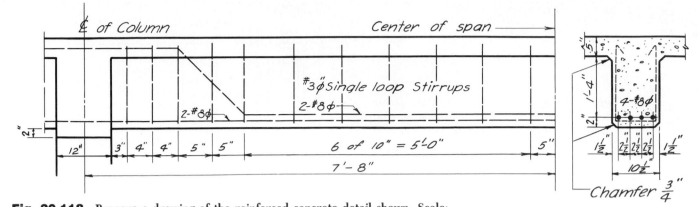

Fig. 20-118. Prepare a drawing of the reinforced concrete detail shown. Scale: 1½″ = 1′-0″.

Fig. 20-119. Prepare a truss detail of Fig. 20-79 at a scale of 1″ = 1′-0″ (1:12). NOTE: Prepare the left half of the symmetrical beam as shown about centerline.

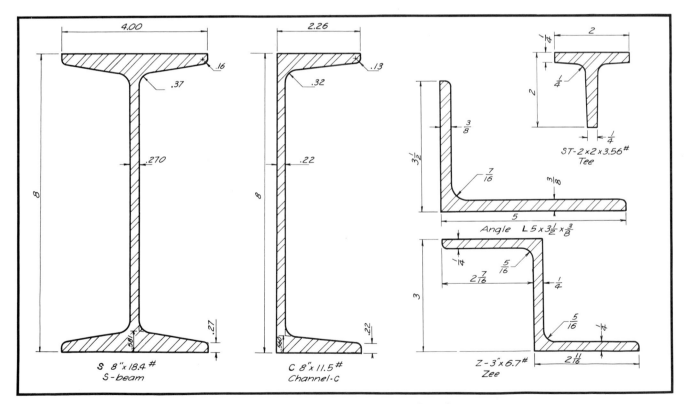

Fig. 20-120. Detail the standard steel shapes shown at an appropriate scale and show sectioning.

21 Map Drafting

MAPPING—A CHANGING INDUSTRY

Mapmakers are people who have been trained to gather information and prepare maps. Mapmakers are also called *cartographers*. The skilled map drafter prepares maps in detail for the civil engineer, scientist, geographer, and geologist.

Mapmaking is a pictorial method of representing facts about the surface of the earth or other bodies in the solar system. Rockets fired from Kennedy Space Center (Fig. 21-1) to the moon or to Mars, as in the Mariner program, carry the necessary equipment for mapping the moon and the planets. A method called *photogrammetry* is used in today's modern mapping industry. It is this method or technique that has been used in mapping the moon and the planets of our solar system.

CAREER OPPORTUNITIES

Many jobs are available for those who are able to prepare maps. The field of civil engineering is always expanding. Railroads, highways, harbor facilities, airports, and space stations are just a few areas in this broad industry where map planning is being used.

The drafter may prepare maps and charts under the direction of the design engineer or cartographer. There may be opportunities for advancement in job areas such as photogrammetry, surveying with laser beams, or research and development projects with the geographer. Additional information about careers in civil engineering and mapmaking is available from:

Association of American
  Geographers
1146 16th Street N.W.
Washington, DC 20036

MAP SIZES

As the world has entered the space age, there are maps that show enormous distances. The scale (size) is quite small on maps that cover large distances. Some maps that indicate ownership of property, such as city plats, must be very accurate. They may be drawn to a large scale in order to note all the information of physical property. Maps showing the geography of states or countries, which show boundary lines, streams, lakes, or coastlines, may use a scale of several miles to the inch.

SCALES USED ON MAPS

The civil engineer's scale is used for map drawings. Distances are given in decimals of a foot or meter, such as tenths, hundredths, and so forth. (See the maps in geography and history books.) Practically all the countries of the world use the metric system. In this system, distances are commonly measured in kilometers instead of miles.

The scale of a map is generally noted as 1 in. equals 500 ft, or 1 part equals 6000 parts, noted as 1:6000. The scale of 1 in. equals 1 mile can also be shown as 1:63 360. Graphic scales should be shown on maps as part of their basic information.

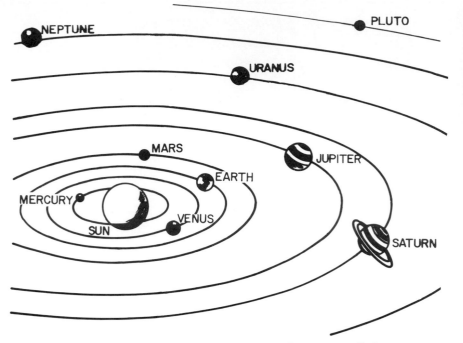

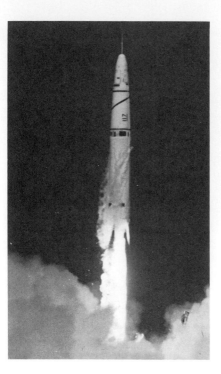

Fig. 21-1. Missiles can make map surveys of our solar system. All planets move in the same direction around the Sun. The Earth is 90 million miles (145 million kilometers) from the Sun. The planet Pluto is 3.56 billion miles (5.6 billion kilometers) from the Sun. (U.S. Air Force photo.)

■ PLATS OF A SURVEY

A map used to show the boundaries of a piece of land and to identify it is called a *plat*. The amount and kind of information presented depend upon the purpose of the map. The plat of a plane survey that was made to accompany the legal description of a property is shown in Fig. 21-2. Accuracy of information on a plat is all-important; it must agree with the legal description.

■ CITY PLAT

Maps of cities are made for the following reasons: to keep a record of street improvements, to show the location of utilities, and to record sizes and location of property for tax purposes. A part of such a city plat is

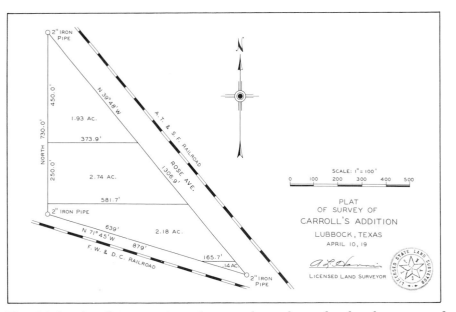

Fig. 21-2. Plat of a survey. Note the parts that make up the plat: the acreage of each part, the iron pipe locating the corners of the graphic scale, the signature of the surveyor, and the official seal. (One acre = 43 560 square feet.)

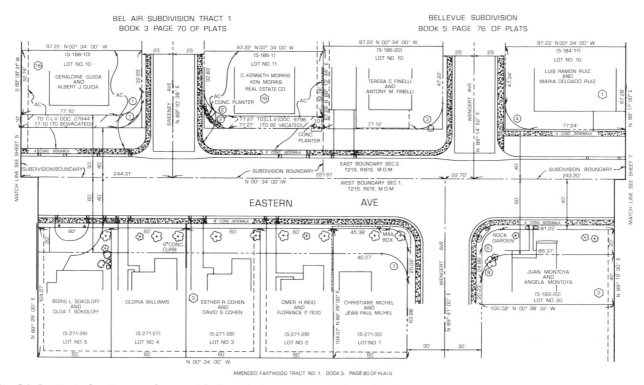

Fig. 21-3. Part of a city map drawn with the aid of a computer. (City of Las Vegas and Calcomp.)

shown in Fig. 21-3. Notice the numbering of the lots and the location of streets, sidewalks, and other details found in a city.

OPERATIONS MAPS

Operations maps are maps that show the relationship between the land's physical features and the operation that is to be performed. Such a map is found in Fig. 21-4. Operations maps can greatly help engineering, management, or government groups in the presentation of a project. A presentation well done will aid in the selling of a program.

CONTOURS

Since maps are one-view drawings, vertical distances and differences in ground level do not show. They can,

however, be shown by lines of constant level called *contours*. This is illustrated in Fig. 21-5. The contour lines represent the height of the ground above sea level. Contour lines that are close together indicate a steeper slope than lines that are far apart. This can be seen by projecting the intersections of the horizontal level lines with the profile section, as shown in Fig. 21-5.

Note that the contour map and the profile correspond to the plan and section of an ordinary drawing. Note the horizontal line, *AA*, or cutting plane, on the contour map. This line shows the position, or line, on which the profile is taken. You can see how the profile would change if the cutting plane were moved toward the ocean or to some other new position. The distance between contour lines will change according to needs and

the scale of maps. A 10-ft distance may be quite satisfactory, while a 5-ft distance may be used on maps that require a larger scale. For close detail work, such as an irrigation project, the contour distances may be reduced to .5, 1, or 2 ft. On small-scale maps with a high degree of relief, distances may be increased from 20 to 200 ft or more. As an aid to reading the map, every fifth contour is usually emphasized by drawing a much heavier line (refer to Fig. 21-7).

A technical pen is satisfactory for inking contour lines. It is replacing the contour pen. The contour pen has blades that swivel so that contour lines can be easily followed.

There are many forms of flexible curves and splines that can be bent to match curves to be drawn. The vinyl, or rubber-covered, curve in

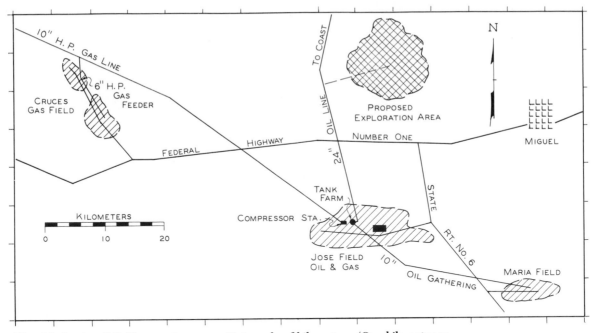

Fig. 21-4. An oil-field operations map. Note scale of kilometers. (One kilometer = 0.621 statute mile.)

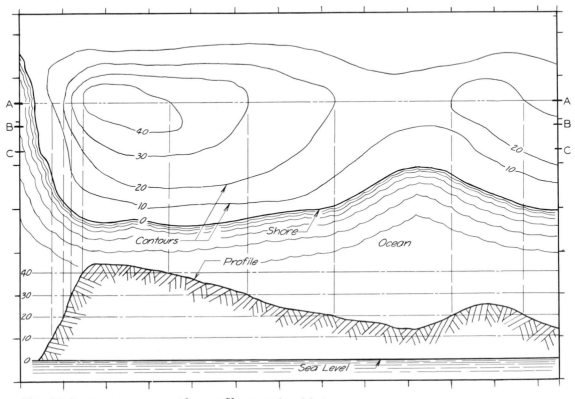

Fig. 21-5. A contour map with a profile at section *AA*.

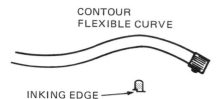

CONTOUR
FLEXIBLE CURVE

INKING EDGE →

Fig. 21-6. A flexible curve assists in plotting irregular curves in contour drafting.

Fig. 21-6 has a flexible core made of a strip of metal. The desired curve will hold without support. It is instantly adjustable and formed to any contour.

A contour map is shown in Fig. 21-7. It uses a *contour distance* (vertical distance between contour lines) of 20 ft. The elevation in feet is marked in a break in each contour line. Notice that the drainage is shown as intermittent streams.

Before a contour map can be drawn, elevations must be obtained in the field for several key points that control the drawing of the contours. The following methods are used: a grid system where all intersection elevations are obtained, along with important elevations on grid lines; points located by transit and stadia rod, with the corresponding elevations figured by plane table; and aerial photographic surveys (Fig. 21-8). The finished map in Fig. 21-8 was produced by photogrammetric methods. The equipment for making photogrammetric maps is shown in Fig. 21-9. The actual drawing of the map is usually done by scribing lines in the coating on a piece of film. An example of scribing on film is shown in Fig. 21-10.

Experience in surveying is needed in all of these mapping methods.

Topographic maps present complete pictorial descriptions of the areas shown. These maps show such information as boundaries, natural

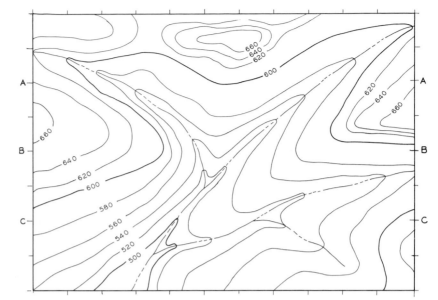

Fig. 21-7. A contour map with intermittent streams indicated by the lines with three dashes.

Fig. 21-8. Portions of an aerial photo and a topographic map of Concepción, Chile, compiled by photogrammetric methods. Actual map sheets were done at a scale of 1:2000 with 1-m contours. Photos from which maps were prepared are at a scale of 1:10 000. (Aerial Service Corp., Philadelphia, a division of Litton Industries.)

features, structures, vegetation, and relief (elevations and depressions).

Symbols are used for many of the features shown on topographic maps. Some of these are given in Fig.

21-11. Maps using topographic symbols can be obtained at a low cost from the Director, U.S. Geological Survey, Department of the Interior, or from the U.S. Coast and Geodetic

Survey, Department of Commerce, Washington, DC. Naval charts (maps) come from the Hydrographic Office, Bureau of Navigation, Department of the Navy.

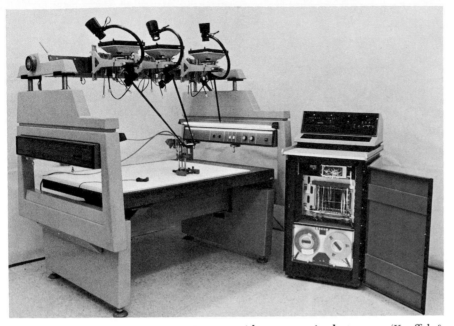

Fig. 21-9. Photogrammetric equipment with computerized storage. (Keuffel & Esser Co.)

Fig. 21-10. Scribing by hand or with a computerized plotter with a jewel point on coated film—a very accurate process. (Keuffel & Esser.)

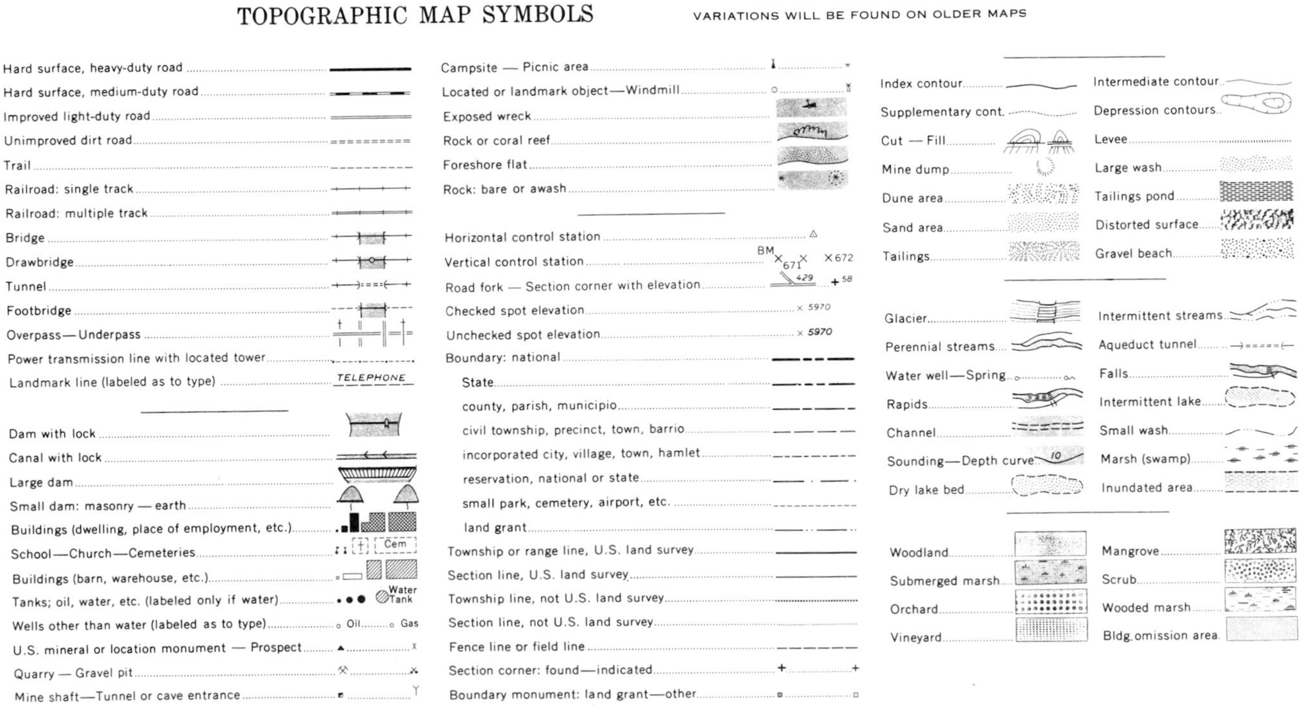

TOPOGRAPHIC MAP SYMBOLS   VARIATIONS WILL BE FOUND ON OLDER MAPS

Hard surface, heavy-duty road

Hard surface, medium-duty road

Improved light-duty road

Unimproved dirt road

Trail

Railroad: single track

Railroad: multiple track

Bridge

Drawbridge

Tunnel

Footbridge

Overpass — Underpass

Power transmission line with located tower

Landmark line (labeled as to type)   TELEPHONE

Dam with lock

Canal with lock

Large dam

Small dam: masonry — earth

Buildings (dwelling, place of employment, etc.)

School — Church — Cemeteries   Cem

Buildings (barn, warehouse, etc.)

Tanks; oil, water, etc. (labeled only if water)   Water Tank

Wells other than water (labeled as to type)   Oil   Gas

U.S. mineral or location monument — Prospect

Quarry — Gravel pit

Mine shaft — Tunnel or cave entrance

Campsite — Picnic area

Located or landmark object — Windmill

Exposed wreck

Rock or coral reef

Foreshore flat

Rock: bare or awash

Horizontal control station

Vertical control station   BM 671   672

Road fork — Section corner with elevation   429   58

Checked spot elevation   5970

Unchecked spot elevation   5970

Boundary: national

State

county, parish, municipio

civil township, precinct, town, barrio

incorporated city, village, town, hamlet

reservation, national or state

small park, cemetery, airport, etc.

land grant

Township or range line, U.S. land survey

Section line, U.S. land survey

Township line, not U.S. land survey

Section line, not U.S. land survey

Fence line or field line

Section corner: found — indicated   +   +

Boundary monument: land grant — other

Index contour   Intermediate contour

Supplementary cont.   Depression contours

Cut — Fill   Levee

Mine dump   Large wash

Dune area   Tailings pond

Sand area   Distorted surface

Tailings   Gravel beach

Glacier   Intermittent streams

Perennial streams   Aqueduct tunnel

Water well — Spring   Falls

Rapids   Intermittent lake

Channel   Small wash

Sounding — Depth curve   10   Marsh (swamp)

Dry lake bed   Inundated area

Woodland   Mangrove

Submerged marsh   Scrub

Orchard   Wooded marsh

Vineyard   Bldg. omission area

Fig. 21-11. Some conventional symbols used on maps. Note the intermittent streams as used in Fig. 21-7. (U.S. Coast and Geodetic Survey.)

Aeronautical maps (Fig. 21-12) use special symbols that need to be understood in order to be read. Some symbols from the U.S. Coast and Geodetic Survey are shown in Fig. 21-13.

■ BLOCK DIAGRAMS

Discussion thus far has been given to mapping in the horizontal plane and the vertical plane by profiles or sections. To help you see the three-dimensional problem, a *block diagram* is also used. It is a three-

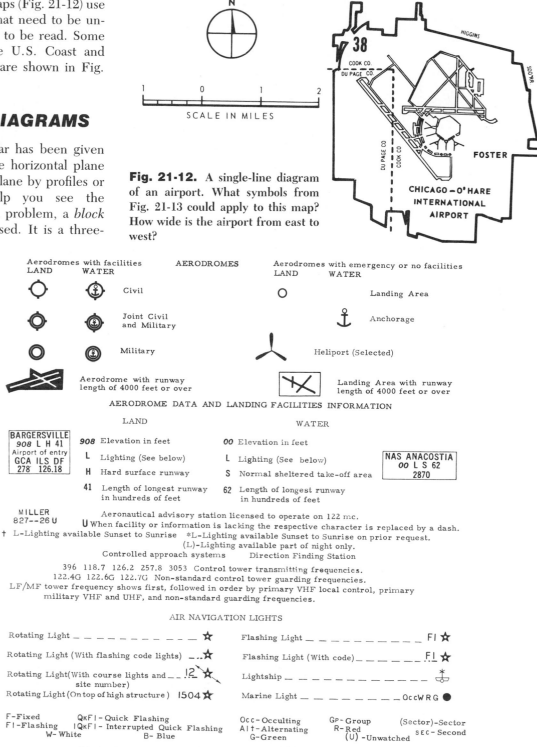

Fig. 21-12. A single-line diagram of an airport. What symbols from Fig. 21-13 could apply to this map? How wide is the airport from east to west?

SCALE IN MILES

AERODROMES

Aerodromes with facilities
LAND   WATER

Civil

Joint Civil and Military

Military

Aerodrome with runway length of 4000 feet or over

Aerodromes with emergency or no facilities
LAND   WATER

Landing Area

Anchorage

Heliport (Selected)

Landing Area with runway length of 4000 feet or over

AERODROME DATA AND LANDING FACILITIES INFORMATION

LAND   WATER

BARGERSVILLE
908 L H 41
Airport of entry
GCA ILS DF
278' 126.18

908 Elevation in feet

L Lighting (See below)

H Hard surface runway

41 Length of longest runway in hundreds of feet

00 Elevation in feet

L Lighting (See below)

S Normal sheltered take-off area

62 Length of longest runway in hundreds of feet

NAS ANACOSTIA
00 L S 62
2870

MILLER
827--26 U

U Aeronautical advisory station licensed to operate on 122 mc.
When facility or information is lacking the respective character is replaced by a dash.

† L-Lighting available Sunset to Sunrise   *L-Lighting available Sunset to Sunrise on prior request.
(L)-Lighting available part of night only.

Controlled approach systems   Direction Finding Station

396 118.7 126.2 257.8 3053 Control tower transmitting frequencies.
122.4G 122.6G 122.7G Non-standard control tower guarding frequencies.
LF/MF tower frequency shows first, followed in order by primary VHF local control, primary
military VHF and UHF, and non-standard guarding frequencies.

AIR NAVIGATION LIGHTS

Rotating Light _ _ _ _ _ _ _ _ _ _ ☆   Flashing Light _ _ _ _ _ _ _ _ _ Fl ☆

Rotating Light (With flashing code lights) _ ..☆   Flashing Light (With code) _ _ _ _ _ _ Fl ☆

Rotating Light (With course lights and _ _ !2 ☆   Lightship _ _ _ _ _ _ _ _ _ _ _ *
site number)

Rotating Light (On top of high structure) 1504 ☆   Marine Light _ _ _ _ _ _ _ _ _ OccWRG ●

F-Fixed   QkFl-Quick Flashing   Occ-Occulting   Gp-Group   (Sector)-Sector
Fl-Flashing   IQkFl-Interrupted Quick Flashing   Alt-Alternating   R-Red   sec-Second
W-White   B-Blue   G-Green   (U)-Unwatched

Marine lights are white unless colors are indicated; alternating lights
are red and white unless otherwise indicated

Fig. 21-13. Some aeronautical symbols. (U.S. Coast and Geodetic Survey.)

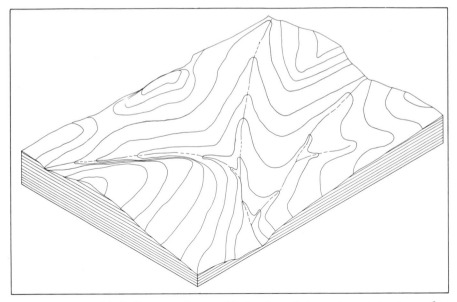

Fig. 21-14. A block diagram shows a block of earth in an isometric view. This model was made from Fig. 21-7.

dimensional projection using the isometric view (Fig. 21-14). This block diagram has been developed from Fig. 21-7. Keep in mind that each contour represents a level plane, similar to a card in a deck of cards. True lengths are measured on the isometric axes.

GEOLOGICAL MAPPING

Geology is the science dealing with the makeup and structure of the earth's surface and interior depths.

The crust of the earth is made up of three groups of rock: igneous or crystalline, sedimentary, and metamorphic. *Igneous* rocks, for purposes of this general discussion, are the basic materials that make up the earth's crustal ring. This rock was once *molten* (melted). It has cooled, but it has not been eroded nor has its makeup changed. *Sedimentary* rocks, as a rule, are deposited in water in layers of different thicknesses similar to the layers of an on-

ion. If the onion is cut perpendicular to its axis, a series of concentric rings will be noted. In a slice of the earth's crust made in a sedimentary area, a similar pattern can be seen. A series of layers that can be identified by texture, color, and material is visible (refer to pictures of the Grand Canyon). *Metamorphic* rocks are generally considered to be sedimentary rocks. These rocks have been deeply buried, heated to high temperatures, and recomposed so that they can no longer be identified as sedimentary rocks.

Nature, being ever-changing, folds, tips, and slices these sedimentary layers in a number of ways. Sometimes sedimentary layers include large areas of crystalline rock. At times, sedimentary layers are formed on top of crystalline rocks. Often, the whole mass may be tipped, perhaps for miles. This tipping raises the mass of rock high above sea level. Sometimes the mass is dropped thousands of feet. The geologist making investigations has

the problem of representing what has happened or what a particular area looks like.

Figure 21-15 is part of a geological surface map. The turquoise lines represent the line of surface exposure of the contact between two *formations* (like the line of two contacting layers of a cut onion). The geologist locates this in the field and notes the observation point with the "T" symbol. The *strike* (direction of this contact) is shown by the top of the T. The *dip* (slope of the contact) is indicated by the figures at the T, such as 23° on the T and the right-hand side and below the section line *XX*. Since the stem of the T, in this case, is pointing to the east, or to the right, the dip is 23° to the east. In other words, this formation slopes 23° below the horizontal and to the east. Another T symbol on the left side shows a 30° dip to the west. The heavy, borken line near the left edge represents a *fault trace* (the line along which the layer broke).

GEOLOGICAL SECTIONS

These sections help in the interpretation of the surface map. An example (Fig. 21-16) shows what the geologist believes the area below the surface is like. This is a section along line *XX* of Fig. 21-15. The dips that the geologist noted are used in developing the curvature of the folds. By means of a typical section of the region, the geologist can determine the various normal thicknesses of each formation or stratum. These values are used in making this section. The fault, as indicated, shows the area at the right to be upthrown. The displacement is easily seen by comparing the position of formation *A* on either side of the fault.

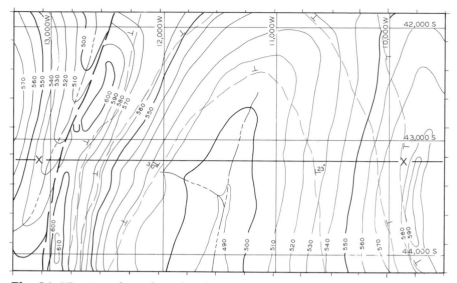

Fig. 21-15. Part of a geological surface map. **Where is the highest elevation on line XX?**

SUBSURFACE MAPPING

This kind of mapping is a means of showing details of strata lying below the surface of the earth. It can show the top or bottom of a given forma-tion or possibly an assumed horizon. Information for constructing such a map is obtained from many sources. These sources may include core holes, electrically recorded logs, seismograph surveys, and so forth. An example of information that was obtained from electrically recorded logs taken in a series of oil wells is shown in Fig. 21-17. The wells are located on a grid pattern. Producing wells are indicated by a solid black circle. Dry holes are indicated by an open circle with outward-extending rays. The top of a producing sand that is cut by a fault on the west is shown. Notice that the contours are numbered with negative values, or depths that are below sea level. The greater the value, the deeper the point below sea level. Section *X-X* shows the thickness of the sand and the level of the *oil-water contact*.

Geological maps and sections are greatly improved by the use of col-ors. In Fig. 21-15, colors may be applied to each of the formations that show between the turquoise forma-tion-contact lines. This can also be done in Fig. 21-16 by applying the same colors to the corresponding for-mations. The use of colors helps in bringing out the three-dimensional relationship of the surface and the shape of the structure. Color also helps in understanding the geology of the area. Paper prints of the trac-ing are colored and then rubbed carefully to give smooth, even color texture. Color is used on U.S Geo-logical Survey maps.

The making of geological maps and drawings is an important part of the extractive minerals industry, partic-ularly for petroleum. With the aid of maps, it is possible to keep proper records and information so that activ-ity in this economic field can be con-tinued. Standards for records are dif-ferent from company to company. However, general standards are well covered in technical literature such as publications of the AIME, petro-leum branch, the AAPG, U.S. Geo-logical Survey, U.S. Bureau of mines, and others.

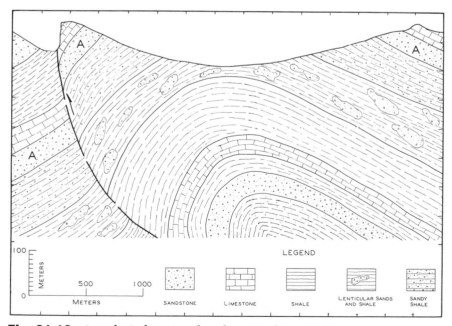

LEGEND

SANDSTONE LIMESTONE SHALE LENTICULAR SANDS AND SHALE SANDY SHALE

Fig. 21-16. A geological section along line *XX* of Fig. 21-15.

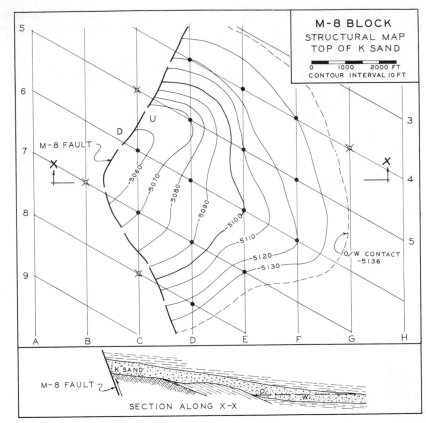

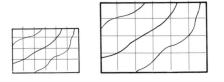

Fig. 21-17. A structural map showing strata details below the surface.

■ NOTES AND DEFINITIONS

Maps are important in understanding the news, in studying geography and history, in making car trips, and in surveying, geology, civil engineering, petroleum engineering, and space exploration.

Brief definitions for some of the terms used in map drafting are listed here. For more complete descriptions, refer to books on surveying and geology.

1. *Bearing*. The direction of a line as shown by its angle with a north-south line (meridian).
2. *Meridian*. North-south line.

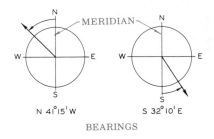

BEARINGS

3. *Grid*. A series of uniformly spaced horizontal and perpendicular lines used to locate points by coordinates, or for enlarging or reducing a figure.

4. *Profile*. A section of the earth on a vertical plane, showing the intersection of the surface of the earth and the plane (see Fig. 21-5).

PROFILE

5. *Cut*. Earth to be removed to prepare for construction, such as a desired level or slope for a road.
6. *Fill*. Earth to be supplied and put in place to prepare for a construction in order to obtain a desired level or slope.
7. *Grade*. A particular level or slope, as a downgrade.

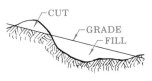

8. *Coordinate system*. A system for locating points by reference to lines that are generally at right angles.

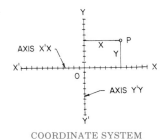

COORDINATE SYSTEM

9. *Contour*. A line of constant level showing where a level plane cuts through the surface of the earth (see Fig. 21-5).
10. *Block diagram*. A pictorial drawing (generally isometric) of

a block of earth, showing profiles and contours (see Fig. 21-14).

LEVEL PLANE

CONTOUR

11. *Fault*. A break in the earth's crust with a movement of one side of the break parallel to the line of the break.

FAULT

12. *Stratum*. A layer of rock, earth, sand, and the like, horizontal or inclined, arranged in flat form clearly different from the matter next to it. (Plural is *strata*.)

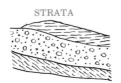

STRATA

13. *Fold*. A bend in a layer or layers of rock brought about by forces acting upon the rock after it has been formed.

FOLD

14. *Dip*. The angle that an inclined stratum, or like geological feature, makes with a horizontal plane. The direction of dip is perpendicular to the strike.

DIP   STRIKE

15. *Strike*. The direction at the surface of the intersection of a stratum with a horizontal plane.

REVIEW

1. If a map covers a large area, the scale is _____.

2. If a map covers a small area, the scale is _____.

3. Define a plat.

4. Define a contour line.

5. Name some reasons for making and using a contour map.

6. What is a topographic map?

7. What is a geological map?

8. What is a cartographer?

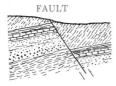

Problems

Fig. 21-18. Plot the map of Chicago on a C-size sheet. Scale: ¾″ = 1 mile, or other suitable scale. Calculate the number of square miles, the number of acres, or the square kilometers, that make up this city. (1 square mile = 2 589 988 square meters.)

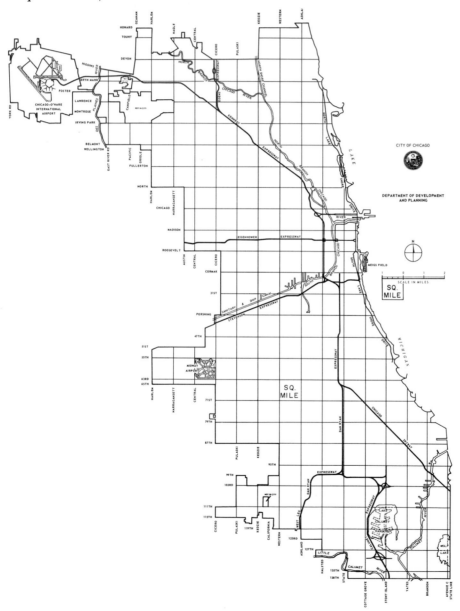

Fig. 21-19. (A) Make a drawing of a plat survey as shown in Fig. 21-2. Use a working space of 11″ × 15″. Scale: 1″ = 100′. Start the lowest point 8¼″ from the left border and 1½″ up from the bottom border. (B) Lay out a residential tract of land and divide it into lots. Plan most lots with 75′ frontage and 150′ or more in depth. Rose Avenue is a main street.

Fig. 21-20. Prepare a drawing of a contour as shown in Fig. 21-5. Use a working space of 8½″ × 11″. Draw 2 times the size shown in the figure. Use dividers or a grid to enlarge the figure (vertical profile scale: 1″ = 20′; horizontal scale: 1″ = 500′). Draw the profile on line *CC*.

Fig. 21-21. Make a city map, using the data provided in Fig. 21-3, showing streets, sidewalks, and lots with dimensions.

Fig. 21-22. Make an operations map as shown in Fig. 21-4. Draw a grid sheet over the map. Short marks along the border are to assist in drawing the grid. Redraw on an 8″ × 15″ working space. How long is the 10″ gas line? How long is the 24″ oil line?

Fig. 21-23. Make a contour map as shown in Fig. 21-7. Use the grid on the border to enlarge the contours. Scale: grid = 1″ squares. Working space: 8″ × 11″.

Fig. 21-24. Prepare a contour map of Fig. 21-15 and a vertical profile through *XX*. The grids are 1″ apart. Scale: 1″ = 100′.

22 Graphic Charts and Diagrams

IMPORTANCE OF GRAPHIC CHARTS AND DIAGRAMS

Graphic charts and diagrams are an important part of technical drawing. They are important to scientists, engineers, mathematicians, and nearly everyone else in everyday life.

Scientists use charts and diagrams to record and study the results of research. Engineers use them to record information about materials and conditions. Mathematicians use them to record facts and numerical information. Doctors use charts to record body temperature, heart action, and other body functions. Most people use and read charts and diagrams to learn about the weather, the stock market, and finances and for many other purposes. Because Fig. 22-1 is a chart, it can readily be seen that driver reaction distance and automobile braking distance increase as speed is increased. Figure 22-2 contains the same information, but it takes more time to read, study, and understand it. You can see and understand the relationship of speed

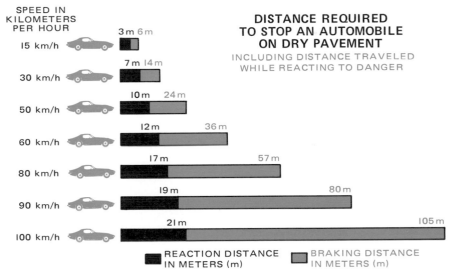

DISTANCE REQUIRED TO STOP AN AUTOMOBILE ON DRY PAVEMENT
INCLUDING DISTANCE TRAVELED WHILE REACTING TO DANGER

NOTE: DISTANCES ARE APPROXIMATE.

Fig. 22-1. Chart showing stopping distances at different speeds for automobiles.

STOPPING DISTANCES AT DIFFERENT SPEEDS FOR AUTOMOBILES			
KILOMETERS PER HOUR	DISTANCE IN METERS		
	REACTION DISTANCE	BRAKING DISTANCE	TOTAL DISTANCE
15	3	3	6
30	7	7	14
50	10	14	24
60	12	24	36
80	17	40	57
90	19	61	80
100	21	84	105

Fig. 22-2. The same information takes longer to read in this form.

and distance more easily in Fig. 22-1 because of its *graphic* (pictorial) presentation.

DEFINITIONS

Graphic charts and diagrams are pictures of numerical information that show the relationship of one thing to another. Sometimes they show trends in such areas as economics. You can tell at a glance whether the cost of living and wages are rising or falling over a period of time. Information like this can be determined quite easily because of the pictorial nature of a graph or chart.

Charts can also show ratios. An example of a chart that shows the ratio of speed versus distance is Fig. 22-1. Charts can show percentages of a whole. This type of chart is called a *bar chart* or a *pie chart*. Charts can also be used to explain information that is not numerical. For instance, a *flowchart* shows sequential information. That means that the chart shows which operation comes first, second, third, and so on.

Graphic charts may also be used to solve various kinds of mathematical problems. You should be able to answer the following questions easily by studying the chart in Fig. 22-3.

1. What is the normal water level in the lake?
2. When does the highest water level occur?
3. At what hours is the water level lowest?

You probably had no trouble answering the questions because charts of this type are easily understood.

The *curve on a chart* is not necessarily a curved line. As shown in Fig. 22-4, a curve on a chart may be a straight line, a curved line, a broken line, or a stepped line. It may also be a straight line or curved line adjusted to plotted points.

The *selection of the proper scales* (vertical and horizontal squares) is important. The vertical and horizontal scales must give a true pictorial *impression* by the angle of slope of the curve. In Fig. 22-5, you will notice that different *impressions* are given by the three charts. Chart A

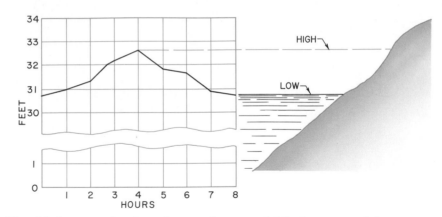

Fig. 22-3. A graphic chart showing the rise and fall of water in a lake.

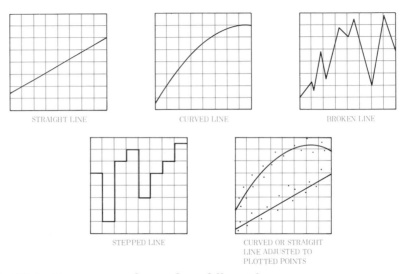

Fig. 22-4. Curves on graphs may have different forms.

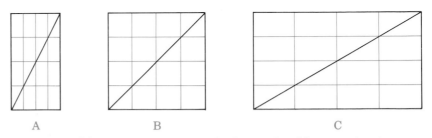

Fig. 22-5. A false impression may result if vertical and horizontal scales are not properly selected.

presents a very abrupt change. Chart B presents a normal change. At C, a very slow or gradual change is indicated. You should choose the scale that gives the most accurate pictorial impression.

Printed grid or *graph paper* is available in many forms. It may be

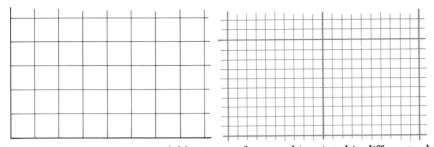

Fig. 22-6. Graph paper is available in many forms and is printed in different colors.

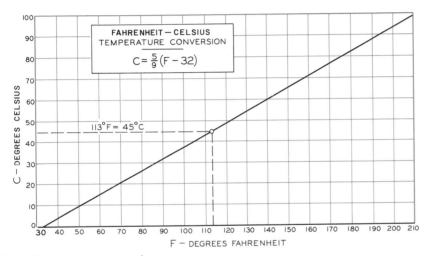

Fig. 22-7. A conversion chart.

FAHRENHEIT – CELSIUS
TEMPERATURE CONVERSION

$$C = \frac{5}{9}(F - 32)$$

$113°F = 45°C$

C – DEGREES CELSIUS

F – DEGREES FAHRENHEIT

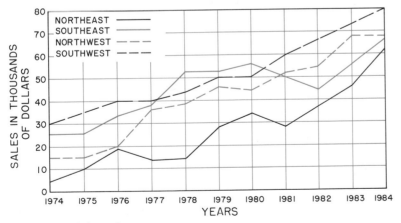

NORTHEAST ———
SOUTHEAST ———
NORTHWEST – – –
SOUTHWEST – – –

SALES IN THOUSANDS OF DOLLARS

YEARS

Fig. 22-8. A multiline chart.

purchased with lines ruled for drafting use at 4, 5, 8, 10, 16, and 20 to the inch, and in many other forms (Fig. 22-6). Metric sizes are also available. Graph paper that has certain lines printed more heavily than others is also available. This type of graph paper makes it easy to plot points and to read the finished chart. Figure 22-6 shows every tenth line in heavy print. In this case, the heavy lines are 1 in. (25 mm) apart. The lines on grid paper may form squares or rectangles, as shown in Fig. 22-5.

■ LINE CHARTS

Line charts are most often used to show *trends* or changes. For example, changes in the weather, ups and downs in sales, or trends in population growth can be plotted and shown graphically on a line chart. A line chart can have one or several curves. A conversion chart with one curve (Fig. 22-7) is often convenient for changing from one value to another. The chart in Fig. 22-7 shows the conversion from Fahrenheit to Celsius degrees. Figure 22-8 contains four curves that compare sales in various parts of the country.

■ TO DRAW A LINE CHART

Follow steps 1 through 9 to make a line chart.

1. Prepare and list the information to be presented (Fig. 22-9).

SCORING INFORMATION EASTERN HIGH SCHOOL	
GAME NUMBER	POINTS SCORED
1	38
2	20
3	50
4	40
5	40
6	30
7	10
8	40
9	55
10	45

Fig. 22-9. Information to be presented in a line chart.

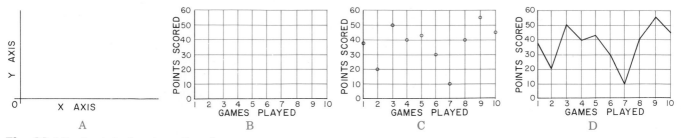

Fig. 22-10. Steps in drawing a line chart.

2. Select plain paper or ready-ruled graph paper.
3. Select a suitable size and proportion for your chart so that the overall design will be effective.
4. Select an appropriate scale.
5. If graph paper is not used, lay off and draw thin horizontal lines (called X-axis, or abscissa, lines). Then draw thin vertical lines (called Y-axis, or ordinate, lines). These steps are shown in Fig. 22-10A. You will notice that the intersection of X and Y is zero.
6. Lay off the scale divisions on the X axis and the Y axis (Fig. 22-10B).
7. Mark the scale values on the X and Y axes (Fig. 22-10B).
8. Plot the points accurately by using the information you have listed (Fig. 22-9). You will find that it is usually better to use small circles, triangles, or squares

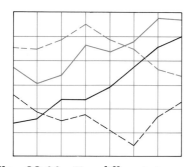

Fig. 22-11. Use different types of lines and different colors to distinguish curves on a multiline chart.

rather than crosses or dots for plotting points (Fig. 22-10C).
9. Connect the points to complete the line chart (Fig. 22-10D).

If you draw more than one curve on a chart, use different types of lines or different colors for each curve (Fig. 22-11). Use a full, continuous line of the brightest color for the most important curve. In general, the scales and other identifying notes, or captions, are placed below the X axis and to the left of the Y axis. For large charts, you may sometimes want to show the Y-axis scale at both

the right and the left sides. You may also want to show the X-axis scale at the top and bottom. This will make the chart more convenient to read.

ENGINEERING CHARTS

Experimental information may be plotted from tests and used to obtain an unknown value. In Fig. 22-12, the results of tests have been plotted. These results show as a straight-line curve when drawn in an adjusted position. (Ω is the Greek letter

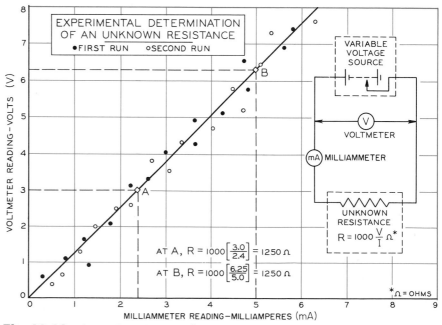

Fig. 22-12. An engineering test chart.

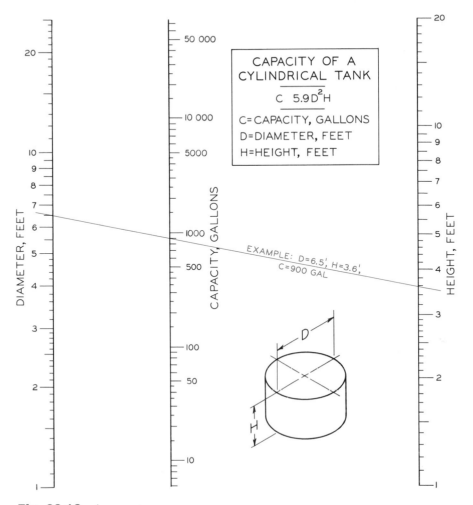

Fig. 22-13. A nomogram.

omega.) Values can be taken from two points and inserted in the formula. In this way, the value of the unknown resistance can be obtained and checked. Notice that tests were made on two occasions, and the results were plotted on the chart. A straight-line curve has been drawn along the center of the path made by the dots.

Nomograms are charts that show the solutions to problems containing three or more variables (kinds of information). Figure 22-13 is an example of this type of chart. A straight line from values on the outside scales will cross the inside scale. The solu-

tion to the equation can be read at this point of intersection. *Nomography* is a special kind of chart construction that requires more than simple mathematics.

BAR CHARTS

A bar chart is probably the most familiar kind of graphic chart. Bar charts are easily read and understood. A bar chart may consist of a single rectangle representing 100 percent (Fig. 22-14F). The chart pictured in Fig. 22-14F represents the total number of games won, lost, and tied.

TO DRAW A ONE-COLUMN BAR CHART

Follow steps A through F to make a one-column bar chart.

A. Prepare and list the information to be presented (Fig. 22-14A).
B. Lay off the long side equal to 100 units. Figure 23-14B shows how this is done.
C. Lay off a suitable width and complete the rectangle (Fig. 22-14C).
D. Lay off the percentage of the parts and draw lines parallel to the base (Fig. 22-14D).

	QUANTITY	PERCENTAGE
GAMES WON	18	60
GAMES LOST	9	30
GAMES TIED	3	10
GAMES PLAYED	30	100

A

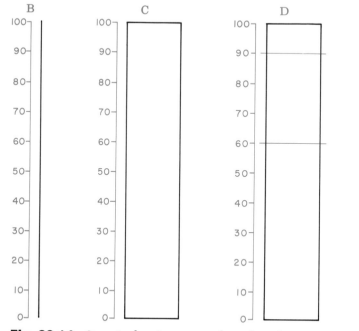

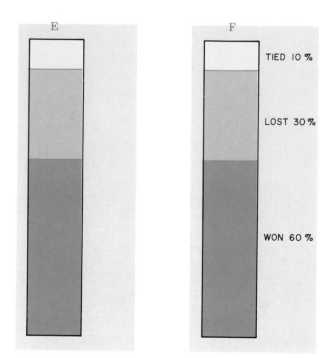

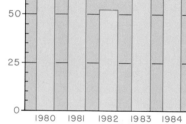

TIED 10 %

LOST 30 %

WON 60 %

Fig. 22-14. Steps in drawing a one-column bar chart.

PERSONAL SAVINGS IN DOLLARS	
1980 —	$ 65.00
1981 —	78.00
1982 —	52.00
1983 —	85.00
1984 —	60.00

A

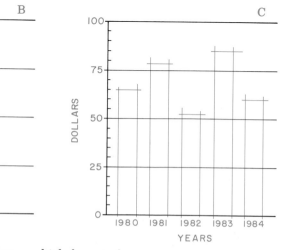

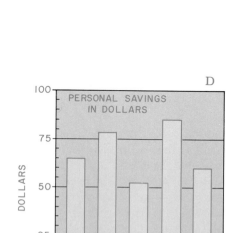

PERSONAL SAVINGS IN DOLLARS

Fig. 22-15. Steps in drawing a multiple-bar graph.

E. Crosshatch, shade, or color the various parts, as shown in Fig. 22-14E.

F. Letter all necessary information in or near the parts so that it can be read easily (Fig. 22-14F).

■ TO DRAW A MULTIPLE-COLUMN BAR CHART

Follow steps A through D to make a multiple-column bar chart.

A. Prepare and list the information to be presented (Fig. 22-15A).

B. Select a suitable scale and lay off the X and Y axes. Lay off the scale divisions (Fig. 22-15B).

C. Block in the bars using the information gathered in Fig. 22-15A. Allow enough space between the bars for all necessary lettering. Make the bars any convenient width so that the overall appearance is pleasing (Fig. 22-15C).

D. Complete the bar chart by adding shading or color to the bars, lettering, and any other lines and information. This will add to the appearance of the chart and the ease with which it can be read and understood. A three-dimensional appearance may also be added, as shown in Fig. 23-15D.

A bar chart with horizontal bars is shown in Fig. 22-16. It gives speed ranges for Caterpillar tractors. Note that the bars do not start at the same line because they show different speed ranges. This kind of chart is called a *progressive chart*.

A *compound-bar chart* is shown in Fig. 22-17, where the total length of the bars is made up of two parts. The black portion is the distance traveled at a given speed before application of

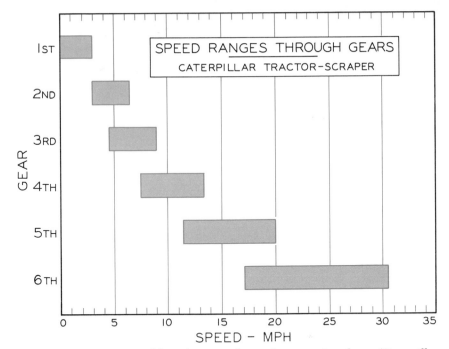

Fig. 22-16. This horizontal-bar chart is a form of progressive chart. (Caterpillar Tractor Co.)

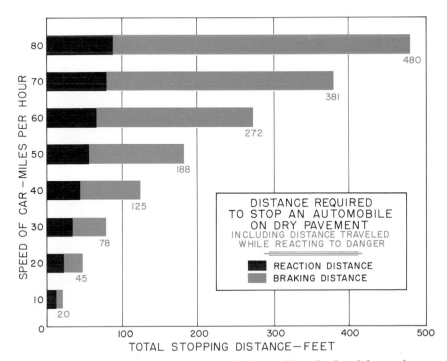

Fig. 22-17. A compound-bar chart in which the total length of each bar is the sum of two parts.

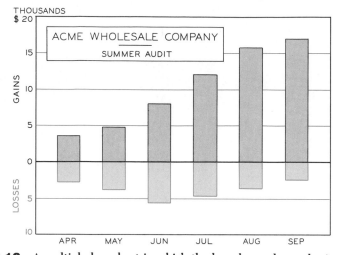

Fig. 22-18. A multiple-bar chart in which the bars have plus and minus values.

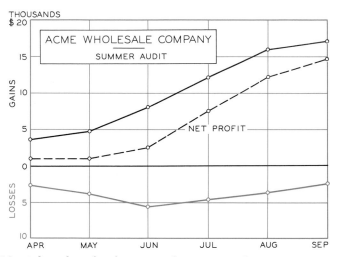

Fig. 22-19. A line chart for the same information as that in Fig. 22-18.

the brakes. The turquoise portion is the distance traveled after applying the brakes. The two portions are added together graphically to give the total distance involved.

The multiple-bar chart shown in Fig. 22-18 has minus values. These values are represented by bars set below the *X* axis. The *Y* axis must have values less than 0 and be drawn past the *X* axis. The same information is given in Fig. 22-19. The vertical scale is the same as in Fig. 22-18. The dashed line shows the net gain. It can be found by laying off the minus values down from the total-gain values.

■ PIE CHARTS

In this type of chart, a circle represents 100 percent. Various sectors represent parts of the whole (Fig. 22-20). These may be drawn flat or in pictorial, as shown.

■ TO DRAW A PIE CHART

Follow steps A through D to make a pie chart.

A. Prepare and list the information to be presented (Fig. 22-21A).

B. Draw a circle of the desired size. Lay off and draw the radial lines

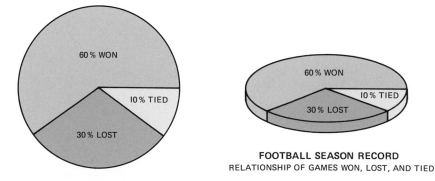

FOOTBALL SEASON RECORD
RELATIONSHIP OF GAMES WON, LOST, AND TIED

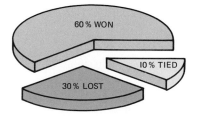

Fig. 22-20. A 100 percent circular chart, or pie chart.

DISTRIBUTION OF CLASS TREASURY		
ITEM	COST	%
DANCE	$ 72.00	40
PARTY	36.00	20
PICNIC	32.40	18
CLASS PLAY	21.60	12
PHOTOGRAPHS	18.00	10
TOTAL	$180.00	100 %

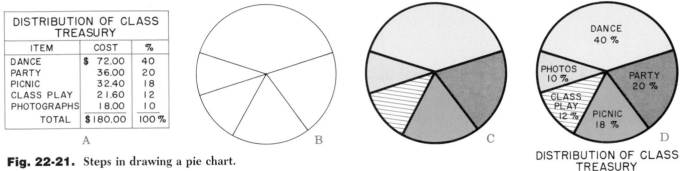

DISTRIBUTION OF CLASS TREASURY

Fig. 22-21. Steps in drawing a pie chart.

representing the amount or percentage for each part on the circumference of the circle (Fig. 22-21B). If a protractor is used, 3.6° = 1 percent. Thirty percent is 30 × 3.6°, or 108°, and so on. If a circle is to be divided into a 24-hour day, each hour represents 15° on the circle.

C. Crosshatch, shade, or color the various parts, as shown in Fig. 22-21C.

D. Complete the pie chart by adding all necessary information (Fig. 22-21D).

■ PICTORIAL CHARTS, OR PICTOGRAPHS

Pictorial charts, or *pictographs*, are similar to bar charts. Pictures or symbols are used instead of bars. Figure 22-22 illustrates a pictorial graphic chart, or pictograph. The chart pictured is a multiple-bar chart. In this chart, each figure represents 100 people. However, it is understood that a symbol or picture may represent any number that the maker of the chart assigns to it. There are three kinds of individuals represented. Adhesive symbols may be used on graphic charts. They are available in many forms (Fig. 22-23). These symbols make pictorial charts easy to construct.

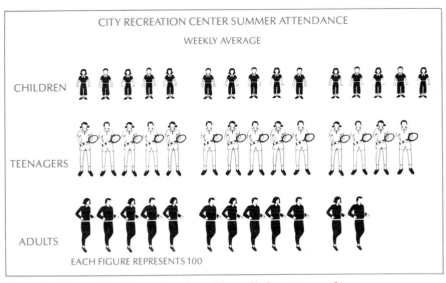

Fig. 22-22. A pictorial graphic chart (also called a *pictograph*).

Fig. 22-23. Many styles of adhesive symbols are available for use on graphic charts. (Chart-Pak, Inc.)

■ ORGANIZATION CHARTS AND FLOWCHARTS

There are many kinds of organization charts. Most, however, have the features of a flowchart. Figure 22-24 is an example. It shows the path, or flow, of drawings from the top engineer to the shop, and the organization of the drafting department.

A flowchart may show the path or series of operations that it takes to manufacture a product or a material.

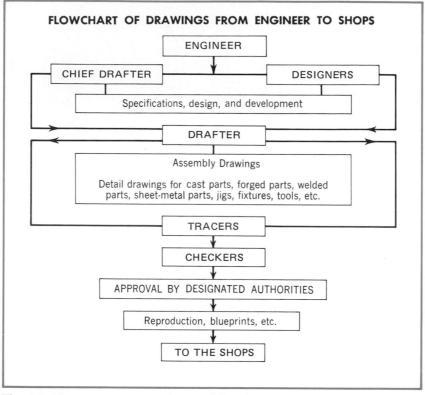

Fig. 22-24. An organization chart and flowchart combined.

An example of this is the flowchart of steelmaking (Fig. 22-25).

■ *TAPE DRAFTING*

Tape drafting is a convenient method of preparing graphic charts. Adhesive tape comes in many colors, designs, and widths. It is applied from a roll dispenser (Fig. 22-26) and pressed onto the chart in the desired position. Tapes of different widths provide a quick and simple way of making bar charts.

■ *THE USE OF COLOR*

Black-and-white charts are often used by scientists and mathematicians for recording information. Black-and-white charts can also be found in newspapers. The use of

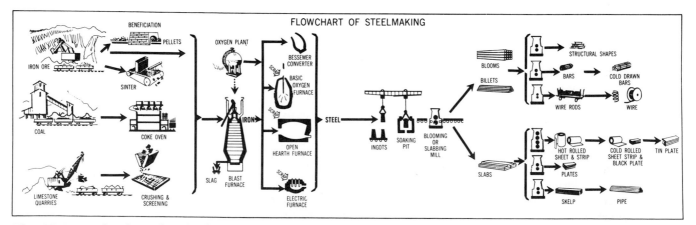

Fig. 22-25. A flowchart of steelmaking. (American Iron and Steel Institute.)

Fig. 22-26. Applying adhesive tape to a graphic chart. (R. J. Capece/McGraw-Hill.)

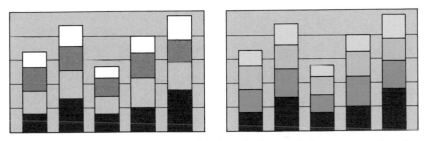

Fig. 22-27. A multiple-bar chart in black and white and the same one in color.

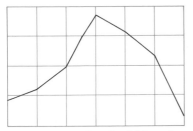

Fig. 22-28. Scales should be selected with care so that the appearance of the curve plotted from the data will aid the understanding of the information.

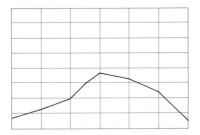

Fig. 22-29. The vertical scale used here is too small. It gives the effect of very little change in values. The movement is slow.

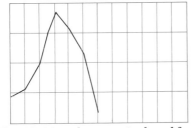

Fig. 22-30. This curve is plotted from the same data used for Figs. 22-28 and 22-29. The horizontal scale is too small (if Fig. 22-28 is a correct picture). The movement is fast.

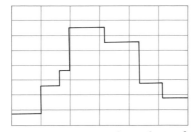

Fig. 22-31. A step chart shows data that remains constant during regular or irregular intervals. This figure might show time periods during which a price remained constant or was raised or was lowered.

color, however, has become quite common in the preparation of charts and diagrams for magazines, books, pamphlets, and various other publications. Color is also used a great deal in making charts for display purposes.

The use of color adds a great deal to the appearance and emphasis of the chart and makes it easier to understand (Fig. 22-27). Color may be added in a variety of ways. Colored pencils, felt-tipped pens, watercolors, or other similar materials are easy to use. Most of these can be found in the drafting room, art room, or at home. Commercially prepared pressure-sensitive materials are available at art and engineering supply stores.

CHARTS AND ELEMENTS OF CHARTS

A great variety of graphic charts can be made for visual communication. The general characteristics and uses of a few types have been discussed in this chapter.

A variety of charts is shown in Figs. 22-28 to 22-36. Some of the elements of chart making and the effects these elements have are presented in these illustrations.

Remember these important points when you make a chart: Every chart should have a suitable title, well lettered and placed near the chart. Every chart should also have a key to tell what the elements represent.

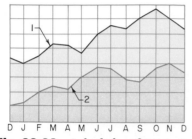

Fig. 22-32. A shaded-surface, or strata, chart uses shaded areas for contrast. This illustration might show the total amount of each of two materials used each month.

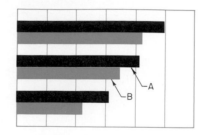

Fig. 22-34. A comparison-bar chart that might be used for two or three values. The illustration might show the amount made (A) and the amount sold (B) of an item by various companies in 1 year or by one company for several years.

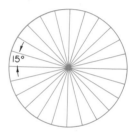

Fig. 22-36. A pie chart representing 24 hours in a day may be used to show time relationships. Each division is 15° and represents 1 hour. Fractional parts of an hour may be estimated.

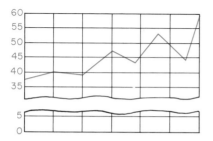

Fig. 22-33. An omission chart may be used for some purposes, as shown, in order to use a larger vertical scale. There are no values below 35, and so a portion of the chart is broken out.

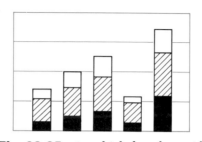

Fig. 22-35. A multiple-bar chart with divided bars. The bars are divided to show the amount of each of three substances that make up the total.

REVIEW

1. What kind of chart is used to show the solutions to problems that contain three or more kinds of information?

2. Is the curve on a chart always a curved line?

3. Refer to the chart in Fig. 22-8. What was the dollar volume of sales in the Northeast region in 1982?

4. Refer to the chart in Fig. 22-17. How far does a car travel during reaction time at a speed of 60 mph?

5. In Fig. 22-17, what is the total stopping distance at a speed of 60 mph? Convert your answer to km/h.

6. Refer to Fig. 22-13. What is the capacity in gallons of a cylindrical tank 10 ft in diameter and 17 ft high? Translate your answer to metric units.

7. In Fig. 22-13, what is the height of a cylindrical tank that holds 100 gal and is 2 ft in diameter?

8. In Fig. 22-13, a tank holds 3000 gal and is 10 ft high. What is its diameter?

9. What is another name for a pictorial graphic chart?

Problems

Fig. 22-37. Draw a pie chart to show the population distribution in the following four regions of the United States:

Northeast	24.2%
North Central	27.8%
South	31.2%
West	16.8%

Use color or various types of crosshatching for contrast.

Fig. 22-38. Draw a pie chart showing that 67.4% of the population of the United States lives within metropolitan areas, while only 32.6% lives outside metropolitan areas. Use different colors or crosshatch areas for contrast.

Fig. 22-39. Make a one-column bar chart to show how your allowance was spent last month.

Fig. 22-40. Make a multiple-bar chart showing a comparison of how you spent your allowance over a period of 4 months.

Fig. 22-41. Assignment 1: The average cost per pound for beef varied over a period of 10 years as follows:

Year	Cost per pound
1975	$1.20
1976	1.10
1977	1.20
1978	1.28
1979	1.15
1980	1.25
1981	1.32
1982	1.28
1983	1.55
1984	2.20

Make a line chart showing this relationship. Assignment 2: The average cost of pork varied somewhat differently. Add a second line to your chart showing a comparison of the cost of beef to the cost of pork over the same period of time. Use colored pencils or different types of lines for contrast.

1975	$1.20
1976	1.30
1977	1.50
1978	1.45
1979	1.38
1980	1.40
1981	1.60
1982	1.70
1983	1.80
1984	2.00

Assignment 3: Add a line to the chart showing the average cost of poultry for the same 10 years.

1975	$0.44
1976	0.52
1977	0.60
1978	0.40
1979	0.46
1980	0.70
1981	0.80
1982	0.66
1983	0.76
1984	0.90

Fig. 22-42. Draw a flowchart showing how to mass-produce a project of your choice in the school shop.

Fig. 22-43. Make an organizational chart showing the administrative structure of your school.

Fig. 22-44. Assignment 1: Draw a pie chart or a one-column bar chart showing a breakdown of the average person's income if 24.5% goes to federal taxes, 5.8% goes to state taxes, and 1.3% goes to local taxes. Be sure the figures total 100%. Assignment 2: Compute the dollar value of each category above for a gross income of $26 500. Mark each on your chart. Be sure the figures total $26 500.

Fig. 22-45. Assignment 1: Make a pictorial chart (pictograph) showing male and female population in your grade in school. Assignment 2: Make a bar chart showing male and female population in your drawing class.

Fig. 22-46. Draw a pictorial chart showing the enrollment of technical drawing classes in your school. Your instructor can supply the information.

Fig. 22-47. Assignment 1: Make a line graph showing the hourly change in outside temperature for a 12-hour period during any day. Assignment 2: Record similar information for several days and make a multiline chart to show a comparison.

Fig. 22-48. Draw a bar chart to show home consumption of electricity for 1 year as follows:

Month	Kilowatt hours used
January	900
February	885
March	800
April	783
May	722
June	600
July	494
August	478
September	525
October	650
November	735
December	820

Fig. 22-49. Plot the batting averages of the players on your favorite baseball team.

Fig. 22-50. The number of cars per mile of road in the United States is growing. Draw a pictorial chart from the data below to show the growth and anticipated increase.

Year	Cars per mile of road	(or)	Cars per kilometer of road
1930	9		2
1950	15		5
1960	20		8
1970	26		12
1980	38		17
1990 (est.)	45		23

Fig. 22-51. Draw a vertical-bar chart showing the following student attendance for a given week of school. The total school enrollment is 925.

Day	Attendance
Monday	625
Tuesday	715
Wednesday	800
Thursday	775
Friday	695

Fig. 22-52. Assignment 1: Compute your daily calorie intake for 1 week. Make a line chart representing this information. Assignment 2: Use the same information and prepare a horizontal-bar chart.

Fig. 22-53. Make a multiline chart representing individual game scores of the top five players on the school basketball team for any given season.

Fig. 22-54. Assignment 1: From the stock-market listings in the newspaper, select any stock and record its daily status for 10 days. Plot the information on a line chart. Assignment 2: Select several stocks and make a multiline chart showing a comparison of growth and decline.

Fig. 22-55. Make a pictorial chart or a pie chart showing a breakdown of the source of each dollar received by the federal government. Use the information given below.

Individual income tax	$0.38
Employment tax	0.26
Corporate income tax	0.14
Borrowing	0.10
Excise tax	0.07
Other (miscellaneous)	0.05

Fig. 22-56. Make a pictorial chart or a pie chart showing a breakdown of the expenditure of each dollar by the federal government. Use the information given below.

National defense	$0.31
Income security	0.27
Interest	0.08
Health	0.07
Commerce, transportation, housing	0.06
Veterans	0.05
Education	0.04
Agriculture	0.03
Other (miscellaneous)	0.09

Fig. 22-57. The following information includes five common foods and the number of calories and grams of carbohydrates in a 4-ounce serving of each. Make a bar chart illustrating these facts.

Food	Calories	Carbo-hydrates
Chocolate ice cream	150	14
Peas	75	14
Pizza	260	29
Milk	85	6
Strawberries	30	6

Fig. 22-58. Assignment 1: The data in the list below represent a percentage breakdown for a family budget. Make a pie chart illustrating this information.

Food	23.1%
Housing	24.0%
Transportation	8.8%
Clothing	10.9%
Medical Care	5.6%
Income Tax	12.5%
Social Security	3.8%
Miscellaneous	11.3%

Assignment 2: Compute the dollar value of each category for a gross income of $25 000. Mark each on your chart. Be sure the figures total $25 000.

Fig. 22-59. Accidents involving children occur in various places. Draw a horizontal bar chart or a pie chart, using the places and percentages given below.

At home	25%
Between home and school	8%
On school grounds	15%
In school buildings	21%
In other places	31%

23 Electrical and Electronics Drafting

POWERFUL PROGRESS

Progress in making electrical power started with Thomas Alva Edison's electricity generating station in New York City in 1882. Since then electricity has become one of humanity's practical servants. It has completely changed the communication, manufacturing, and utility industries. Electricity and electronics are a powerful team for our space shuttles, computers, communications systems, automated machinery, and everyday appliances.

Imagine miniature circuits (microelectronics) on a chip of silicon that's smaller than a fingernail. The newest communications system in America has a digital (number) switching system that uses a single silicon chip to replace 150 000 transistors. This communications control system using a single chip can make a million decisions in one second. Of course, a few years ago the transistor took the place of the vacuum tube.

The space shuttle utilizes fuel cells for electric power. Designers are capable of utilizing the power for the shuttle during ascent, descent, and in-orbit operations with carefully calculated energy requirements (Fig. 23-1).

CAREER OPPORTUNITIES

The electronics industry is one of the biggest manufacturing industries in the country. This industry offers special opportunities to young women and men who can make the freehand and formal drawings it needs. Electronic drafting is based upon the same basic rules of all other drafting. However, electronic drafting also requires the preparation of schematic diagrams, block diagrams, and technical illustrations.

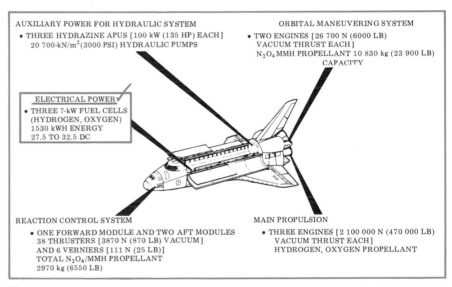

AUXILIARY POWER FOR HYDRAULIC SYSTEM
- THREE HYDRAZINE APUS [100 kW (135 HP) EACH]
  20 700-kN/m^2(3000 PSI) HYDRAULIC PUMPS

ORBITAL MANEUVERING SYSTEM
- TWO ENGINES [26 700 N (6000 LB)
  VACUUM THRUST EACH]
  N$_2$O$_4$ MMH PROPELLANT 10 830 kg (23 900 LB)
  CAPACITY

ELECTRICAL POWER
- THREE 7-kW FUEL CELLS
  (HYDROGEN, OXYGEN)
  1530 kWH ENERGY
  27.5 TO 32.5 DC

REACTION CONTROL SYSTEM
- ONE FORWARD MODULE AND TWO AFT MODULES
  38 THRUSTERS [3870 N (870 LB) VACUUM]
  AND 6 VERNIERS [111 N (25 LB)]
  TOTAL N$_2$O$_4$/MMH PROPELLANT
  2970 kg (6550 LB)

MAIN PROPULSION
- THREE ENGINES [2 100 000 N (470 000 LB)
  VACUUM THRUST EACH]
  HYDROGEN, OXYGEN PROPELLANT

Fig. 23-1. The power subsystem can provide as much as 7000 watts (average) to 12 000 watts (peak) for major energy-consuming payloads. (NASA.)

PREPARING FOR OPPORTUNITIES

Industry provides training programs in electrical and electronics drafting. However, it helps if you have learned drafting and electronics in high school before entering such programs. A few courses at technical college or junior college in electronic drafting and in *electromechanical systems* (moving things with electric power) will help if you are looking for a technician rating on the design team. To learn more about careers, write to the following groups:

Institute of Electrical and Electronic
    Engineers (IEEE)
345 East 47 Street
New York, NY 10017

Electronic Industries Association
    (EIA)
2001 I Street N.W.
Washington, DC 20036

THE ELECTRICAL OR ELECTRONIC DRAFTER

An electrical or electronic drafter on a design-engineering team must always develop new skills. The drafter must always be learning new things about the field he or she works in. The drafter must show good judgment, skill, and originality when working from design sketches or written instructions. Future electronic developments will be more spectacular than those of today. Scientists, along with electrical engineers, designers, technicians, and drafters, seem to be building an electrical and electronic world. The industrial robot shown in Fig. 23-2 is an example of electromechanical progress.

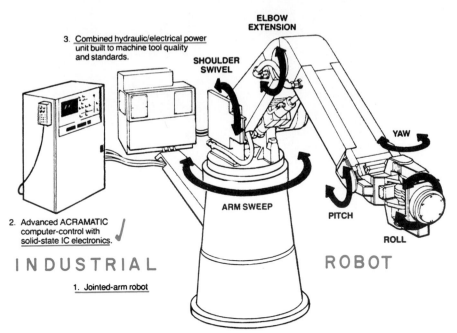

3. Combined hydraulic/electrical power unit built to machine tool quality and standards.

ELBOW EXTENSION

SHOULDER SWIVEL

YAW

ARM SWEEP

PITCH

ROLL

2. Advanced ACRAMATIC computer-control with solid-state IC electronics.

INDUSTRIAL ROBOT

1. Jointed-arm robot

Fig. 23-2. By definition, a *robot* is a reprogrammable machine designed to handle materials or tools for a variety of tasks. (Cincinnati Milacron.)

ELECTRONIC ENVIRONMENT

People have always tried to control their *environment* (surroundings). Today, electronic devices control the air we breathe at home and at school. Electronic devices have been made that control the cooling, heating, lighting, and sound systems for different ways of living.

The electronic room shown in Fig. 23-3 is inside a 25-ft (7620-mm) Fiberglas dome. The room has a recessed living area in the center. This arrangement makes it easier to relax. The room also has an electronic sight and sound system. *Mood lighting* (lighting that changes brightness or color) responds to the remote-controlled color television, the television recorder, and the sound system.

The pictorial sketch (Fig. 23-4) shows the major *components* (parts) of the electronic environment. The center table at 1 is the control center for the components of the room. The rectangle at 2 is a color television set. The rectangle at 3 is an electronic video recorder (EVR) that tapes television shows. The rectangles numbered 4 are speaker towers. They also contain the mood lighting that changes with the sound. The center table rises mechanically to show the controls for the sound system, color television, and EVR. The overhead lighting is *kinetic* (moving). It changes with the beat of the music. You can see three of the four sound towers that hold speakers and special lighting. Each tower has three speakers, one each for high-, low-, and middle-range sound.

The electric circuits for this room must be planned as a wiring diagram. The electrical design engineer makes this diagram. The electronic components for the room must be designed by the electronic design team.

Fig. 23-3. A room especially designed for an electronic environment. (Motorola, Inc.)

ELECTRICAL AND ELECTRONIC DRAFTING

Students who have had a basic course in electricity will find it easier to understand and make electrical or electronic drawings. This chapter will introduce electrical and electronic symbols, wiring diagrams, and circuit diagrams. The chapter will also relate these to electrical and electronic drafting. For students who have not had a basic course in electricity, the paragraphs that follow give a brief explanation of the subject.

ELECTRICITY

The source of electrical energy is the tiny *atom*. All atoms are made up of many kinds of *particles*. One of these particles is the *electron*. The electron is the most important particle in the study of electricity and electronics.

The electrons in an atom rotate around the *nucleus*, or center, of the atom in definite paths. These paths are called *orbits* (Fig. 23-5). All electrons in all atoms are the same. Each electron has what is called a *negative charge* of electricity.

Different kinds of atoms have different numbers of electrons and other particles. When atoms have the same number of electrons and the same number of *protons* (a particle in the nucleus), they are the same *element*. Copper, gold, and lead are examples of elements.

When atoms join together, they form *molecules*. When different kinds of atoms join, they form *compounds*. Water, acids, and salt are common compounds.

VOLTAGE AND CURRENT

Sometimes, electrons can be made to leave their "parent" atoms, the atoms of which they were originally part. This happens, for example, when a piece of wire is connected across the *terminals* (electrical connections) of a battery. The battery produces an electrical pressure called *voltage*. The symbol for voltage is V. The voltage causes a steady stream of electrons to flow through the wire. Connect a light bulb (load)

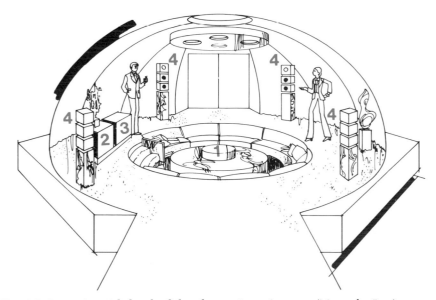

Fig. 23-4. A pictorial sketch of the electronic equipment. (Motorola, Inc.)

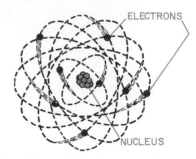

Fig. 23-5. The electron in the structure of an atom.

to the wire. Electrons will move through the *lamp filament* (thin wire in the bulb) from the battery (power source) (Fig. 23-6). The energy of the moving electrons is changed into heat energy as the filament becomes white hot. The glow of the filament produces the light.

The electron pathway is formed by the battery, the wire, and the lamp filament. This is a simple form of *electric circuit*. In other circuits, electrical energy is changed into other kinds of energy. Some of these kinds of energy are magnetism, sound, and light.

A *direct current* (dc) is a flow of electrons through a circuit in one direction only (Fig. 23-7). An *alternat-*

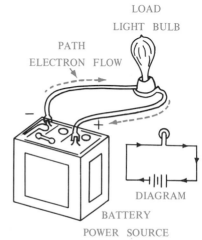

Fig. 23-6. A simple electric circuit.

ing current (ac) is a flow of electrons in one direction during a fixed time period, and then in the opposite direction during a similar time period (Fig. 23-8). One complete alternation is called a *cycle*. The number of times this cycle is repeated in one second is called the *frequency* of the alternating current, such as 60 cycle. We measure frequency in *hertz*, for which the symbol is Hz. Current is measured in amperes, and the symbol for current is *I*.

■ RESISTANCE

Electrons can move through some materials more easily than through other materials. Electric current will flow more easily through a copper wire than through a steel wire of the same size. We say that the steel offers more *resistance* than the copper. Materials with small resistance to the flow of electrons are called

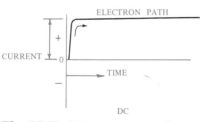

Fig. 23-7. Direct current attains magnitude and keeps it as long as the circuit is complete.

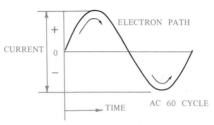

Fig. 23-8. Alternating current builds up from zero to a maximum in a positive direction, falls to zero, then builds to a maximum in a negative direction and falls back to zero.

conductors. Silver is the best conductor known. However, it costs too much for general use. Copper and aluminum are good conductors. They are the most widely used. Materials through which electrons will not flow easily are called *insulators*. The insulators used most often are glass, porcelain, plastics, and rubber compounds. We measure resistance in *ohms*, and the symbol for resistance is *R*.

■ ELECTRICITY AND ELECTRONICS

Electricity has to do with the flow of electrons moving through wires or other metal conductors. Common examples are house wiring systems, generators, and transformers. Electricity refers to an energy source.

Electronics has to do with the flow of electrons moving through metal conductors and conductors other than metals. Some of these other conductors are gases, vacuums, and materials called *semiconductors*. The most common semiconductors are transistors and diodes made of germanium or silicon. Electronics refers to devices that make use of electricity.

Both electricity and electronics deal with electrons flowing through circuits. These circuits carry energy for a definite purpose.

■ BASIC ELECTRICAL UNITS

As we have seen, volts are used to measure pressure, ohms to measure resistance, and amperes to measure current. In addition, we use *watts* (symbol: W) to measure power. You can often tell the value of a unit used to show an electrical *quantity* (amount) by looking at the *prefix* (beginning) of the word. Take, for example, the unit kilovolt. *Kilo* means a thousand. Thus, 1 kilovolt (kV) equals 1000 V. Similarly, 1 kilowatt (kW) equals 1000 W, and 1 kilohm (kΩ) equals 1000 Ω. Another prefix is *milli*, which means thousandth. Thus, 1 milliampere (mA) equals 1 one-thousandth (0.001) A. For other unit prefixes, see Appendix B on the metric system.

■ BASIC FORMULAS

The amounts of voltage, current, and resistance in a circuit are related. This relationship is called *Ohm's law*. Ohm's law may be expressed as shown in the 12 formulas of Fig. 23-9. V = volts (pressure), I = amperes (current), and R = ohms (resistance). These letters are used in mathematical formulas to mean *unknown amounts*. They are not International System (SI) symbols. The SI unit symbols are V for electrical pressure, A for current, and R for resistance. These SI unit symbols should only be used with *known amounts*.

■ BASIC ELECTRIC CIRCUITS

These include series circuits, parallel circuits, and series and parallel circuits together. These terms are explained in the following paragraphs.

Series Circuits

Series circuits are those where the current flows from the source (battery, generator, and so forth) through one resistance (lamp, motor, and so forth) after another. This is shown in Figs. 23-10, 23-11, and 23-12.

In Fig. 23-10, a bell (*A*) gets its power from a battery (*C*) when the

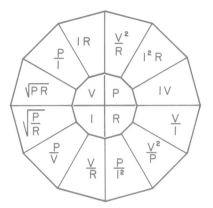

Fig. 23-9. Power expressed as formulas in Ohm's law.

circuit is closed (when the electron pathway is complete). The circuit is normally open. The pushbutton (*B*) closes the circuit.

In Fig. 23-11, a buzzer (*A*) gets its power by the current from the transformer (*C*). What is item *B*? What does it do in this circuit?

In Fig. 23-12, four lamps (*C*, *D*, *E*, and *F*) get power from a generator (*A*) when the fused switch (*B*) is closed. All the lights must be on. If any one is not, the circuit will be open. Some strings of Christmas tree lights go out completely when just one lamp burns out. They are connected in series.

Parallel Circuits

Parallel circuits let the current flow through more than one path. This is shown in Figs. 23-13 and 23-14.

There are three separate branches, or paths (*C*, *D*, and *E*), with lamps in Fig. 23-13. Each lamp is separate from the others. If one lamp is burned out, the others will still work. With a parallel string of lights on a Christmas tree, the remaining lights will still burn if some are missing, loose, or burned out.

A siren is shown in Fig. 23-14. It can be turned on by any of the push-

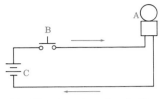

Fig. 23-10. A series-circuit diagram.

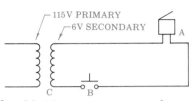

Fig. 23-11. A series-circuit diagram for a buzzer.

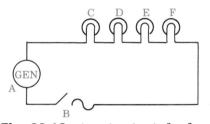

Fig. 23-12. A series-circuit for four lamps.

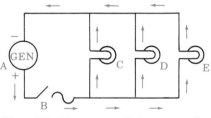

Fig. 23-13. A parallel circuit for three lamps.

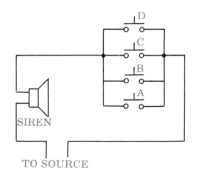

Fig. 23-14. A parallel-circuit diagram with four pushbuttons.

buttons *A*, *B*, *C*, or *D*. These buttons are all connected in parallel. A good use of this would be in an alarm system to warn of an attempted holdup in a store. The pushbuttons, connected in parallel, would be under counters and in the cashier's office.

Notice that the symbol for the siren is the same as for a loudspeaker. So that they will not be confused, the note SIREN has been added.

Combination Circuits

Combining series and parallel circuits permits many different arrangements. In Fig. 23-15, lamps *C* and *D* are in series. Lamps *E* and *F* are in parallel. Both lamps *C* and *D* must be on if switch *A* is closed, since they are in a series. When switches *A* and *B* are closed, all the lamps (*C*, *D*, *E*, and *F*) are lighted. Lamps *E* and *F* will work separately. If one fails, the other will stay lighted. Lamps *E* and *F* will work separately. If one fails, the other will stay lighted because they are in parallel. However, because lamps *C* and *D* are in series, when one fails, the other will not light, as we have learned from the Christmas tree lights.

■ ELECTRICAL INSTRUMENTS

Many kinds of electrical instruments can be used to measure electricity. Two main ones are the ammeter and the voltmeter. The *ammeter* measures electric current in amperes. To measure the amount of current flowing through a resistance (light, motor, and so on), connect the ammeter directly in series with the resistance you want to measure. This is shown in Fig. 23-16.

The *voltmeter* measures the electromotive force (pressure) in volts. Connect the voltmeter in parallel with the part of a circuit where you want to measure the voltage. This is shown in Fig. 23-17.

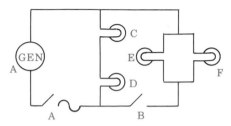

Fig. 23-15. A combination series and parallel circuit.

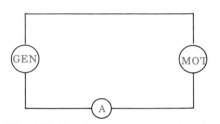

Fig. 23-16. Ammeter connection in series in a circuit.

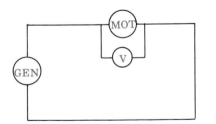

Fig. 23-17. Voltmeter connection in parallel in a circuit.

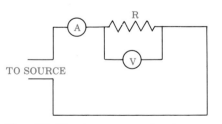

Fig. 23-18. Ammeter and voltmeter connections.

Figure 23-18 shows both an ammeter and a voltmeter connected in a circuit. The ammeter measures the current flowing through the resistance *R*. The voltmeter measures voltage flowing across the resistance. The amperes and the volts are then measured. Resistance can be measured with an ohmmeter.

■ GRAPHIC SYMBOLS

We use graphic symbols on electrical and electronic diagrams to show the components and workings in a circuit. Symbols can be drawn quickly and easily with templates (Fig. 23-19).

The graphic symbols in Fig. 23-20 and the following sections about using graphic symbols are adapted from the *American National Standard Graphic Symbols for Electrical and Electronics Diagrams* (ANSI Y32.2) by permission of the Institute of Electrical and Electronic Engineers, Inc. (IEEE).

Graphic symbols for electrical engineering are a shorthand way to show through drawings how a circuit works or how the parts of the circuit are connected. A graphic symbol shows what a part in the circuit does. Drafters use graphic symbols on single-line (one-line) diagrams, on schematic diagrams, or on connection or wiring diagrams. You can relate graphic symbols with parts lists, descriptions, or instructions by marking the symbol.

Drafting and Graphic Symbols

1. A symbol is made up of all its various parts.
2. The direction a symbol is facing on a drawing does not change its

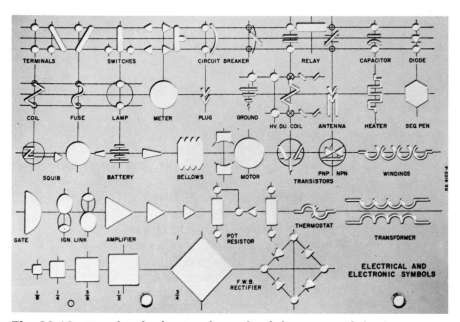

Fig. 23-19. Template for drawing electrical and electronic symbols. (RapiDesign, Inc.)

mcaning. This is true even if the symbol is drawn backwards.

3. The width of a line does not affect the meaning of the symbol. Sometimes, however, a wider line can be used to show that something is important.

4. Each symbol shown in this standard is the right size related to all the other symbols. That is, if one of these symbols is drawn twice as big as shown in the standard, any other symbol should also be drawn twice as big.

5. A symbol can be drawn any size needed. Symbols are not drawn to scale. However, their size must fit in with the rest of the drawing.

6. The arrowhead of a symbol can be drawn closed → or open →.

7. The standard symbol for a *terminal* (○) can be added to any one of the graphic symbols used where connecting lines are at-

tached. These are not part of the graphic symbol unless the terminal symbol is part of the symbol shown in this standard.

8. To make a diagram simpler, a symbol for a device may be drawn in parts. If this is done, the relationship of the parts must be shown.

9. Most of the time, the angle of a line connected to a graphic symbol does not matter. Generally, lines are drawn horizontally and vertically.

10. Sometimes it may be desirable to draw paths and equipment that will be added to the circuit later, or that are connected to the circuit but are not part of it. This is done by drawing lines made up of short dashes: - - - -.

11. If details of type, impedance, rating, and so on, are needed, they may be drawn next to a symbol. The abbreviations used should be from the *American*

National Standard Abbreviations for Use on Drawings (Y1.1). Letters that are joined together and used as parts of graphic symbols are not abbreviations.

■ CIRCUIT COMPONENTS

Figure 23-21 shows and names some of the electrical and electronic components most often used. A symbol for a component should look like that component. You should know what a component does and how it works.

■ ELECTRICAL DIAGRAMS

There are many kinds of electrical diagrams. Each kind of diagram suits its purpose. The definitions that follow are from the *American National Standard Drafting Manual, Electri-*

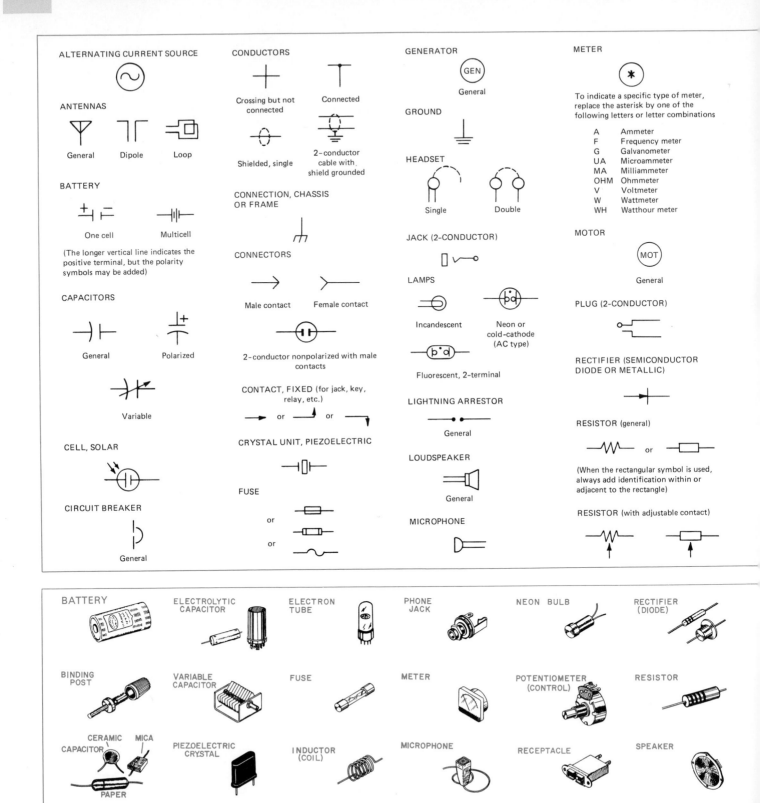

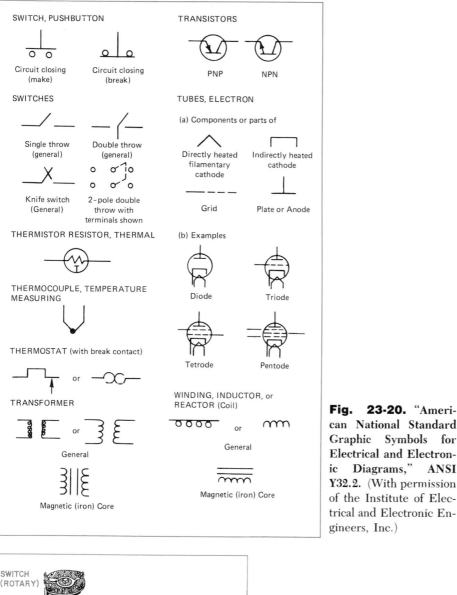

Fig. 23-20. "American National Standard Graphic Symbols for Electrical and Electronic Diagrams," ANSI Y32.2. (With permission of the Institute of Electrical and Electronic Engineers, Inc.)

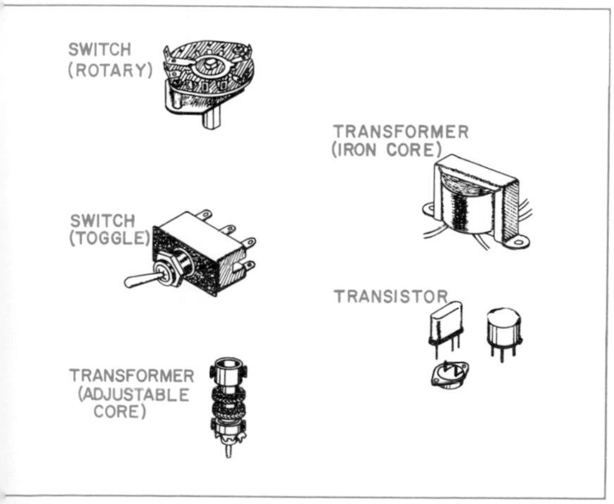

cal Diagrams (ANSI Y14–15) with the permission of the publisher, the American Society of Mechanical Engineers, 345 East 47 Street, New York, NY.

Single-Line (One-Line) Diagram. This diagram shows, using single lines and graphic symbols, the course of an electric circuit and the parts of the circuit.

Schematic, or Elementary, Diagram. This diagram shows, using graphic symbols, the ways a circuit is connected and what the circuit does. The schematic diagram does not have to show the size or shape of the parts of the circuit. It does not have to show where the parts of the circuit actually are.

Connection or Wiring Diagram. This diagram shows how the components of a circuit are connected. It may cover connections inside or outside the components. It has as much detail as is needed to make or trace connections. The connection diagram usually shows how a component looks and where it is placed.

Interconnection Diagram. This is a kind of connection or wiring diagram that shows only connections outside a component. An interconnection diagram shows connections between components. The connections inside the component are usually left out.

■ LINE CONVENTIONS AND LETTERING

Sometimes the size of a drawing is changed. When this is done, remember to choose a line thickness and letter size that will still let people understand the drawing after it is made larger or smaller. A guide for draw-

Fig. 23-21. Some electrical and electronic components, with their names and appearance. (Heath Company, Inc.)

LINE APPLICATION	LINE THICKNESS
FOR GENERAL USE	MEDIUM
MECHANICAL CONNECTION, SHIELDING, & FUTURE CIRCUITS LINE	MEDIUM
BRACKET-CONNECTING DASH LINE	MEDIUM

USE OF THESE LINE THICKNESSES OPTIONAL

BRACKETS, LEADER LINES, ETC.	THIN
BOUNDARY OF MECHANICAL GROUPING	THIN
FOR EMPHASIS	THICK

Fig. 23-22. Line conventions for electrical diagrams.

ing lines on electrical diagrams is shown in Fig. 23-22.

Draw lines of medium thickness for general use on electrical diagrams. Use thin lines for brackets, leader lines, and so on. When something special needs to be set off, such as main or transmission paths, use a line thick enough to show the difference. Line thickness and lettering used with electrical diagrams should conform with American National Standard Y14.2 (latest issue) and local needs. This is so that microfilm of the diagrams can be made.

SYMBOLS AND LAYOUTS

A symbol can be drawn any size needed. However, its size must fit in with the rest of the drawing. Keep in mind whether the drawing will be made larger or smaller. For most electrical diagrams meant to be used for manufacturing, or for use in a smaller form, draw symbols about 1.5 times the size of those shown in American National Standard Y32.2.

Layout of Electrical Diagrams. Lay out electrical diagrams so that the main parts are easily seen. The

parts of the diagram should have space between them. This is so that there will be an even balance between blank spaces and lines. Allow enough blank area around symbols so that notes or reference information will not be crowded. Avoid larger spaces, however. Only allow large spaces if circuits will be added there later.

SINGLE-LINE DIAGRAMS

The single-line diagram (Fig. 23-23) tells in a basic way how a circuit works. It leaves out much of the detailed information usually shown on schematic or connection diagrams. Single-line diagrams make it possible to draw complex circuits in a simple

way. You can draw diagrams of communications or power systems where a single line means a multiconductor communications or power circuit.

Most of the time, single-line diagrams can be drawn with the same methods used to draw schematic diagrams.

SCHEMATIC DIAGRAMS

Following are some guidelines for making schematic diagrams.

Layout. Use a layout that follows the circuit, signal, or transmission path either from input to output, from power source to load, or in the order that the equipment works. Do not use long interconnecting lines between parts of the circuit.

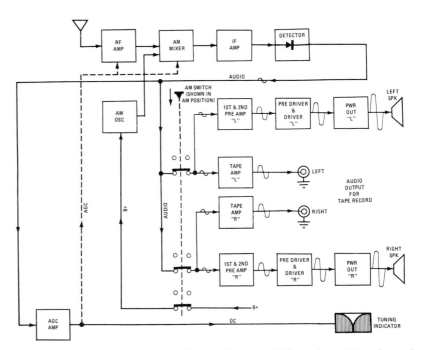

Fig. 23-23. A block diagram for the complete signal flow of an AM radio. A loop antenna picks up the signal and presents it to the RF amplifier. The mixer receives both the selected AM and oscillator signals. At the AM detector, the audio is recovered and applied to the audio channels. The stereo channels are connected in parallel to the AM switch, providing monaural operation. (Motorola, Inc.)

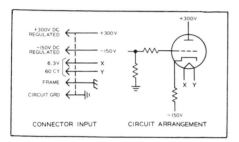

Fig. 23-24. Identification of interrupted lines. At left, a group of interrupted lines on the diagram. At right, single lines interrupted on the diagram.

Connecting Lines. It is better to draw connecting lines horizontally or vertically. Use as few bends and crossovers as possible. Do not connect four or more lines at one point if it can just as easily be drawn another way. When connecting lines are drawn parallel (side-by-side), the spacing between lines after making the drawing smaller should be no less than 1/16 in. (2 mm). Draw parallel lines in groups. It is best to draw them in groups of three. Allow double spacing between groups of lines. When you group parallel lines, group them together according to what they do.

Interrupted Single Lines. When a single line is interrupted, show where the line is going in the same place you *identify* (name) it. This is shown in Fig. 23-24 for the power and filament circuit paths. The following section on interrupted group lines tells how to identify grouped and bracketed lines. Do the same for single interrupted lines.

Interrupted Grouped Lines. When interrupted lines are grouped and bracketed, identify the lines as shown in Fig. 23-25. You can show at the brackets where lines are meant to go or where they are meant to be connected. Do this by using notes

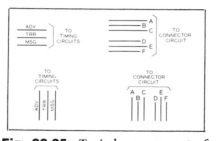

Fig. 23-25. Typical arrangement of line identifications and circuit destinations.

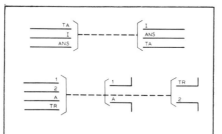

Fig. 23-26. Typical interrupted lines interconnected by dashed lines. The dashed line shows the interrupted paths that are to be connected. Individual line identifications indicate matching connections.

outside the brackets, as shown in Fig. 23-25, or by using a dashed line between brackets, as shown in Fig. 23-26. When using the dashed line to connect brackets, draw it so that it will not be mistaken for part of one of the bracketed lines. Begin the dashed line in one bracket and end it in no more than two brackets. When drawing schematics, carefully follow the rules given above.

■ SCHEMATIC DIAGRAMS FOR ELECTRONICS AND COMMUNICATIONS

The following tells in detail about schematic diagrams used with electronic and communications equipment. Use this material with the general standards of schematic diagrams already learned.

Layout. In general, lay out schematic diagrams so that they can be read from left to right (input on left and output on right). Complex diagrams should generally be laid out to read from upper left to lower right. They may be laid out in two or more layers. Each layer should be read from left to right. The circuit layout should follow the signal, or transmission path, from input to output, or in the order that the circuit works. Where possible, draw endpoints for outside connections at the outer edges of the circuit layout.

Ground Symbols. Use the ground symbol ⊥ or ⏚ only when the circuit ground is at a potential level equivalent to that of earth potential. Use the symbol ⌿ when you do not get an earth potential from connecting the ground wire to the structure that houses or supports the circuit parts. This is known as a *chassis ground.*

■ SOME ELECTRICAL CIRCUITS

Look at Fig. 23-27. The bell (*C*) and the buzzer (*E*) get power separately from the same battery (*A*) by the pushbuttons (*B* and *D*).

In Fig. 23-28, the current is from an outside source. Three-way

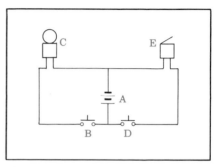

Fig. 23-27. A bell and buzzer circuit with pushbuttons.

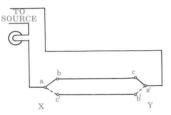

Fig. 23-28. A three-way switch.

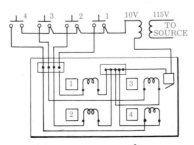

Fig. 23-29. Annunciator diagram.

switches (*X* and *Y*) are used. This is done so that the light can be turned off or on by either of the switches. Switch *X* might be at the garage. Switch *Y* might be at the house. Each switch has three terminals. If either switch is opened, the light will be turned off. However, the light can be turned on by the switch at the other end.

The circuit diagram in Fig. 23-29 is for an *annunciator* (a device for signaling from different places to a station or post). When any of the buttons is pushed, a buzzer in the annunciator sounds. Each button releases, or allows, a tab to drop down. The tab shows the place where the button is pushed. Trace the circuits that begin with each of the buttons. The source of the current is a transformer.

Some single-line graphic symbols are shown in Fig. 23-30. A single-line, or one-line, diagram is shown in Fig. 23-31. This diagram shows the component parts of a circuit. It uses single lines and graphic symbols.

The single lines mean two or more conductors. The diagram shows all you usually need to know about how the circuit works. However, it is not as detailed as a schematic diagram. A single-line diagram is a simpler way to show circuits. It is often used in the fields of communications and electrical power *transmission* (sending) (Fig. 23-31). Drafters usually draw the highest voltage at the top or draw the highest voltage at the top or left side of the diagram. They draw the lower-voltage lines in the order of their value below or on the right side of the diagram. Drafters give other information at the proper places on the diagram. Such information includes where lines are, what equipment is used, and so on. A wiring diagram for a motor starter is shown at A in Fig. 23-32. A schematic, or one-line, diagram for the

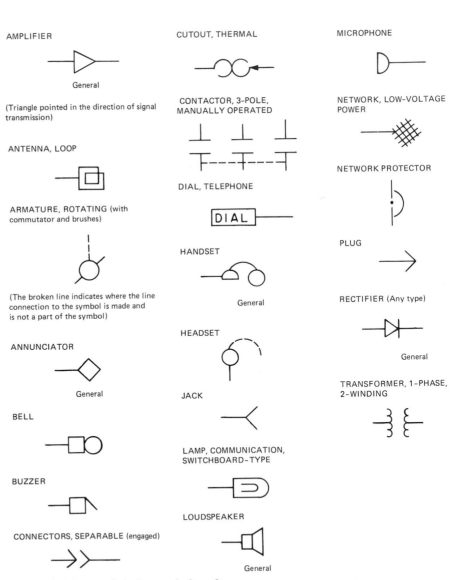

Fig. 23-30. Symbols for single-line diagrams.

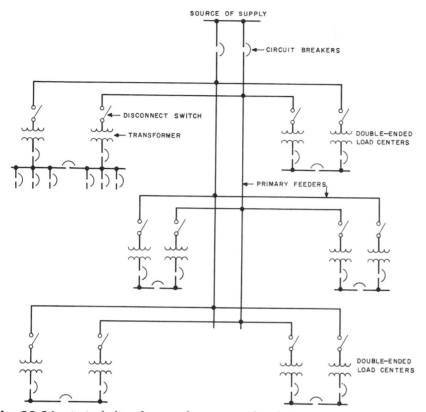

Fig. 23-31. A single-line diagram for a power distribution system.

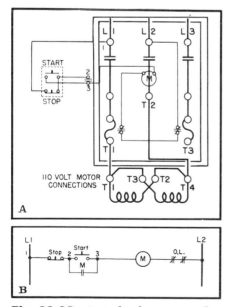

Fig. 23-32. A single-phase starter for a three-motor system.

same circuit is shown at B. The uses of a motor starter are given below:

1. To protect the circuit against burnouts caused by long *overloads* (too much power on the circuit). This is called *thermal overload protection*. Ordinary fuses cannot protect the circuit in this way.
2. To allow remote control by start-stop buttons or automatic devices such as thermostats.
3. To allow sequence control so that the order in which equipment is started can be controlled.

■ COLOR CODES

Color codes are an easy way to show information when drawing circuit di-

agrams. Color codes are also used on the actual wiring of the circuit. In electrical and electronic work, drafters use a color code to show certain characteristics of components, to identify wire leads, and to show where wires are connected. A color code may be included on the diagram (Fig. 23-33).

When using a color code, you should look up the Electronic Industries Association (EIA) standards. Also look up any other codes that might be needed. Note the different code used in Fig. 23-33. It is not the same code as the EIA standard code used in Fig. 23-34.

Color codes are used to give exact information about electrical equipment. You can find out about such uses in published standards and in textbooks.

■ BLOCK DIAGRAMS

A block diagram (Fig. 23-35) is usually made up of squares or rectangles, or "blocks." They are joined by single lines. The blocks show how the components or stages are related when the circuit is working. Note the arrowheads at the terminal ends of the lines. These arrowheads show which way the signal path travels from input to output, reading the diagram from left to right.

The identification of the stage is lettered within the block or just outside it. The blocks are often drawn along with symbols and a schematic diagram (see Figs. 23-40 through 23-44).

Engineers often draw or sketch block diagrams as a first step in designing new equipment. Because blocks are easy to sketch, the engineer can try many different layouts before deciding which to use. The

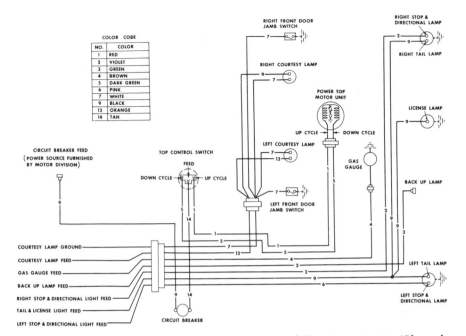

COLOR CODE	
NO.	COLOR
1	RED
2	VIOLET
3	GREEN
4	BROWN
5	DARK GREEN
6	PINK
7	WHITE
9	BLACK
13	ORANGE
14	TAN

Fig. 23-33. Wire color scheme used on an automobile wiring circuit. (Chevrolet Motor Division, General Motors Corp.)

overlay method of sketching discussed in Chap. 2 can be used with great success in drawing or sketching block diagrams.

Block diagrams are used in catalogs, descriptive folders, and advertisements for electrical equipment.

They are also used in technical-service literature to aid in the repair of equipment. The size of a block diagram is generally determined by the amount of information lettered on the components.

ELECTRICAL LAYOUTS FOR BUILDINGS

Figure 23-36 shows the usual way an architect indicates where electric outlets and switches are to be

placed. This plan only shows where the lights, base plugs, and switches are to be placed. You cannot build a good electrical system using this diagram. For this, a complete and detailed set of electrical drawings is needed. These must be made by someone who knows the engineering needs of the system. A list of the symbols used is shown in Fig. 23-37.

THE INTERCONNECTION DIAGRAM

Figure 23-38 is a connection, or wiring, diagram. It shows the electrical connections between the different components of an electrical or electronics system. Generally, the connections inside each component are not shown. The name of each component is given. Each component is shown on the diagram by a rectangle.

PRINTED CIRCUIT DRAWINGS

These are used in making actual printed circuit boards. These boards are used in electronic equipment. The drawings are exact layouts of the pattern of the circuit needed. The drawing is made actual size or larger. If drawn larger, it can be made

EIA Standard Color Code		
Color	Abbre-viation	Number
Black	BLK	0
Brown	BRN	1
Red	RED	2
Orange	ORN	3
Yellow	YEL	4
Green	GRN	5
Blue	BLU	6
Violet	VIO	7
Gray	GRA	8
White	WHT	9

Fig. 23-34. The EIA color code standard.

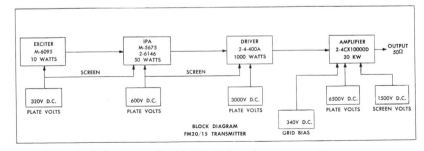

Fig. 23-35. Block diagram of a 20 000-W broadcast transmitter.

SYMBOLS

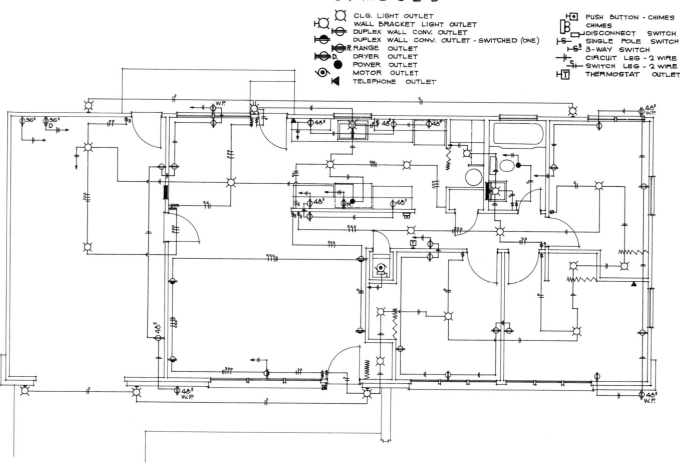

ELECTRICAL PLAN

Fig. 23-36. Electrical plan for a ranch house. (See Chap. 20.)

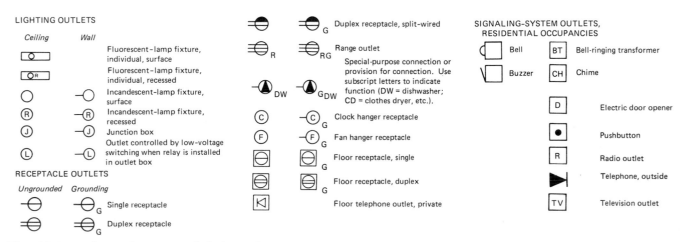

Fig. 23-37. Electrical wiring symbols for architectural layout. (Continued on next page.)

SWITCH OUTLETS

S   Single-pole switch

S₂   Double-pole switch

S₃   Three-way switch

S_P   Switch and pilot lamp

S_L   Switch for low-voltage switching
           system

S_D   Door switch

⊖S   Switch and single receptacle

⊜S   Switch and double receptacle

(S)   Ceiling pull switch

CIRCUITING (wiring method identification by notation
on drawing or in specifications)

———————— Wiring concealed in ceiling or wall

———/// ——— 3 wires

———//// ——— 4 wires, etc.

— — — — — — Wiring concealed in floor

- - - - - - - Wiring exposed

Branch circuit home run to panel
board. Number of arrows indicates
number of circuits. A numeral at
each arrow may be used to identify
circuit number. Any circuit without
further identification indicates a two-
wire circuit. For a greater number of
wires, indicate with cross lines.

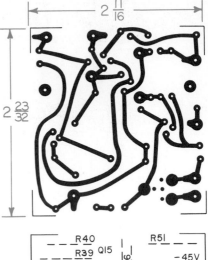

Fig. 23-37. Electrical wiring symbols for architectural layout. (Continued.)

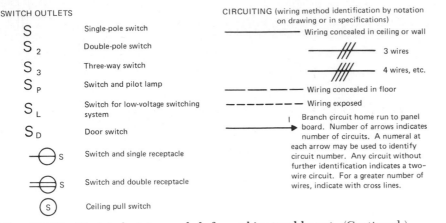

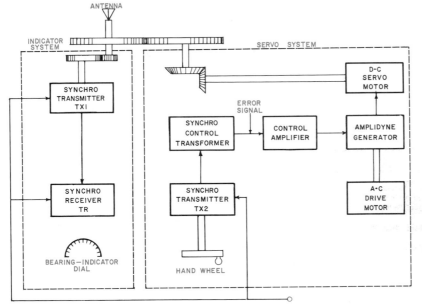

Fig. 23-38. Interconnection diagram showing the different units of a typical dc servo system used for rotating a search radar system.

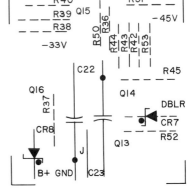

Fig. 23-39. A printed circuit and a component identification overlay.

Etching is one way to remove the copper from all areas of the insulating base except for the circuits needed.

The components may be shown on the printed circuit board by using symbols or other markings. This information is transferred to the printed circuit diagram from a component identification overlay (Fig. 23-39).

smaller by photography. The lines (conductors) on the pattern should be at least ⅟₃₂ in. (1 mm) wide. They should be spaced at least ⅟₃₂ in. (1 mm) apart. The circuit layout pattern is transferred to a copper-clad insulating base. This is done by photography or in some other way.

REVIEW

1. Describe a series circuit in technical terms.
2. Draw a simple parallel circuit.

3. Explain how a block diagram and a schematic diagram are different.

4. How are voltage and amperage related in Ohm's law?

Problems

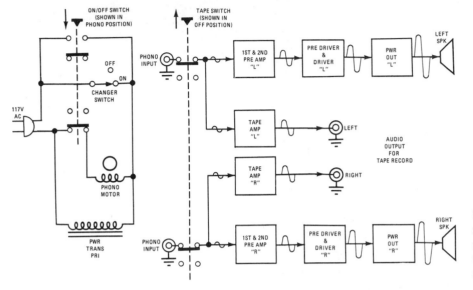

Fig. 23-40. Prepare a complete signal-flow diagram for an AM-FM stereo unit as shown. Estimate sizes and draw twice the size. Use a template if one is available.

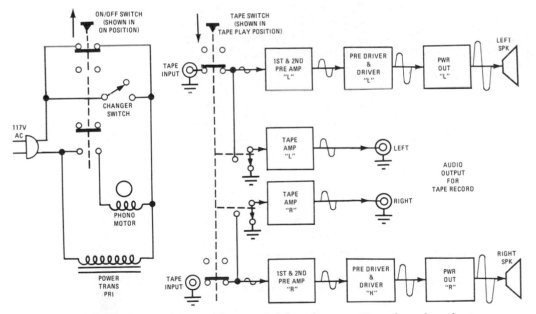

Fig. 23-41. Draw the tape-player signal-flow diagram. Note that when the tape button is depressed, all other functions (AM, FM, and phono) are disabled. Also, the changer switch cannot turn on the phono motor. Since there is no outlet on the left and right audio-output jacks, no direct tape recording can take place. The tape preamp is grounded.

24 Aerospace Drafting

THE AEROSPACE INDUSTRY

New ways of making airplanes, missiles, and spaceships are always being found. This means challenging opportunities exist for young men and women interested in aerospace design and drafting (Fig. 24-1). Aero-

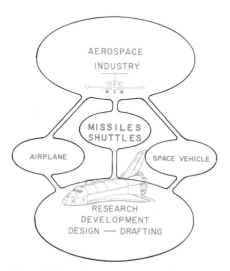

Fig. 24-1. The aerospace industry is a high-tech working force that paces transportation and discovery.

space design and drafting deals with all kinds of flying vehicles. These vehicles fly at all speeds and at all *altitudes* (heights). The basic aerospace team is made up of hundreds of scientists, engineers, designers, drafters, and technicians and thousands of skilled workers. The government works with industry to develop aircraft. The aircraft industry has many interests. Some of these are hovering helicopters, stunt biplanes, and vehicles that can fly faster than sound for thousands of miles. The aerospace industry even makes spacecraft that orbit the earth (Fig. 24-2) or fly millions of miles to the planets Jupiter and Saturn.

TESTING AND RESEARCH

The federal government, through many different agencies and the National Aeronautics and Space Administration (NASA), helps industry with vehicle testing and research. *Aeronautics* is the science of designing, building, and operating aircraft. Air-

craft experiments explore how far we can travel into space. Today's leisure aircraft normally fly at *subsonic* (below the speed of sound) *Mach numbers*. Mach 1 is the speed of sound, about 1195 km/h or 742 mph at sea level. Tomorrow's leisure aircraft may be able to fly at *supersonic speeds* (faster than the speed of sound) and *hypersonic speeds* (5 times the speed of sound).

CAREER OPPORTUNITIES

The aviation industry has grown with the speed of the jet. New careers and jobs have developed that did not exist thirty years ago. The knowledge needed for aircraft design and drafting is learned in school and on the job. Those who work in this field usually begin as technicians or engineering aides. If interested, you should study science and mathematics in high school and junior college. You should also take a course in technical drawing or engineering graphics. As a member of a design team,

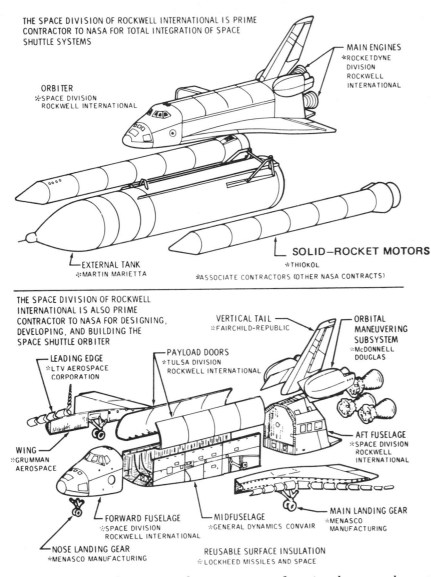

THE SPACE DIVISION OF ROCKWELL INTERNATIONAL IS PRIME CONTRACTOR TO NASA FOR TOTAL INTEGRATION OF SPACE SHUTTLE SYSTEMS

MAIN ENGINES
*ROCKETDYNE DIVISION ROCKWELL INTERNATIONAL

ORBITER
*SPACE DIVISION ROCKWELL INTERNATIONAL

SOLID—ROCKET MOTORS
*THIOKOL
*ASSOCIATE CONTRACTORS (OTHER NASA CONTRACTS)

EXTERNAL TANK
*MARTIN MARIETTA

THE SPACE DIVISION OF ROCKWELL INTERNATIONAL IS ALSO PRIME CONTRACTOR TO NASA FOR DESIGNING, DEVELOPING, AND BUILDING THE SPACE SHUTTLE ORBITER

VERTICAL TAIL
*FAIRCHILD-REPUBLIC

ORBITAL MANEUVERING SUBSYSTEM
*McDONNELL DOUGLAS

PAYLOAD DOORS
*TULSA DIVISION ROCKWELL INTERNATIONAL

LEADING EDGE
*LTV AEROSPACE CORPORATION

AFT FUSELAGE
*SPACE DIVISION ROCKWELL INTERNATIONAL

WING
*GRUMMAN AEROSPACE

MAIN LANDING GEAR
*MENASCO MANUFACTURING

FORWARD FUSELAGE
*SPACE DIVISION ROCKWELL INTERNATIONAL

MIDFUSELAGE
*GENERAL DYNAMICS CONVAIR

NOSE LANDING GEAR
*MENASCO MANUFACTURING

REUSABLE SURFACE INSULATION
*LOCKHEED MISSILES AND SPACE

Fig. 24-2. An aircraft company relies on many manufacturing plants to make parts for an ever-changing industry. (NASA.)

Many *specialists* (people with special skills) are needed in aerospace industries, from the person on the drawing board to the person testing models in the wind tunnel. The big aircraft companies usually rely on many other manufacturing plants across the country to make parts for different aircraft. The largest companies are those that make the power plants and *framework* (skeleton) of the aircraft. Thousands of parts are made by smaller manufacturers.

AIRCRAFT MATERIALS

Today's high-speed and supersonic aircraft have parts made of many different materials. Aluminum, magnesium, and lightweight steel are usually strong enough for the inside, or substructure, parts. The superstructure (skeleton) and skins become very hot during supersonic flight. For this reason, they must be made of materials that are strong and can stand the heat. Designers today are interested in making lightweight structures. They are also interested in electronic equipment that works

the drafting technician (Fig. 24-3) may make drawings of mechanical or electrical systems. These are drawn from the designer's sketch.

CAREER ADVANCEMENT

Drafting technicians may move up as they gain experience in a certain area. With some schooling and on-

the-job training, they may become design technicians. Design technicians have more responsibilities than drafting technicians.

To learn more about careers, write to the group listed below:

American Institute of Aeronautics and Astronautics
1290 Avenue of the Americas
New York, NY 10019

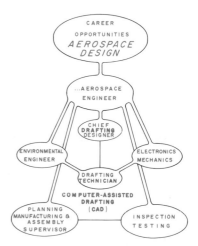

CAREER OPPORTUNITIES
AEROSPACE DESIGN

...AEROSPACE ENGINEER

CHIEF DRAFTING DESIGNER

ENVIRONMENTAL ENGINEER

ELECTRONICS MECHANICS

DRAFTING TECHNICIAN

COMPUTER-ASSISTED DRAFTING (CAD)

PLANNING MANUFACTURING & ASSEMBLY SUPERVISOR

INSPECTION TESTING

Fig. 24-3. The aerospace industry offers many drafting opportunities.

faster and better than the human mind.

MAJOR AIRCRAFT COMPONENTS

Figure 24-4 shows the McDonnell Douglas F-4C Phantom jet. It has 62 basic parts. The *fuselage* (or central body), part 14, holds the pilot and passenger compartments. The fuselage has three parts. The fuselage structure is made up of a set of shaped *bulkheads* (walls) and rings. The fuselage also has *longitudinal* (lengthwise) members. Together these form a strong framework. The outer skin is made of sheet metal. It is fastened to the framework with rivets or machine-screw fasteners.

The Wings and Airfoil

The wings are made up of ribs. These ribs form the shape of the wing. The ribs are connected to *spars* (beams). The spars run toward the fuselage and away from it. The wing skins are attached to the ribs and spars with metal fasteners. Besides the wings, other parts control the aircraft. These are called *airfoils*. They include ailerons, rudders, stabilizers, flaps, and tabs. A *chord line* is a straight line between the leading edge and the trailing edge of an airfoil. The general two-view drawing shows the sleek form of an aircraft that is made to be flown at high speed and high altitude (Fig. 24-5).

What Makes an Airplane Fly?

Figure 24-6 shows why an airplane can fly. Any wing passing through the air at an angle pushes the air downward. This causes an equal and opposite upward push. Lift (flying), then, is caused by the angle of the wing as it moves through air. Look at the shape and position of the wing as you try to figure out how well an aircraft can fly.

The Landing Gear

The landing gear is raised and lowered by *hydraulics* (fluid under pressure) or electricity. Hydraulic shock absorbers ease the shock of landing (Fig. 24-7).

The Power Plant

Piston or jet engines are often placed in the lower part of the fuselage. They can also be placed under the wings.

Many other systems are also found on aircraft. These include air conditioning, compartment pressurization (to keep normal air pressure), radar, radio (communications), hydraulic, electronic, and plumbing systems. Many trades and industries help make the complex aerospace designs of today.

The parts and systems ready to be put together are generally shown on a plan. This plan has many details. Many kinds of drawings are needed before the aircraft is built. To understand how to build an aircraft from these drawings, you must understand multiview and pictorial drawings (Fig. 24-8).

AIRCRAFT DRAFTING PRACTICES

The bigger aircraft companies have engineering manuals for their workers. The manuals are carefully made to save time and money in manufacturing. A designer and drafter must know the company's manual. They must also know the standards used by all companies. The Society of Automotive Engineers (SAE) publishes a book of aerospace-automotive drawing standards.

METHODS OF UNDIMENSIONED DRAWINGS

The undimensioned drawing is a way of drafting in the aerospace industry. It frees designers and drafters from the detail dimensioning needed for most finished layouts. Automatic drafting equipment makes undimensioned drafting possible.

The drawings are made with devices that control drawing accuracy. Some methods are listed for full-size drawings.

1. Lines are drawn on matte-surface (dull and rough) plastic film.
2. Drawings are made with a technical pen that controls line widths.
3. Drawing is accurate. The accuracy needed depends on the part being drawn.

Undimensioned drawings are useful for several reasons.

1. Less time is needed to make finished drawings.
2. Drawings are used to check fit. Full-size drawings are a natural place to test whether parts fit. If parts do not fit on full-size drawings, they will probably not fit on the final product.
3. Flat-pattern drawings can be easily checked. A flat pattern gives a design team time to check out problems.
4. A contact photo makes the master layout. The drawing can be transferred with a photograph.
5. Changes caused by mistakes are less common. By using undimen-

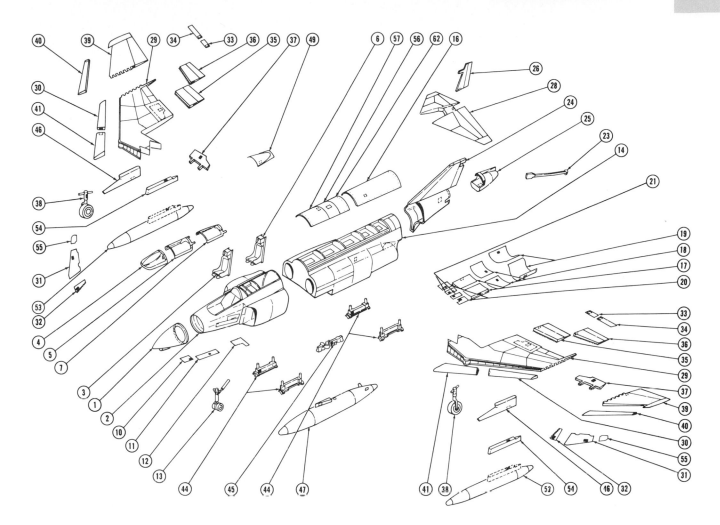

1. Radome
2. Forward fuselage
3. Pilot seat
4. Windshield
5. Forward canopy
6. Radar operation seat
7. Aft canopy
10. Nose landing gear door, forward
11. Nose landing gear door, aft
12. Hydraulic compartment access door
13. Nose landing gear shock strut
14. Center fuselage
16. Fuel tank door
17. Engine access door
18. Engine access door
19. Engine access door
20. Engine access door
21. Auxiliary engine air door
23. Arresting hook
24. Aft fuselage
25. Tail cone
26. Rudder
28. Stabilator
29. Center section wing
30. Leading edge flap
31. Main landing gear strut door
32. Main landing gear inboard door
33. Inboard spoiler
34. Outboard spoiler
35. Flap
36. Aileron
37. Speed brake
38. Main landing gear shock strut
39. Outer wing
40. Leading edge flap, outboard
41. Leading edge flap, inboard
44. Missile rack
45. Bomb rack
46. Missile pylon
47. External centerline fuel tank
49. Data link access door
53. External wing fuel tank
54. External wing fuel tank pylon
55. Landing gear door, outboard
56. Boom IFR receptacle access door
57. Fuel cell access door
62. Fuel cell access door

Fig. 24-4. An exploded view helps to explain the parts of the McDonnell Douglas F-4C Phantom jet. (McDonnell Douglas Corp.)

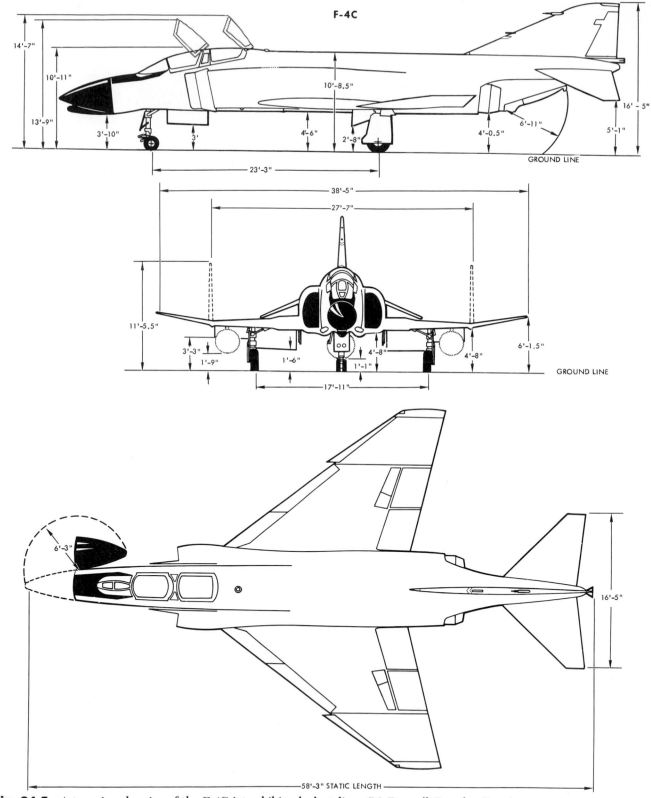

Fig. 24-5. A two-view drawing of the F-4C jet exhibits sleek styling. (McDonnell Douglas Corp.)

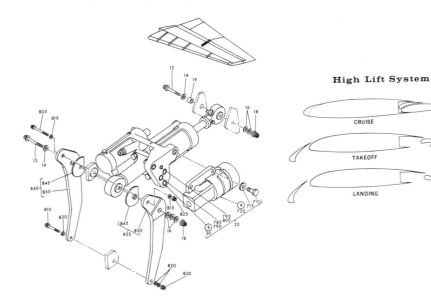

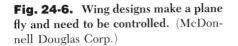

Fig. 24-6. Wing designs make a plane fly and need to be controlled. (McDonnell Douglas Corp.)

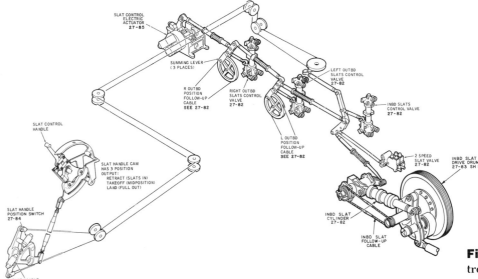

Fig. 24-7. The landing gear is controlled by a hydraulic system. (McDonnell Douglas Corp.)

sioned drawings with full-size drawings, and by not including measurements that may be wrong or unclear, drawing mistakes have been lowered by 40 percent.

6. They may be drawn by automation. *Automation* means electronic equipment doing jobs that people once did. In the future age of automation, the full-size undimensioned drawing will be ideal. The instructions for the drawing may be put onto a computer tape. Then the tape can be played back and changed to make new, modified drawings. The program will be used for the numerically con-

trolled machine in the areas where parts are made automatically.

The kinds of drawings that can be made in the undimensioned way are the following:

1. Parts that have flat patterns that can be made in a hydropress,

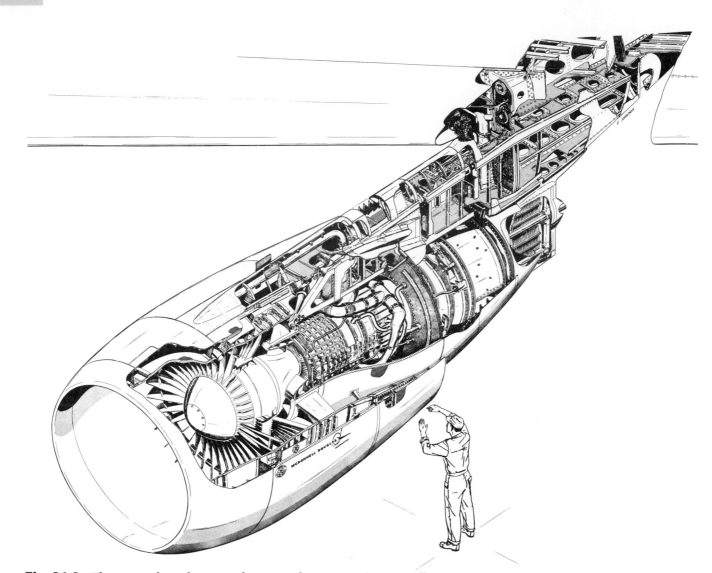

Fig. 24-8. The power plant of a jet may be mounted on a wing. (McDonnell Douglas Corp.)

power-press brake, or stretch brake. You can see an example of flat-pattern development in Fig. 24-9.

2. Machined parts that are drawn as undimensioned drawings on a piece must be made by profile machine methods (Fig. 24-10).

3. Parts needing a plaster pattern with a three-dimensional form.

4. Drawings needing artwork layout. Circuits and lighting panels are natural for undimensioned drawing. The artwork is drawn twice the size. Then it is photoreduced to the right size for making the needed part. Photoreduction makes lines clean and sharp and meets engineering standards.

■ DRAWINGS FOR LARGE AIRCRAFT

Examine some of the drawings prepared for a passenger jet. The assembly drawing in Fig. 24-11 is used to show the inboard profile of the DC-10. The cutaway pictorial of a wing engine in Fig. 24-8 shows the huge size of a power plant compared

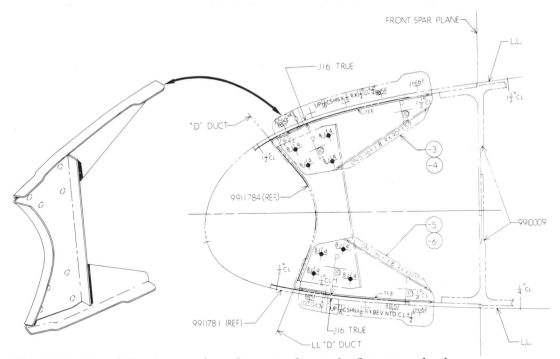

Fig. 24-9. Aircraft drawings may be undimensioned, as in this flat-pattern development. (McDonnell Douglas Corp.)

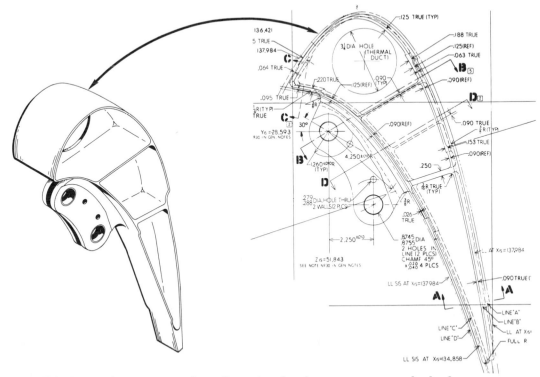

Fig. 24-10. Machine parts may be undimensioned and meet accurate standards of the aircraft company. (McDonnell Douglas Corp.)

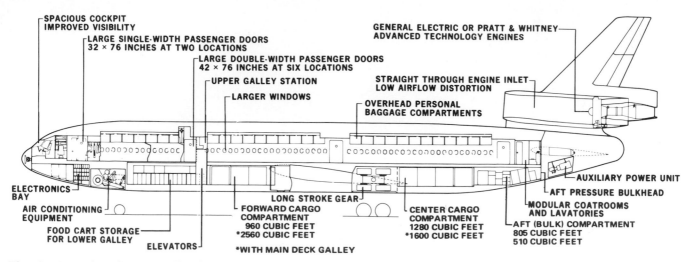

SPACIOUS COCKPIT IMPROVED VISIBILITY

LARGE SINGLE-WIDTH PASSENGER DOORS 32 × 76 INCHES AT TWO LOCATIONS

LARGE DOUBLE-WIDTH PASSENGER DOORS 42 × 76 INCHES AT SIX LOCATIONS

UPPER GALLEY STATION

LARGER WINDOWS

GENERAL ELECTRIC OR PRATT & WHITNEY ADVANCED TECHNOLOGY ENGINES

STRAIGHT THROUGH ENGINE INLET LOW AIRFLOW DISTORTION

OVERHEAD PERSONAL BAGGAGE COMPARTMENTS

AUXILIARY POWER UNIT

ELECTRONICS BAY

AIR CONDITIONING EQUIPMENT

FOOD CART STORAGE FOR LOWER GALLEY

ELEVATORS

LONG STROKE GEAR

FORWARD CARGO COMPARTMENT 960 CUBIC FEET *2560 CUBIC FEET

***WITH MAIN DECK GALLEY**

CENTER CARGO COMPARTMENT 1280 CUBIC FEET *1600 CUBIC FEET

AFT PRESSURE BULKHEAD

MODULAR COATROOMS AND LAVATORIES

AFT (BULK) COMPARTMENT 805 CUBIC FEET 510 CUBIC FEET

Fig. 24-11. The inboard profile of a commercial jet transport reveals the use of space in the main body. (McDonnell Douglas Corp.)

to a human. Figure 24-12 shows the giant transport in a three-view orthographic projection. In Fig. 24-6, the exploded pictorial of the outboard aileron damper shows its parts. These kinds of drawings are needed to completely understand aircraft design, building, and flight checkout.

■ SMALLER AIRCRAFT

The new *navigational* (direction-finding) equipment used today has meant challenges for the pilots of small aircraft. The design today includes a Bendix automatic pilot and

flight director system. This system, with communications and navigational equipment, lets the aircraft cruise safely from 175 to 200 mph (280 to 320 km/h).

The Piper Aircraft Corporation designed and made the *Navajo* (Fig. 24-13) as an all-metal superstructure. However, Fiberglas parts are used where the toughness of such a material is needed. Such places are the nose cone, wing, rudder and vertical fin tips, door frame, and windshield channel.

The wing of this plane has a stepped-down main spar, a front and rear spar, *lateral* (crosswise) stringers, *longitudinal* (lengthwise) ribs, and stressed skin sheets. The wings are joined together with heavy steel plates. This makes a main spar that runs unbroken from wing tip to wing tip. Besides the sturdy splice joint, the wing is attached to the fuselage at the front, center, and trailing edge. Flat rivets are used forward of the main spar. This allows the air to flow more smoothly over the wing. The wing root fillet and the swept leading edge between the fuselage and the

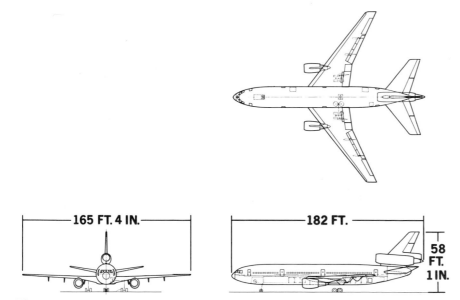

165 FT. 4 IN.

182 FT.

58 FT. 1 IN.

Fig. 24-12. Three-view drawing of a DC-10 jet transport. (McDonnell Douglas Corp.)

Fig. 24-13. This advanced small aircraft has an all-metal superstructure with Fiberglas components. (Piper Aircraft Corp.)

Fig. 24-14. An aircraft made of unique plastic materials. (Windecker Industries, Inc.)

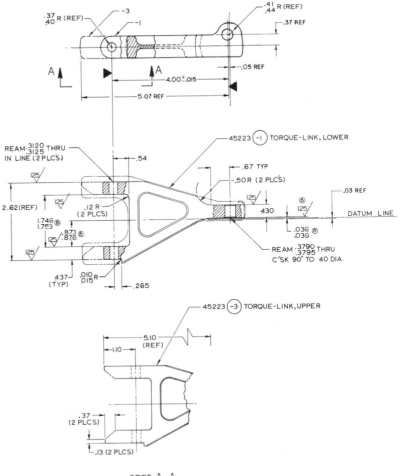

Fig. 24-15. A typical dimensioned aircraft part to be machined. (North American Rockwell Corp.)

nacelle (which holds the engine) also smooth the airflow.

New Material for Small Aircraft

A new material has been used in small leisure aircraft in recent years. Windecker Industries, Inc., has made a plastic plane of a material called Fibaloy. The *Eagle* (Fig. 24-14) is built of plastic made stronger with Fiberglas. The aircraft has no riveted sections or seams formed by lapping skins. The aircraft becomes stronger after it is made because of a curing process inside the plastic. A chemical reaction causes the plastic to become hard like steel.

Typical Drawings

Some of the common engineering drawings for a smaller aircraft are

shown. A torque link from a small airplane is a common dimensioned machine drawing (Fig. 24-15). The forging blank drawing is shown in Fig. 24-16. This drawing tells all that the forger and inspector of the blank need to know. A forging machine drawing gives information for the machinist, set-up person, inspector, and others who may put together bushings, bearings, etc.

Lofting Layouts

Full-size drawings for large jobs are made by *lofting*. This word comes

from the ship loft, where the exact lines of the shapes of ships are worked out. Lofting is important to aircraft design layout. Contours of wing sections are drawn without mistakes by lofting. The curves are *faired* (adjusted or smoothed out) to get smooth surfaces. Templates may be made when needed. The drawing board is generally too small for such work. Layouts are made on special loft floors that have all the space needed. Ribs are drawn with trace chords and a loft line. The drawings of ribs are undimensional. This

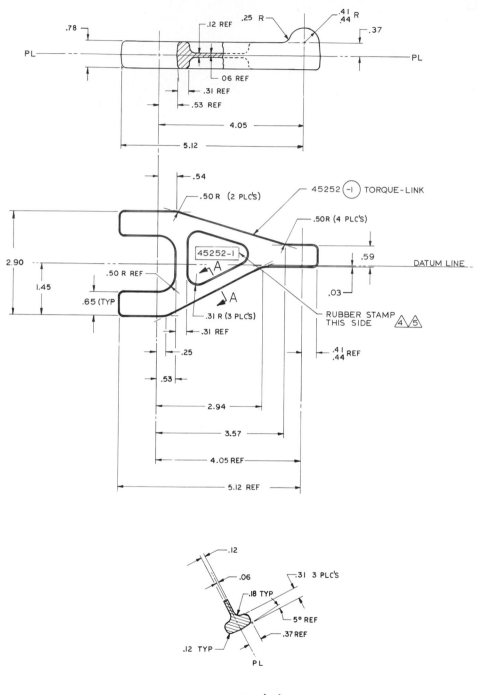

Fig. 24-16. A forged part dimensioned for accurate mechanical service. A torque link arm. (North American Rockwell Corp.)

method is used for sheet aluminium and steel parts. The drawing is transferred to the material by a photo process.

Light Jet Helicopter

Another small aircraft is the light jet helicopter. This craft set records for speed, distance, climbing ability, and altitude. The three-view drawing (Fig. 24-17) shows the structural form. The overall measurements are in U.S. customary and metric measure. The electronic components for navigation and communications (called *avionics*) let the helicopter make many different movements.

■ BUSINESS JETS

The aircraft manufacturers have designed a few business jets that are each one of a kind. They can travel at speeds from 508 to 548 mph (820 to 880 km/h). The design has a wing with eight spars. The wing has twice the needed strength and is designed not to fail. A spar is the principal part of the framework in an airplane wing. It runs from tip to tip or from root to tip. The highly polished aluminum skin brings out the sleekly designed lines of the jet.

The three-view drawing of the business jet in Fig. 24-18 shows another style. The assembly-line photo (Fig. 24-19) shows the final stages of assembly.

The basic air-frame designs for business jets are made to pass federal testing.

■ THE SPACE SHUTTLE

The newest kind of spacecraft in the NASA space program is the space

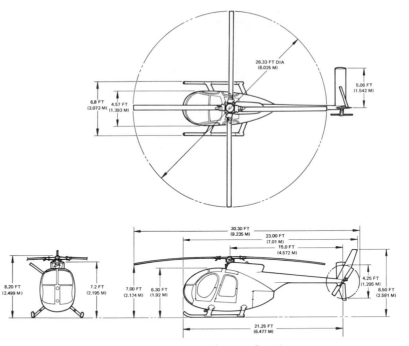

Fig. 24-17. A small jet helicopter. (Hughes Tool Co.)

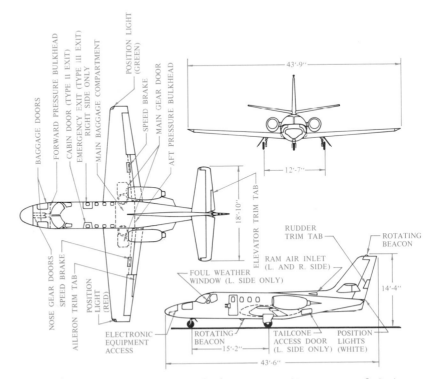

Fig. 24-18. A three-view drawing of a business jet. (Cessna Aircraft Co.)

Fig. 24-19. The final assembly line prepares the plane for tests. (Cessna Aircraft Co.)

shuttle (Fig. 24-20). It consists of a vehicle called the *orbiter* and two solid rocket boosters to lift it into orbit. It also includes a large external liquid *propellant* (fuel) tank. The space shuttle is designed for scientific, civilian, or military use.

The shuttle orbiter is different from older spacecraft in that when its mission in space is over, it can land back on Earth like an airplane. It can then be made ready for other missions in the Earth's orbit. The orbiter is about the size of a small jet passenger airplane.

◼ KINDS OF AIRCRAFT DRAWINGS

Many different kinds of drawings are made for many different kinds of aircraft. For helicopters, leisure aircraft, military jets, jet transports, biplanes, and space shuttles, drawings are needed. Drawings are used for castings, forgings, sheet-metal layout, schematics, and lofting. Sketching is used for changes and new designs. Of these, one of the important drafting areas is in forming sheet metal.

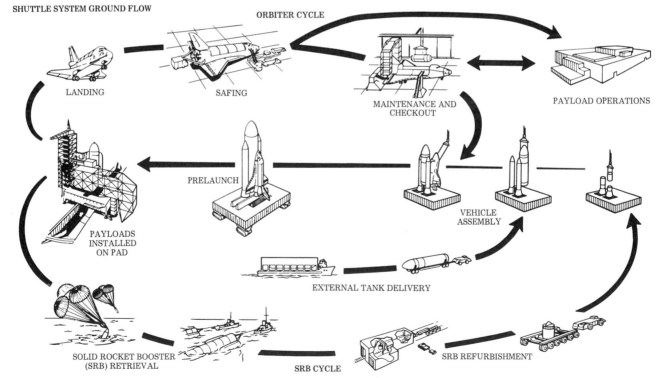

SHUTTLE SYSTEM GROUND FLOW

ORBITER CYCLE

LANDING

SAFING

MAINTENANCE AND CHECKOUT

PAYLOAD OPERATIONS

PRELAUNCH

PAYLOADS INSTALLED ON PAD

VEHICLE ASSEMBLY

EXTERNAL TANK DELIVERY

SOLID ROCKET BOOSTER (SRB) RETRIEVAL

SRB CYCLE

SRB REFURBISHMENT

Fig. 24-20. The space shuttle is an experimental laboratory that will follow the service diagram illustrated. (NASA.)

REVIEW

1. Describe the basic structure of an airplane fuselage and wing.

2. Name some good points of the undimensioned drawing.

3. Define the word "aeronautics."

4. What makes an airplane fly?

5. What are the basic components of a space shuttle?

6. How are the landing gear and ailerons controlled?

Problems

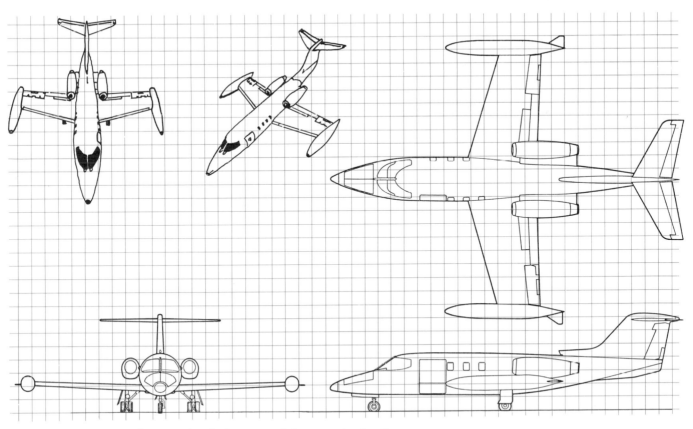

Fig. 24-21. Using grid paper, sketch the views of the executive jet. Prepare one pictorial view of the Lear Jet.

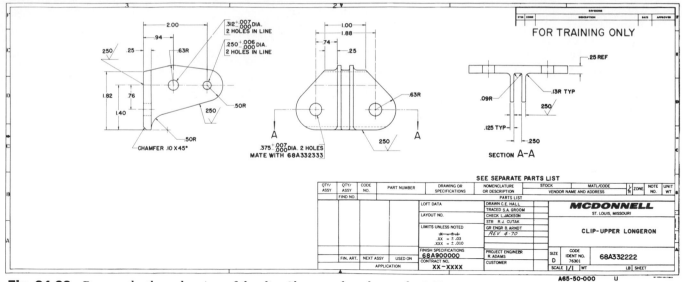

Fig. 24-22. Prepare the three drawings of the clip. Change order: change the 2.00 dimension in zone F-3 to 2.625.

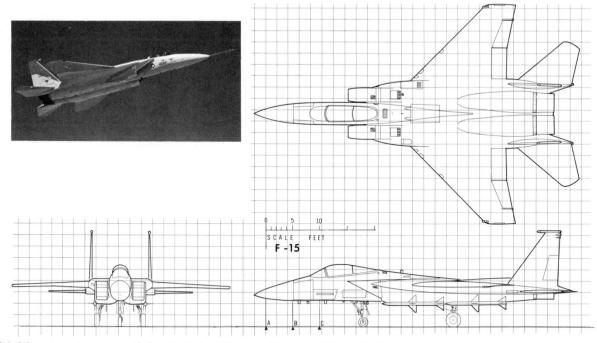

Fig. 24-23. Draw one view of the F-15 aircraft on grid paper. Using dividers, locate the contours accurately.

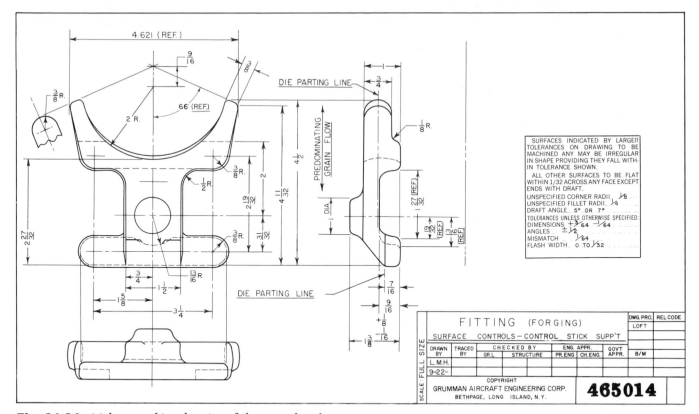

Fig. 24-24. Make a working drawing of the control-stick support.

Fig. 24-25. Prepare a working drawing of the link in Fig. 24-15. Dimension completely.

Fig. 24-26. Prepare a working drawing of the fork link in Fig. 24-16 and complete dimensions.

Reference Tables

AMERICAN NATIONAL STANDARDS

A few standards that are useful for reference are listed below. Standards are subject to revisions and latest issues should be consulted. A catalog with prices is published by the American National Standards Institute, Inc., 1430 Broadway, New York, NY 10018.

American National Drafting Standards Manual

TABLE A-1. FRACTIONAL-INCH, DECIMAL-INCH, AND MILLIMETER EQUIVALENT CHART

4ths	8ths	16ths	32nds	64ths	To 2 places	To 3 places	To 4 places	To 1 place	To 2 places	To 3 places
				1/64	.02	.016	.0156	0.4	.40	.397
			1/32		.03	.031	.0312	0.8	.80	.794
				3/64	.05	.047	.0469	1.2	1.20	1.191
		1/16	...	...	.06	.062	.0625	1.6	1.59	1.588
				5/64	.08	.078	.0781	2.0	1.98	1.984
			3/32		.09	.094	.0938	2.4	2.38	2.381
				7/64	.11	.109	.1094	2.8	2.78	2.778
	1/8	...	...	...	.12	.125	.1250	3.2	3.18	3.175
				9/64	.14	.141	.1406	3.6	3.57	3.572
			5/32		.16	.156	.1562	4.0	3.97	3.969
				11/64	.17	.172	.1719	4.4	4.37	4.366
		3/16	...	...	.19	.188	.1875	4.8	4.76	4.762
				13/64	.20	.203	.2031	5.2	5.16	5.159
			7/32		.22	.219	.2188	5.6	5.56	5.556
				15/64	.23	.234	.2344	6.0	5.95	5.953
1/4	...	...	...	...	.25	.250	.2500	6.4	6.35	6.350
				17/64	.27	.266	.2656	6.8	6.75	6.747
			9/32		.28	.281	.2812	7.1	7.14	7.144
				19/64	.30	.297	.2969	7.5	7.54	7.541
		5/16	...	...	.31	.312	.3125	7.9	7.94	7.938
				21/64	.33	.328	.3281	8.3	8.33	8.334
			11/32		.34	.344	.3438	8.7	8.73	8.731
				23/64	.36	.359	.3594	9.1	9.13	9.128
	3/8	...	...	...	.38	.375	.3750	9.5	9.52	9.525
				25/64	.39	.391	.3906	9.9	9.92	9.922
			13/32		.41	.406	.4062	10.3	10.32	10.319
				27/64	.42	.422	.4219	10.7	10.72	10.716
		7/16	...	...	.44	.438	.4375	11.1	11.11	11.112
				29/64	.45	.453	.4531	11.5	11.51	11.509
			15/32		.47	.469	.4688	11.9	11.91	11.906
				31/64	.48	.484	.4844	12.3	12.30	12.303
					.50	.500	.5000	12.7	12.70	12.700
				33/64	.52	.516	.5156	13.1	13.10	13.097
			17/32		.53	.531	.5312	13.5	13.49	13.494
				35/64	.55	.547	.5469	13.9	13.89	13.891
		9/16	...	...	.56	.562	.5625	14.3	14.29	14.288
				37/64	.58	.578	.5781	14.7	14.68	14.684
			19/32		.59	.594	.5938	15.1	15.08	15.081
				39/64	.61	.609	.6094	15.5	15.48	15.478
	5/8	...	...	...	.62	.625	.6250	15.9	15.88	15.875
				14/64	.64	.641	.6406	16.3	16.27	16.272
			21/32		.66	.656	.6562	16.7	16.67	16.669

TABLE A-1. (*CONTINUED*)

4ths	8ths	16ths	32nds	64ths	To 2 places	To 3 places	To 4 places	To 1 place	To 2 places	To 3 places
					Fractions		**Decimal inches**			**Millimeters**
				43/64	.67	.672	.6719	17.1	17.07	17.066
		11/16	. . .	. . .	.69	.688	.6875	17.5	17.46	17.462
				45/64	.70	.703	.7031	17.9	17.86	17.859
			23/32	. . .	.72	.719	.7188	18.3	18.26	18.256
				47/64	.73	.734	.7344	18.6	18.65	18.653
3/4	. . .	. . .	. . .	. . .	.75	.750	.7500	19.1	19.05	19.050
				49/64	.77	.766	.7656	19.4	19.45	19.447
			25/32	. . .	.78	.781	.7812	19.8	19.84	19.844
				51/64	.80	.797	.7969	20.2	20.24	20.241
		13/16	. . .	. . .	.81	.812	.8125	20.6	20.64	20.638
				53/64	.83	.828	.8281	21.0	21.03	21.034
			27/32	. . .	.84	.844	.8438	21.4	21.43	21.431
				55/64	.86	.859	.8594	21.8	21.83	21.828
	7/8	. . .	. . .	. . .	.88	.875	.8750	22.2	22.22	22.225
				57/64	.89	.891	.8906	22.6	22.62	22.622
			29/32	. . .	.91	.906	.9062	23.0	23.02	23.019
				59/64	.92	.922	.9219	23.4	23.42	23.416
		15/16	. . .	. . .	.94	.938	.9375	23.8	23.81	23.812
				61/64	.95	.953	.9531	24.2	24.21	24.209
			31/32	. . .	.97	.969	.9688	24.6	24.61	24.606
				63/64	.98	.984	.9844	25.0	25.00	25.003
					1.00	1.000	1.0000	25.4	25.40	25.400

TABLE A-2. AMERICAN NATIONAL STANDARD UNIFIED AND AMERICAN THREAD SERIES[a]

Threads per inch for coarse, fine, extra-fine, 8-thread, 12-thread, and 16-thread series[b]
[tap-drill sizes for approximately 75 percent depth of thread (not American Standard)]

Nominal size (basic major dia.)	Coarse-thd. series UNC and NC[c] in classes 1A, 1B, 2A, 2B, 3A, 3B, 2, 3		Fine-thd. series UNF and NF[c] in classes 1A 1B, 1A, 2B, 3A, 3B, 2, 3		Extra-fine thd. series UNEF and NEF[d] in classes 2A, 2B, 2, 3		8-thd. series 8N[c] in classes 2A, 2B, 2, 3		12-thd. series 12UN and 12N[d] in classes 2A, 2B, 2, 3		16-thd. series 16UN and 16N[d] in classes 2A, 2B, 2, 3	
	Thd. /in.	Tap drill	Thd. /in.	Tap drill	Thd. /in.	Tap drill	Thd. /in.	Tap drill	Thd. /in.	Tap drill	Thd. /in.	Tap drill
0(0.060)	. . . .	. . . . .	80	3/64								
1(0.073)	64	No. 53	72	No. 53								
2(0.086)	56	No. 50	64	No. 50								

TABLE A-2. (CONTINUED)

Nominal size (basic major dia.)	Coarse-thd. series UNC and NC[c] in classes 1A, 1B, 2A, 2B, 3A, 3B, 2, 3		Fine-thd. series UNF and NF[c] in classes 1A, 1B, 1A, 2B, 3A, 3B, 2, 3		Extra-fine thd. series UNEF and NEF[d] in classes 2A, 2B, 2, 3		8-thd. series 8N[c] in classes 2A, 2B, 2, 3		12-thd. series 12UN and 12N[d] in classes 2A, 2B, 2, 3		16-thd. series 16UN and 16N[d] in classes 2A, 2B, 2, 3	
	Thd. /in.	Tap drill	Thd. /in.	Tap drill	Thd. /in.	Tap drill	Thd. /in.	Tap drill	Thd. /in.	Tap drill	Thd. /in.	Tap drill
3(0.099)	48	No. 47	56	No. 45								
4(0.112)	40	No. 43	48	No. 42								
5(0.125)	40	No. 38	44	No. 37								
6(0.138)	32	No. 36	40	No. 33								
8(0.164)	32	No. 29	36	No. 29								
10(0.190)	24	No. 25	32	No. 21								
12(0.216)	24	No. 16	28	No. 14	32	No. 13						
1/4	20	N0. 7	28	No. 3	32	7/32						
5/16	18	Let. F	24	Let. I	32	9/32						
3/8	16	5/16	24	Let. Q	32	11/32						
7/16	14	Let. U	20	25/64	28	13/32						
1/2	13	27/64	20	29/64	28	15/32	..	....	12	27/64		
9/16	12	31/64	18	33/64	24	33/64	..	....	12	31/64		
5/8	11	17/32	18	37/64	24	37/64	..	....	12	35/64		
11/16	....	.....	..	.....	24	41/64	..	....	12	39/64		
3/4	10	21/32	16	11/16	20	45/64	..	....	12	43/64	16	11/16
13/16	....	.....	..	.....	20	49/64	..	....	12	47/64	16	3/4
7/8	9	49/64	14	13/16	20	53/64	..	....	12	51/64	16	13/16
15/16	....	.....	..	.....	20	57/64	..	....	12	55/64	16	7/8
1	....	....	14	15/16	..	.....	8	7/8				
1	8	7/8	12	59/64	20	61/64	..	....	12	59/64	16	15/16
1 1/16	....	.....	..	.....	18	1	..	....	12	63/64	16	1
1 1/8	7	63/64	12	1 3/64	18	1 5/64	8	1	12	1 3/64	16	1 1/16
1 3/16	....	.....	..	.....	18	1 9/64	..	....	12	1 7/64	16	1 1/8
1 1/4	7	1 7/64	12	1 11/64	18	1 3/16	8	1 1/8	12	1 11/64	16	1 3/16
1 5/16	....	.....	..	.....	18	1 17/64	..	....	12	1 15/64	16	1 1/4
1 3/8	6	1 7/32	12	1 19/64	18	1 5/16	8	1 1/4	12	1 19/64	16	1 5/16
1 7/16	....	.....	..	.....	18	1 3/8	..	....	12	1 23/64	16	1 3/8
1 1/2	6	1 11/32	12	1 27/64	18	1 7/16	8	1 3/8	12	1 27/64	16	1 7/16
1 9/16	....	.....	..	.....	18	1 1/2	..	....	..	.....	16	1 1/2
1 5/8	....	.....	..	.....	18	1 9/16	8	1 1/2	12	1 35/64	16	1 9/16
1 11/16	....	.....	..	.....	18	1 5/8	..	....	..	.....	16	1 5/8
1 3/4	5	1 9/16	..	.....	16	1 11/16	8[e]	1 5/8	12	1 43/64	16	1 11/16
1 13/16	....	.....	..	.....	..	.....	..	....	..	.....	16	1 3/4

TABLE A-2. (CONTINUED)

Nominal size (basic major dia.)	Coarse-thd. series UNC and NC[c] in classes 1A, 1B, 2A, 2B, 3A, 3B, 2, 3		Fine-thd. series UNF and NF[c] in classes 1A, 1B, 2A, 2B, 3A, 3B, 2, 3		Extra-fine thd. series UNEF and NEF[d] in classes 2A, 2B, 2, 3		8-thd. series 8N[c] in classes 2A, 2B, 2, 3		12-thd. series 12UN and 12N[d] in classes 2A, 2B, 2, 3		16-thd. series 16UN and 16N[d] in classes 2A, 2B, 2, 3	
	Thd./in.	Tap drill	Thd./in.	Tap drill	Thd./in.	Tap drill	Thd./in.	Tap drill	Thd./in.	Tap drill	Thd./in.	Tap drill
1 7/8	....	.....	..	.....	..	.....	8	1 3/4	12	1 51/64	16	1 13/16
1 15/16	....	.....	..	.....	..	.....	..	...	..	.....	16	1 7/8
2	4 1/2	1 25/32	..	.....	16	1 15/16	8[e]	1 7/8	12	1 59/64	16	1 15/16

[a]ANSI B1.1. Dimensions are in inches. [b]See ANSI B1.1 for 20-thread, 28-thread, and 32-thread series.
[c]Limits of size for classes are based on a length of engagement equal to the nominal diameter.
[d]Limits of size for classes are based on length of engagement equal to nine times the pitch.
[e]These sizes, with specified limits of size, based on a length of engagement of 9 threads in classes 2A and 2B, are designated UN.

Note. If a thread is in both the 8-, 12-, or 16-thread series and the coarse, fine, or extra-fine-thread series, the symbols and tolerances of the latter series apply.

TABLE A-3. ISO METRIC SCREW THREADS

Nominal size dia (mm)		Series with graded pitches				Series with constant pitches																	
		Coarse		Fine		4		3		2		1.5		1.25		1		0.75		0.5		0.35	
Preferred		Thread pitch	Tap drill size	Thread pitch	Tap drill size	Thread pitch	Tap drill size	Thread pitch	Tap drill size	Thread pitch	Tap drill size	Thread pitch	Tap Drill size	Thread pitch	Tap drill size	Thread pitch	Tap drill size	Thread pitch	Tap drill size	Thread pitch	Tap drill size	Thread pitch	Tap drill size
1.6		0.35	1.25																				
	1.8	0.35	1.45																				
2		0.4	1.6																				
	2.2	0.45	1.75																				
2.5		0.45	2.05																		0.35	2.15
3		0.5	2.5																		0.35	2.65
	3.5	0.6	2.9																		0.35	3.15
4		0.7	3.3															0.5	3.5			
	4.5	0.75	3.7															0.5	4.0			
5		0.8	4.2															0.5	4.5			
6		1	5.0														0.75	5.2					
8		1.25	6.7	1	7.0											1	7.0	0.75	7.2				
10		1.5	8.5	1.25	8.7									1.25	8.7	1	9.0	0.75	9.2				
12		1.75	10.2	1.25	10.8							1.5	10.5	1.25	10.7	1	11						
	14	2	12	1.5	12.5							1.5	12.5	1.25	12.7	1	13						
16		2	14	1.5	14.5							1.5	14.5			1	15						
	18	2.5	15.5	1.5	16.5					2	16	1.5	16.5			1	17						
20		2.5	17.5	1.5	18.5					2	18	1.5	18.5			1	19						
	22	2.5	19.5	1.5	20.5					2	20	1.5	20.5			1	21						
24		3	21	2	22					2	22	1.5	22.5			1	23						

TABLE A-3. (CONTINUED)

Nominal size dia (mm) Preferred	Series with graded pitches — Coarse Thread pitch	Coarse Tap drill size	Fine Thread pitch	Fine Tap drill size	4 Thread pitch	4 Tap drill size	3 Thread pitch	3 Tap drill size	2 Thread pitch	2 Tap drill size	1.5 Thread pitch	1.5 Tap Drill size	1.25 Thread pitch	1.25 Tap drill size	1 Thread pitch	1 Tap drill size	0.75 Thread pitch	0.75 Tap drill size	0.5 Thread pitch	0.5 Tap drill size	0.35 Thread pitch	0.35 Tap drill size
27	3	24	2	25					2	25	1.5	25.5			1	26						
30	3.5	26.5	2	28					2	28	1.5	28.5			1	29						
33	3.5	29.5	2	31					2	31	1.5	31.5										
36	4	32	3	33					2	34	1.5	34.5										
39	4	35	3	36					2	37	1.5	37.5										
42	4.5	37.5	3	39	4	38	3	39	2	40	1.5	40.5										
45	4.5	39	3	42	4	41	3	42	2	43	1.5	43.5										
46	5	43	3	45	4	44	3	45	2	46	1.5	46.5										

TABLE A-4. SIZES OF NUMBERED AND LETTERED DRILLS

No.	Size	No.	Size	No.	Size	Letter	Size
80	0.0135	53	0.0595	26	0.1470	A	0.2340
79	0.0145	52	0.0635	25	0.1495	B	0.2380
78	0.0160	51	0.0670	24	0.1520	C	0.2420
77	0.0180	50	0.0700	23	0.1540	D	0.2460
76	0.0200	49	0.0730	22	0.1570	E	0.2500
75	0.0210	48	0.0760	21	0.1590	F	0.2570
74	0.0225	47	0.0785	20	0.1610	G	0.2610
73	0.0240	46	0.0810	19	0.1660	H	0.2660
72	0.0250	45	0.0820	18	0.1695	I	0.2720
71	0.0260	44	0.0860	17	0.1730	J	0.2770
70	0.0280	43	0.0890	16	0.1770	K	0.2810
69	0.0292	42	0.0935	15	0.1800	L	0.2900
68	0.0310	41	0.0960	14	0.1820	M	0.2950
67	0.0320	40	0.0980	13	0.1850	N	0.3020
66	0.0330	39	0.0995	12	0.1890	O	0.3160
65	0.0350	38	0.1015	11	0.1910	P	0.3230
64	0.0360	37	0.1040	10	0.1935	Q	0.3320
63	0.0370	36	0.1065	9	0.1960	R	0.3390
62	0.0380	35	0.1100	8	0.1990	S	0.3480
61	0.0390	34	0.1110	7	0.2010	T	0.3580
60	0.0400	33	0.1130	6	0.2040	U	0.3680
59	0.0410	32	0.1160	5	0.2055	V	0.3770
58	0.0420	31	0.1200	4	0.2090	W	0.3860
57	0.0430	30	0.1285	3	0.2130	X	0.3970
56	0.0465	29	0.1360	2	0.2210	Y	0.4040
55	0.0520	28	0.1405	1	0.2280	Z	0.4130
54	0.0550	27	0.1440				

TABLE A-5. ACME AND STUB ACME THREADS*

ANSI preferred diameter-pitch combinations.

Nominal (major) dia.	Threads /in.	Nominal (major) dia.	Threads /in.	Nominal (major) dia.	Threads /in.	Nominal (major) dia.	Threads /in.
1/4	16	3/4	6	1 1/2	4	3	2
5/16	14	7/8	6	1 3/4	4	3 1/2	2
3/8	12	1	5	2	4	4	2
7/16	12	1 1/8	5	2 1/4	3	4 1/2	2
1/2	10	1 1/4	5	2 1/2	3	5	2
5/8	8	1 3/8	4	2 3/4	3		

*ANSI B1.5 and B1.8. Diameters in inches.

TABLE A-6. METRIC TWIST DRILL SIZES

Metric drill sizes (mm)[a]		Decimal equivalent in inches	Metric drill sizes (mm)[a]		Decimal equivalent in inches
Preferred	Available	(ref)	Preferred	Available	(ref)
	0.40	.0157	0.95		.0374
	0.42	.0165		0.98	.0386
	0.45	.0177	1.00		.0394
	0.48	.0189		1.03	.0406
0.50		.0197	1.05		.0413
	0.52	.0205		1.08	.0425
0.55		.0217	1.10		.0433
	0.58	.0228		1.15	.0453
0.60		.0236	1.20		.0472
	0.62	.0244	1.25		.0492
0.65		.0256	1.30		.0512
	0.68	.0268		1.35	.0531
0.70		.0276	1.40		.0551
	0.72	.0283		1.45	.0571
0.75		.0295	1.50		.0591
	0.78	.0307		1.55	.0610
0.80		.0315	1.60		.0630
	0.82	.0323		1.65	.0650
0.85		.0335	1.70		.0669
	0.88	.0346		1.75	.0689
0.90		.0354	1.80		.0709
	0.92	.0362		1.85	.0728
			1.90		.0748

TABLE A-6. (CONTINUED)

Metric drill sizes (mm)[a]		Decimal equivalent in inches	Metric drill sizes (mm)[a]		Decimal equivalent in inches
Preferred	Available	(ref)	Preferred	Available	(ref)
	1.95	.0768		3.50	.1378
2.00		.0787	3.60		.1417
	2.05	.0807			
2.10		.0827		3.70	.1457
	2.15	.0846	3.80		.1496
				3.90	.1535
2.20		.0866	4.00		.1575
	2.30	.0906		4.10	.1614
2.40		.0945			
2.50		.0984	4.20		.1654
2.60		.1024		4.40	.1732
			4.50		.1772
	2.70	.1063		4.60	.1811
2.80		.1102	4.80		.1890
	2.90	.1142			
3.00		.1181	5.00		.1969
	3.10	.1220		5.20	.2047
			5.30		.2087
3.20		.1260		5.40	.2126
	3.30	.1299	5.60		.2205
3.40		.1339		5.80	.2283

[a]Metric drill sizes listed in the "Preferred" column are based on the R 40 series of preferred numbers shown in the ISO standard R497. Those listed in the "Available" column are based on the R 80 series from the same document.

TABLE A-7. AMERICAN NATIONAL STANDARD REGULAR HEXAGON BOLTS

Diameter	Flats	Height			Diameter	Flats	Height		
		Unfinished	Semi-finished	Finished			Unfinished	Semi-finished	Finished
1/4	7/16	11/64	5/32	5/32	7/8	1 5/16	37/64	35/64	35/64
5/16	1/2	7/32	13/64	13/64	1	1 1/2	43/64	39/64	39/64
3/8	9/16	1/4	15/64	15/64	1 1/8	2 11/16	3/4	11/16	11/16
7/16	5/8	19/64	9/32	9/32	1 1/4	1 7/8	27/32	25/32	25/32
1/2	3/4	11/32	5/16	5/16	1 3/8	1 1/16	29/32	27/32	27/32
9/16	13/16	....	....	23/64	1 1/2	2 1/4	1	15/16	15/16
5/8	15/16	27/64	25/64	25/64	1 3/4	2 5/8	1 5/32	1 3/32	1 3/32
3/4	1 1/8	1/2	15/32	15/32	2	3	1 11/32	1 7/32	1 7/32

TABLE A-8. REGULAR METRIC HEXAGON BOLTS

Nominal size (mm)	Width across flats	Thickness	Nominal size (mm)	Width across flats	Thickness
1.6	3.2	1.1	16	24	10
2	4	1.4	18	27	12
2.5	5	1.7	20	30	13
3	5.5	2	22	32	14
4	7	2.8	24	36	15
5	8	3.5	27	41	17
6	10	4	30	46	19
8	13	5.5	33	50	21
10	17	7	36	55	23
12	19	8	39	60	25
14	22	9			

TABLE A-9. AMERICAN NATIONAL STANDARD REGULAR HEXAGON NUTS

Diameter	Unfinished		Semifinished		Finished	
	Flats	Thickness	Flats	Thickness	Flats	Thickness
$\frac{1}{4}$	$\frac{7}{16}$	$\frac{7}{32}$	$\frac{7}{16}$	$\frac{13}{64}$	$\frac{7}{16}$	$\frac{7}{32}$
$\frac{5}{16}$	$\frac{9}{16}$	$\frac{17}{64}$	$\frac{9}{16}$	$\frac{1}{4}$	$\frac{1}{2}$	$\frac{17}{64}$
$\frac{3}{8}$	$\frac{5}{8}$	$\frac{21}{64}$	$\frac{5}{8}$	$\frac{5}{16}$	$\frac{9}{16}$	$\frac{21}{64}$
$\frac{7}{16}$	$\frac{3}{4}$	$\frac{3}{8}$	$\frac{3}{4}$	$\frac{23}{64}$	$\frac{11}{16}$	$\frac{3}{8}$
$\frac{1}{2}$	$\frac{13}{16}$	$\frac{7}{16}$	$\frac{13}{16}$	$\frac{27}{64}$	$\frac{3}{4}$	$\frac{7}{16}$
$\frac{9}{16}$	$\frac{7}{8}$	$\frac{1}{2}$	$\frac{7}{8}$	$\frac{31}{64}$	$\frac{7}{8}$	$\frac{31}{64}$
$\frac{5}{8}$	1	$\frac{35}{64}$	1	$\frac{17}{32}$	$\frac{15}{16}$	$\frac{35}{64}$
$\frac{3}{4}$	$1\frac{1}{8}$	$\frac{21}{32}$	$1\frac{1}{8}$	$\frac{41}{64}$	$1\frac{1}{8}$	$\frac{41}{64}$
$\frac{7}{8}$	$1\frac{5}{16}$	$\frac{49}{64}$	$1\frac{5}{16}$	$\frac{3}{4}$	$1\frac{5}{16}$	$\frac{3}{4}$
1	$1\frac{1}{2}$	$\frac{7}{8}$	$1\frac{1}{2}$	$\frac{55}{64}$	$1\frac{1}{2}$	$\frac{55}{64}$
$1\frac{1}{8}$	$1\frac{11}{16}$	1	$1\frac{11}{16}$	$\frac{31}{32}$	$1\frac{11}{16}$	$\frac{31}{32}$
$1\frac{1}{4}$	$1\frac{7}{8}$	$1\frac{3}{32}$	$1\frac{7}{8}$	$1\frac{1}{16}$	$1\frac{7}{8}$	$1\frac{1}{16}$
$1\frac{3}{8}$	$2\frac{1}{16}$	$1\frac{13}{64}$	$2\frac{1}{16}$	$1\frac{11}{64}$	$2\frac{1}{16}$	$1\frac{11}{64}$
$1\frac{1}{2}$	$2\frac{1}{4}$	$1\frac{5}{16}$	$2\frac{1}{4}$	$1\frac{9}{32}$	$2\frac{1}{4}$	$1\frac{9}{32}$
$1\frac{5}{8}$	...	...	$2\frac{7}{16}$	$1\frac{25}{64}$		
$1\frac{3}{4}$	...	...	$2\frac{5}{8}$	$1\frac{1}{2}$	$2\frac{5}{8}$	$1\frac{1}{2}$
$1\frac{7}{8}$	...	...	$2\frac{13}{16}$	$1\frac{39}{64}$		
2	...	...	3	$1\frac{23}{32}$	3	$1\frac{23}{32}$

TABLE A-10. REGULAR METRIC HEXAGON NUTS

Nominal size (mm)	Distance across flats	Thickness			Nominal size (mm)	Distance across flats	Thickness		
		Regular	Jamb	Thick			Regular	Jamb	Thick
1.6	3.2	1.3			16	24	13	8.0	16
2	4.0	1.6	1.2		18	27	15	9.0	18.5
2.5	5.0	2.0			20	30	16	9.0	20
3	5.5	2.4	1.6	4.0	22	32	18	10	22
4	7.0	3.2	2.0	5.0	24	36	19	10	24
5	8.0	4.0	2.5	5.0					
6	10	5.0	3.0	6.0	27	41	22	12	27
8	13	6.5	5.0	8.0	30	46	24	12	30
					33	50	26		
10	17	8.0	6.0	10	36	55	29		
12	19	10	7.0	12					
14	22	11	8.0	14	39	60	31		

TABLE A-11. AMERICAN NATIONAL STANDARD REGULAR SQUARE BOLTS AND NUTS

Diameter	Bolthead		Nut	
	Flats	Height of head	Flats	Thickness of nut
$1/4$	$3/8$	$11/64$	$7/16$	$7/32$
$5/16$	$1/2$	$13/64$	$9/16$	$17/64$
$3/8$	$9/16$	$1/4$	$5/8$	$21/64$
$7/16$	$5/8$	$19/64$	$3/4$	$3/8$
$1/2$	$3/4$	$21/64$	$13/16$	$7/16$
$5/8$	$15/16$	$27/64$	1	$35/64$
$3/4$	$1 1/8$	$1/2$	$1 1/8$	$21/32$
$7/8$	$1 5/16$	$19/32$	$1 5/16$	$49/64$
1	$1 1/2$	$21/32$	$1 1/2$	$7/8$
$1 1/8$	$1 11/16$	$3/4$	$1 11/16$	1
$1 1/4$	$1 7/8$	$27/32$	$1 7/8$	$1 3/32$
$1 3/8$	$2 1/16$	$29/32$	$2 1/16$	$1 13/64$
$1 1/2$	$2 1/4$	1	$2 1/4$	$1 5/16$
$1 5/8$	$2 7/16$	$1 3/32$		

TABLE A-12. AMERICAN NATIONAL STANDARD HEAVY HEXAGON BOLTS

Diameter	Flats — Unfinished semifinished finished	Height of head — Unfinished	Height of head — Semifinished finished	Diameter	Flats — Unfinished semifinished finished	Height of head — Unfinished	Height of head — Semifinished
$1/2$	$7/8$	$7/16$	$13/32$	$1\,5/8$	$2\,9/16$	$1\,9/32$	$1\,7/32$
$5/8$	$1\,1/16$	$17/32$	$1/2$	$1\,3/4$	$2\,3/4$	$1\,3/8$	$1\,5/16$
$3/4$	$1\,1/4$	$5/8$	$19/32$	$1\,7/8$	$2\,15/16$	$1\,15/32$	$1\,13/32$
$7/8$	$1\,7/16$	$23/32$	$11/16$	2	$3\,1/8$	$1\,9/16$	$1\,7/16$
1	$1\,5/8$	$13/16$	$3/4$	$2\,1/4$	$3\,1/2$	$1\,3/4$	$1\,5/8$
$1\,1/8$	$1\,13/16$	$29/32$	$27/32$	$2\,1/2$	$3\,7/8$	$1\,15/16$	$1\,13/16$
$1\,1/4$	2	1	$15/16$	$2\,3/4$	$4\,1/4$	$2\,1/8$	2
$1\,3/8$	$2\,3/16$	$1\,3/32$	$1\,1/32$	3	$4\,5/8$	$2\,5/16$	$2\,3/16$
$1\,1/2$	$2\,3/8$	$1\,3/16$	$1\,1/8$				

Note: $1\,5/8$, $1\,7/8$ not in unfinished or semifinished bolts.

TABLE A-13. AMERICAN NATIONAL STANDARD HEAVY NUTS, SQUARE AND HEXAGON

Diameter	Flats — Unfinished semifinished square and hexagon	Thickness of nut — Unfinished square and hexagon	Thickness of nut — Semifinished hexagon	Diameter	Flats — Unfinished semifinished square and hexagon	Thickness of nut — Unfinished square and hexagon	Thickness of nut — Semifinished hexagon
$1/4$	$1/2$	$1/4$	$15/64$	1	$1\,5/8$	1	$63/64$
$5/16$	$9/16$	$5/16$	$19/64$	$1\,1/8$	$1\,13/16$	$1\,1/8$	$1\,7/64$
$3/8$	$11/16$	$3/8$	$23/64$	$1\,1/4$	2	$1\,1/4$	$1\,7/32$
$7/16$	$3/4$	$7/16$	$27/64$	$1\,3/8$	$2\,3/16$	$1\,3/8$	$1\,11/32$
$1/2$	$7/8$	$1/2$	$31/64$	$1\,1/2$	$2\,3/8$	$1\,1/2$	$1\,15/32$
$9/16$	$15/16$	....	$35/64$	$1\,5/8$	$2\,9/16$	....	$1\,19/32$
$5/8$	$1\,1/16$	$5/8$	$39/64$	$1\,3/4$	$2\,3/4$	$1\,3/4$	$1\,23/32$
$3/4$	$1\,1/4$	$3/4$	$47/64$	$1\,7/8$	$2\,15/16$	....	$1\,27/32$
$7/8$	$1\,7/16$	$7/8$	$55/64$	2	$3\,1/8$	2	$1\,31/32$

Note: $9/16$, $1\,5/8$, $1\,7/8$ not in unfinished nuts.

D	H nom.	R nom.	K max.	U min.	V max.	C nom.	J nom.	P max.	Q
10 (0.190)	9/64	15/32	0.145	0.083	0.027	3/32	0.141	0.127	0.090
12 (0.216)	5/32	35/64	0.162	0.091	0.029	7/64	0.156	0.144	0.110
1/4	3/16	5/8	0.185	0.100	0.032	1/8	0.188	0.156	0.125
5/16	15/64	25/32	0.240	0.111	0.036	11/64	0.234	0.203	0.156
3/8	9/32	15/16	0.294	0.125	0.041	13/64	0.281	0.250	0.188
7/16	21/64	1 3/32	0.345	0.143	0.046	15/64	0.328	0.297	0.219
1/2	3/8	1 1/4	0.400	0.154	0.050	9/32	0.375	0.344	0.250
9/16	27/64	1 13/32	0.454	0.167	0.054	5/16	0.422	0.391	0.281
5/8	15/32	1 9/16	0.507	0.182	0.059	23/64	0.469	0.469	0.313
3/4	9/16	1 7/8	0.620	0.200	0.065	7/16	0.563	0.563	0.375
7/8	21/32	2 3/16	0.731	0.222	0.072	33/64	0.656	0.656	0.438
1	3/4	2 1/2	0.838	0.250	0.081	19/32	0.750	0.750	0.500
1 1/8	27/32	2 13/16	0.939	0.283	0.092	43/64	0.844	0.844	0.562
1 1/4	15/16	3 1/8	1.064	0.283	0.092	3/4	0.938	0.938	0.625
1 3/8	1 1/32	3 7/16	1.159	0.333	0.109	53/64	1.031	1.031	0.688
1 1/2	1 1/8	3 3/4	1.284	0.333	0.109	29/32	1.125	1.125	0.750

TABLE A-14. AMERICAN NATIONAL STANDARD SQUARE-HEAD SETSCREWS AND POINTS

Note: Threads may be coarse-, fine-, or 8-threaded series, class 2A. Coarse thread normally used on 1/4 in. and larger. When length equals nominal diameter or less $Y = 118°$. When length exceeds nominal diameter $Y = 90°$.

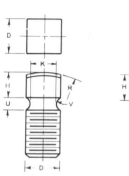

OVAL (ROUND) POINT

CONE POINT

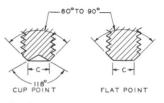

CUP POINT   FLAT POINT

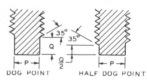

DOG POINT   HALF DOG POINT

TABLE A-15. AMERICAN NATIONAL STANDARD SLOTTED-HEAD CAPSCREWS

Nominal diameter D	A max.	B max.	C max.	E max.	F max.	G max.	H average	I max.	J max.	K max.	M max.
1/4	0.375	0.216	0.172	0.075	0.097	0.500	0.140	0.068	0.437	0.191	0.117
5/16	0.437	0.253	0.203	0.084	0.115	0.625	0.177	0.086	0.562	0.245	0.151
3/8	0.562	0.314	0.250	0.094	0.142	0.750	0.210	0.103	0.625	0.273	0.168
7/16	0.625	0.368	0.297	0.094	0.168	0.812	0.210	0.103	0.750	0.328	0.202
1/2	0.750	0.413	0.328	0.106	0.193	0.875	0.210	0.103	0.812	0.354	0.218
9/16	0.812	0.467	0.375	0.118	0.213	1.000	0.244	0.120	0.937	0.409	0.252
5/8	0.875	0.521	0.422	0.133	0.239	1.125	0.281	0.137	1.000	0.437	0.270
3/4	1.000	0.612	0.500	0.149	0.283	1.375	0.352	0.171	1.250	0.546	0.338
7/8	1.125	0.720	0.594	0.167	0.334	1.625	0.423	0.206			
1	1.312	0.803	0.656	0.188	0.371	1.875	0.494	0.240			
1 1/8	....	....	....	0.196	....	2.062	0.529	0.257			
1 1/4	....	....	....	0.211	....	2.312	0.600	0.291			
1 3/8	....	....	....	0.226	....	2.562	0.665	0.326			
1 1/2	....	....	....	0.258	....	2.812	0.742	0.360			

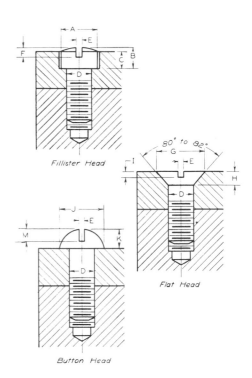

Fillister Head

Flat Head

Button Head

TABLE A-16. AMERICAN NATIONAL STANDARD MACHINE SCREWS

Diameter		A	B	C	E	F	G	H	I	J	K	M
Nominal	Max.											
0	0.060	0.119	0.035	0.056	0.113	0.053	0.096	0.045	0.059	....	....	0.023
1	0.073	0.146	0.043	0.068	0.138	0.061	0.118	0.053	0.071	....	....	0.026
2	0.086	0.172	0.051	0.080	0.162	0.069	0.140	0.062	0.083	0.167	0.053	0.031
3	0.099	0.199	0.059	0.092	0.187	0.078	0.161	0.070	0.095	0.193	0.060	0.035
4	0.112	0.225	0.067	0.104	0.211	0.086	0.183	0.079	0.107	0.219	0.068	0.039
5	0.125	0.252	0.075	0.116	0.236	0.095	0.205	0.088	0.120	0.245	0.075	0.043
6	0.138	0.279	0.083	0.128	0.260	0.103	0.226	0.096	0.132	0.270	0.082	0.048
8	0.164	0.332	0.100	0.152	0.309	0.120	0.270	0.113	0.156	0.322	0.096	0.054
10	0.190	0.385	0.116	0.176	0.359	0.137	0.313	0.130	0.180	0.373	0.110	0.060
12	0.216	0.438	0.132	0.200	0.408	0.153	0.357	0.148	0.205	0.425	0.125	0.067
1/4	0.250	0.507	0.153	0.232	0.472	0.175	0.414	0.170	0.237	0.492	0.144	0.075
5/16	0.3125	0.635	0.191	0.290	0.590	0.216	0.518	0.211	0.295	0.615	0.178	0.084
3/8	0.375	0.762	0.230	0.347	0.708	0.256	0.622	0.253	0.355	0.740	0.212	0.094
7/16	0.4375	0.812	0.223	0.345	0.750	0.328	0.625	0.265	0.368	....	....	0.094
1/2	0.500	0.875	0.223	0.354	0.813	0.355	0.750	0.297	0.412	....	....	0.106
9/16	0.5625	1.000	0.260	0.410	0.938	0.410	0.812	0.336	0.466	....	....	0.118
5/8	0.625	1.125	0.298	0.467	1.000	0.438	0.875	0.375	0.521	....	....	0.133
3/4	0.750	1.375	0.372	0.578	1.250	0.547	1.000	0.441	0.612	....	....	0.149

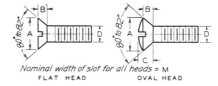

Nominal width of slot for all heads = M
FLAT HEAD   OVAL HEAD

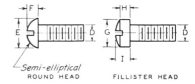

Semi-elliptical
ROUND HEAD   FILLISTER HEAD

PAN HEAD

TABLE A-17. METRIC CAPSCREWS

Nominal size	Hexagon head		Socket head			Flat head		Fillister head		Pan head	
	A	H	A	H	Key size	A	H	A	H	A	H
M3	5.5	2	5.5	3	2.5	5.6	1.6	6	2.4	5.6	1.9
4	7	2.8	7	4	3	7.5	2.2	8	3.1	7.5	2.5
5	8.5	3.5	9	5	4	9.2	2.5	10	3.8	9.2	3.1
6	10	4	10	6	5	11	3	12	4.6	11	3.8
8	13	5.5	13	8	6	14.5	4	16	6	14.5	5
10	17	7	16	10	8	18	5	20	7.5	18	6.2
12	19	8	18	12	10						
14	22	9	22	14	12						
16	24	10	24	16	14						

HEXAGON HEAD SOCKET HEAD FLAT HEAD FILLISTER HEAD PAN HEAD

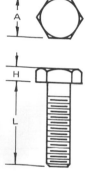
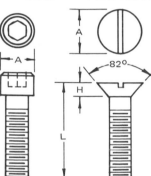
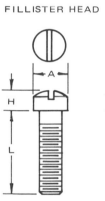
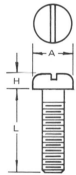

TABLE A-18. AMERICAN NATIONAL STANDARD PLAIN WASHERS[a]

Nominal washer size[b]		Inside dia.	Outside dia.	Nominal thickness
		A	B	C
....	....	0.078	0.188	0.020
....	....	0.094	0.250	0.020
...	....	0.125	0.312	0.032
No. 6	0.138	0.156	0.375	0.049
No. 8	0.164	0.188	0.438	0.049
No. 10	0.190	0.219	0.500	0.049
3/16	0.188	0.250	0.562	0.049
No. 12	0.216	0.250	0.562	0.065
1/4	0.250 N[c]	0.281	0.625	0.065

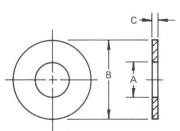

TABLE A-18. (CONTINUED)

Nominal washer size[b]			Inside dia.	Outside dia.	Nominal thickness
			A	B	C
1/4	0.250	W	0.312	0.734	0.065
5/16	0.312	N	0.344	0.688	0.065
5/16	0.312	W	0.375	0.875	0.083
3/8	0.375	N	0.406	0.812	0.065
3/8	0.375	W	0.438	1.000	0.083
7/16	0.438	N	0.469	0.922	0.065
7/16	0.438	W	0.500	1.250	0.083
1/2	0.500	N	0.531	1.062	0.095
1/2	0.500	W	0.562	1.375	0.109
9/16	0.562	N	0.594	1.156	0.095
9/16	0.562	W	0.625	1.469	0.095
5/8	0.625	N	0.656	1.312	0.095
5/8	0.625	W	0.688	1.750	0.134
3/4	0.750	N	0.812	1.469	0.134
3/4	0.750	W	0.812	2.000	0.148
7/8	0.875	N	0.938	1.750	0.134
7/8	0.875	W	0.938	2.250	0.165
1	1.000	N	1.062	2.000	0.134
1	1.000	W	1.062	2.500	0.165
1 1/8	1.125	N	1.250	2.250	0.134
1 1/8	1.125	W	1.250	2.750	0.165
1 1/4	1.250	N	1.375	2.500	0.165
1 1/4	1.250	W	1.375	3.000	0.165
1 3/8	1.375	N	1.500	2.750	0.165
1 3/8	1.375	W	1.500	3.250	0.180
1 1/2	1.500	N	1.625	3.000	0.165
1 1/2	1.500	W	1.625	3.500	0.180
1 5/8	1.625		1.750	3.750	0.180
1 3/4	1.750		1.875	4.000	0.180
1 7/8	1.875		2.000	4.250	0.180
2	2.000		2.125	4.500	0.180
2 1/4	2.250		2.375	4.750	0.220
2 1/2	2.500		2.625	5.000	0.238
2 3/4	2.750		2.875	5.250	0.259
3	3.000		3.125	5.500	0.284

[a]ANSI B18.22.1
[b]Nominal washer sizes are intended for use with comparable nominal screw or bolt sizes
[c]N = narrow; W = wide.

TABLE A-17. METRIC CAPSCREWS

Nominal size	Hexagon head A	Hexagon head H	Socket head A	Socket head H	Socket head Key size	Flat head A	Flat head H	Fillister head A	Fillister head H	Pan head A	Pan head H
M3	5.5	2	5.5	3	2.5	5.6	1.6	6	2.4	5.6	1.9
4	7	2.8	7	4	3	7.5	2.2	8	3.1	7.5	2.5
5	8.5	3.5	9	5	4	9.2	2.5	10	3.8	9.2	3.1
6	10	4	10	6	5	11	3	12	4.6	11	3.8
8	13	5.5	13	8	6	14.5	4	16	6	14.5	5
10	17	7	16	10	8	18	5	20	7.5	18	6.2
12	19	8	18	12	10						
14	22	9	22	14	12						
16	24	10	24	16	14						

HEXAGON HEAD   SOCKET HEAD   FLAT HEAD   FILLISTER HEAD   PAN HEAD

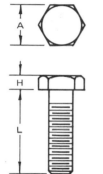
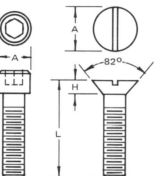
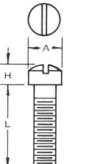
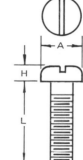

82°

TABLE A-18. AMERICAN NATIONAL STANDARD PLAIN WASHERS[a]

Nominal washer size[b]		Inside dia. A	Outside dia. B	Nominal thickness C
....	....	0.078	0.188	0.020
....	....	0.094	0.250	0.020
....	....	0.125	0.312	0.032
No. 6	0.138	0.156	0.375	0.049
No. 8	0.164	0.188	0.438	0.049
No. 10	0.190	0.219	0.500	0.049
3/16	0.188	0.250	0.562	0.049
No. 12	0.216	0.250	0.562	0.065
1/4	0.250 N[c]	0.281	0.625	0.065

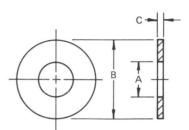

TABLE A-18. (*CONTINUED*)

Nominal washer size[b]			Inside dia.	Outside dia.	Nominal thickness
			A	B	C
¼	0.250	W	0.312	0.734	0.065
5/16	0.312	N	0.344	0.688	0.065
5/16	0.312	W	0.375	0.875	0.083
3/8	0.375	N	0.406	0.812	0.065
3/8	0.375	W	0.438	1.000	0.083
7/16	0.438	N	0.469	0.922	0.065
7/16	0.438	W	0.500	1.250	0.083
½	0.500	N	0.531	1.062	0.095
½	0.500	W	0.562	1.375	0.109
9/16	0.562	N	0.594	1.156	0.095
9/16	0.562	W	0.625	1.469	0.095
5/8	0.625	N	0.656	1.312	0.095
5/8	0.625	W	0.688	1.750	0.134
¾	0.750	N	0.812	1.469	0.134
¾	0.750	W	0.812	2.000	0.148
7/8	0.875	N	0.938	1.750	0.134
7/8	0.875	W	0.938	2.250	0.165
1	1.000	N	1.062	2.000	0.134
1	1.000	W	1.062	2.500	0.165
1⅛	1.125	N	1.250	2.250	0.134
1⅛	1.125	W	1.250	2.750	0.165
1¼	1.250	N	1.375	2.500	0.165
1¼	1.250	W	1.375	3.000	0.165
1⅜	1.375	N	1.500	2.750	0.165
1⅜	1.375	W	1.500	3.250	0.180
1½	1.500	N	1.625	3.000	0.165
1½	1.500	W	1.625	3.500	0.180
1⅝	1.625		1.750	3.750	0.180
1¾	1.750		1.875	4.000	0.180
1⅞	1.875		2.000	4.250	0.180
2	2.000		2.125	4.500	0.180
2¼	2.250		2.375	4.750	0.220
2½	2.500		2.625	5.000	0.238
2¾	2.750		2.875	5.250	0.259
3	3.000		3.125	5.500	0.284

[a]ANSI B18.22.1
[b]Nominal washer sizes are intended for use with comparable nominal screw or bolt sizes
[c]N = narrow; W = wide.

TABLE A-19. AMERICAN NATIONAL STANDARD LOCK WASHERS[a]

Nominal washer size[b]	Inside dia. min.	Regular		Extra duty		Hi-collar	
		Outside dia. max.	Thick-ness min.	Outside dia. max.	Thick-ness min.	Outside dia. max.	Thick-ness min.
No. 2	0.086	0.088	0.172	0.020	0.208	0.027	
No. 3	0.099	0.101	0.195	0.025	0.239	0.034	
No. 4	0.112	0.115	0.209	0.025	0.253	0.034	0.173	0.022
No. 5	0.125	0.128	0.236	0.031	0.300	0.045	0.202	0.030
No. 6	0.138	0.141	0.250	0.031	0.314	0.045	0.216	0.030
No. 8	0.164	0.168	0.293	0.040	0.375	0.057	0.267	0.047
No. 10	0.190	0.194	0.334	0.047	0.434	0.068	0.294	0.047
No. 12	0.216	0.221	0.377	0.056	0.497	0.080		
1/4	0.250	0.255	0.489	0.062	0.535	0.084	0.365	0.078
5/16	0.312	0.318	0.586	0.078	0.622	0.108	0.460	0.093
3/8	0.375	0.382	0.683	0.094	0.741	0.123	0.553	0.125
7/16	0.438	0.446	0.779	0.109	0.839	0.143	0.647	0.140
1/2	0.500	0.509	0.873	0.125	0.939	0.162	0.737	0.172
9/16	0.562	0.572	0.971	0.141	1.041	0.182		
5/8	0.625	0.636	1.079	0.156	1.157	0.202	0.923	0.203
11/16	0.688	0.700	1.176	0.172	1.258	0.221		
3/4	0.750	0.763	1.271	0.188	1.361	0.241	1.111	0.218
13/16	0.812	0.826	1.367	0.203	1.463	0.261		
7/8	0.875	0.890	1.464	0.219	1.576	0.285	1.296	0.234
15/16	0.938	0.954	1.560	0.234	1.688	0.308		
1	1.000	1.017	1.661	0.250	1.799	0.330	1.483	0.250
1 1/16	1.062	1.080	1.756	0.266	1.910	0.352		
1 1/8	1.125	1.144	1.853	0.281	2.019	0.375	1.669	0.313
1 3/16	1.188	1.208	1.950	0.297	2.124	0.396		
1 1/4	1.250	1.271	2.045	0.312	2.231	0.417	1.799	0.313
1 5/16	1.312	1.334	2.141	0.328	2.335	0.438		
1 3/8	1.375	1.398	2.239	0.344	2.439	0.458	2.041	0.375
1 7/16	1.438	1.462	2.334	0.359	2.540	0.478		
1 1/2	1.500	1.525	2.430	0.375	2.638	0.496	2.170	0.375

[a]ANSI B18.21.1.
[b]Nominal washer sizes are intended for use with comparable nominal screw or bolt sizes.

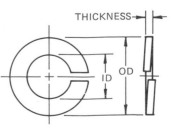

THICKNESS

ID   OD

TABLE A-20. AMERICAN NATIONAL STANDARD COTTER PINS

A nominal	B min.	Hole sizes recommended	A nominal	B min.	Hole sizes recommended
0.031	1/16	3/64	0.188	3/8	13/64
0.047	3/32	1/16	0.219	7/16	15/64
0.062	1/8	5/64	0.250	1/2	17/64
0.078	5/32	3/32			
0.094	3/16	7/64	0.312	5/8	5/16
0.109	7/32	1/8	0.375	3/4	3/8
			0.438	7/8	7/16
0.125	1/4	9/64	0.500	1	1/2
0.141	9/32	5/32	0.625	1 1/4	5/8
0.156	5/16	11/64	0.750	1 1/2	3/4

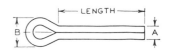

Design of head may vary but outside diameters should be adhered to.

TABLE A-21. AMERICAN NATIONAL STANDARD TAPER PINS

TAPER PINS *Maximum length for which standard reamers are available. Taper 1/4 in. per ft.

Size No.	0000000	000000	00000	0000	000	00	0
Size (large end)	0.0625	0.0780	0.0940	0.1090	0.1250	0.1410	0.1560
Maximum length*	0.625	0.750	1.000	1.000	1.000	1.250	1.250
Size No.	1	2	3	4	5	6	7
Size (large end)	0.1720	0.1930	0.2190	0.2500	0.2890	0.3410	0.4090
Maximum length*	1.250	1.500	1.750	2.000	2.250	3.000	3.750
Size No.	8	9	10	11	12	13	14
Size (large end)	0.4920	0.5910	0.7060	0.8600	1.032	1.241	1.523
Maximum length*	4.500	5.250	6.000	(Special sizes. Special lengths.)			

TABLE A-22. AMERICAN NATIONAL STANDARD SQUARE- AND FLAT-STOCK KEYS AND SHAFT DIAMETERS

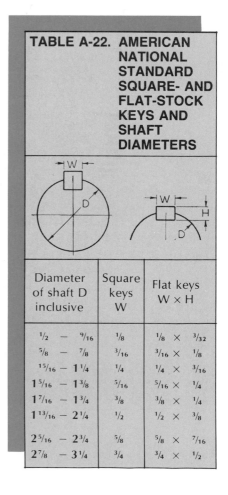

Diameter of shaft D inclusive	Square keys W	Flat keys W × H
1/2 — 9/16	1/8	1/8 × 3/32
5/8 — 7/8	3/16	3/16 × 1/8
15/16 — 1 1/4	1/4	1/4 × 3/16
1 5/16 — 1 3/8	5/16	5/16 × 1/4
1 7/16 — 1 3/4	3/8	3/8 × 1/4
1 13/16 — 2 1/4	1/2	1/2 × 3/8
2 5/16 — 2 3/4	5/8	5/8 × 7/16
2 7/8 — 3 1/4	3/4	3/4 × 1/2

TABLE A-23. WOODRUFF KEYS

Key No.	Nominal			Maximum		
	A	B	E	C	D	H
204	$1/16$	$1/2$	$3/64$	0.203	0.194	0.1718
304	$3/32$	$1/2$	$3/64$	0.203	0.194	0.1561
305	$3/32$	$5/8$	$1/16$	0.250	0.240	0.2031
404	$1/8$	$1/2$	$3/64$	0.203	0.194	0.1405
405	$1/8$	$5/8$	$1/16$	0.250	0.240	0.1875
406	$1/8$	$3/4$	$1/16$	0.313	0.303	0.2505
505	$5/32$	$5/8$	$1/16$	0.250	0.240	0.1719
506	$5/32$	$3/4$	$1/16$	0.313	0.303	0.2349
507	$5/32$	$7/8$	$1/16$	0.375	0.365	0.2969
606	$3/16$	$3/4$	$1/16$	0.313	0.303	0.2193
607	$3/16$	$7/8$	$1/16$	0.375	0.365	0.2813
608	$3/16$	1	$1/16$	0.438	0.428	0.3443
609	$3/16$	$1 1/8$	$5/64$	0.484	0.475	0.3903
807	$1/4$	$7/8$	$1/16$	0.375	0.365	0.2500
808	$1/4$	1	$1/16$	0.438	0.428	0.3130
809	$1/4$	$1 1/8$	$5/64$	0.484	0.475	0.3590
810	$1/4$	$1 1/4$	$5/64$	0.547	0.537	0.4220
811	$1/4$	$1 3/8$	$3/32$	0.594	0.584	0.4690
812	$1/4$	$1 1/2$	$7/64$	0.641	0.631	0.5160
1008	$5/16$	1	$1/16$	0.438	0.428	0.2818

Note: Nominal dimensions are indicated by the key number in which the last two digits give the diameter (B) in eighths and the ones in front of them give the width (A) in thirty-seconds. For example, No. 809 means B = $9/8$ or $1 1/8$ and A = $8/32$ or $1/4$.

TABLE A-24. WIRE AND SHEET-METAL GAGES

Dimensions in Decimal Parts of an Inch

No. of wire gage	American, or Brown & Sharpe	Birming-ham, or Stubs wire	Washburn & Moen or American Steel & Wire Co.	W. & M. steel music wire	New American S. & W. Co. music wire gage	Imperial wire gage	U.S. Standard gage for sheet and plate iron and steel
00000000	....	.....	....	0.0083			
0000000	....	.....	....	0.0087			
000000	....	.....	....	0.0095	0.004	0.464	0.46875
00000	....	.....	....	0.010	0.005	0.432	0.4375
0000	0.460	0.454	0.3938	0.011	0.006	0.400	0.40625
000	0.40964	0.425	0.3625	0.012	0.007	0.372	0.375
00	0.3648	0.380	0.3310	0.0133	0.008	0.348	0.34375
0	0.32486	0.340	0.3065	0.0144	0.009	0.324	0.3125
1	0.2893	0.300	0.2830	0.0156	0.010	0.300	0.28125
2	0.25763	0.284	0.2625	0.0166	0.011	0.276	0.265625
3	0.22942	0.259	0.2437	0.0178	0.012	0.252	0.250
4	0.20431	0.238	0.2253	0.0188	0.013	0.232	0.234375
5	0.18194	0.220	0.2070	0.0202	0.014	0.212	0.21875
6	0.16202	0.203	0.1920	0.0215	0.016	0.192	0.203125
7	0.14428	0.180	0.1770	0.023	0.018	0.176	0.1875
8	0.12849	0.165	0.1620	0.0243	0.020	0.160	0.171875
9	0.11443	0.148	0.1483	0.0256	0.022	0.144	0.15625
10	0.10189	0.134	0.1350	0.027	0.024	0.128	0.140625
11	0.090742	0.120	0.1205	0.0284	0.026	0.116	0.125
12	0.080808	0.109	0.1055	0.0296	0.029	0.104	0.109375
13	0.071961	0.095	0.0915	0.0314	0.031	0.092	0.09375
14	0.064084	0.083	0.0800	0.0326	0.033	0.080	0.078125
15	0.057068	0.072	0.0720	0.0345	0.035	0.072	0.0703125
16	0.05082	0.065	0.0625	0.036	0.037	0.064	0.0625
17	0.045257	0.058	0.0540	0.0377	0.039	0.056	0.05625
18	0.040303	0.049	0.0475	0.0395	0.041	0.048	0.050
19	0.03589	0.042	0.0410	0.0414	0.043	0.040	0.04375
20	0.031961	0.035	0.0348	0.0434	0.045	0.036	0.0375
21	0.028462	0.032	0.03175	0.046	0.047	0.032	0.034375
22	0.025347	0.028	0.0286	0.0483	0.049	0.028	0.03125
23	0.022571	0.025	0.0258	0.051	0.051	0.024	0.028125
24	0.0201	0.022	0.0230	0.055	0.055	0.022	0.025
25	0.0179	0.020	0.0204	0.0586	0.059	0.020	0.021875
26	0.01594	0.018	0.0181	0.0626	0.063	0.018	0.01875
27	0.014195	0.016	0.0173	0.0658	0.067	0.0164	0.0171875
28	0.012641	0.014	0.0162	0.072	0.071	0.0149	0.015625
29	0.011257	0.013	0.0150	0.076	0.075	0.0136	0.0140625
30	0.010025	0.012	0.0140	0.080	0.080	0.0124	0.0125
31	0.008928	0.010	0.0132	......	0.085	0.0116	0.0109375
32	0.00795	0.009	0.0128	......	0.090	0.0108	0.01015625

TABLE A-25. AMERICAN NATIONAL STANDARD WELDED AND SEAMLESS STEEL PIPE

Nominal pipe size	Nominal wall thickness				Tap drill
	Outside dia.	Standard wall	Extra strong wall	Double extra strong wall	
$\frac{1}{8}$	0.405	0.068	0.095	.....	$11/32$
$\frac{1}{4}$	0.540	0.088	0.119	.....	$7/16$
$\frac{3}{8}$	0.675	0.091	0.126	.....	$37/64$
$\frac{1}{2}$	0.840	0.109	0.147	0.294	$23/31$
$\frac{3}{4}$	1.050	0.113	0.154	0.308	$59/64$
1	1.315	0.133	0.179	0.358	$15/32$
$1\frac{1}{4}$	1.660	0.140	0.191	0.382	$1\frac{1}{2}$
$1\frac{1}{2}$	1.900	0.145	0.200	0.400	$147/64$
2	2.375	0.154	0.218	0.436	$27/32$
$2\frac{1}{2}$	2.875	0.203	0.276	0.552	$25/8$
3	3.500	0.216	0.300	0.600	$3\frac{1}{4}$
$3\frac{1}{2}$	4.000	0.226	0.318	.....	$3\frac{3}{4}$
4	4.500	0.237	0.337	0.674	$4\frac{1}{4}$
5	5.563	0.258	0.375	0.750	$55/16$
6	6.625	0.280	0.432	0.864	$65/16$
8	8.625	0.322	0.500	0.875	$75/16$

Note: To find the inside diameter, subtract twice the wall thickness from the outside diameter. Schedule numbers have been set up for wall thicknesses for pipe and the standard should be consulted for complete information. Standard wall thicknesses are for Schedule 40 up to and including nominal size 10. Extra strong walls are Schedule 80 up to and including size 8, and Schedule 60 for size 10.

AMERICAN NATIONAL STANDARD LIMITS AND FITS

The following tables are designed for use in the basic hole system of limits and fits described in Chap. 6, Dimensioning. Information for these tables is adapted from ANSI B4.1, "Preferred Limits and Fits for Cylindrical Parts." For larger sizes and additional information, refer to the standard.

There are five distinct classes of fits:

RC   Running or sliding clearance fits
LC   Locational clearance fits
LT   Transition clearance or interference fits
LN   Locational interference fits
FN   Force or shrink fits

These five classes of fits are placed in three general categories as follows.

Running and Sliding Fits (Table A-26).   These fits provide a similar running performance, with suitable lubrication allowance, throughout the range of sizes. The clearances for the first two classes, used chiefly as slide fits, increase more slowly than for the other classes, so that accurate location is maintained, even when this is at the expense of free relative motion.

RC 1: *Close sliding fits* accurately locate parts that must assemble without perceptible play.

RC 2: *Sliding fits* are for accurate location, but with greater maximum clearance than RC 1. Parts move and turn easily but do not run freely, and in the larger sizes may seize with small temperature changes.

TABLE A-26. American National Standard Running and Sliding Fits[a]

Basic hole system. Limits are in thousandths of an inch.

Nominal size range, inches Over To	Class RC 1 close sliding fit			Class RC 2 sliding fit			Class RC 3 precision running fit			Class RC 4 close running fit		
	Limits of clearance	Standard limits		Limits of clearance	Standard limits		Limits of clearance	Standard limits		Limits of clearance	Standard limits	
		Hole	Shaft		Hole	Shaft		Hole	Shaft		Hole	Shaft
0–0.12	0.1	+0.2	−0.1	0.1	0.25	+0.1	−0.3	+0.4	−0.3	0.3	+0.6	−0.3
	0.45	−0	−0.25	0.55	−0	−0.3	0.95	−0	−0.55	1.3	−0	−0.7
0.12–0.24	0.15	+0.2	−0.15	0.15	+0.3	−0.15	0.4	+0.5	−0.4	0.4	+0.7	−0.4
	0.5	−0	−0.3	0.65	−0	−0.35	1.12	−0	−0.7	1.6	−0	−0.9
0.24–0.40	0.2	+0.25	−0.2	0.2	+0.4	−0.2	0.5	+0.6	−0.5	0.5	+0.9	−0.5
	0.6	−0	−0.35	0.85	−0	−0.45	1.5	−0	−0.9	2.0	−0	−1.1
0.40–0.71	0.25	+0.3	−0.25	0.25	+0.4	−0.25	0.6	+0.7	−0.6	0.6	+1.0	−0.6
	0.75	−0	−0.45	0.95	−0	−0.55	1.7	−0	−1.3	2.8	−0	−1.3
0.71–1.19	0.3	+0.4	−0.3	0.3	+0.5	−0.3	0.8	+0.8	−0.8	0.8	+1.2	−0.8
	0.95	−0	−0.55	1.2	−0	−0.7	2.1	−0	−1.3	2.8	−0	−1.6
1.19–1.97	0.4	+0.4	−0.4	0.4	+0.6	−0.4	1.0	+1.0	−1.0	1.0	+1.6	−1.0
	1.1	−0	−0.7	1.4	−0	−0.8	2.6	−0	−1.6	3.6	−0	−2.0
1.97–3.15	0.4	+0.5	−0.4	0.4	+0.7	−0.4	1.2	+1.2	−1.2	1.2	+1.8	−1.2
	1.2	−0	−0.7	1.6	−0	−0.9	3.1	−0	−1.9	4.2	−0	−2.4

[a]ANSI B4.1.

TABLE A-26. (CONTINUED)

Nominal size range, inches	Class RC 5 Medium running fits			Class RC 6			Class RC 7			Class RC 8 Loose running fits			Class RC 9		
	Limits of clearance	Standard limits		Limits of clearance	Standard limits		Limits of clearance	Standard limits		Limits of clearance	Standard limits		Limits of clearance	Standard limits	
Over To		Hole	Shaft		Hole	Shaft		Hole	Shaft		Hole	Shaft		Hole	Shaft
0–0.12	0.6	+0.6	−0.6	0.6	+1.0	−0.6	1.0	+1.0	−1.0	2.5	+1.6	−2.5	4.0	+2.5	− 4.0
	1.6	−0	−1.0	2.2	−0	−1.2	2.6	−0	+1.6	5.1	−0	−3.5	8.1	−0	− 5.6
0.12–0.24	0.8	+0.7	−0.8	0.8	+1.2	−0.8	1.2	+1.2	−1.2	2.8	+1.8	−2.8	4.5	+3.0	− 4.5
	2.0	−0	−1.3	2.7	−0	−1.5	3.1	−0	−1.9	5.8	−0	−4.0	9.0	−0	− 6.0
0.24–0.40	1.0	+0.9	−1.0	1.0	+1.4	−1.0	1.6	+1.4	−1.6	3.0	+2.2	−3.0	5.0	+3.5	− 5.0
	2.5	−0	−1.6	3.3	−0	−1.9	3.9	−0	−2.5	6.6	−0	−4.4	10.7	−0	− 7.2
0.40–0.71	1.2	+1.0	−1.2	1.2	+1.6	−1.2	2.0	+1.6	−2.0	3.5	+2.8	−3.5	6.0	+4.0	− 6.0
	2.9	−0	−1.9	3.8	−0	−2.2	4.6	−0	−3.0	7.9	−0	−5.1	12.8	−0	− 8.8
0.71–1.19	1.6	+1.2	−1.6	1.6	+2.0	−1.6	2.5	+2.0	−2.5	4.5	+3.5	−4.5	7.0	+5.0	− 7.0
	3.6	−0	−2.4	4.8	−0	−2.8	5.7	−0	−3.7	10.0	−0	−6.5	15.5	−0	−10.5
1.19–1.97	2.0	+1.6	−2.0	2.0	+2.5	−2.0	3.0	+2.5	−3.0	5.0	+4.0	−5.0	8.0	+6.0	− 8.0
	4.6	−0	−3.0	6.1	−0	−3.6	7.1	−0	−4.6	11.5	−0	−7.5	18.0	−0	−12.0
1.97–3.15	2.5	+1.8	−2.5	2.5	+3.0	−2.5	4.0	+3.0	−4.0	6.0	+4.5	−6.0	9.0	+7.0	− 9.0
	5.5	−0	−3.7	7.3	−0	−4.3	8.8	−0	−5.8	13.5	−0	−9.0	20.5	−0	−13.5

RC 3: *Precision running fits* are about the closest fits expected to run freely, and are for precision work at slow speeds and light journal pressures. They are not suitable under appreciable temperature differences.

RC 4: *Close running fits* are chiefly for running fits on accurate machinery with moderate surface speeds and journal pressures, where accurate location and minimum play are desired.

RC 5 and RC 6: *Medium running fits* are for higher running speeds, heavy journal pressures, or both.

RC 7: *Free running fits* are for use where accuracy is not essential, where large temperature variations are likely, or under both these conditions.

RC 8 and RC 9: *Loose running fits* are for materials such as cold-rolled shafting and tubing, made to commercial tolerances.

Locational Fits (Tables A-27 through A-29). These fits determine only the location of mating parts and may provide rigid or accurate location—as in interference fits—or some freedom of location—as in clearance fits. They fall into the following three groups.

LC: *Locational clearance fits* are for normally stationary parts that can be freely assembled or disassembled. They run from snug fits for parts requiring accuracy of location, through the medium clearance fits for parts such as spigots, to the looser fastener fits where freedom of assembly is of prime importance.

LT: *Transition locational fits* fall between clearance and interference fits for application where accuracy of location is important, but a small amount of clearance or interference is permissible.

LN: *Locational interference fits* are used where accuracy of location is of prime importance, and for parts needing rigidity and alignment with no special requirements for bore pressure. Such fits are not for parts that transmit frictional loads from one part to another by virtue of the tightness of fit; these conditions are met by force fits.

TABLE A-27. American National Standard Clearance Locational Fits[a]

Basic hole system. Limits are in thousandths of an inch.

Nominal size range, inches Over To	Class LC 1 Limits of clearance	Class LC 1 Hole	Class LC 1 Shaft	Class LC 2 Limits of clearance	Class LC 2 Hole	Class LC 2 Shaft	Class LC 3 Limits of clearance	Class LC 3 Hole	Class LC 3 Shaft	Class LC 4 Limits of clearance	Class LC 4 Hole	Class LC 4 Shaft	Class LC 5 Limits of clearance	Class LC 5 Hole	Class LC 5 Shaft
0–0.12	0	+0.25	+0	0	+0.4	+0	0	+0.6	+0	0	+1.6	+0	0.1	+0.4	−0.1
	0.45	−0	−0.2	0.65	−0	−0.25	1	−0	−0.4	2.6	−0	−1.0	0.75	−0	−0.35
0.12–0.24	0	+0.3	+0	0	+0.5	+0	0	+0.7	+0	0	+1.8	+0	0.15	+0.5	−0.15
	0.5	−0	−0.2	0.8	−0	−0.3	1.2	−0	−0.5	3.0	−0	−1.2	0.95	−0	−0.45
0.24–0.40	0	+0.4	+0	0	+0.6	+0	0	+0.9	+0	0	+2.2	+0	0.2	+0.6	−0.2
	0.65	−0	−0.25	1.0	−0	−0.4	1.5	−0	−0.6	3.6	−0	−1.4	1.2	−0	−0.6
0.40–0.71	0	+0.4	+0	0	+0.7	+0	0	+1.0	+0	0	+2.8	+0	0.25	+0.7	−0.25
	0.7	−0	−0.3	1.1	−0	−0.4	1.7	−0	−0.7	4.4	−0	−1.6	1.35	−0	−0.65
0.71–1.19	0	+0.5	+0	0	+0.8	+0	0	+1.2	+0	0	+3.5	+0	0.3	+0.8	−0.3
	0.9	−0	−0.4	1.3	−0	−0.5	2	−0	−0.8	5.5	−0	−2.0	1.6	−0	−0.8
1.19–1.97	0	+0.6	+0	0	+1.0	+0	0	+1.6	+0	0	+4.0	+0	0.4	+1.0	−0.4
	1.0	−0	−0.4	1.6	−0	−0.6	2.6	−0	−1	6.5	−0	−2.5	2.0	−0	−1.0
1.97–3.15	0	+0.7	+0	0	+1.2	+0	0	+1.8	+0	0	+4.5	+0	0.4	+1.2	−0.4
	1.2	−0	−0.5	1.9	−0	−0.7	3	−0	−1.2	7.5	−0	−3	2.3	−0	−1.1

Nominal size range, inches Over To	Class LC 6 Limits of clearance	Class LC 6 Hole	Class LC 6 Shaft	Class LC 7 Limits of clearance	Class LC 7 Hole	Class LC 7 Shaft	Class LC 8 Limits of clearance	Class LC 8 Hole	Class LC 8 Shaft	Class LC 9 Limits of clearance	Class LC 9 Hole	Class LC 9 Shaft	Class LC 10 Limits of clearance	Class LC 10 Hole	Class LC 10 Shaft	Class LC 11 Limits of clearance	Class LC 11 Hole	Class LC 11 Shaft
0–0.12	0.3	+1.0	−0.3	0.6	+1.6	−0.6	1.0	+1.6	1.0	2.5	+2.5	− 2.5	4	+ 4	− 4	5	+ 6	− 5
	1.9	−0	−0.9	3.2	−0	−1.6	3.6	−0	−2.0	6.6	−0	− 4.1	12	− 0	− 8	17	− 0	−11
0.12–0.24	0.4	+1.2	−0.4	0.8	+1.8	−0.8	1.2	+1.8	−1.2	2.8	+3.0	− 2.8	4.5	+ 5	− 4.5	6	+ 7	− 6
	2.3	−0	−1.1	3.8	−0	−2.0	4.2	−0	−2.4	7.6	−0	− 4.6	14.5	− 0	− 9.5	20	− 0	−13
0.24–0.40	0.5	+1.5	−0.5	1.0	+2.2	−1.0	1.6	+2.2	−1.6	3.0	+3.5	− 3.0	5	+ 6	− 5	7	+ 9	− 7
	2.8	−0	−1.4	4.6	−0	−2.4	5.2	−0	−3.0	8.7	−0	− 5.2	17	− 0	−11	25	− 0	−16
0.40–0.71	0.6	+1.6	−0.6	1.2	+2.8	−1.2	2.0	+2.8	−2.0	3.5	+4.0	− 3.5	6	+ 7	− 6	8	+10	− 8
	3.2	−0	−1.6	5.6	−0	−2.8	6.4	−0	−3.6	10.3	−0	− 6.3	20	− 0	−13	28	− 0	−18
0.71–1.19	0.8	+2.0	−0.8	1.6	+3.5	−1.6	2.5	+3.5	−2.5	4.5	+5.0	− 4.5	7	+ 8	− 7	10	+12	−10
	4.0	−0	−2.0	7.1	−0	−3.6	8.0	−0	−4.5	13.0	−0	− 8.0	23	− 0	−15	34	− 0	−22
1.19–1.97	1.0	+2.5	−1.0	2.0	+4.0	−2.0	3.0	+4.0	−3.0	5	+6	− 5	8	+10	− 8	12	+16	−12
	5.1	−0	−2.6	8.5	−0	−4.5	9.5	−0	−5.5	15	−0	− 9	28	− 0	−18	44	− 0	−28
1.97–3.15	1.2	+3.0	−1.2	2.5	+4.5	−2.5	4.0	+4.5	−4.0	6	+7	− 6	10	+12	−10	14	+18	−14
	6.0	−0	−3.0	10.0	−0	−5.5	11.5	−0	−7.0	17.5	−0	−10.5	34	− 0	−22	50	− 0	−32

[a]From ANSI B4.1.

TABLE A-28. AMERICAN NATIONAL STANDARD TRANSITION LOCATIONAL FITS[a]

Basic hole system. Limits are in thousandths of an inch.

Nominal size range, inches Over To	Class LT 1 Fit	Standard limits Hole	Shaft	Class LT 2 Fit	Standard limits Hole	Shaft	Class LT 3 Fit	Standard limits Hole	Shaft	Class LT 4 Fit	Standard limits Hole	Shaft	Class LT 5 Fit	Standard limits Hole	Shaft	Class LT 6 Fit	Standard limits Hole	Shaft
0–0.12	−0.10 +0.50	+0.4 −0	+0.10 −0.10	−0.2 +0.8	+0.6 −0	+0.2 −0.2	... ...	... ...	... ...	... ...	... ...	... ...	−0.5 +0.15	+0.4 −0	+0.5 +0.25	−0.65 +0.15	+0.4 −0	+0.65 +0.25
0.12–0.24	−0.15 +0.65	+0.5 −0	+0.15 −0.15	−0.25 +0.95	+0.7 −0	+0.25 −0.25	... ...	... ...	... ...	... ...	... ...	... ...	−0.6 +0.2	+0.5 −0	+0.6 +0.3	−0.8 +0.2	+0.5 −0	+0.8 +0.3
0.24–0.40	−0.2 +0.8	+0.6 −0	+0.2 −0.2	−0.3 +1.2	+0.9 −0	+0.3 −0.3	−0.5 +0.5	+0.6 −0	+0.5 +0.1	−0.7 +0.8	+0.9 −0	+0.7 +0.1	−0.8 +0.2	+0.6 −0	+0.8 +0.4	−1.0 +0.2	+0.6 −0	+1.0 +0.4
0.40–0.71	−0.2 +0.9	+0.7 −0	+0.2 −0.2	−0.35 +1.35	+1.0 −0	+0.35 −0.35	−0.5 +0.6	+0.7 −0	+0.5 +0.1	−0.8 +0.9	+1.0 −0	+0.8 +0.1	−0.9 +0.2	+0.7 −0	+0.9 +0.5	−1.2 +0.2	+0.7 −0	+1.2 +0.5
0.71–1.19	−0.25 +1.05	+0.8 −0	+0.25 −0.25	−0.4 +1.6	+1.2 −0	+0.4 −0.4	−0.6 +0.7	+0.8 −0	+0.6 +0.1	−0.9 +1.1	+1.2 −0	+0.9 +0.1	−1.1 +0.2	+0.8 −0	+1.1 +0.6	−1.4 +0.2	+0.8 −0	+1.4 +0.6
1.19–1.97	−0.3 +1.3	+1.0 −0	+0.3 −0.3	−0.5 +2.1	+1.6 −0	+0.5 −0.5	−0.7 +0.9	+1.0 −0	+0.7 +0.1	−1.1 +1.5	+1.6 −0	+1.1 +0.1	−1.3 +0.3	+1.0 −0	+1.3 +0.7	−1.7 +0.3	+1.0 −0	+1.7 +0.7
1.97–3.15	−0.3 +1.5	+1.2 −0	+0.3 −0.3	−0.6 +2.4	+1.8 −0	+0.6 −0.6	−0.8 +1.1	+1.2 −0	+0.8 +0.1	−1.3 +1.7	+1.8 −0	+1.3 +0.1	−1.5 +0.4	+1.2 −0	+1.5 +0.8	−2.0 +0.4	+1.2 −0	+2.0 +0.8

[a]ANSI B4.1.

TABLE A-29. AMERICAN NATIONAL STANDARD INTERFERENCE LOCATIONAL FITS[a]

Basic hole system. Limits are in thousandths of an inch.

Nominal size range, inches Over To	Class LN 1 Limits of interference	Standard limits Hole	Shaft	Class LN 2 Limits of interference	Standard limits Hole	Shaft	Class LN 3 Limits of interference	Standard limits Hole	Shaft
0–0.12	0 0.45	+0.25 −0	+0.45 +0.25	0 0.65	+0.4 −0	+0.65 +0.4	0.1 0.75	+0.4 −0	+0.75 +0.5
0.12–0.24	0 0.5	+0.3 −0	+0.5 +0.3	0 0.8	+0.5 −0	+0.8 +0.5	0.1 0.9	+0.5 0	+0.9 +0.6
0.24–0.40	0 0.65	+0.4 −0	+0.65 +0.4	0 1.0	+0.6 −0	+1.0 +0.6	0.2 1.2	+0.6 −0	+1.2 +0.8
0.40–0.71	0 0.8	+0.4 −0	+0.8 +0.4	0 1.1	+0.7 −0	+1.1 +0.7	0.3 1.4	+0.7 −0	+1.4 +1.0
0.71–1.19	0 1.0	+0.5 −0	+1.0 +0.5	0 1.3	+0.8 −0	+1.3 +0.8	0.4 1.7	+0.8 −0	+1.7 +1.2
1.19–1.97	0 1.1	+0.6 −0	+1.1 +0.6	0 1.6	+1.0 −0	+1.6 +1.0	0.4 2.0	+1.0 −0	+2.0 +1.4
1.97–3.15	0.1 1.3	+0.7 −0	+1.3 +0.7	0.2 2.1	+1.2 −0	+2.1 +1.4	0.4 2.3	+1.2 −0	+2.3 +1.6

[a]ANSI B4.1.

Force Fits (Table A-30). A force fit is a special type of interference fit, normally characterized by maintenance of constant bore pressures throughout the range of sizes. Thus the interference varies almost directly with diameter, and, to maintain the resulting pressures within reasonable limits, the difference between its minimum and maximum value is small.

FN 1: *Light drive fits* require light assembly pressures and produce more or less permanent assemblies. They are suitable for thin sections, long fits, or cast-iron external members.

FN 2: *Medium drive fits* are for ordinary steel parts or for shrink fits on light sections. They are about the tightest fits that can be used with high-grade cast-iron external members.

FN 3: *Heavy drive fits* are suitable for heavier steel parts or for shrink fits in medium sections.

FN 4 and FN 5: *Force fits* are for parts that can be highly stressed, or for shrink fits where heavy pressing forces are impractical.

In each of the five tables on limits and fits, the range of nominal sizes (left column) is given in inches. All other sizes or values are given in thousandths of an inch to save space on the tables and must be converted from thousandths of an inch to inches. This is done simply by moving the decimal point three places to the left. For example, 0.2 in the tables becomes .0002 in.

The basic diameter is given in the first column, "Nominal size range, inches." For example, a basic diameter of 2.5000 in. falls between 1.97 and 3.15 on the tables. Read

TABLE A-30. AMERICAN NATIONAL STANDARD FORCE AND SHRINK FITS[a]

Basic hole system. Limits are in thousandths of an inch.

Nominal size range, inches Over To	Class FN 1 Light drive fit Limits of interference	Standard limits Hole	Shaft	Class FN 2 Medium drive fit Limits of interference	Standard limits Hole	Shaft	Class FN 3 Heavy drive fit Limits of interference	Standard limits Hole	Shaft	Class FN 4 Force fit Limits of interference	Standard limits Hole	Shaft	Class FN 5 Force fit Limits of interference	Standard limits Hole	Shaft
0–0.12	0.05	+0.25	+0.5	0.2	+0.4	+0.85	. . .	. . .	. . .	0.3	+0.4	+ 0.95	0.3	+0.6	+1.3
	0.5	−0	+0.3	0.85	−0	+0.6	. . .	. . .	. . .	0.95	−0	+ 0.7	1.3	−0	+0.9
0.12–0.24	0.1	+0.3	+0.6	0.2	+0.5	+1.0	. . .	. . .	. . .	0.4	+0.5	+ 1.2	0.5	+0.7	+1.7
	0.6	−0	+0.4	1.0	−0	+0.7	. . .	. . .	. . .	1.2	−0	+ 0.9	1.7	−0	+1.2
0.24–0.40	0.1	+0.4	+0.75	0.4	+0.6	+1.4	. . .	. . .	. . .	0.6	+0.6	+ 1.6	0.5	+0.9	+2.0
	0.75	−0	+0.5	1.4	−0	+1.0	. . .	. . .	. . .	1.6	−0	+ 1.2	2.0	−0	+1.4
0.40–0.56	0.1	+0.4	+0.8	0.5	+0.7	+1.6	. . .	. . .	. . .	0.7	+0.7	+ 1.8	0.6	+1.0	+2.3
	0.8	−0	+0.5	1.6	−0	+1.2	. . .	. . .	. . .	1.8	−0	+ 1.4	2.3	−0	+1.6
0.56–0.71	0.2	+0.4	+0.9	0.5	+0.7	+1.6	. . .	. . .	. . .	0.7	+0.7	+ 1.8	0.8	+1.0	+2.5
	0.9	−0	+0.6	1.6	−0	+1.2	. . .	. . .	. . .	1.8	−0	+ 1.4	2.5	−0	+1.8
0.71–0.95	0.2	+0.5	+1.1	0.6	+0.8	+1.9	. . .	. . .	. . .	0.8	+0.8	+ 2.1	1.0	+1.2	+3.0
	1.1	−0	+0.7	1.9	−0	+1.4	. . .	. . .	. . .	2.1	−0	+ 1.6	3.0	−0	+2.2
0.95–1.19	0.3	+0.5	+1.2	0.6	+0.8	+1.9	0.8	+0.8	+ 2.1	1.0	+0.8	+ 2.3	1.3	+1.2	+3.3
	1.2	−0	+0.8	1.9	−0	+1.4	2.1	−0	+ 1.6	2.3	−0	+ 1.8	3.3	−0	+2.5
1.19–1.58	0.3	+0.6	+1.3	0.8	+1.0	+2.4	1.0	+1.0	+ 2.6	1.5	+1.0	+ 3.1	1.4	+1.6	+4.0
	1.3	−0	+0.9	2.4	−0	+1.8	2.6	−0	+ 2.0	3.1	−0	+ 2.5	4.0	−0	+3.0
1.58–1.97	0.4	+0.6	+1.4	0.8	+1.0	+2.4	1.2	+1.0	+ 2.8	1.8	+1.0	+ 3.4	2.4	+1.6	+5.0
	1.4	−0	+1.0	2.4	−0	+1.8	2.8	−0	+ 2.2	3.4	−0	+ 2.8	5.0	−0	+4.0
1.97–2.56	0.6	+0.7	+1.8	0.8	+1.2	+2.7	1.3	+1.2	+ 3.2	2.3	+1.2	+ 4.2	3.2	+1.8	+6.2
	1.8	−0	+1.3	2.7	−0	+2.0	3.2	−0	+ 2.5	4.2	−0	+ 3.5	6.2	−0	+5.0

[a]ANSI B4.1.

across the top row in Table A-26 to the next major heading, "Class RC 1, close sliding fit." If an RC 1 fit is desired, the limits would be determined as follows: Read across from 1.97–3.15 to the next column, which shows that the minimum clearance, or allowance (tightest permissible condition), is .0004 in. and that the maximum clearance (loosest permissible condition) is .0012 in.

Under the "Standard limits" column are the applicable limits for the "Hole" and the "Shaft." Using the example above, add .0005 to 2.5000 in. to obtain the upper limit or size of the hole. Since the lower limit for the hole is zero, the lower limit and the basic size are the same, or 2.5000 in. Therefore, on the drawing the hole would be dimensioned as

$$
\begin{array}{ccc}
2.5005 & & +.0005 \\
2.0000 & \text{or} & 2.5000 \quad 0
\end{array}
$$

Limit dimensions applied to the shaft would be done in the same way, as follows:

$$
\begin{array}{lr}
\text{Basic shaft size} & 2.5000 \\
\text{Upper limit} & -.0004 \\
\hline
 & 2.4996
\end{array}
$$

$$
\begin{array}{lr}
\text{Basic shaft size} & 2.5000 \\
\text{Lower limit} & -.0007 \\
\hline
 & 2.4993
\end{array}
$$

Therefore, on the drawing the shaft would be dimensioned as

$$
\begin{array}{ccc}
2.4996 & & 0 \\
2.4993 & \text{or} & 2.4996 \quad -.0003
\end{array}
$$

The allowance (minimum clearance) is 2.5000 (minimum hole size) minus 2.4996 (maximum shaft size), or .0004 in. The maximum clearance is 2.5005 (maximum hole size) minus 2.4993 (minimum shaft size), or .0012 in.

TABLE A-31. STEEL-WIRE NAILS

American Steel & Wire Company Gage

Size	Length	Common wire nails and brads		Casing nails		Finishing nails	
		Gage, dia.	No. to pound	Gage, dia.	No. to pound	Gage, dia.	No. to pound
2d	1	15	876	15½	1010	16½	1351
3d	1¼	14	568	14½	635	15½	807
4d	1½	12½	316	14	473	15	584
5d	1¾	12½	271	14	406	15	500
6d	2	11½	181	12½	236	13	309
7d	2¼	11½	161	12½	210	13	238
8d	2½	10¼	106	11½	145	12½	189
9d	2¾	10¼	96	11½	132	12½	172
10d	3	9	69	10½	94	11½	121
12d	3¼	9	64	10½	87	11½	113
16d	3½	8	49	10	71	11	90
20d	4	6	31	9	52	10	62
30d	4½	5	24	9	46		
40d	5	4	18	8	35		
50d	5½	3	14				
60d	6	2	11				

TABLE A-32. AMERICAN NATIONAL STANDARD LARGE RIVETS

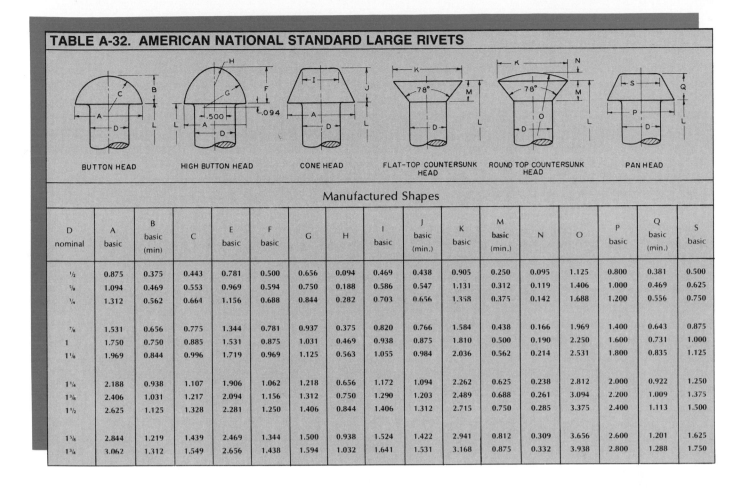

BUTTON HEAD   HIGH BUTTON HEAD   CONE HEAD   FLAT-TOP COUNTERSUNK HEAD   ROUND TOP COUNTERSUNK HEAD   PAN HEAD

Manufactured Shapes

D nominal	A basic	B basic (min)	C	E basic	F basic	G	H	I basic	J basic (min.)	K basic	M basic (min.)	N	O	P basic	Q basic (min.)	S basic
½	0.875	0.375	0.443	0.781	0.500	0.656	0.094	0.469	0.438	0.905	0.250	0.095	1.125	0.800	0.381	0.500
⅝	1.094	0.469	0.553	0.969	0.594	0.750	0.188	0.586	0.547	1.131	0.312	0.119	1.406	1.000	0.469	0.625
¾	1.312	0.562	0.664	1.156	0.688	0.844	0.282	0.703	0.656	1.358	0.375	0.142	1.688	1.200	0.556	0.750
⅞	1.531	0.656	0.775	1.344	0.781	0.937	0.375	0.820	0.766	1.584	0.438	0.166	1.969	1.400	0.643	0.875
1	1.750	0.750	0.885	1.531	0.875	1.031	0.469	0.938	0.875	1.810	0.500	0.190	2.250	1.600	0.731	1.000
1⅛	1.969	0.844	0.996	1.719	0.969	1.125	0.563	1.055	0.984	2.036	0.562	0.214	2.531	1.800	0.835	1.125
1¼	2.188	0.938	1.107	1.906	1.062	1.218	0.656	1.172	1.094	2.262	0.625	0.238	2.812	2.000	0.922	1.250
1⅜	2.406	1.031	1.217	2.094	1.156	1.312	0.750	1.290	1.203	2.489	0.688	0.261	3.094	2.200	1.009	1.375
1½	2.625	1.125	1.328	2.281	1.250	1.406	0.844	1.406	1.312	2.715	0.750	0.285	3.375	2.400	1.113	1.500
1⅝	2.844	1.219	1.439	2.469	1.344	1.500	0.938	1.524	1.422	2.941	0.812	0.309	3.656	2.600	1.201	1.625
1¾	3.062	1.312	1.549	2.656	1.438	1.594	1.032	1.641	1.531	3.168	0.875	0.332	3.938	2.800	1.288	1.750

TABLE A-33.   AMERICAN NATIONAL STANDARD SLOTTED–HEAD WOOD SCREWS

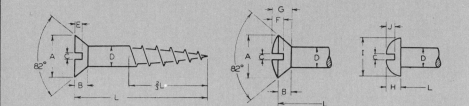

Maximum Dimensions

Nominal size	D	A	B	C	E	F	G	H	I	J	Number threads per inch
0	0.060	0.119	0.035	0.023	0.015	0.030	0.056	0.053	0.113	0.039	32
1	0.073	0.146	0.043	0.026	0.019	0.038	0.068	0.061	0.138	0.044	28
2	0.086	0.172	0.051	0.031	0.023	0.045	0.080	0.069	0.162	0.048	26
3	0.099	0.199	0.059	0.035	0.027	0.052	0.092	0.078	0.187	0.053	24
4	0.112	0.225	0.067	0.039	0.030	0.059	0.104	0.086	0.211	0.058	22
5	0.125	0.252	0.075	0.043	0.034	0.067	0.116	0.095	0.236	0.063	20
6	0.138	0.279	0.083	0.048	0.038	0.074	0.128	0.103	0.260	0.068	18
7	0.151	0.305	0.091	0.048	0.041	0.081	0.140	0.111	0.285	0.072	16
8	0.164	0.332	0.100	0.054	0.045	0.088	0.152	0.120	0.309	0.077	15
9	0.177	0.358	0.108	0.054	0.049	0.095	0.164	0.128	0.334	0.082	14
10	0.190	0.385	0.116	0.060	0.053	0.103	0.176	0.137	0.359	0.087	13
12	0.216	0.438	0.132	0.067	0.060	0.117	0.200	0.153	0.408	0.096	11
14	0.242	0.491	0.148	0.075	0.068	0.132	0.224	0.170	0.457	0.106	10
16	0.268	0.544	0.164	0.075	0.075	0.146	0.248	0.187	0.506	0.115	9
18	0.294	0.597	0.180	0.084	0.083	0.160	0.272	0.204	0.555	0.125	8
20	0.320	0.650	0.196	0.084	0.090	0.175	0.296	0.220	0.604	0.134	8
24	0.372	0.756	0.228	0.094	0.105	0.204	0.344	0.254	0.702	0.154	7

TABLE A-34. AMERICAN WELDING SOCIETY STANDARD WELDING SYMBOLS

Basic Welding Symbols and Their Location Significance

Location Significance	Fillet	Plug or Slot	Spot or Projection	Seam	Back or Backing	Surfacing	Flange Edge	Flange Corner
Arrow Side						Not used		
Other Side						Not used	Not used	Not used
Both Sides		Not used	Not used	Not used	Not used	Not used	Not used	Not used
No Arrow Side or Other Side Significance	Not used	Not used			Not used	Not used	Not used	Not used

Location Significance	Scarf for Brazed Joint	Square	V	Bevel	Groove U	Groove J	Flare-V	Flare-Bevel	Location Significance
									Arrow Side
									Other Side
	Not used								Both Sides
	Not used	Not used	Not used	Not used	Not used	Not used	Not used	Not used	No Arrow Side or Other Side Significance

Supplementary Symbols Used with Welding Symbols

Supplementary Symbols

	Weld-All-Around	Field Weld	Melt-Thru	Backing, Spacer	Contour Flush	Contour Convex	Contour Concave

Typical Welding Symbols

Square-Groove Welding Symbol

Flare-V and Flare-Bevel-Groove Welding Symbols

Edge- and Corner- Flange Welding Symbols

Chain Intermittent Fillet Welding Symbol

Surfacing Welding Symbol Indicating Built-up Surface

Staggered Intermittent Fillet Welding Symbol

Single-V Groove Welding Symbol Indicating Root Penetration

Back or Backing Welding Symbol

Double-Bevel-Groove Welding Symbol

Projection Welding Symbol

Double-Fillet Welding Symbol

Spot Welding Symbol

Seam Welding Symbol

Welding Symbols for Combined Welds

Plug Welding Symbol

Slot Welding Symbol

Backgouging Welding Symbol

Flash or Upset Welding Symbol

Location of Elements of a Welding Symbol

Basic Joints—Identification of Arrow Side and Other Side of Joint

Butt Joint

Corner Joint

T-Joint

Lap Joint

Edge Joint

Process Abbreviations

Where process abbreviations are to be included in the tail of the welding symbol, reference is made to Table A, Designation of Welding and Allied Processes by Letters, of AWS 2.4-79, 71.

A2.1-79

© 1979 by American Welding Society

AMERICAN WELDING SOCIETY, INC.
2501 N.W. 7th Street, Miami, Florida 33125

APPENDIX B
The Metric System

DEVELOPMENT OF THE METRIC SYSTEM

The metric system of measurement was developed in France in the late 1700s. Within a century, most other countries had adopted it. This helped their trade, because they now manufactured to the same standards and could use each other's goods more easily. The main holdout countries were Great Britain and the United States. Great Britain converted to metrics during the 1960s and 1970s. The United States is now doing the same.

The metric system is a very practical system built around units of 10. It was to be a *natural measurement* based on the Earth's own dimensions. The meter was originally defined as one ten-millionth of the distance from the North Pole to the equator. Measures of volume and weight were related to linear distance. No such unified system had ever been made before. Today the meter is more precisely defined as a wavelength of the red-orange light of krypton 86.

In 1960, the International Organization of Weights and Measures, representing countries from all over the world, gave the metric system its formal title of Système International d-Unités, which is abbreviated SI. The basic SI units are shown in Fig. B-1.

USING THE METRIC SYSTEM

The most common measurements of the metric system are the *meter*, for length; the *gram*, for mass or weight; and the *liter*, for volume.

The parts and multiples of these units are based on the number 10. This is similar to our money system and our number system, which are also based on 10. It is easier to figure with units of 10 than to divide by 12 (inches) or by 16 (ounces).

Special names are given to certain multiples and divisions of the basic unit. These names become a *prefix* to the name of that unit. The prefixes apply to any of the unit names— meter, gram, or liter. The prefixes and their corresponding amounts are shown in Fig. B-2.

Unit	Name	Symbol
Length	Meter	m
Mass	Kilogram	kg
Time	Second	s
Electric current	Ampere	A
Temperature	Kelvin	K
Luminous intensity	Candela	cd

Fig. B-1. The basic SI units.

Prefix	Amount	Fraction	Decimal
Milli	One thousandth	$\frac{1}{1000}$	0.001
Centi	One hundredth	$\frac{1}{100}$	0.01
Deci	One-tenth	$\frac{1}{10}$	0.1
Deka	Ten	10	10.0
Hecto	Hundred	100	100.0
Kilo	Thousand	1000	1000.0

Fig. B-2. Metric prefixes and their corresponding amounts.

Adding the prefixes to the word *meter* gives the following:

millimeter (mm) = one thousandth of a meter

centimeter (cm) = one hundredth of a meter

decimeter (dm) = one tenth of a meter

dekameter (dam) = ten meters

hectometer (hm) = one hundred meters

kilometer (km) = one thousand meters

Adding the prefixes to the word *gram* gives the following:

milligram (mg) = one thousandth of a gram

centigram (cg) = one hundredth of a gram

decigram (dg) = one tenth of a gram

dekagram (dag) = ten grams

hectogram (hg) = one hundred grams

kilogram (kg) = one thousand grams

Adding the prefixes to the word *liter* gives the following:

milliliter (ml) = one thousandth of a liter

centiliter (cl) = one hundredth of a liter

deciliter (dl) = one tenth of a liter

dekaliter (dal) = ten liters

hectoliter (hl) = one hundred liters

kiloliter (kl) = one thousand liters

As a basis for comparison, approximate U.S. Customary and metric equivalents to remember are: one liter is a little larger than a quart; one kilogram is a little greater than two pounds; one meter is a little longer than a yard; and one kilometer is about five eighths of a mile. Figure B-3 is a list of metric system equivalents.

Figure B-4 shows the actual size as well as the relationship between metric measures of length, volume, and mass.

■ THE METRIC SYSTEM IN DRAFTING

In drafting, the primary dimension is length. On drawings, the millimeter is used for dimensions. This is true whether the drawing is prepared with all metric dimensions or whether it is dual-dimensioned, with sizes expressed in both metric and U.S. Customary units. In cases where drawings are not dual-dimensioned, a metric equivalent chart (Fig. B-5) may be used to make the conversions. However, not all dimensions are found on the chart. Therefore, it is useful to know how to convert mathematically from one system to the other.

■ INCHES TO MILLIMETERS

One inch is equal to 25.4 mm. Multiplying an inch dimension by 25.4 will give an answer in millimeters. To simplify the calculation, the inch dimension should be in decimal form.

Example: How many millimeters equal 2.62 in.? Multiply 2.62 by 25.4:

$$
\begin{array}{r}
2.62 \\
25.4 \\
\hline
1048 \\
1310 \\
524 \\
\hline
66.548 \text{ mm}
\end{array}
$$

■ MILLIMETERS TO INCHES

One millimeter is equal to .0394 in. Multiplying a millimeter dimension by .0394 will give the answer in inches.

Example: How many inches equal 66.5 mm? Multiply 66.5 by .0394:

$$
\begin{array}{r}
66.5 \\
.0394 \\
\hline
2660 \\
5958 \\
1995 \\
\hline
2.62010 \text{ in., or } 2.62 \text{ in.}
\end{array}
$$

■ ROUNDING OFF

It is sometimes necessary to round numbers to fewer digits. For example, 25.63220 mm rounded to three decimal places would be 25.632 mm. The same number rounded to two places would be 25.63 mm. The procedure is as follows:

A. When the first number (digit) dropped is less than 5, the last number kept does not change. For example, 15.232 rounded to

Length

Centimeter	= 0.3937 inch
Meter	= 3.28 feet
Meter	= 1.094 yards
Kilometer	= 0.621 statute mile
Kilometer	= 0.5400 nautical mile
Inch	= 2.54 centimeters
Foot	= 0.3048 meter
Yard	= 0.9144 meter
Statute mile	= 1.61 kilometers
Nautical mile	= 1.852 kilometers

Area

Square centimeter	= 0.155 square inch
Square meter	= 10.76 square feet
Square meter	= 1.196 square yards
Hectare	= 2.47 acres
Square kilometer	= 0.386 square miles
Square inch	= 6.45 square centimeters
Square foot	= 0.0929 square meter
Square yard	= 0.836 square meter
Acre	= 0.405 hectare
Square mile	= 2.59 square kilometers

Volume

Cubic centimeter	= 0.0610 cubic inch
Cubic meter	= 35.3 cubic feet
Cubic meter	= 1.308 cubic yards
Cubic inch	= 16.39 cubic centimeters
Cubic foot	= 0.0283 cubic meter
Cubic yard	= 0.765 cubic meter

Capacity

Milliliter	= 0.0338 U.S. fluid ounce
Liter	= 1.057 U.S. liquid quarts
Liter	= 0.908 U.S. dry quart
U.S. fluid ounce	= 29.57 milliliters
U.S. liquid quart	= 0.946 liter
U.S. dry quart	= 1.101 liters

Mass or weight

Gram	= 15.43 grains
Gram	= 0.0353 avoirdupois ounce
Kilogram	= 2.205 avoirdupois pounds
Metric ton	= 1.102 short, or net, tons
Grain	= 0.0648 gram
Avoirdupois ounce	= 28.35 grams
Avoirdupois pound	= 0.4536 kilogram
Short, or net, ton	= 0.907 metric ton

Fig. B-3. Metric system equivalents.

two decimal places would be 15.23.

B. When the first number dropped is greater than 5, the last number kept should be increased by 1. For example, 6.436 rounded to two decimal places would be 6.44.

C. When the first number dropped is 5 followed by at least one number greater than zero, the last number kept should be increased by 1. For example, 8.4253 rounded to two decimal places would be 8.43.

D. When the first number dropped is 5 followed by zeros, the last number kept is increased by 1 if it is an odd number, no change is made. For example, 3.23500 rounded to two places would be 3.24. However, 3.22500 would be 3.22.

 **ANGULAR DIMENSIONS**

Angular dimensions do not need to be converted. Angles dimensioned in degrees, minutes, and seconds or in decimal degrees are common to both systems.

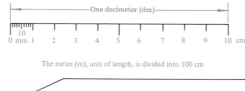

The meter (m), unit of length, is divided into 100 cm

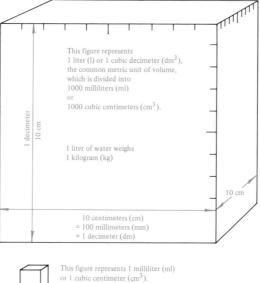

This figure represents
1 liter (l) or 1 cubic decimeter (dm^3),
the common metric unit of volume,
which is divided into
1000 milliliters (ml)
or
1000 cubic centimeters (cm^3).

1 liter of water weighs
1 kilogram (kg)

1 decimeter
10 cm

10 cm

10 centimeters (cm)
= 100 millimeters (mm)
= 1 decimeter (dm)

This figure represents 1 milliliter (ml)
or 1 cubic centimeter (cm^3).
1 ml of water weighs 1 gram (g)

Fig. B-4. Relationship of metric measures of length, volume, and mass. (Metric Association.)

mm	in.*	mm	in.	in.	mm†	in.	mm
1	= 0.0394	17	= 0.6693	1/32 (0.03125)	= 0.794	17/32 (0.53125)	= 13.493
2	= 0.0787	18	= 0.7087	1/16 (0.0625)	= 1.587	9/16 (0.5625)	= 14.287
3	= 0.1181	19	= 0.7480	3/32 (0.9375)	= 2.381	19/32 (0.59375)	= 15.081
4	= 0.1575	20	= 0.7874	1/8 (0.1250)	= 3.175	5/8 (0.6250)	= 15.875
5	= 0.1969	21	= 0.8268	5/32 (0.15625)	= 3.968	21/32 (0.65625)	= 16.668
6	= 0.2362	22	= 0.8662	3/16 (0.1875)	= 4.762	11/16 (0.6875)	= 17.462
7	= 0.2756	23	= 0.9055	7/32 (0.21875)	= 5.556	23/32 (0.71875)	= 18.256
8	= 0.3150	24	= 0.9449	1/4 (0.2500)	= 6.349	3/4 (0.7500)	= 19.050
9	= 0.3543	25	= 0.9843	9/32 (0.28125)	= 7.144	25/32 (0.78125)	= 19.843
10	= 0.3937	26	= 1.0236	5/16 (0.3125)	= 7.937	13/16 (0.8125)	= 20.637
11	= 0.4331	27	= 1.0630	11/32 (0.34375)	= 8.731	27/32 (0.84375)	= 21.431
12	= 0.4724	28	= 1.1024	3/8 (0.3750)	= 9.525	7/8 (0.8750)	= 22.225
13	= 0.5118	29	= 1.1418	13/32 (0.40625)	= 10.319	29/32 (0.90625)	= 23.018
14	= 0.5512	30	= 1.1811	7/16 (0.4375)	= 11.112	15/16 (0.9375)	= 23.812
15	= 0.5906	31	= 1.2205	15/32 (0.46875)	= 11.906	31/32 (0.96875)	= 24.606
16	= 0.6299	32	= 1.2599	1/2 (0.5000)	= 12.699	1 (1.0000)	= 25.400

*Calculated to *nearest* fourth decimal place. †Calculated to *nearest* third decimal place.

Fig. B-5. Inch-millimeter equivalent chart.

Glossary

(v) = *Verb*   (n) = *Noun*   (adj) = *Adjective*

acme *(n)*   A screw-thread form.

addendum *(n)*   The distance a gear tooth extends above the pitch circle.

airbrush *(n)*   A miniature spray gun that sprays a solution (usually watercolor) to give shading effects.

allowance *(n)*   The smallest space permitted between mating parts.

alloy *(n)*   Two or more metals combined to form a new metal.

anneal *(v)*   to heat slowly to a critical temperature, then gradually cool. Used to soften and to remove internal stresses.

assembly drawing *(n)*   A drawing of several parts to show how they are put together.

automation *(n)*   The automatic operation of equipment that increases productivity without additional expenditure of human energy.

auxiliary view *(n)*   An additional view, usually of a slanted surface, showing that surface in true shape.

azimuth *(n)*   In descriptive geometry, the measure of how far a line is off due north.

babbitt or **babbitt metal** *(n)*   A friction-bearing metal, invented by Isaac Babbitt. Composed of antimony, tin, and copper.

bearing *(n)*   Any part that bears up, or supports, another part. In particular, the support for a turning shaft.

bearing *(n)*   In descriptive geometry, the angle a line in a top view makes with a north-south line.

bevel *(n)*   A surface slanted to another surface. Called a *miter* when the angle is 45°.

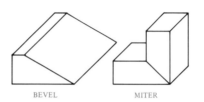

BEVEL   MITER

blueprint. *(n)*   A copy of a drawing made by a machine. It has white lines on a blue background.

bolt circle *(n)*   A circular centerline locating the centers of holes arranged in a circle.

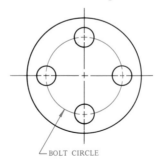

BOLT CIRCLE

bore *(v)*   To enlarge or to finish a hole with a cutting tool called a *boring bar*, used in a boring mill or lathe.

boss *(n)*   A raised circular surface as used on a casting or forging.

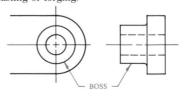

BOSS

brass *(n)*   An alloy of copper and zinc or of copper, zinc, and lead.

braze *(v)*   To join two pieces of metal by using hard solder, such as brass or zinc.

broach *(v)*   To machine (finish) and change the forms of holes or outside surfaces to a desired shape, generally other than round. *(n)* A tool with a series of chisel edges used to broach.

bronzes *(n)*   Alloys of copper and tin in varying proportions, mostly copper. Sometimes other metals, such as zinc, are added.

buff *(v)*   To polish a surface on a fabric wheel by rubbing it with some rough material.

burnish *(v)*   To smooth or polish by a sliding or rolling motion.

burr *(n)*   A rough or jagged edge caused by cutting or punching.

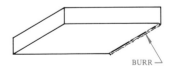

BURR

bushing *(n)*   A hollow cylindrical sleeve used as a bearing or as a guide for drills or other tools.

cabinet drawing *(n)*   A kind of oblique pictorial drawing in which receding edges are drawn to one half their proportional length.

CAD *(n)*   Computer-aided drafting or computer-aided design and drafting.

CADD *(n)*   Computer-aided design drafting or computer-aided design and drafting.

caliper *(n)* A measuring device with two adjustable legs. It is used for measuring thickness or diameters.

cam *(n)* A machine part mounted on a turning shaft, used to change rotary (turning) motion into back-and-forth motion. See Chap. 19.

CAM *(n)* Computer-aided manufacturing.

caseharden (**carbonize** or **carburize**) *(v)* To heat-treat steel. The outside surface is hardened by making it absorb carbon. It is then cooled rapidly by quenching in oil.

casting *(n)* A part formed by pouring molten metal into a hollow form (mold) of the desired shape, then allowing it to harden.

cavalier drawing *(n)* A kind of oblique pictorial drawing in which receding edges are drawn to their full proportional length.

center drill *(n)* A special combination drill and countersink. It is used to drill the ends of stock to be turned between centers on a lathe.

chamfer *(v)* To bevel an edge. *(n)* An edge that has been beveled.

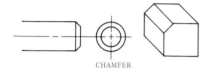

CHAMFER

chip *(n)* A small slice of material (such as silicon) containing electronic devices (transistors, diodes, resistors, and/or capacitors) which perform functions within a computer.

circular pitch *(n)* The distance from a point on one gear tooth to the same point on the next tooth measured along the pitch circle.

clearance *(n)* The space that a moving part needs in order to function. On gears, the distance between the top of a tooth and the bottom of the mating space.

core *(v)* To form the hollow part of a casting. A part made of sand and shaped like the hollow part (called a *core*) is placed in the mold (see **casting**). The core is broken up and removed after the casting is cool.

counterbore *(v)* To make a hole deeper and wider. *(n)* A counterbored hole.

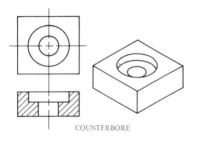

COUNTERBORE

countersink *(v)* To drill a cone shape at the end of a hole. *(n)* A countersunk hole.

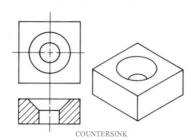

COUNTERSINK

crown *(n)* The contour of the face of a belt pulley, rounded or angular, used to keep the belt in place. The belt tends to climb to the highest place.

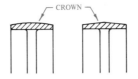

CROWN

CRT *(n)* Cathode-ray tube. Similar to a television screen, this screen allows the production of a drawing (visual display) without the use of vellum or Mylar.

cutting plane *(n)* An imaginary plane passed through an object to cut away part of it.

dependum *(n)* The distance a gear tooth extends below the pitch circle.

detail drawing *(n)* A drawing of one part of a machine or structure.

development *(n)* A drawing of all the surfaces of an object in true shape.

diametral pitch *(n)* The number of gear teeth per inch of pitch diameter.

diazo *(n)* A print with dark lines on a white background. The opposite of a blueprint.

die *(n)* A hardened metal block used to make a desired shape by cutting or pressing. Also, a tool used to cut external screw threads.

die casting *(n)* A casting made by pouring molten alloy (or plastic composition) into a metal mold or die, generally forced under pressure. Die castings are smooth and accurate.

die stamping *(n)* A piece that has been formed or cut from sheet material, generally metal.

dimension line *(n)* A line with arrowheads at either end to show the distance between two points.

display *(n)* A visual output as seen on the screen of a CRT. It may be a drawing or alphanumeric data.

dividers *(n)* A tool with two legs joined at the top. Each leg ends in a needle point.

Dividers are used to find and lay off measurements.

dowel *(n)* A cylindrical-shaped pin used to fasten parts together.

draft *(n)* The tapered sides on a foundry pattern that allow it to be easily pulled from the sand.

drill *(v)* To make a cylindrical hole using a revolving tool with cutting edges, called a *drill*, generally a twist drill.

drop forging *(n)* A piece formed between dies while hot, using pressure or a drop hammer.

elevation *(n)* A drawing of a facade of a structure.

exploded view *(n)* A drawing showing how an object would look if an "explosion" had propelled its component parts away from each other. Used to show how the parts relate to each other.

face *(v)* To machine (finish) a flat surface on a lathe with the surface perpendicular to the axis of rotation.

FAO Finished all over.

file *(v)* To smooth, finish, or shape with a file.

fillet *(n)* The rounded-in corner between two surfaces.

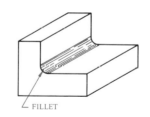

FILLET

fit *(n)* The tightness or looseness between two mating parts.

fixture *(n)* A device used to hold a workpiece during machining (finishing).

flange *(n)* A rim extension, as at the end of a pipe.

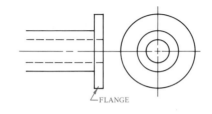

FLANGE

forge *(v)* To form hot metal into a desired shape by hammering or by pressure.

French curve *(n)* A tool used to draw curved lines that are not arcs. Also called an *irregular curve*.

functional drawing *(n)* A drawing that uses the fewest number of views and the fewest number of lines to give the exact information required.

gage *(n)* A device to find whether a specified dimension on an object is within specified limits.

gage line *(n)* A line along which rivets are spaced.

galvanize *(v)* To give a coating of zinc or zinc and lead.

gasket *(n)* A thin piece of material placed between two surfaces to produce a tight joint.

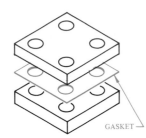

gear *(n)* A toothed wheel used to transmit power or motion from one shaft to another. A machine part used to transmit motion or force.

grind *(v)* To use an *abrasive* (rough-surfaced) wheel to polish or to finish a surface.

hardware *(n)* Physical equipment.

heat-treat *(v)* To change the properties of metal by carefully controlled heating and cooling.

IC *(n)* Integrated circuit. See **Chip.**

interactive *(adj)* Refers to the need for a human to initiate communication between different parts of a computer system.

isometric drawing *(n)* A kind of pictorial drawing based on height, width, and depth axes in equal 120° angles with each other.

jig *(n)* A special device used to guide a cutting tool; it may also hold the workpiece.

kerf *(n)* a slot or groove made by a cutting tool.

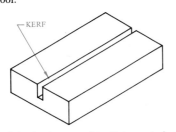

key *(n)* A piece used to fasten a hub to a shaft, or for a similar purpose.

keyway or **keyseat** *(n)* A groove or slot in a shaft or hub into which a key fits.

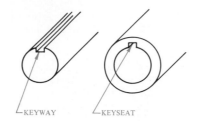

knurl *(v)* To roughen or indent a rounded surface so that it can be held or turned by hand.

lead *(n)* The distance a screw moves along its axis and against a fixed mating part when given one full turn.

limit *(n)* A boundary. Indication of only the largest and smallest permissible dimensions. See Chap. 6.

load *(n)* The weight or pressure borne by a structure.

lug *(n)* An "ear" forming a part of, and extending from, a part.

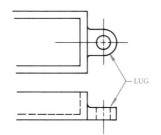

mallable casting *(n)* A casting toughened by annealing.

matte *(adj)* Dull and rough, as a surface.

micrometer caliper *(n)* A device used to find exact measurements of diameter thickness.

microprocessor *(n)* The central processing unit of a microcomputer.

mill *(v)* To machine (finish) a part on a milling machine, using a rotating toothed cutter.

neck *(v)* To cut a groove around a cylindrical part, generally at a change in diameter.

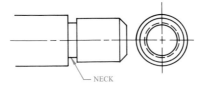

oblique drawing *(n)* A kind of pictorial drawing that shows one face of the object in true shape, but other faces on a distorted angle.

operator *(n)* A CAD user. Not all users are designers and drafters. A designer or drafter, however, may be an operator.

orthographic projection *(n)* A system of showing an object in several views.

overlay *(n)* A translucent sheet placed over a drawing. It can be used to trace the drawing. It can also contain additional material to be superimposed on the drawing.

pattern *(n)* See **development.**

PC *(n)* Printed circuit. An electronic assembly comprised of a nonconductive board, a conductive copper pattern, and electronic components.

peen *(v)* To stretch or bend material with the peen, or ball, end of a machinist's hammer.

perspective drawing *(n)* A kind of pictorial drawing that shows objects as they look to the eye.

photodrawing *(n)* A drawing prepared from a photograph, or a photograph on which dimensions, changes, or additions have been drawn.

pickle *(v)* To clean a metal object by using a weak sulfuric acid bath.

pictorial drawing *(n)* A drawing that looks like a picture.

pitch *(n)* On a screw, the distance from one point on the thread form to the corresponding point on the next form, measured in line with the axis of the screw. Also, on any slanting structure, the degree of slant.

plate *(v)* To electrochemically coat a metal object with another metal.

polish *(v)* To smooth with a very fine *abrasive* (rough surface).

program *(n)* A detailed set of coded instructions that are logically ordered. These are used for the operation of a computer.

programmer *(n)* An individual who designs the sets of instructions, or programs, for the CAD system.

projector *(n)* In pictorial drawing, a line along a receding axis.

punch *(v)* To make a hole in thin metal by pressing a tool of the desired shape through it.

ream *(v)* To make a hole the exact size by finishing with a turning cutting tool that has many cutting edges.

rendering *(n)* Surface shading used in a drawing.

rib *(n)* A thin, flat part of an object used to brace or strengthen another part.

rise *(n)* In a slanting structure, such as a roof or staircase, the vertical distance from the bottom to the top.

rivet *(n)* A metal rod with a preformed head on one end. A rivet is used to join the metal parts together permanently.

round *(n)* The rounded-over corner of two surfaces.

ROUND

run *(n)* In a slanting structure, such as a roof or staircase, the horizontal distance from the bottom to the top.

schematic *(n)* Usually, an electric wiring diagram using symbols.

sectional view *(n)* A drawing of an object as if part of it were cut away to show the inside.

section lines *(n)* In a sectional view, thin, evenly spaced lines that mark the cut surface.

shaft *(n)* Round stock to which gears, pulleys, or turning pieces are attached for support or to transmit power.

shear *(v)* To cut material between two blades.

shim *(n)* A thin plate of metal used between two surfaces to adjust the distance between them.

skew lines *(n)* Lines that are not parallel and do not intersect.

software *(n)* The programs that are input for the computer. The program gives instructions concerning the operational sequence.

specifications *(n)* In construction and manufacturing, a detailed list of facts concerning materials and measurements.

spline *(n)* A metal strip that drafters use to trace a curve.

spline *(n)* A long keyway.

spotface *(v)* To finish a circular spot slightly below a rough surface on a casting. This provides a smooth, flat seat for a bolthead or other fastener.

state of the art *(n)* The latest technical advancement

steel casting *(v)* A part made of cast iron to which scrap steel has been added when melted.

tap *(v)* To cut threads in a hole with a threading tool called a *tap*. *(n)* A hardened screw, fluted to provide cutting edges.

taper pin *(v)* A piece of round stock that narrows gradually and uniformly in diameter.

technical illustration *(v)* A pictorial drawing made to simplify and interpret technical information.

technical pen *(n)* A pen with a point of a specific size, to draw lines of a specific width.

temper *(v)* To make hardened steel less brittle by heating it in various ways, as in a bath of oil, salt, sand, or lead, to a specified temperature, and then cooling.

template or **templet** *(n)* A flat form or pattern of full size, used to lay out a shape and to locate holes or other features.

tolerance *(n)* The distance allowed away from the exact size before the size is wrong. See Chap. 6.

turn *(v)* To machine (finish) a piece on a lathe by turning the piece against a cutting tool (as when forming a cylindrical surface).

turnkey *(n)* A complete CAD system.

upset *(v)* To make an enlarged section or shoulder on a rod, bar, or similar piece while forging.

vanishing point *(n)* In perspective drawing, the point at which receding axes converge.

vernier *(n)* A small auxiliary scale to find fractional parts of a major scale.

washer *(n)* A ring of metal used as a seat for a bolt or nut.

web *(n)* A thin, flat part of an object used to brace or strengthen another part.

weld *(v)* To join pieces of metal by heating them to the melting point, then pressing or hammering them together. See Chap. 17.

working drawing *(n)* A drawing that contains all the information needed to make an object.

Index

BUTLER PUBLIC LIBRARY
BUTLER, PENNSYLVANIA 16001

604.2 MEC

Mechanical drawing / Thomas
E. French... [et al.].
10th ed.